Atomic and Molecular Collision Theory

NATO ADVANCED STUDY INSTITUTES SERIES

A series of edited volumes comprising multifaceted studies of contemporary scientific issues by some of the best scientific minds in the world, assembled in cooperation with NATO Scientific Affairs Division.

Series B: Physics

Recent Volumes in this Series

Volume 69 – Photovoltaic and Photoelectrochemical Solar Energy Conversion
edited by F. Cardon, W. P. Gomes, and W. Dekeyser

Volume 70 – Current Topics in Elementary Particle Physics
edited by K. H. Mütter and K. Schilling

Volume 71 – Atomic and Molecular Collision Theory
edited by Franco A. Gianturco

Volume 72 – Phase Transitions: *Cargèse 1980*
edited by Maurice Lévy, Jean-Claude Le Guillou, and Jean Zinn-Justin

Volume 73 – Scattering Techniques Applied to Supramolecular and Nonequilibrium Systems
edited by Sow-Hsin Chen, Benjamin Chu, and Ralph Nossal

Volume 74 – Rigorous Atomic and Molecular Physics
edited by G. Velo and A. S. Wightman

Volume 75 – Nonlinear Phenomena in Physics and Biology
edited by Richard H. Enns, Billy L. Jones, Robert M. Miura, and Sadanand S. Rangnekar

Volume 76 – Metal Hydrides
edited by Gust Bambakidis

Volume 77 – Nonlinear Phenomena at Phase Transitions and Instabilities
edited by T. Riste

This series is published by an international board of publishers in conjunction with NATO Scientific Affairs Division

A Life Sciences B Physics	Plenum Publishing Corporation London and New York
C Mathematical and Physical Sciences	D. Reidel Publishing Company Dordrecht, Boston, and London
D Behavioral and Social Sciences E Applied Sciences	Sijthoff & Noordhoff International Publishers Alphen aan den Rijn, The Netherlands, and Germantown, U.S.A.

Atomic and Molecular Collision Theory

Edited by

Franco A. Gianturco
Cittá Universitaria
Rome, Italy

PLENUM PRESS • NEW YORK AND LONDON
Published in cooperation with NATO Scientific Affairs Division

Library of Congress Cataloging in Publication Data

NATO Advanced Study Institute on Atomic and Molecular Collision Theory (1980: Cortona, Italy).
Atomic and molecular collision theory.

(NATO advanced study institutes series. Series B, Physics; v. 71)
"Published in cooperation with NATO Scientific Affairs Division."
"Proceedings of a NATO Advanced Study Institute on Atomic and Molecular Collision Theory held at Il Palazzone di Cortona, Arezzo, Italy, on September 15-26, 1980" —T.p. verso.
Includes bibliographies and indexes.
1. Collisions (Nuclear physics)—Congresses. 2. Atoms—Congresses. 3. Molecules—Congresses. I. Gianturco, Franco A., 1938- . II. North Atlantic Treaty Organization. Division of Scientific Affairs. III. Title. IV. Series.
QC794.6.C6 N38 1980 539.7′54 81-15362
ISBN 0-306-40807-4 AACR2

Proceedings of a NATO Advanced Study Institute on Atomic and Molecular Collision Theory held at Il Palazzone di Cortona, Arezzo, Italy, on September 15-26, 1980

A Division of Plenum Publishing Corporation
233 Spring Street, New York, N.Y. 10013

Printed in the United States of America

PREFACE

Until recently, the field of atomic and molecular collisions was left to a handful of practitioners who essentially explored it as a branch of atomic physics and gathered their experimental results mainly from spectroscopy measurements in bulk. But in the past ten years or so, all of this has dramatically changed, and we are now witnessing the rapid growth of a large body of research that encompasses the simplest atoms as well as the largest molecules, that looks at a wide variety of phenomena well outside purely spectroscopic observation, and that finds applications in an unexpectedly broad range of physico-chemical and physical processes. The latter are in turn surprisingly close to very important sectors of applied research, such as the modeling of molecular lasers, the study of isotope separation techniques, and the energy losses in confined plasmas, to mention just a few of them.

As a consequence of this healthy state of affairs, greatly diversified research pathways have developed; however, their specialized problems are increasingly at risk of being viewed in isolation, although they are part of a major and extended branch of physics or chemistry. This is particularly true when it comes to the theory of this work -- where well-established methods and models of one subfield are practically unknown to researchers in other subfields -- and, consequently, the danger of wasteful duplication arising is quite real.

The scope of the Advanced Study Institute from which this volume originated was, therefore, to bring together under the same heading some of the most significant areas of the theory of atomic and molecular collision processes. The different chapters that follow here are meant to outline the relevant theoretical tools and reveal their essential unity while, at the same time, describe in detail the many varied forms that their application can take. Because of the obvious limit of time imposed by the duration of the Institute, not all the current possible sidepaths could be examined; all those that have recently been significantly spurred by concurrent experimental progress, however, have been taken into consideration and this in the hope of providing newcomers with a compact view of the field as well as a self-contained source from which to begin a systematic study. To this end, the bibliography in each chapter is as exhaustive as possible, as far as published results are concerned, thus enabling the interested student to go back to the most up-to-date sources.

Many people have contributed to make this effort come true, and while it would be impossible to thank them all individually the following assistance must not go unnoticed. The friendly atmosphere provided by the staff of the Palazzone and the Oasi Residence in Cortona encouraged student participation at classes, thereby affording them the opportunity of pointing out inconsistencies in the first draft of the lecture notes. The generous and professional help of the secretarial staff in Professor Toennies' laboratory in Göttingen improved the overall presentation of several of the chapters. The unflagging support of our secretary, before, during and after The Institute took place, allowed the coming into being of this volume in the best possible spirit. Finally the financial aid of the Italian National Research Council (CNR) which made possible the completion of this volume is also gratefully acknowledged.

Franco A. Gianturco
Professor of Quantum Chemistry
Istituto di Chimica Fisica
Città Universitaria
00189 Rome, Italy
Director of the Adv.Study Inst.

Rome, November 1980

CONTENTS

COLLISIONS INVOLVING ELECTRONS AND PHOTONS

Electron and Photon Collisions with Atoms and Ions........... 1
C.J. Joachain

Electron and Photon Collisions with Molecules................ 69
P.G. Burke

Recombination Processes in Atomic and Molecular Physics...... 123
J.N. Bardsley

Charge Exchange and Ionization in Ion-Atom Collisions........ 165
R. McCarroll

Atoms in Astrophysical Plasmas............................... 233
H. Van Regemorter

COLLISIONS OF ATOMS AND MOLECULES

Low Energy Atom-Atom Collisions.............................. 279
M.S. Child

Internal Energy Transfers in Molecular Collisions............ 315
F.A. Gianturco

Chemical Photophysics.. 393
R.D. Levine

COLLISIONS UNDER SPECIAL CONDITIONS

Collisions of Charged Particles with Highly Excited Atoms.... 431
I.C. Percival

Electron-Atom Scattering in the Field of a Laser............ 461
M.H. Mittleman

Lecturers.. 489

Participants... 492

Index.. 501

PART I

COLLISIONS INVOLVING ELECTRONS AND PHOTONS

ELECTRON AND PHOTON COLLISIONS WITH ATOMS AND IONS

C.J. JOACHAIN

Physique Théorique, Faculté des Sciences

Université Libre de Bruxelles, Belgium

Abstract

The theoretical methods used in recent investigations of electron and photon collisions with atoms and ions are reviewed.

1. INTRODUCTION
2. ELECTRON COLLISIONS WITH ATOMS AND IONS
   2.1. Potential scattering.
   2.2. Multichannel scattering. General features.
   2.3. Elastic scattering and excitation at low energies.
   2.4. Electron-atom (ion) collisions at intermediate and high energies.
3. PHOTON COLLISIONS WITH ATOMS AND IONS
   3.1. Semi-classical theory of the interaction of atoms with electromagnetic radiation.
   3.2. Photoionization of atoms and ions.

1. INTRODUCTION

The theoretical study of electron and photon collisions with atomic systems has attracted a considerable amount of interest in recent years. This is due to several reasons. Firstly, there is an increasing demand for electron and photon collision cross sections in other fields such as astrophysics, laser physics and plasma physics. Secondly, a number of important advances have occured on the experimental side. These include absolute measurements of cross sections as well as experiments using tunable dye lasers, synchrotron radiation, polarized beams or targets, and coincidence techniques. Many of these experiments provide very stringent tests of the theory and have stimulated the development of new theoretical approaches. Finally, the availability of increasingly more powerful computers has made it possible to perform calculations which would have been impossible to carry out a decade ago.

In these lectures I shall outline the theory of electron and photon collisions with atoms and ions. I shall start by discussing non-relativistic electron-atom (ion) scattering and then proceed to survey photon collisions, in particular photo-ionization. Atomic units will be used, except where otherwise stated.

2. ELECTRON COLLISIONS WITH ATOMS AND IONS

2.1. Potential Scattering

In this section we shall review the non-relativistic scattering of a spinless particle by a potential field $V(\mathbf{r})$. This will allow us to introduce in a simple way some of the basic ideas which are required in the analysis of electron scattering by atoms and ions. A detailed account of potential scattering may be found for example in the books by Bransden[1], Joachain[2] and Burke[3].

We start from the time-independent Schrödinger equation, which we write in atomic units as

$$[-\frac{1}{2}\nabla^2 + V(\mathbf{r})]\psi(\mathbf{r}) = E\,\psi(\mathbf{r}) \tag{2.1}$$

where the mass of the particle has been set equal to one, and E is its total energy. We denote respectively by $\mathbf{k}_i$ and $\mathbf{k}_f$ the initial and final momenta of the particle, with $k = |\mathbf{k}_i| = |\mathbf{k}_f|$ and $E = k^2/2$. It is also convenient to introduce the "reduced" potential $U(\mathbf{r}) = 2V(\mathbf{r})$. For a potential vanishing faster than r^{-1} at large distances the stationary scattering wave function $\psi_{\mathbf{k}_i}^{(+)}$ representing a plane wave incident in the z-direction (which we choose along $\hat{\mathbf{k}}_i$) and an outgoing (+) spherical wave has the asymptotic form

$$\psi_{\mathbf{k}_i}^{(+)}(\mathbf{r}) \underset{r\to\infty}{\to} A(k)\,[e^{i\mathbf{k}_i\cdot\mathbf{r}} + f(k,\theta,\phi)\,\frac{e^{ikr}}{r}] \tag{2.2}$$

where $f(k,\theta,\phi)$ is the scattering amplitude corresponding to scattering in the direction $\Omega \equiv (\theta,\phi)$, and the coefficient A is independent of $\mathbf{r}$.

Cross sections

The cross section of a given type of event occuring in a collision process is defined as the ratio of the number of events of this type per unit time and per unit scatterer, to the relative flux of the incident particles with respect to the target.

Applying this general definition to the simple case of potential scattering, we first introduce the differential cross section $d\sigma/d\Omega$. It is given by the outgoing flux of particles scattered through the spherical surface $r^2 d\Omega$ (with $r \to \infty$) per unit solid angle, divided by the incident flux. The flux of particles passing through an area $d\mathbf{S}$ is given by

$$\mathbf{j}\cdot d\mathbf{S} = \frac{1}{2i}(\psi^*\nabla\psi - \psi\nabla\psi^*).d\mathbf{S} \tag{2.3}$$

where $\mathbf{j}$ is the probability current density associated with the Schrödinger equation (2.1). Except for the case of scattering in the forward direction, the term $\mathbf{j}_{int}$ arising from the interference between the incoming plane wave and the outgoing spherical wave may be ignored. Substituting the second term of (2.2) into (2.3), we then find that the radial flux passing through an area $r^2 d\Omega$, due to the outgoing spherical wave alone, is given for large r by

$$\mathbf{j}_{out} \cdot d\mathbf{S} = |A|^2 k |f(k,\theta,\phi)|^2 d\Omega \tag{2.4}$$

The flux of the incoming plane wave A exp(ikz) through a unit surface perpendicular to the z-direction is given by $|A|^2 k$, so that the differential cross section is

$$\frac{d\sigma}{d\Omega} = |f(k,\theta,\phi)|^2 \tag{2.5}$$

The total cross section is obtained by integrating over all scattering angles,

$$\sigma_{tot}(k) = \int_0^{2\pi} d\phi \int_0^{\pi} d\theta \sin\theta \, |f(k,\theta,\phi)|^2 \tag{2.6}$$

In the forward direction the interference term $\mathbf{j}_{int}$ cannot be neglected ; it is found by using the conservation of the probability flux that

$$\sigma_{tot} = \frac{4\pi}{k} \mathrm{Im}\, f(k, \theta=o) \tag{2.7}$$

This important relation is known as the optical theorem.

Partial wave analysis

Let us restrict ourselves to the case of a spherically symmetric potential V(r). We can then expand the wave function $\psi^{(+)}_{\mathbf{k}_i}(\mathbf{r})$ in partial waves,

$$\psi^{(+)}_{\mathbf{k}_i}(\mathbf{r}) = \sum_{\ell=o}^{\infty} c_\ell(k)\, R_\ell(k,r)\, P_\ell(\cos\theta) \tag{2.8}$$

where ℓ is the orbital angular momentum quantum number of the particle, $P_\ell(\cos\theta)$ are the Legendre polynomials and the coefficients $c_\ell(k)$ must be determined so that (2.2) is satisfied. It is convenient to set

$$u_\ell(k,r) = r\ R_\ell(k,r) \tag{2.9}$$

so that the function $u_\ell(k,r)$ satisfies the radial Schrödinger equation

$$\left[\frac{d^2}{dr^2} - \frac{\ell(\ell+1)}{r^2} + k^2 - U(r)\right]u_\ell(k,r) = 0 \tag{2.10}$$

Provided that $U(r)$ is less singlular than r^{-2} at the origin, the indicial equation for (2.10) gives two solutions which for $r \to 0$ behave like

$$u_\ell(k,r) \underset{r\to o}{\sim} r^{\ell+1} \tag{2.11}$$

and

$$u_\ell(k,r) \underset{r\to o}{\sim} r^{-\ell} \tag{2.12}$$

Only the first one satisfies the physical condition of regularity. Since $U(r)$ vanishes faster than r^{-1} for large r, one finds that

$$u_\ell(k,r) \underset{r\to\infty}{\to} A_\ell(k)\ \sin(kr - \frac{\ell\pi}{2} + \delta_\ell(k)) \tag{2.13}$$

where $A_\ell(k)$ is an arbitrary normalization factor, and the quantity $\delta_\ell(k)$ is known as the phase shift. In the absence of interaction, the regular solution of (2.10) is proportional to $\sin(kr - \ell\pi/2)$ for larger r, so that the phase shift is a measure of the departure of the radial wave function from the form it has when the potential is zero.

In order to determine the scattering amplitude, we consider the partial wave expansion (2.8) of the wave function for large r, and compare it with its asymptotic form (2.2). Using (2.9), (2.13) and the expansion of a plane wave in Legendre polynomials, we first obtain the quantities $c_\ell(k)$ by equating the coefficients of the ingoing spherical wave $\exp[-i(kr - \ell\pi/2)]$ in (2.2) and (2.8). Next, by matching the coefficients of the outgoing spherical wave, we find that

$$f(k,\theta) = \frac{1}{2ik}\sum_{\ell=o}^{\infty}(2\ell+1)[e^{2i\delta_\ell(k)} - 1]P_\ell(\cos\theta) \quad (2.14)$$

The scattering amplitude is independent of ϕ since the potential is spherically symmetric. The differential cross section is $d\sigma/d\Omega = |f(k,\theta)|^2$ and the total cross section is

$$\sigma_{tot}(k) = \frac{4\pi}{k^2}\sum_{\ell=o}^{\infty}(2\ell+1)\sin^2\delta_\ell(k) = \sum_{\ell=o}^{\infty}\sigma_\ell(k) \quad (2.15)$$

where the quantities $\sigma_\ell(k)$ are called the partial wave cross sections. Comparing (2.14) and (2.15), and remembering that $P_\ell(1) = 1$, we retrieve the optical theorem (2.7). More generally, it is a simple matter to prove the "unitarity relation"

$$\mathrm{Im}\, f(k,\theta) = \frac{k}{4\pi}\int f^*(k,\theta')\, f(k,\theta_o)d\Omega' \quad (2.16)$$

where θ_o is the angle between the directions (θ,o) and (θ',ϕ').

It is of interest to rewrite (2.13) in two alternative forms

$$u_\ell(k,r) \underset{r\to\infty}{\to} N_1(k)[\sin(kr - \frac{\ell\pi}{2}) + \cos(kr - \frac{\ell\pi}{2})K_\ell(k)] , \quad (2.17)$$

$$u_\ell(k,r) \underset{r\to\infty}{\to} N_2(k)[e^{-i(kr - \frac{\ell\pi}{2})} - e^{i(kr - \frac{\ell\pi}{2})}S_\ell(k)] \quad (2.18)$$

where N_1 and N_2 are normalization factors. The K-matrix and S-matrix elements introduced above are related to the phase shift by

$$K_\ell(k) = \tan\delta_\ell(k) \quad (2.19)$$

$$S_\ell(k) = e^{2i\delta_\ell(k)} \quad (2.20)$$

and we note that

$$S_\ell(k) = \frac{1 + i\,K_\ell(k)}{1 - i\,K_\ell(k)} \quad (2.21)$$

It is also convenient to introduce the T-matrix element

$$T_\ell(k) = S_\ell(k) - 1 \qquad (2.22)$$

If the potential is real then eq. (2.10) is real for real energies and the radial functions $u_\ell(k,r)$ can therefore be chosen to be real. As a result, the phase shift is also real and $|S_\ell| = 1$. We also note that for scattering by spherically symmetric potentials the S, K and T matrices are diagonal in the angular momentum representation.

For most potentials of interest in atomic physics the phase shifts must be determined by integrating numerically eq. (2.10), subject to the boundary conditions (2.11) and (2.13). In practice, because of the centrifugal barrier term $\ell(\ell+1)/r^2$ in (2.10), only the first few phase shifts are appreciably different from zero at low energies. If the potential can be neglected for $r > a$, one can divide the domain of the variable r into an internal region ($r \leqslant a$) and an external region ($r \geqslant a$). The exterior solution may be written for example as

$$u_\ell(k,r) = A_\ell(k)\, kr[\cos\delta_\ell(k) j_\ell(kr) - \sin\delta_\ell(k) n_\ell(kr)], \quad r \geqslant a \qquad (2.23)$$

where j_ℓ and n_ℓ are spherical Bessel and Neumann functions, respectively. The phase shift can then be obtained from the boundary conditions that both u_ℓ and du_ℓ/dr - or the logarithmic derivative $u_\ell^{-1}(du_\ell/dr)$ - be continuous[4] at $r = a$.

In general the phase shift - and therefore the cross sections - are slowly varying functions of the incident particle energy and of the strength of the potential. However, under certain circumstances - in particular for the case of low-energy scattering by a deep attractive potential well - it may happen that a phase shift δ_ℓ suffers a rapid variation for a given potential strength and in a certain energy interval ΔE. In this case, δ_ℓ may be written in this energy interval as

$$\delta_\ell \simeq \xi_\ell + \delta_\ell^r \qquad (2.24)$$

where $\xi_\ell = \tan^{-1}[j_\ell(ka)/n_\ell(ka)]$ is a phase shift corresponding to scattering by a "hard sphere" of radius a, and

$$\delta_\ell^r = \tan^{-1}\frac{\Gamma/2}{E_r - E}, \qquad \Gamma \ll \Delta E \tag{2.25}$$

Outside the interval $(E_r-\Gamma/2,\ E_r+\Gamma/2)$, the phase shift δ_ℓ is dominated by the "hard sphere" contribution ξ_ℓ, and the incident wave does not penetrate deeply in the scattering region. On the contrary, within the small energy region $(E_r-\Gamma/2,\ E_r+\Gamma/2)$, the quantity δ_ℓ^r increases rapidly through an odd multiple of $\pi/2$ and governs the behaviour of δ_ℓ. If the "hard sphere" scattering in the partial wave ℓ can be neglected, together with the scattering contributed by the other partial waves, one has

$$\frac{d\sigma}{d\Omega} \simeq \frac{d\sigma_\ell}{d\Omega} = \frac{(2\ell+1)^2}{k^2}\,\frac{\Gamma^2/4}{(E_r-E)^2 + \Gamma^2/4}\,P_\ell^2(\cos\theta) \tag{2.26}$$

and we see that for any angle θ the quantity $d\sigma/d\Omega$ exhibits a sharp peak of width Γ about the value $E = E_r$. Moreover, in the neighbourhood of E_r the angular distribution does not depend on the energy, but on the angular momentum ℓ. This behaviour is characteristic of a pure, narrow resonance. From (2.26) the pure resonance total cross section is

$$\sigma_{tot} \simeq \sigma_\ell = \frac{4\pi(2\ell+1)}{k^2}\,\frac{\Gamma^2/4}{(E_r-E)^2 + \Gamma^2/4} \tag{2.27}$$

which is known as the Breit-Wigner formula. At $E = E_r$ the pure resonance total cross section (2.27) reaches its maximum possible value $\sigma_\ell^{max} = 4\pi(2\ell+1)/k^2$. The energy E_r is known as the position and Γ as the width of the resonance. In general, departures from the Lorentz shape (2.27) arise, due for example to "hard sphere" scattering in the partial wave ℓ, or to the scattering contributed by other partial waves.

The physical significance of a narrow resonance may be inferred by looking at the radial wave function $u_\ell(k,r)$ near $E = E_r$. One finds that the probability of finding the scattered particle within the potential well becomes large near the energy $E = E_r$ for which a resonance appears in the ℓ th partial wave, so that the particle is nearly bound in the well. Thus a resonance corresponds to a metastable state whose lifetime τ is much longer than a typical collision time. As a result, there is a long time delay in the scattering. Using Heisenberg's uncertainty relation $\Delta E\ \Delta t > \hbar$, with $\Delta E \approx \Gamma$ and $\Delta t \simeq \tau$, we find that $\tau \simeq \hbar/\Gamma$.

The integral equation of potential scattering

For potentials which vanish faster than r^{-1} at large r the stationary scattering wave function $\psi^{(+)}_{\mathbf{k}_i}$ has been defined above as a solution of the Schrödinger equation (2.1) satisfying the boundary condition (2.2). It can be shown[1-3] that $\psi^{(+)}_{\mathbf{k}_i}$ is also a solution of an equivalent integral equation - the Lippmann-Schwinger equation - which directly takes into account the boundary condition (2.2). That is

$$\psi^{(+)}_{\mathbf{k}_i}(\mathbf{r}) = \Phi_{\mathbf{k}_i}(\mathbf{r}) + \int G_0^{(+)}(\mathbf{r},\mathbf{r}')\, U(\mathbf{r}')\, \psi^{(+)}_{\mathbf{k}_i}(\mathbf{r}')d\mathbf{r}' \qquad (2.28)$$

where $\Phi_{\mathbf{k}_i}(\mathbf{r}) = \langle \mathbf{r}|\mathbf{k}_i\rangle = (2\pi)^{-3/2}\exp(i\mathbf{k}_i\cdot\mathbf{r})$ is a plane wave corresponding to the incident momentum $\mathbf{k}_i$ and the free Green's function $G_0^{(+)}(\mathbf{r},\mathbf{r}')$ is given by

$$G_0^{(+)}(\mathbf{r},\mathbf{r}') = -(2\pi)^{-3}\int \frac{e^{i\mathbf{k}'\cdot(\mathbf{r}-\mathbf{r}')}}{k'^2-k^2-i\varepsilon}\, d\mathbf{k}', \qquad \varepsilon \to o^+ \qquad (2.29)$$

or

$$G_0^{(+)}(\mathbf{r},\mathbf{r}') = -\frac{1}{4\pi}\frac{e^{ik|\mathbf{r}-\mathbf{r}'|}}{|\mathbf{r}-\mathbf{r}'|} \qquad (2.30)$$

By looking at the asymptotic behaviour of (2.28) one can show that the scattering amplitude f is given by the integral representation

$$f = -2\pi^2 \langle \Phi_{\underset{\sim}{k}_f} | U | \psi^{(+)}_{\underset{\sim}{k}_i} \rangle = -(2\pi)^2 \langle \Phi_{\underset{\sim}{k}_f} | V | \psi^{(+)}_{\underset{\sim}{k}_i} \rangle = -(2\pi)^2 T_{fi} \tag{2.31}$$

where $\Phi_{\underset{\sim}{k}_f}(\underset{\sim}{r}) = \langle \underset{\sim}{r} | \underset{\sim}{k}_f \rangle = (2\pi)^{-3/2} \exp(i\underset{\sim}{k}_f \cdot \underset{\sim}{r})$ is a plane wave corresponding to the final momentum $\underset{\sim}{k}_f$, and $T_{fi} = \langle \Phi_{\underset{\sim}{k}_f} | V | \psi^{(+)}_{\underset{\sim}{k}_i} \rangle$ is called the transition matrix element. We remark that our plane waves have been "normalised" here in such a way that $\langle \Phi_{\underset{\sim}{k}_f} | \Phi_{\underset{\sim}{k}_i} \rangle = \langle \underset{\sim}{k}_f | \underset{\sim}{k}_i \rangle = \delta(\underset{\sim}{k}_i - \underset{\sim}{k}_f)$.

If the potential is spherically symmetric we readily find by analyzing (2.28) in partial waves that the radial functions $R_\ell(k,r)$, defined in (2.8), and normalized in such a way that

$$R_\ell(k,r) \underset{r\to\infty}{\to} j_\ell(kr) - n_\ell(kr)\, K_\ell(k) \tag{2.32}$$

satisfy the integral equations

$$R_\ell(k,r) = j_\ell(kr) + \int_0^\infty G_\ell(r,r')\, U(r')\, R_\ell(k,r') r'^2\, dr' \tag{2.33}$$

Here

$$G_\ell(r,r') = k\, j_\ell(kr_<)\, n_\ell(kr_>) \tag{2.34}$$

where $r_<$ is the lesser and $r_>$ the greater of r and r'. Moreover, we have for $\tan \delta_\ell(k)$ the integral representation

$$\tan \delta_\ell = -k \int_0^\infty j_\ell(kr)\, U(r)\, R_\ell(k,r)\, r^2\, dr \tag{2.35}$$

Scattering by a Coulomb potential

The discussion of the above paragraphs must be modified when the potential behaves asymptotically as r^{-1}. We start by considering the scattering by a pure Coulomb potential acting between two particles having respectively the charges Z_1 and Z_2, namely

$$V_c(r) = \frac{Z_1 Z_2}{r} \tag{2.36}$$

Setting $\gamma = Z_1Z_2/k$, it may be shown[5] by using parabolic coordinates that the Schrödinger equation admits the regular solution

$$\psi_c = C\, e^{ikz}\, {}_1F_1(-\,i\gamma;\, 1;\, ik(r-z)) \qquad (2.37)$$

where C is a constant. Moreover, one has[5]

$$\psi_c \underset{|r-z|\to\infty}{\longrightarrow} C\, \frac{e^{\pi\gamma/2}}{\Gamma(1+i\gamma)} \left[e^{i[kz+\gamma \log kr(1-\cos\theta)]} \left(1 + \frac{\gamma^2}{ikr(1-\cos\theta)} + \ldots\right) + f_c(k,\theta)\, \frac{e^{i(kr-\gamma\log 2kr)}}{r} \left(1 + \frac{(1+i\gamma)^2}{ikr(1-\cos\theta)} + \ldots\right)\right] \qquad (2.38)$$

where

$$f_c(k,\theta) = -\,\frac{\gamma}{2k\sin^2\frac{1}{2}\theta}\, e^{-i\gamma\log(\sin^2\frac{1}{2}\theta) + 2i\sigma_0} \qquad (2.39)$$

with $\sigma_0 = \arg\Gamma(1+i\gamma)$. We see that both the incident wave and the outgoing scattered wave are distorted by logarithmic phase factors; this is a direct consequence of the long range nature of the Coulomb potential. The differential cross section is given by

$$\frac{d\sigma_c}{d\Omega} = |f_c(k,\theta)|^2 = \frac{\gamma^2}{4k^2\sin^4\frac{1}{2}\theta} = \frac{(Z_1Z_2)^2}{16E^2\sin^4\frac{1}{2}\theta} \qquad (2.40)$$

a result first obtained by Rutherford using classical mechanics. We remark that the differential cross section (2.40) diverges in the forward direction and that the total Coulomb cross section obtained by integrating (2.40) over all scattering angles is infinite.

For most electron collision problems, the interaction is not a pure Coulomb potential. For example, in the case of electron scattering by a positive or negative ion, the screening due to the target electrons leads to a deviation from the form (2.36) at short distances. In this case, it is useful to make a partial wave expansion of the wave equation. The radial equation now reads

$$\left[\frac{d^2}{dr^2} - \frac{\ell(\ell+1)}{r^2} + k^2 - U_c(r) - U_s(r)\right]\chi_\ell(k,r) = 0 \tag{2.41}$$

where $U_c(r) = 2V_c(r) = 2\gamma k/r$ and $U_s(r) = 2V_s(r)$, V_s being a short range potential. For r larger than a value a chosen so that $U_s(r)$ can be neglected compared with $U_c(r)$ and $\ell(\ell+1)/r^2$, the equation (2.41) reduces to the radial Schrödinger equation for the Coulomb potential, namely

$$\left[\frac{d^2}{dr^2} - \frac{\ell(\ell+1)}{r^2} + k^2 - \frac{2\gamma k}{r}\right]w_\ell(k,r) = 0 \tag{2.42}$$

The regular and an irregular solution of this equation are the spherical Coulomb functions $F_\ell(k,r)$ and $G_\ell(k,r)$ such that[5]

$$F_\ell(k,r) \underset{r\to\infty}{\to} \sin(kr - \frac{\ell\pi}{2} - \gamma \log 2kr + \sigma_\ell(k)) \tag{2.43}$$

and

$$G_\ell(k,r) \underset{r\to\infty}{\to} - \cos(kr - \frac{\ell\pi}{2} - \gamma \log 2kr + \sigma_\ell(k)) \tag{2.44}$$

where $\sigma_\ell = \arg \Gamma(\ell+1+i\gamma)$ is called the Coulomb phase shift. In analogy with (2.23), the regular solution of (2.41) can be written for $r \geqslant a$ as a linear combination of the spherical Coulomb functions $F_\ell(k,r)$ and $G_\ell(k,r)$. That is,

$$\chi_\ell(k,r) = A_\ell(k)[\cos \hat{\delta}_\ell(k)\, F_\ell(k,r) - \sin \hat{\delta}_\ell(k)\, G_\ell(k,r)], \quad r \geqslant a \tag{2.45}$$

so that

$$\chi_\ell(k,r) \underset{r\to\infty}{\to} A_\ell(k) \sin(kr - \frac{\ell\pi}{2} - \gamma \log 2kr + \sigma_\ell(k) + \hat{\delta}_\ell(k)) \tag{2.46}$$

The quantity $\hat{\delta}_\ell$ is the phase shift due to the short range potential V_s in the presence of the Coulomb potential V_c. We remark that $\hat{\delta}_\ell$ vanishes when V_s is not present. The scattering amplitude is

$$f(k,\theta) = f_c(k,\theta) + f_s(k,\theta) \tag{2.47}$$

where f_c is the Coulomb amplitude (2.42) and the amplitude f_s arising from the additional short range potential is

$$f_s(k,\theta) = \frac{1}{2ik}\sum_{\ell=o}^{\infty}(2\ell+1)\, e^{2i\sigma_\ell(k)}[e^{2i\hat{\delta}_\ell(k)} - 1]P_\ell(\cos\theta) \quad (2.48)$$

The differential cross section is given by

$$\frac{d\sigma}{d\Omega} = |f_c(k,\theta) + f_s(k,\theta)|^2$$
$$= |f_c(k,\theta)|^2 + |f_s(k,\theta)|^2 + 2\mathrm{Re}\left[f_c^*(k,\theta)\, f_s(k,\theta)\right] \quad (2.49)$$

and we see that it contains an interference term between the Coulomb amplitude and the amplitude f_s.

The Born series

We shall now discuss briefly a few important approximation methods. We begin by the Born series, which is obtained if one elects to solve the Lippmann-Schwinger equation (2.28) by perturbation theory. Starting from the "unperturbed" incident plane wave $\Phi_{\underset{\sim}{k}_i}(\underset{\sim}{r})$, we then generate for $\psi^{(+)}_{\underset{\sim}{k}_i}$ the Born series

$$\psi^{(+)}_{\underset{\sim}{k}_i}(\underset{\sim}{r}) = \sum_{n=o}^{\infty}\phi_n(\underset{\sim}{r}) \quad (2.50)$$

with

$$\phi_o(\underset{\sim}{r}) = \Phi_{\underset{\sim}{k}_i}(\underset{\sim}{r}) , \qquad \phi_n(\underset{\sim}{r}) = \int K_n(\underset{\sim}{r},\underset{\sim}{r}')\, \phi_o(\underset{\sim}{r}')d\underset{\sim}{r}' , \ n \geqslant 1$$

and

$$K_1(\underset{\sim}{r},\underset{\sim}{r}') = G_0^{(+)}(\underset{\sim}{r},\underset{\sim}{r}')U(\underset{\sim}{r}'), \quad K_n(\underset{\sim}{r},\underset{\sim}{r}') = \int K_1(\underset{\sim}{r},\underset{\sim}{r}'')K_{n-1}(\underset{\sim}{r}'',\underset{\sim}{r}')d\underset{\sim}{r}''$$
$$n \geqslant 2 \quad (2.51)$$

We see that the Born series (2.50) is a perturbation series in powers of the interaction potential. Substituting the series (2.50) into the expression (2.31) we obtain the Born series for the scattering amplitude, namely

$$f = \sum_{n=1}^{\infty} \bar{f}_{Bn} \tag{2.52}$$

where

$$\bar{f}_{B1} = -2\pi^2 \langle \Phi_{\mathbf{k}_f} | U | \Phi_{\mathbf{k}_i} \rangle = -\frac{1}{4\pi} \int e^{i\boldsymbol{\Delta}\cdot\mathbf{r}} U(\mathbf{r}) d\mathbf{r} \tag{2.53}$$

and

$$\bar{f}_{Bn} = -2\pi^2 \langle \Phi_{\mathbf{k}_f} | U G_0^{(+)} U \ldots G_0^{(+)} U | \Phi_{\mathbf{k}_i} \rangle , \quad n \geqslant 2 \tag{2.54}$$

In (2.53) we have introduced the momentum transfer $\boldsymbol{\Delta} = \mathbf{k}_i - \mathbf{k}_f$, while in (2.54) the potential appears n times and the Green's function (n-1) times. It is apparent from the above equations that the Born series (2.52) may be pictured as a multiple scattering series in which the particle interacts repeatedly with the potential and propagates freely between two successive interactions.

Except for the first Born term $\bar{f}_{B1}$, the evaluation of the quantities $\bar{f}_{Bn}$ is in general difficult. For screened Coulomb (Yukawa) potentials $U(r) = U_0 \exp(-\alpha r)/r$, superposition of such potentials or "Yukawa-type" potentials

$$U(r) = \sum_{i=1}^{N} U_{o,i}\, r^{n_i}\, e^{-\alpha_i r}, \qquad (n_1=-1,\ n_i \geqslant -1,\ i \geqslant 2,\ \alpha_i > 0) \tag{2.55}$$

which play an important role in electron-atom collisions, it is convenient to work in momentum space. The basic integrals to be considered are then Dalitz integrals[6] of the type

$$I_{m,n}(\alpha,\beta;\mathbf{k}_i,\mathbf{k}_f,k) = \int d\boldsymbol{\kappa} \frac{1}{\kappa^2-k^2-i\varepsilon} \frac{1}{(\alpha^2+|\boldsymbol{\kappa}-\mathbf{k}_i|^2)^m(\beta^2+|\boldsymbol{\kappa}-\mathbf{k}_f|^2)^n}$$

$$(m,n = 1,2,\ldots) \tag{2.56}$$

A Born series for partial waves may also be obtained by solving the radial integral equations (2.33) by successive iterations, starting from the zero-order approximation $R_\ell^{(o)}(k,r)=j_\ell(kr)$. In particular, the first Born approximation for tan δ_ℓ is given by

$$(\tan \delta_\ell)_{B1} = -k \int_0^\infty [j_\ell(kr)]^2 \, U(r) r^2 dr \tag{2.57}$$

The eikonal approximation.

Originally used in optics[7], the eikonal method was first introduced in quantum collision theory by Molière[8] and considerably developed by Glauber[9], who proposed a very fruitful generalization of the method. Detailed discussions of the properties of the eikonal approximation and related methods can be found in Joachain[2] and Byron and Joachain[10].

Let us assume that the "short wavelength" (semi-classical) condition $ka \gg 1$ is satisfied (where a is the range of the potential), together with the "high-energy" requirement $|V_0|/E = |U_0|/k^2 \ll 1$ (where $|V_0|$ is a typical strength of the potential and $|U_0| = 2|V_0|$). If these two conditions are satisfied, the free Green's function (2.29) may, when used in the Lippmann-Schwinger equation (2.28) be written approximately as

$$G_0^{(+)}(\mathbf{r},\mathbf{r}') \simeq -(2\pi)^{-3} e^{i\mathbf{k}_i\cdot(\mathbf{r}-\mathbf{r}')} \int \frac{e^{i\mathbf{Q}\cdot(\mathbf{r}-\mathbf{r}')}}{2\mathbf{k}_i\cdot\mathbf{Q} - i\varepsilon} d\mathbf{Q} \tag{2.58}$$

where we have set $\mathbf{Q} = \boldsymbol{\kappa} - \mathbf{k}_i$ and linearized the denominator of the integrand by neglecting the Q^2 term. Upon insertion in the Lippmann-Schwinger equation, the linearized propagator (2.58) leads to the eikonal wave function

$$\psi_E(\mathbf{r}) = (2\pi)^{-3/2} \exp[\, i\mathbf{k}_i\cdot\mathbf{r} - \frac{i}{2k} \int_{-\infty}^{z} U(\mathbf{b},z')dz'] \tag{2.59}$$

In writing (2.59) we have adopted a cylindrical coordinate system, with $\mathbf{r} = \mathbf{b} + z\hat{\mathbf{k}}_i$ and the integral is evaluated along a straight line parallel to $\mathbf{k}_i$. In fact, since the actual phase of the scattering wave function should be evaluated in the semi-classical limit along a curved trajectory, it is reasonable to expect that an improvement may be achieved by performing the z' integration in the phase along a straight line parallel to the bisector of the scattering angle θ (i.e. perpendicular to the momentum transfer $\boldsymbol{\Delta}$). This suggestion, first made by Glauber[9], does indeed improve the scattering amplitude. Working in a cylindrical coordinate system such that $\mathbf{r} = \mathbf{b} + z\hat{\mathbf{n}}$, where $\hat{\mathbf{n}}$ is perpendicular to $\boldsymbol{\Delta}$, we then obtain the eikonal scattering amplitude

$$f_E = \frac{k}{2\pi i}\int e^{i\boldsymbol{\Delta}\cdot\mathbf{b}}\,[e^{i\chi(k,\mathbf{b})} - 1]\,d^2\mathbf{b} \qquad (2.60)$$

where the eikonal phase shift function $\chi(k,\mathbf{b})$ is given by

$$\chi(k,\mathbf{b}) = \frac{1}{k}\chi_o(\mathbf{b}), \qquad \chi_o(\mathbf{b}) = -\frac{1}{2}\int_{-\infty}^{+\infty} U(\mathbf{b},z)\,dz \qquad (2.61)$$

For potentials which possess azimuthal symmetry, (2.60) reduces to

$$f_E = \frac{k}{i}\int_0^{\infty} J_0(\Delta b)\,[e^{i\chi(k,b)} - 1]\,b\,db \qquad (2.62)$$

where J_0 is an ordinary Bessel function or order zero.

Unlike the first Born approximation, the eikonal approximation satisfies unitarity. The eikonal method may be generalized to complex potentials, in which case the phase shift function $\chi(k,\mathbf{b})$ becomes complex.

By analogy with the Born series (2.52), we may define an eikonal series by expanding the eikonal amplitude f_E as a power series in the potential strength. Thus we write

$$f_E = \sum_{n=1}^{\infty} \bar{f}_{En} \qquad (2.63)$$

where

$$\bar{f}_{En} = k^{1-n}\frac{i^{n-1}}{2\pi n!}\int e^{i\boldsymbol{\Delta}\cdot\mathbf{b}}\,[\chi_o(\mathbf{b})]^n\,d^2\mathbf{b} \qquad (2.64)$$

and we note that $\bar{f}_{E1} = \bar{f}_{B1}$. Remarkable relationships exist between the higher terms ($n \geqslant 2$) of the eikonal and Born series[2,10,11]. We consider real; central potentials so that the terms $\bar{f}_{En}$ are alternately real (n odd) and imaginary (n even). Furthermore, we restrict our attention to "Yukawa-type" interactions. Table 1 shows the dependence of the Born terms $\bar{f}_{Bn}$ (for large k) and of the eikonal terms $\bar{f}_{En}$ as a function of k and of the magnitude Δ of the momentum transfer. We remark that for large Δ the results quoted in Table 1 are given to leading order in k^{-1}, Δ^{-1} and $\log \Delta$. It is worth noting that at large Δ the convergence of the Born series is slower than at small Δ. It is apparent from Table 1 that at all Δ the dominant contribution (which is framed) is given by the first Born term.

TABLE 1

Dependence of various terms of the Born and eikonal series for scattering by Yukawa-type potentials, as a function of (large) k and Δ. The dominant contributions are framed. The terms located above the dashed line contribute through order k^{-2} to the differential cross section

Order of Pert. Theory	Term	Small Δ ($\Delta < 1$)	Large Δ ($\Delta > k$)
First	$\bar{f}_{B1} = \bar{f}_{E1}$	$\boxed{1}$	$\boxed{\Delta^{-2}}$
Second	Re $\bar{f}_{B2}$	k^{-2}	$k^{-2}\ \Delta^{-2}$
	Re $\bar{f}_{E2}$	0	0
	Im $\bar{f}_{B2}$	k^{-1}	$k^{-1}\ \Delta^{-2}\ \log \Delta$
	Im $\bar{f}_{E2}$	k^{-1}	$k^{-1}\ \Delta^{-2}\ \log \Delta$
Third	Re $\bar{f}_{B3}$	k^{-2}	$k^{-2}\ \Delta^{-2}\ \log^2 \Delta$
	Re $\bar{f}_{E3}$	k^{-2}	$k^{-2}\ \Delta^{-2}\ \log^2 \Delta$
	- - - - - -	- - - - - -	- - - - - -
	Im $\bar{f}_{B3}$	k^{-3}	$k^{-3}\ \Delta^{-2}\ \log \Delta$
	Im $\bar{f}_{E3}$	0	0
n^{th} ($n > 3$)	$\bar{f}_{Bn}$	$(ik)^{1-n}$	$(ik)^{1-n}\ \Delta^{-2}\ \log^{n-1} \Delta$
	$\bar{f}_{En}$	$(ik)^{1-n}$	$(ik)^{1-n}\ \Delta^{-2}\ \log^{n-1} \Delta$

We also remark from Table 1 that a consistent calculation of the scattering amplitude and of the differential cross section through order k^{-2} requires the inclusion of the terms $\bar{f}_{B1}$, Re $\bar{f}_{B2}$, Im $\bar{f}_{B2}$ (or Im $\bar{f}_{E2}$) and Re $\bar{f}_{B3}$ (or $\bar{f}_{E3}$) which lie above the dashed line. It is worth stressing that neither the second Born amplitude $f_{B2}=\bar{f}_{B1}+\bar{f}_{B2}$ nor the eikonal amplitude f_E are correct to order k^{-2}. On the other hand the "eikonal-Born series" (EBS) amplitude[12]

$$f_{EBS} = \bar{f}_{B1} + \bar{f}_{B2} + \bar{f}_{E3} \tag{2.65}$$

is consistent through order k^{-2}. This is particularly interesting since the term $\bar{f}_{E3}$ (which for large k reproduces the quantity Re $\bar{f}_{B3}$ at all angles) is much easier to calculate than Re $\bar{f}_{B3}$.

Variational methods

Although bound state problems have been studied by means of variational methods since the early days of quantum theory, it was only much later that variational principles were proposed to analyze scattering problems[13-16]. In what follows we shall only consider the Hulthén-Kohn method, which has been widely applied to electron-atom scattering problems.

Let u_ℓ^t be a trial function satisfying the same boundary and asymptotic conditions as the exact radial function u_ℓ (normalized according to (2.17), with $N_1 = 1$), and let $I_\ell[u_\ell^t]$ be the functional

$$I_\ell[u_\ell^t] = \int_0^\infty u_\ell^t \, L_\ell[u_\ell^t]\, dr \tag{2.66}$$

where $L_\ell \equiv [d^2/dr^2 + k^2 - \ell(\ell+1)/r^2 - U(r)]$. If δI_ℓ and δK_ℓ are small changes in I_ℓ and $K_\ell = \tan \delta_\ell$ resulting from a small change δu_ℓ in u_ℓ, one has

$$\delta[I_\ell + k K_\ell] = 0 \tag{2.67}$$

This is known as the Hulthén-Kohn variational principle. It can be used to obtain an approximate value of the phase shift and an approximation to the radial function, by choosing a trial function $u_\ell^t(c_1,c_2,\ldots c_n,K_\ell^t,r)$ depending on n parameters $c_1,c_2,\ldots c_n$ and on a trial K-matrix element $K_\ell^t = \tan\delta_\ell^t$. The function u_ℓ^t is substituted in (2.66) and the n+1 parameters $c_1,c_2,\ldots c_n,K_\ell^t$ are determined from (2.67) by requiring that

$$\frac{\partial I_\ell}{\partial c_i} = 0 \quad (i = 1,2,\ldots n), \qquad \frac{\partial I_\ell}{\partial K_\ell^t} = -k \tag{2.68}$$

A variational estimate of K_ℓ, accurate to the second order, is then given by

$$[K_\ell] = K_\ell^t + k^{-1} I_\ell \tag{2.69}$$

A similar Hulthén-Kohn variational principle may be written down for the scattering amplitude. A comprehensive account of the properties of the Hulthén-Kohn and related variational methods has been given by Nesbet[17].

Analytic properties of scattering amplitudes

To conclude this outline of potential scattering, we remark that scattering amplitudes can be studied as functions of complex wave numbers, complex energies and even complex angular momenta. Detailed discussions may be found for example in the books by de Alfaro and Regge[18], Newton[19], Joachain[2] and Burke[3]. In particular, we mention that for a large class of potentials the scattering amplitude satisfies the forward dispersion relation

$$\mathrm{Re}\, f(E) = \bar{f}_{B1} + \sum_{i=1}^{N} \frac{R_i}{E-E_i} + \frac{P}{\pi}\int_0^\infty \frac{\mathrm{Im}\, f(E')}{E'-E}\, dE' \tag{2.70}$$

where E_i are the bound state energies, R_i the corresponding residues, and the symbol P means that the principal value of the integral must be taken.

2.2. Multichannel Scattering. General Features

We now turn to the analysis of electron collisions with atoms and ions, taking into account the internal structure of the target and the fact that electrons are spin-1/2 particles. This problem is much more complicated than the situation discussed in Section 2.1. Indeed, if the incident electron has enough energy, the target can be excited or ionized during the collision, so that a number (finite or infinite) of channels will be open. In addition, since the incident and target electrons are identical, the total scattering wave function must satisfy the Pauli exclusion principle. This gives rise to exchange effects which are in general difficult to handle.

As we have seen in Section 2.1, the scattering amplitude provides a meeting ground between theory and experiment. One of the major developments which have taken place recently in electron-atom collisions has been towards "perfect scattering experiments", in which the complete scattering amplitudes are determined (apart from an unobservable phase factor). In potential scattering there is one scattering amplitude to be determined for a given energy and scattering direction, but in electron-atom (ion) collisions this only happens in special cases, such as the elastic scattering of electrons by spin-zero atoms where the spin-orbit interaction is negligible (for example in electron-helium scattering). In many other electron-atom scattering processes, more than one scattering amplitude is required for a complete description of the collision and several independent measurements must be performed, involving for example polarized electron beams and (or) targets, or electron-photon coincidence experiments.

Polarized electrons[20]

As a first example, let us consider the elastic scattering of electrons by spin-zero atoms, including spin-orbit effects. In this case there are two scattering amplitudes, the "non-spin-flip" amplitude f and the spin-flip amplitude g. Using the density matrix formalism, one finds that the differential cross section for an incident electron beam with polarization $\mathbf{P}_i$ is given by[20,21]

$$\frac{d\sigma}{d\Omega} = \left(\frac{d\sigma}{d\Omega}\right)_{unpol} [1 + S(\theta)\, \mathbf{P}_i \cdot \hat{\mathbf{n}}] \tag{2.71}$$

where

$$\left(\frac{d\sigma}{d\Omega}\right)_{unpol} = |f|^2 + |g|^2 \tag{2.72}$$

is the differential cross section for an unpolarized beam,

$$S(\theta) = \frac{f^* g + f g^*}{|f|^2 + |g|^2} = \frac{2\mathrm{Re}(f g^*)}{(d\sigma/d\Omega)_{unpol}} \tag{2.73}$$

is called the Sherman function, and $\hat{\mathbf{n}} = (\mathbf{k}_i \times \mathbf{k}_f)/|\mathbf{k}_i \times \mathbf{k}_f|$ is the unit vector perpendicular to the scattering plane. The result (2.71) provides the basic equation for the measurement of electron polarization by "Mott scattering". It is worth noting that an initially unpolarized electron beam ($\mathbf{P}_i = 0$) is polarized by the scattering, the polarization $\mathbf{P}_f$ of the scattered beam being given in this case by $\mathbf{P}_f = S(\theta)\hat{\mathbf{n}}$. More generally, for an incident electron beam with polarization $\mathbf{P}_i$, the polarization $\mathbf{P}_f$ after scattering is given by[20]

$$\mathbf{P}_f = \frac{[\mathbf{P}_i \cdot \hat{\mathbf{n}} + S(\theta)]\hat{\mathbf{n}} + T(\theta)[\mathbf{P}_i - (\mathbf{P}_i \cdot \hat{\mathbf{n}})\hat{\mathbf{n}}] + U(\theta)[\hat{\mathbf{n}} \times \mathbf{P}_i]}{1 + S(\theta)\, \mathbf{P}_i \cdot \hat{\mathbf{n}}} \tag{2.74}$$

where $S(\theta)$ is the Sherman function (2.73), while

$$T(\theta) = \frac{|f|^2 - |g|^2}{|f|^2 + |g|^2}, \qquad U(\theta) = i\, \frac{f^* g - f g^*}{|f|^2 + |g|^2} \tag{2.75}$$

The function $S(\theta)$ can be determined by performing a double scattering experiment, while a triple scattering experiment is needed to measure the quantities $T(\theta)$ and $U(\theta)$. These experiments are discussed in detail by Kessler[20]. Since $S^2 + T^2 + U^2 = 1$, the measurement of S, T and U only yields two independent results. A third independent result is obtained by measuring the differential cross section for an unpolarized beam, given by (2.72). Three of the four parameters of the two complex amplitudes $f = |f| \exp(i\lambda)$ and $g = |g| \exp(i\mu)$ can thus be measured, namely $|f|$, $|g|$ and the phase difference $(\lambda-\mu)$. As a result, the two amplitudes f and g are known, except for an unobservable phase factor.

Another simple, but interesting example is provided by the elastic scattering of electrons by spin-1/2 atoms, in the absence of spin-orbit interaction (e.g. in atomic hydrogen or in the light alkalis). We begin by considering the case in which the incident electrons and the target atoms are completely polarized. Denoting by A↑ the target atoms having a given spin direction, the scattering of electrons with spin parallel or antiparallel to the atomic spin leads to the following possibilities[20]

Process	Cross Section	
$e\downarrow + A\uparrow \to e\downarrow + A\uparrow$	$\|f(\theta)\|^2$	(2.76)
$e\downarrow + A\uparrow \to e\uparrow + A\downarrow$	$\|g(\theta)\|^2$	(2.77)
$e\uparrow + A\uparrow \to e\uparrow + A\uparrow$	$\|f(\theta) - g(\theta)\|^2$	(2.78)

The scattering amplitude $f(\theta)$ is called the direct amplitude, while $g(\theta)$ is the exchange amplitude. It is worth stressing that this exchange amplitude arises as a consequence of the Pauli exclusion principle and should not confused with the spin-orbit amplitude introduced in the previous example. We note that the use of spin-polarized electron beams and targets enables the separation of direct and exchange scattering. In fact, it is sufficient for the observation of the individual cross sections listed in (2.76)-(2.78) to perform simpler experiments, in which either the electrons or the target atoms are initially unpolarised[20]. We also remark that the exchange interaction can produce a spin polarization in electron scattering, for example if free electrons are fired on alkali atoms whose valence electrons all have a given spin direction. The differential cross section for elastic scattering with unpolarized particles is given by

$$\left(\frac{d\sigma}{d\Omega}\right)_{unpol} = \frac{1}{2}|f|^2 + \frac{1}{2}|g|^2 + \frac{1}{2}|f-g|^2$$
$$= \frac{1}{4}|f^+|^2 + \frac{3}{4}|f^-|^2 \qquad (2.79)$$

where $f^+ = f + g$ and $f^- = f - g$ are respectively the singlet (S=0) and triplet (S=1) scattering amplitudes. Thus we see that two scattering amplitudes (f and g, or f^+ and f^-) are necessary to describe the elastic scattering of electrons by spin-1/2 atoms, if spin-orbit interactions are neglected[22]. Thus three independent measurements must be performed at each energy and scattering angle to determine these amplitudes within an unobservable phase. Polarization experiments which have been carried out to measure the quantities $|f(\theta)|$ and $|g(\theta)|$ are discussed for example in the book by Kessler[20], where an analysis of polarization effects arising in inelastic electron-atom collisions (either through the spin-orbit interaction or the exchange interaction) can also be found.

Electron-photon coincidences and angular correlations.

Let us now consider electron-photon coincidence experiments, in which the atom is excited in the collision, and the angular distribution of the radiation emitted in the decay of the excited state is observed in coincidence with the scattered electron. The theory of such measurements has been developed extensively in recent years[23-27]. As an example, we shall analyze the case of the $1^1S - 2^1P$ transition in helium, for which the first electron-photon angular correlation has been measured. In a collision at intermediate energies, the excited 2^1P states are populated in a time of the order of 10^{-15} s, which is much shorter than any characteristic lifetime of these states. Neglecting spin-orbit and spin-spin interactions, the excitation into the 2^1P states can therefore be described as a coherent superposition of excitations into degenerate magnetic sub-levels[23], and immediately after the collision the eigenstate $\psi(2^1P)$ of the excited atom may be written as

$$\psi(2^1P) = a(1)|11\rangle + a(o)|10\rangle + a(-1)|1-1\rangle \tag{2.80}$$

where the amplitudes a(M) are functions of the incident electron energy E and the electron scattering angles (θ,ϕ) and describe the excitation of the magnetic sub-levels $|LM\rangle$ (with L = 1). The dependence of ϕ can be factored out as follows

$$a(M) = a_M(E,\theta)\ e^{-iM\phi} \tag{2.81}$$

and the mirror symmetry of the electron-atom scattering process in the plane of scattering implies that $a_{-1} = -a_1$. Hence, there are two independent scattering amplitudes, so that again three independent measurements are required at each energy and scattering angle. Setting $a_o = |a_o| \exp(i\alpha_o)$ and $a_1 = |a_1| \exp(i\alpha_1)$ these measurements are conveniently parametrized by the three quantities

$$\sigma = \sigma_o + 2\sigma_1, \qquad \lambda = \sigma_o/\sigma, \qquad \chi = \alpha_1 - \alpha_o \tag{2.82}$$

where $\sigma_o = |a_o|^2$ and $\sigma_1 = |a_1|^2$ are partial differential cross sections for exciting the magnetic sub-levels $|10\rangle$ and $|1\pm1\rangle$, respectively. We note that σ is the differential cross section for excitation of the 2^1P state and that $0 \leqslant \lambda \leqslant 1$, while $-\pi \leqslant \chi \leqslant \pi$.

The joint probability density for scattering of the electron in the direction (θ_e,ϕ_e) in any 2^1P excitation, with subsequent emission of the photon in the direction $(\theta_\gamma,\phi_\gamma)$, summed over photon polarizations, can be written as[23,28]

$$\frac{d^2P_c}{d\Omega_e\, d\Omega_\gamma} = \frac{\sigma}{\sigma_{tot}} \frac{dP_c}{d\Omega_\gamma} \tag{2.83}$$

where σ_{tot} is the total (integrated) cross section for excitation of the 2^1P level at energy E, and $dP_c/d\Omega_\gamma$ is the conditional probability density for photon emission after electron scattering in a particular direction, which is given by

$$\frac{dP_c}{d\Omega_\gamma} = \frac{3}{8\pi} \{\lambda \sin^2\theta_\gamma + (\frac{1-\lambda}{2})(1 + \cos^2\theta_\gamma) - (\frac{1-\lambda}{2})\sin^2\theta_\gamma \cos 2(\phi_\gamma - \phi_e)$$
$$+ [\lambda(1-\lambda)]^{1/2} \cos \chi \sin 2\theta_\gamma \cos(\phi_\gamma - \phi_e)\} \qquad (2.84)$$

The remarkably simple dependence on θ_γ and ϕ_γ exhibited by this formula entirely follows from the quantum theory of radiation. On the other hand, the (complicated) dependence on the electron scattering angle θ_e, which results from electron-atom collision dynamics, is contained in λ and χ.

In the experiment of Eminyan et al[28] the electron beam is incident in the Z-direction on the target located at the origin of the coordinate system (see Fig. 1). The scattered electrons are

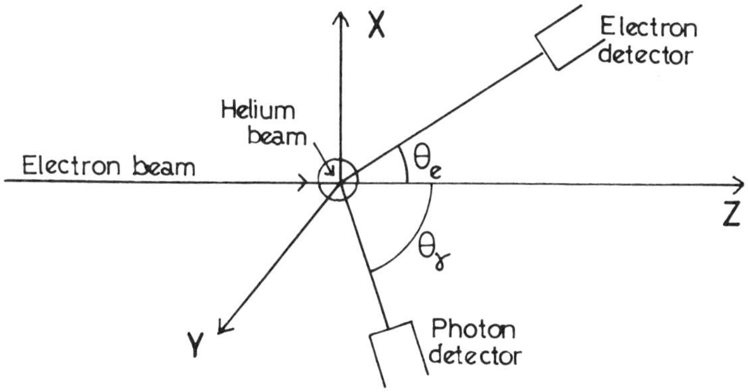

Figure 1. The geometry of the electron-photon coincidence experiment of Eminyan et al[28]. The Y axis is perpendicular to the scattering (XZ) plane. The coordinate axes XYZ define the "collision frame"

recorded by a detector whose position defines the scattering plane, which is taken to be the XZ plane, so that $\phi_e = 0$. Photons are counted without regard to polarization by a detector placed in the XZ plane on the opposite side of the electron beam from the electron detector, so that $\phi_\gamma = \pi$. In this case (2.84) takes the form $dP_c/d\Omega_\gamma = (3/8\pi)N_c$, where the angular correlation function N_c is given by

$$N_c = \lambda \sin^2\theta_\gamma + (1-\lambda)\cos^2\theta_\gamma - 2[\lambda(1-\lambda)]^{1/2}\cos\chi \sin\theta_\gamma \cos\theta_\gamma \tag{2.85}$$

The two parameters λ and $|\chi|$ can be extracted from the observed angular correlation by fitting the data to the theoretical expression (2.85). They can be related to expectation values of the atomic orbital angular momentum in the excited P state as

$$\begin{aligned} \langle L_x \rangle &= 0 & \langle L_x^2 \rangle &= \lambda \\ \langle L_y \rangle &= -2[\lambda(1-\lambda)]^{1/2} \sin\chi & \langle L_y^2 \rangle &= 1 \\ \langle L_z \rangle &= 0 & \langle L_z^2 \rangle &= 1-\lambda \end{aligned} \tag{2.86}$$

where $\underset{\sim}{L}^2 = L_x^2 + L_y^2 + L_z^2$ has eigenvalues $L(L+1)$, with $L = 1$ in the present case.

The above formulation of the electron-photon coincidence problem follows the treatment of Macek and Jaecks[23]. Fano and Macek[25] have developed a different approach in which the excited state population produced by electron impact can be described by an orientation vector $\underset{\sim}{O}^{col}$, proportional to the average angular momentum of the target atom, and by an alignment tensor A^{col} whose components are proportional to the mean values of expressions quadratic in the components of the target angular momentum. The non-vanishing components of both quantities, defined in the "collision frame" (see Fig. 1) are given in the present case by

$$\begin{aligned} O_{1-}^{col} &= \langle L_y \rangle [L(L+1)]^{-1} = -[\lambda(1-\lambda)]^{1/2} \sin\chi \\ A_0^{col} &= \langle 3L_z^2 - \underset{\sim}{L}^2 \rangle [L(L+1)]^{-1} = \tfrac{1}{2}(1-3\lambda) \\ A_{1+}^{col} &= \langle L_x L_z + L_z L_x \rangle [L(L+1)]^{-1} = [\lambda(1-\lambda)]^{1/2} \cos\chi \\ A_{2+}^{col} &= \langle L_x^2 - L_y^2 \rangle [L(L+1)]^{-1} = \lambda - 1 \end{aligned} \tag{2.87}$$

The most extensive electron-photon coincidence experiments have been carried out thus far on ^1P excitations in helium[28-33]. Recently, however, measurements have also been performed for the excitation of the 2p state of atomic hydrogen[34-37]. Data provided by electron-photon coincidence experiments have given very sensitive tests of theoretical approximations developed to calculate the scattering amplitudes which describe the electron-atom excitation process.

In addition to the experiments with polarized electrons and (or) targets, and the electron-photon coincidence measurements considered above, various other types of experiments are very useful in the analysis of electron-atom collision processes. These include the absolute measurement of differential cross sections[38], electron impact polarization studies of line radiation[39] and electron scattering from laser excited atoms[40,41].

2.3. Elastic scattering and excitation at low energies

We shall now discuss various theoretical methods[42] which have proved to be particularly useful in studying the dynamics of electron-atom (ion) collision processes. In this section we shall assume that the incident electron energy is low enough so that only a few target states can be excited (i.e. only a few channels are open). It is then possible to represent explicitly all open channels in the total electron-atom scattering wave function, and we shall consider shortly several methods which exhibit this feature. Before we do this, however, it is convenient to introduce a representation in which the total electron-atom wave function is expanded in a complete set of target atom eigenfunctions. For the sake of illustration we shall consider first the simple case of scattering by a one-electron atom or ion of nuclear charge Z. Choosing the nucleus of the atom as the origin of our coordinate system, and denoting by $\mathbf{r}_1$ and $\mathbf{r}_2$ the position vectors of the two electrons, we expand the spatial part of the full wave function as

$$\Psi^{\pm}(\mathbf{r}_1,\mathbf{r}_2) = \sum_n [F_n^{\pm}(\mathbf{r}_1)\,\psi_n(\mathbf{r}_2) \pm F_n^{\pm}(\mathbf{r}_2)\,\psi_n(\mathbf{r}_1)] \tag{2.88}$$

where the superscripts + and - refer to the singlet (S = 0) and triplet (S = 1) cases, respectively, and $\psi_n(\mathbf{r}) \equiv \langle \mathbf{r}|n\rangle$ is an eigenfunction of the target corresponding to an eigenenergy w_n, namely

$$(-\frac{1}{2}\nabla^2 - \frac{Z}{r} - w_n)\,\psi_n(\mathbf{r}) = 0 \tag{2.89}$$

The Schrödinger equation satisfied by $\Psi^{\pm}(\mathbf{r}_1,\mathbf{r}_2)$ is

$$(-\frac{1}{2}\nabla_1^2 - \frac{1}{2}\nabla_2^2 - \frac{Z}{r_1} - \frac{Z}{r_2} + \frac{1}{r_{12}} - E)\,\Psi^{\pm}(\mathbf{r}_1,\mathbf{r}_2) = 0 \tag{2.90}$$

By projecting this equation with the basis functions ψ_n and using (2.89) we find that the functions $F_n^{\pm}$ must satisfy the infinite set of coupled integro-differential equations

$$(\nabla^2 + k_n^2)\,F_n^{\pm}(\mathbf{r}) = \sum_m U_{nm}^{\pm}(\mathbf{r})\,F_m^{\pm}(\mathbf{r}) \tag{2.91}$$

with $k_n^2 = 2(E-w_n)$. The potential operators $U_{nm}^{\pm}$ are such that

$$U_{nm}^{\pm}(\mathbf{r})F_m^{\pm}(\mathbf{r}) = 2[\,V_{nm}(\mathbf{r})F_m^{\pm}(\mathbf{r}) \pm \int W_{nm}(\mathbf{r},\mathbf{r}')F_m^{\pm}(\mathbf{r}')d\mathbf{r}'] \tag{2.92}$$

Here V_{nm} is a local direct potential given by

$$V_{nm}(\mathbf{r}) = -\frac{Z}{r}\,\delta_{nm} + \int \psi_n^{*}(\mathbf{r}')\,\frac{1}{|\mathbf{r}-\mathbf{r}'|}\,\psi_m(\mathbf{r}')d\mathbf{r}' \tag{2.93}$$

and $W_{nm}(\mathbf{r},\mathbf{r}')$ is a non-local exchange kernel such that

$$W_{nm}(\mathbf{r},\mathbf{r}') = \psi_n^{*}(\mathbf{r}')\,\psi_m(\mathbf{r})\,[\frac{1}{|\mathbf{r}-\mathbf{r}'|} + w_n + w_m - E] \tag{2.94}$$

For scattering by a neutral atom which is initially in the state $|i\rangle$ the coupled equations (2.91) must be solved subject to the boundary conditions

$$F_n(\underset{\sim}{r}) \underset{r\to\infty}{\to} e^{i\underset{\sim}{k}_n\cdot\underset{\sim}{r}}\,\delta_{ni} + f_{no}(\theta,\phi)\,\frac{e^{ik_n r}}{r} \tag{2.95}$$

for the open channels ($k_n^2 > 0$). In the closed channels $F_n^{\pm}(\underset{\sim}{r})$ vanishes asymptotically. If the target is an ion one must modify the exponents in (2.95) by including the logarithmic phase factors due to the distortion by the Coulomb potential [see (2.38)]. For an unpolarized system we find after averaging over initial spins and summing over final ones that the differential cross section for elastic scattering or for discrete excitation to a final state $|j\rangle$ is given by

$$\frac{d\sigma_{ji}}{d\Omega} = \frac{k_j}{k_i}\left(\frac{1}{4}\,|f_{ji}^{+}|^2 + \frac{3}{4}\,|f_{ji}^{-}|^2\right) \tag{2.96}$$

The above results can readily be generalized to the case of an electron incident on an N-electron atom or ion. The Schrödinger equation then reads

$$H\,\Psi_a^{(+)} = E\,\Psi_a^{(+)} \tag{2.97}$$

where H is the Hamiltonian describing the target atom or ion (having nuclear charge Z and containing N electrons) plus an electron. We shall consider here light atomic systems, so that we may neglect all but Coulomb interactions. Thus

$$H = \sum_{i=1}^{N+1}\left(-\frac{1}{2}\nabla_i^2 - \frac{Z}{r_i}\right) + \sum_{i>j=1}^{N+1}\frac{1}{r_{ij}} \tag{2.98}$$

The Schrödinger equation (2.97) is to be solved subject to the boundary conditions of an incoming wave in channel a and an outgoing wave in all other open channels. Thus the full wave function $\Psi_a^{(+)}$, including spin variables, describing an electron incident on an atom in state $|i\rangle$ and scattered leaving the atom in any state $|j\rangle$ has the asymptotic form (for open channels)

$$\Psi_a^{(+)} \underset{r\to\infty}{\to} e^{i\mathbf{k}_i\cdot\mathbf{r}} \chi_{m_i} \psi_i + \sum_j f_{ji}(\theta,\phi) \frac{e^{ik_j r}}{r} \chi_{m_j} \psi_j \qquad (2.99)$$

where ψ_i and ψ_j are respectively the initial and final target functions, χ_{m_i} and χ_{m_j} are electron spin-1/2 functions, and the superscript (+) indicates an outgoing wave behaviour. The differential cross section for elastic scattering or excitation is then given by

$$\frac{d\sigma_{ji}}{d\Omega} = \frac{k_j}{k_i} |f_{ji}|^2 \qquad (2.100)$$

and the total cross section σ_{tot} $(i \to j)$ is obtained by integrating over the scattering angles. For the case of scattering by ions[43] the appropriate logarithmic phase factors must be included in (2.99). We also note parenthetically that in this case the quantity of interest for applications to astrophysical and laboratory plasmasis the rate coefficient $\langle v\, \sigma_{tot}\, (i \to j)\rangle$, where v is the electron velocity and the average is over a Maxwellian distribution.

Partial wave analysis

In order to carry out the partial wave decomposition[44] we make use of the fact that in the absence of spin-dependent terms in H, the operators $\mathbf{L}^2$, L_z, $\mathbf{S}^2$ and S_z (where $\mathbf{L}$ is the total orbital angular momentum and $\mathbf{S}$ the total spin) commute with H. We can therefore define the initial channel by $a \equiv (\alpha_i\, L_i\, S_i\, k_i\, \ell_i\, s_i\, L\, S\, M_L\, M_S\, \pi)$ where L_i and S_i refer to the atomic state, ℓ_i is the orbital angular momentum of the incident electron and s_i (= 1/2) its spin, π is the parity of the system and α_i specifies any other quantum numbers necessary to define the atomic state. Other channels (n) can be defined in a similar way. Denoting by $q_j \equiv (\mathbf{r}_j, \sigma_j)$ the combined space and spin coordinates of the jth electron, we form the eigenchannel functions $\phi_n(q_1, q_2, \ldots q_N, \hat{\mathbf{r}}_{N+1}, \sigma_{N+1})$ by coupling the target eigenstates with the spin-angle functions of the scattered electron. The total wave function is then expanded in a complete set of these eigenchannel functions,

$$\Psi_a^{(+)} = \mathcal{A} \sum_n r_{N+1}^{-1} u_{na}(r_{N+1}) \phi_n(q_1, q_2, \ldots q_N, \hat{\mathbf{r}}_{N+1}, \sigma_{N+1}) \tag{2.101}$$

where u_{na} is the radial function for an electron in channel n with an incident electron in channel a only, and $\mathcal{A}$ is the antisymmetrization operator. By projecting the Schrödinger equation (2.97) onto the eigenchannel functions ϕ_n, we obtain for the functions u_{na} an infinite set of coupled radial equations, which must be solved subject to the boundary conditions

$$u_{na}(r) \underset{r\to 0}{\sim} r^{\ell_n+1} \tag{2.102}$$

and

$$u_{na}(r) \underset{r\to\infty}{\sim} k_n^{-1/2} \left[\delta_{na} \sin \Theta_n + K_{na} \cos \Theta_n \right] \tag{2.103}$$

for open channels ($k_n^2 > 0$) while u_{na} vanishes asymptotically for closed channels ($k_n^2 < 0$). The quantity Θ_n which appears in (2.103) is given by

$$\Theta_n = k_n r - \frac{\ell_n \pi}{2} + \frac{\mathfrak{z}}{k_n} \log 2 k_n r + \sigma_n \tag{2.104}$$

where $\mathfrak{z} = Z - N$ is the "residual charge" and $\sigma_n = \arg \Gamma(\ell_n + 1 - i\mathfrak{z}/k_n)$ is the Coulomb phase shift. The matrix K_{na} is the multichannel generalization of (2.19); it can be shown to be real (from unitarity) and symmetric (from time reversal invariance).

The S and T-matrix are defined, in analogy with the equations (2.21) and (2.22), by the matrix equations

$$\underset{\sim}{S} = \underset{\sim}{I} + \underset{\sim}{T} = \frac{\underset{\sim}{I} + i \underset{\sim}{K}}{\underset{\sim}{I} - i \underset{\sim}{K}} \tag{2.105}$$

The S-matrix is unitary and symmetric. Its matrix elements can readily be related to the scattering amplitudes f_{ji}. The total cross section for the transition $\alpha_i L_i S_i \rightarrow \alpha_j L_j S_j$ is obtained by integrating (2.100) over the scattering angles, averaging over initial spin directions and summing over final spin directions. It is given by

$$\sigma_{tot}(\alpha_i L_i S_i \to \alpha_j L_j S_j) = \frac{\pi}{k_i^2} \sum_{LS\ell_i\ell_j\pi} \frac{(2L+1)(2S+1)}{2(2L_i+1)(2S_i+1)} |T_{ji}|^2 \quad (2.106)$$

The close-coupling approximation

Let us return to electron scattering by one-electron atoms or ions. In practice, only a few target eigenfunctions ψ_n corresponding to discrete levels can be retained in the expansion (2.88). Since we are dealing here with a low-energy situation for which only a few channels are open, it is reasonable to retain in (2.88) all the open channels and possibly also some closely coupled closed channels. If the eigenfunctions ψ_n kept in (2.88) are labelled from n = 1 to n = M, so that

$$\Psi^{\pm}(\mathbf{r}_1,\mathbf{r}_2) = \sum_{n=1}^{M} [F_n^{\pm}(\mathbf{r}_1)\, \psi_n(\mathbf{r}_2) \pm F_n^{\pm}(\mathbf{r}_2)\, \psi_n(\mathbf{r}_1)] \quad (2.107)$$

we obtain the system of M coupled integro-differential equations

$$(\nabla^2 + k_n^2)\, F_n^{\pm}(\mathbf{r}) = \sum_{m=1}^{M} U_{nm}^{\pm}(\mathbf{r})\, F_m^{\pm}(\mathbf{r}), \qquad n = 1,2,..M \quad (2.108)$$

These equations are called the close-coupling equations; they can be analyzed in partial waves by using the method outline above.

For electron scattering by an N-electron atom (ion) the close-coupling equations (in partial wave form) may be obtained by starting from a total electron-atom wave function of the type [see (2.101)]

$$\Psi_a^{(+)} = \mathcal{A} \sum_{n=1}^{M} F_{na}(r_{N+1})\, \phi_n(q_1,q_2,\ldots q_N,\hat{\mathbf{r}}_{N+1},\sigma_{N+1}) \quad (2.109)$$

with $F_{na}(r) = r^{-1}\, u_{na}(r)$, and projecting the Schrödinger equation (2.97) onto the eigenchannel functions ϕ_n.

The close-coupling equations may be solved either by direct numerical integration[45], or by using variational or R-matrix methods, to which we shall return shortly. The close-coupling approximation is reliable for strong transitions between low-lying states whose energies are well separated from other states. In other cases the convergence is often slow and the method must be improved by adding terms to the expansion (2.109). In order to study this problem, it is convenient to use the Hulthén-Kohn variational method.

The variational method

Let Ψ_a^t and Ψ_b^t be trial functions, which satisfy boundary conditions of the same type as the correct wave functions, but with approximate values of the scattering amplitudes (or of the K-matrix elements). Defining

$$I_{ab} = \langle \Psi_a^t | H-E | \Psi_b^t \rangle \qquad (2.110)$$

the generalization of the Hulthén-Kohn variational principle (2.67) reads

$$\delta(\underset{\sim}{I} - \frac{1}{2} \underset{\sim}{K}) = 0 \qquad (2.111)$$

Thus, given approximate trial values $\underset{\sim}{K}^t$, a variational estimate of $\underset{\sim}{K}$, accurate to second order is given by

$$[\underset{\sim}{K}] = \underset{\sim}{K}^t - 2\,\underset{\sim}{I} \qquad (2.112)$$

It is a simple matter to show that the close-coupling equations result from the Hulthén-Kohn variational principle upon choosing trial functions of the form (2.107) [or (2.109)] and varying each of the unknown functions $F_n^{\pm}$ (or F_{na}). Of course, the Hulthén-Kohn variational principle may be used with other trial functions which are particularly suitable to describe certain important features of the collision process. For example, we may include in the trial function some terms which explicitly represent polarization effects, in the spirit of the polarized orbital method of Temkin[46]. These polarization effects, which are characteristic of the interaction between a charged particle and a neutral polarizable system, give rise in the elastic channel to an attractive long range potential varying like r^{-4} for large r. Together with exchange effects, they play a central role in low-energy electron-atom scattering and remain also important at higher energies.

As in the case of bound state problems, one may take advantage of the variational principle to perform systematic Rayleigh-Ritz-type calculations, in which square integrable (L^2) functions are included in the trial function. This approach can yield excellent results if enough L^2 functions are retained. For example, in the case of e^--H scattering, where Hylleraas-type L^2 functions can be used, very accurate elastic s-wave phase shifts have been obtained by Schwartz[47].

We have pointed out above that the close-coupling equations are obtained if the truncated expansion (2.109) is used as a trial function in the Hulthén-Kohn variational principle. A more general trial function is obtained by adding L^2 functions to the expansion (2.109). The resulting set of mixed integro-differential and algebraic equations[48,49] may be solved either by direct numerical integration[49] or by using variational techniques such as the matrix variational method[17]. This approach, which is known as the correlation approximation, is particularly useful when the remaining strongly coupled channels, which are not represented in (2.109), are closed, because in that case they can be well represented by L^2 functions. We remark, however, that polarization effects are not always accuretely taken into account in the correlation approximation, especially for atoms where a sizable fraction of the dipole polarizability arises from continuum intermediate states. In this case it is necessary to represent these polarization effects by including additional terms in the trial function.

Modified close-coupling expansions. Pseudostates

The foregoing discussion indicates that a general electron-atom collision wave function, suitable for the low-energy range, may be written as

$$\Psi_a^{(+)} = \mathcal{A} \sum_{n=1}^{M} F_{na}(r_{N+1})\ \phi_n(q_1,q_2,\ldots q_N,\hat{\underset{\sim}{r}}_{N+1},\sigma_{N+1})$$
$$+ \mathcal{A} \sum_{n=M+1}^{M+P} F_{na}(r_{N+1})\ \bar{\phi}_n(q_1,q_2,\ldots q_N,\hat{\underset{\sim}{r}}_{N+1},\sigma_{N+1})$$
$$+ \sum_{n=1}^{r} c_n\ \chi_n(q_1,q_2,\ldots q_{N+1}) \qquad (2.113)$$

The first expansion on the right of this equation is the close-coupling expansion (2.109). The second expansion is written in terms of pseudochannel functions $\bar{\phi}_n$, which are obtained by coupling the spin-angle functions of the scattered electron with pseudostates $\bar{\psi}_n$. These pseudostates are constructed so that they represent polarization effects, and can be determined by using methods[50,51] which generalize the polarized orbital approximation. Finally, the third expansion in (2.113) is made in terms of L^2 functions. The approach which retains terms in all three expansions of (2.113) is called the polarized pseudostate approximation.

Integro-differential equations for the functions $F_{na}(r)$, coupled to linear simultaneous equations for the coefficients c_n can be derived either by using $\Psi_a^{(+)}$ as a trial function in the Hulthén-Kohn variational principle, or by projecting the Schrödinger equation (2.97) onto the functions ϕ_n, $\bar{\phi}_n$ and χ_n. These equations may then be solved by using direct numerical integration, variational techniques[52] or the R-matrix method, which we shall now discuss.

The R-matrix method

The R-matrix method was originally introduced by Wigner and Eisenbud[53] in the theory of nuclear reactions. It was first applied to atomic collisions by Burke et al[54,55] and has been developed extensively in recent years to describe a broad range of atomic and molecular processes[56].

The basic idea is to realize that the dynamics of the projectile-target system differs depending on the relative distance r between the two colliding particles, so that configuration space for the system may be divided into an internal region ($r < a$) and and external region ($r > a$). For electron-atom (ion) collisions r is the radial distance between the electron and the nucleus of the atom (ion) and a is chosen in such a way that the charge distribution of the target states of interest is contained within the sphere $r = a$. It is clear that in the internal region the electron-target interaction is strong and difficult to handle, since both electron exchange and electron correlation effects are important. On the other hand, in the external region exchange between the scattered electron and the target may be neglected, and the solution to the collision problem can be readily obtained.

The R-matrix is now defined as follows. We begin by imposing logarithmic boundary conditions on the surface of the internal region, namely

$$\frac{a}{F_n(a)} \left. \frac{d F_n(r)}{dr} \right|_{r=a} = b_n \qquad (2.114)$$

where we have used the simplified notation $F_n(r) \equiv F_{na}(r)$ to denote the radial functions describing the motion of the scattered electron in channels n, and b_n are arbitrary constants which may depend on the channel quantum numbers. With these boundary conditions the spectrum of the Hamiltonian describing the electron-target system in the internal region consists of discrete energy levels E_k, and the corresponding eigenstates Ψ_k form in this region a complete set which may be used to expand the wave function Ψ_E for any energy (for $r < a$) as

$$\Psi_E = \sum_k A_{Ek} \Psi_k \tag{2.115}$$

By substituting this expansion into the Schrödinger equation (2.97) and using the boundary conditions (2.114) satisfied by the radial parts of the functions Ψ_k, one finds that

$$F_n(a) = \sum_m R_{nm}(E) \left(a \frac{dF_m}{dr} - b_m F_m\right)_{r=a} \tag{2.116}$$

where

$$R_{nm}(E) = \frac{1}{2a} \sum_k \frac{w_{nk}(a)\, w_{mk}(a)}{E_k - E} \tag{2.117}$$

is the R-matrix. Here the surface amplitudes $w_{nk}(a)$ are the values of the radial parts of the functions Ψ_k in channel n on the boundary. The central problem in the R-matrix method is therefore to calculate the surface amplitudes $w_{nk}(a)$ and the eigenenergies E_k. The K-matrix (or S-matrix) and cross sections are then related to the R-matrix (2.117) through the solution in the external region.

An interesting feature of the R-matrix theory is that the functions Ψ_k are energy-independent, and can be obtained by standard bound state procedures. In this respect the R-matrix method is similar to L^2 approaches to scattering problems[56-60]. Another important property of the R-matrix is that it is a real meromorphic function of the energy with simple poles lying on the real axis, so that $R_{nm}(E)$ does not contain any branch points. This fact provides the basis of the quantum defect theory[61], in which analytically known properties of electrons moving in a pure Coulomb field are used to describe electron-ion collisions in terms of a few parameters.

The Feshbach projection operator formalism

We shall now outline a formalism due to Feshbach[62], which is very convenient to study various aspects of low-energy electron-atom collisions, in particular resonance scattering. Let P and Q be projection operators, where P projects onto the subspace spanned by the open channel terms included in a close-coupling expansion and Q = 1 - P projects onto the orthogonal subspace. The Schrödinger equation (2.97) may then be rewritten [dropping the superscript (+)]

$$(H_{PP} - E)\, P\, \Psi_a + H_{PQ}\, Q\, \Psi_a = 0 \qquad (2.118\ a)$$

$$(H_{QQ} - E)\, Q\, \Psi_a + H_{QP}\, P\, \Psi_a = 0 \qquad (2.118\ b)$$

where $H_{PP} = PHP$, $H_{QQ} = QHQ$, $H_{PQ} = PHQ$ and $H_{QP} = QHP$. Solving (2.118 b) for $Q\Psi_a$ and substituting into (2.118 a) gives

$$[H_{PP} - H_{PQ}(E - H_{QQ})^{-1} H_{QP} - E]\, P\, \Psi_a = 0 \qquad (2.119)$$

We note that in this notation the close-coupling equations read

$$(H_{PP} - E)\, P\, \Psi_a = 0 \qquad (2.120)$$

The difference between the equations (2.119) and (2.120) is contained in the object

$$\mathcal{V} = H_{PQ}\, (E - H_{QQ})^{-1}\, H_{QP} \qquad (2.121)$$

which is closely related to the optical potential[63].

Let us now consider the eigenvalue problem

$$H_{QQ}\, \chi_\nu = \varepsilon_\nu\, \chi_\nu \qquad (2.122)$$

The operator H_{QQ} may have a discrete spectrum of eigenvalues as well as a continuous spectrum. In the neighborhood of an isolated discrete eigenvalue ε_s, we may write (2.119) in the form

$$[H_{PP} + \sum_{\nu \neq s} H_{PQ} \frac{|\chi_\nu\rangle\langle\chi_\nu|}{E-\varepsilon_\nu} H_{QP} - E]\, P\Psi_a = - H_{PQ} \frac{|\chi_s\rangle\langle\chi'_s|}{E-\varepsilon_s} H_{QP}\, P\Psi_a \qquad (2.123)$$

where the rapidly varying part of $\mathcal{V}$ has been placed on the right-hand side of (2.123). This part may be shown to give rise to resonances due to the fact that a compound state of the (N+1)-electron system (incident electron plus target) coincides with an electron in the continuum of the N-electron system. Such resonances are called Feshbach resonances. It is worth noting that other resonances, known as shape or potential resonances, may also arise when the interaction potential has the right shape and strength to induce a resonant state; the shape resonances are therefore similar to the resonances discussed in Section 2.1 for potential scattering.

Many-body methods

Before leaving the subject of low-energy electron-atom (ion) scattering, we mention that the many-body Green's function methods developed by Taylor et al[64] have also been applied successfully to a variety of electron-atom scattering processes. For elastic scattering an approximation scheme called the "generalized random phase approximation" has been proposed to calculate in a self-consistent way the response function for the target. For inelastic scattering the approach developed by Taylor et al. is similar to the distorted wave approximation considered below.

2.4. Electron-atom (ion) collisions at intermediate and high energies

We shall now consider various methods which have been proposed for the case where the energy of the projectile electron is higher than the first ionization energy of the target atom or ion. Detailed discussions of several of these methods may be found in the review articles of Joachain and Quigg[65], Bransden and McDowell[66] and Byron and Joachain[10]. It is convenient to still distinguish two regions of energy : a "high-energy" domain which extends from a few times the first ionization threshold upwards[67], and an "intermediate energy" region ranging from the low-energy region considered above to a few times the ionization energy of the target.

The Born series

Let us begin by considering the high-energy region, for which it is reasonable to try an approach based on perturbation theory. We shall first discuss direct scattering. We denote respectively by $\mathbf{k}_i$ and $\mathbf{k}_f$ the initial and final momentum of the projectile, with $|\mathbf{k}_i| = k$. The quantity $\boldsymbol{\Delta} = \mathbf{k}_i - \mathbf{k}_f$ is the momentum transfer. The free motion of the colliding particles before the collision is described by the direct arrangement channel Hamiltonian $H_d = K + h$, where K is the kinetic energy operator of the projectile and h the internal target Hamiltonian, such that $h|n\rangle = w_n|n\rangle$. The full Hamiltonian of the system is $H = H_d + V_d$, where V_d is the interaction between the electron and the target in the initial (direct) arrangement channel. For the sake of illustration we shall assume here that the target is a neutral atom of atomic number Z, in which case we have

$$V_d = -\frac{Z}{r} + \sum_{j=1}^{Z} \frac{1}{|\mathbf{r}-\mathbf{r}_j|} \qquad (2.124)$$

Here $\mathbf{r}$ denotes the coordinates of the projectile and $\mathbf{r}_j$ those of the target electrons.

We now write the Born series for the direct scattering amplitude as[68]

$$f = \sum_{n=1}^{\infty} \bar{f}_{Bn} \qquad (2.125)$$

where the nth Born term $\bar{f}_{Bn}$ contains n times the interaction V_d and (n-1) times the direct Green's operator $G_d^{(+)} = (E - H_d + i\varepsilon)^{-1}$, $\varepsilon \to 0^+$.

The first term $\bar{f}_{B1}$ is the familiar first Born amplitude, which has been calculated for a large number of electron-atom elastic, excitation and ionization scattering processes[69]. One of the basic goals of the theory is to obtain systematic improvements over the first Born approximation. To this end, let us first consider the second Born term $\bar{f}_{B2}$. For the direct transition $|\underset{\sim}{k}_i,o\rangle \rightarrow |\underset{\sim}{k}_f,m\rangle$, this term reads[68]

$$\bar{f}_{B2} = 8\pi^2 \int d\underset{\sim}{q} \sum_n \frac{\langle \underset{\sim}{k}_f,m|V_d|\underset{\sim}{q},n\rangle\langle \underset{\sim}{q},n|V_d|\underset{\sim}{k}_i,o\rangle}{q^2 - k^2 + 2(w_n - w_o) - i\varepsilon} \;, \quad \varepsilon \rightarrow o^+ \quad (2.126)$$

A useful approximation for $\bar{f}_{B2}$ at sufficiently high energies may be obtained[70] by replacing the energy differences (w_n-w_o) by an average excitation energy $\bar{w}$, so that the sum on the intermediate target states $|n\rangle$ can be done by closure. An improvement over this approximation consists in evaluating exactly the first few terms in the sum, while treating the remaining states by closure[71]. This method has been widely used in recent years, along with further improvements[72]. It is also worth noting that the sum on intermediate states in (2.24) has been evaluated "exactly" (without using closure) for elastic scattering[73,74] and 1s-2s excitation[74] in $(e^-$-H) scattering at zero momentum transfer.

Let us now examine the behaviour of the terms $\bar{f}_{Bn}$ for large values of k, the wave number of the projectile electron. Looking at Table 2, we see that for direct elastic scattering at small momentum transfers the quantity Re $\bar{f}_{B2}$ (which is governed by polarization effects) gives the dominant contribution (of order k^{-1}) to the first Born differential cross section. The situation at large Δ is easily understood since in this limit the terms $\bar{f}_{Bn}$ $(n \geqslant 2)$ are dominated by processes in which the atom remains in its initial state $|o\rangle$ in all intermediate states. This is equivalent to scattering by the static potential $V_{st} = \langle o|V_d|o\rangle$ of the atom. It is worth noting that for direct elastic scattering the dominant contribution at large k is given by the first Born term $\bar{f}_{B1}$ at all momentum transfers.

TABLE 2

Dependence of various terms of the Born and Glauber multiple scattering series for the direct elastic scattering amplitude, as a function of (large) k and Δ. The dominant contributions are framed. The terms located above the dashed line contribute through order k^{-2} to the differential cross section[75].

Order of Pert. Theory	Term	Small ($\Delta < k^{-1}$)	Interm. Δ ($k^{-1} < \Delta < 1$)	Large Δ ($\Delta > k$)
First	$\bar{f}_{B1} = \bar{f}_{G1}$	$\boxed{1}$	$\boxed{1}$	$\boxed{\Delta^{-2}}$
Second	Re $\bar{f}_{B2}$	k^{-1}	k^{-2}	$k^{-2}\ \Delta^{-2}$
	Re $\bar{f}_{G2}$	0	0	0
	Im $\bar{f}_{B2}$	$k^{-1}\ \log k$	k^{-1}	$k^{-1}\ \Delta^{-2}\ \log \Delta$
	Im $\bar{f}_{G2}$	$k^{-1}\ \log \Delta$	k^{-1}	$k^{-1}\ \Delta^{-2}\ \log \Delta$
Third	Re $\bar{f}_{B3}$	k^{-2}	k^{-2}	$k^{-2}\ \Delta^{-2}\ \log^2 \Delta$
	Re $\bar{f}_{G3}$	k^{-2}	k^{-2}	$k^{-2}\ \Delta^{-2}\ \log^2 \Delta$
	- - - - -	- - - - -	- - - - -	- - - - -
	Im $\bar{f}_{B3}$	k^{-3}	k^{-3}	$k^{-3}\ \Delta^{-2}\ \log \Delta$
	Im $\bar{f}_{G3}$	0	0	0
n^{th} ($n > 3$)	$\bar{f}_{Bn}$	$(ik)^{1-n}$	$(ik)^{1-n}$	$(ik)^{1-n}\Delta^{-2}\log^{n-1} \Delta$
	$\bar{f}_{Gn}$	$(ik)^{1-n}$	$(ik)^{1-n}$	$(ik)^{1-n}\Delta^{-2}\log^{n-1} \Delta$

The situation is different for direct inelastic collisions where for large Δ the first Born term falls off rapidly and the Born series is dominated by the second Born term $\bar{f}_{B2}$. This is illustrated in Table 3 for the case of inelastic s-s transitions.

TABLE 3

Dependence of various terms of the Born and Glauber multiple scattering series for the direct scattering amplitude corresponding to inelastic (s-s) transitions, as a function of (large) k and Δ. The dominant contributions are framed. The terms located above the dashed line contribute through order k^{-2} to the differential cross section[75].

Order of Pert. Theory	Term	Small Δ ($\Delta < k^{-1}$)	Interm. Δ ($k^{-1} < \Delta < 1$)	Large Δ ($\Delta > k$)
First	$\bar{f}_{B1} = \bar{f}_{G1}$	[1]	[1]	Δ^{-6}
Second	Re $\bar{f}_{B2}$	k^{-1}	k^{-2}	$k^{-2}\,\Delta^{-2}$
	Re $\bar{f}_{G2}$	0	0	0
	Im $\bar{f}_{B2}$	$k^{-1}\log k$	k^{-1}	[$k^{-1}\,\Delta^{-2}$]
	Im $\bar{f}_{G2}$	$k^{-1}\log \Delta$	k^{-1}	[$k^{-1}\,\Delta^{-2}$]
Third	Re $\bar{f}_{B3}$	k^{-2}	k^{-2}	$k^{-2}\,\Delta^{-2}\log \Delta$
	Re $\bar{f}_{G3}$	k^{-2}	k^{-2}	$k^{-2}\,\Delta^{-2}\log \Delta$
	- - - - -	- - - - -	- - - - -	- - - - -
	Im $\bar{f}_{B3}$	k^{-3}	k^{-3}	$k^{-3}\,\Delta^{-2}\log \Delta$
	Im $\bar{f}_{G3}$	0	0	0
n^{th} ($n > 3$)	$\bar{f}_{Bn}$	$(ik)^{1-n}$	$(ik)^{1-n}$	$(ik)^{1-n}\,\Delta^{-2}\log^{n-2}\Delta$
	$\bar{f}_{Gn}$	$(ik)^{1-n}$	$(ik)^{1-n}$	$(ik)^{1-n}\,\Delta^{-2}\log^{n-2}\Delta$

The fact that $\bar{f}_{B2}$ falls off more slowly than $\bar{f}_{B1}$ at large momentum transfers is due to the possibility of off-shell elastic scattering in intermediate states, where the projectile can experience the

Coulomb potential of the nucleus. We remark that since the values $\Delta < 1$ correspond to angles $\theta < k^{-1}$, the angular domain in which the first Born approximation is valid shrinks as the energy increases. However, because the dominant contribution to the integrated cross section comes precisely from the region $\Delta < 1$, the first Born values for integrated inelastic cross sections should be reliable at high energies.

The terms $\bar{g}_{Bn}$ of the Born series for exchange scattering are much more difficult to analyze than the direct Born terms $\bar{f}_{Bn}$ we have considered above. We simply mention here that for large k the term $\bar{g}_{B2}$ falls off more slowly than $\bar{g}_{B1}$, except for elastic exchange scattering at small Δ, where the Ochkur amplitude g_{Och} (which is the leading piece of $\bar{g}_{B1}$) is of order k^{-2}.

Our discussion of the Born series has been restricted thus far to elastic and inelastic processes. For ionization collisions, it is very difficult to perform calculations of the Born terms beyond first order, although the second Born term has been recently evaluated in the closure approximation for the ionization of atomic hydrogen[76]. By generalizing the arguments given above for inelastic scattering, it is also possible to show[77] that in certain kinematical situations the second Born term $\bar{f}_{B2}$ again falls off more slowly than $\bar{f}_{B1}$ for large k.

The Glauber approximation

The Glauber method is a many-body generalization of the eikonal approximation which we discussed in Section 2.1. It was first proposed[9] to study high-energy hadron-nucleus collisions, but has also been applied in recent years to analyze atomic collision processes. For a direct collision leading from an initial target state $|o\rangle$ to a final state $|n\rangle$ the Glauber scattering amplitude is given by[9,68]

$$f_G = \frac{k}{2\pi i} \int d^2\underset{\sim}{b}\; e^{i\underset{\sim}{\Delta}.\underset{\sim}{b}} \langle m|\{e^{i\chi_G(k,\underset{\sim}{b},X)} - 1\}|o\rangle \tag{2.127}$$

where the symbol X denotes the ensemble of the target coordinates, and we use a cylindrical coordinate system, with $\underset{\sim}{r} = \underset{\sim}{b} + z\,\underset{\sim}{\hat{z}}$. The Glauber phase shift function χ_G is given in terms of the direct interaction (2.124) between the projectile and the target by

$$\chi_G(k,\underset{\sim}{b},X) = -\frac{1}{k}\int_{-\infty}^{+\infty} V_d(\underset{\sim}{b},z,X)\, dz \tag{2.128}$$

the integration being performed along a z-axis perpendicular to $\underset{\sim}{\Delta}$.

Detailed discussions of the Glauber approximation and related methods may be found in the review articles of Joachain and Quigg[65] and Byron and Joachain[10]. We shall only mention here a few important points concerning the Glauber approach. Firstly, it may be viewed as an eikonal approximation to a "frozen target" model proposed by Chase[78], in which closure is used with an average excitation energy $\overline{w} = 0$. Secondly, considerable insight into the properties of the Glauber method may be gained by expanding the Glauber amplitude (2.127) in powers of V_d, namely

$$f_G = \sum_{n=1}^{\infty} \overline{f}_{Gn} \tag{2.129}$$

where

$$\overline{f}_{Gn} = \frac{k}{2\pi i}\frac{i^n}{n!}\int d^2\underset{\sim}{b}\; e^{i\underset{\sim}{\Delta}.\underset{\sim}{b}} \langle m|[\chi_G(k,\underset{\sim}{b},X)]^n|o\rangle \tag{2.130}$$

and comparing the terms $\overline{f}_{Gn}$ with those of the Born series $\overline{f}_{Bn}$. We note at once that $\overline{f}_{B1} = \overline{f}_{G1}$ because of our choice of z-axis. We also remark that the terms $\overline{f}_{Gn}$ are alternately real or purely imaginary, while the corresponding Born terms $\overline{f}_{Bn}$ are complex for $n \geqslant 2$. This special feature of the Glauber amplitude leads to several defects such as i) the absence of the important term Re $\overline{f}_{B2}$ for elastic scattering and ii) identical cross sections for electron-and positron-atom scattering. Other deficiencies of the Glauber amplitude (2.127) include a logarithmic divergence for elastic scattering in the forward direction (which is due to the choice $\bar{w} = 0$ made in obtaining (2.127) and may be traced to the behaviour of $\bar{f}_{G2}$ at $\Delta = 0$, as shown in Table 2) and a poor description of inelastic collisions involving non-spherically symmetric states. Despite these limitations, the Glauber approximation has been applied to a variety of atomic scattering processes. However, its major role in atomic collision theory has been to stimulate interest in eikonal methods[10] such as the "eikonal-Born series" (EBS) theory[79] which we shall now discuss.

The eikonal-Born series method

The basic idea of the EBS approach consists in analyzing the terms of the Born series (2.125) and the Glauber series (2.129) with the aim of obtaining a consistent expansion of the scattering amplitude in powers of k^{-1}. The main results are summarized in Tables 2 and 3 for elastic and inelastic (s-s) transitions, respectively. We note that for these processes the Glauber term $\bar{f}_{Gn}$ gives in each order of perturbation theory the leading piece of the corresponding Born term (for large k) for all momentum transfers, except in second order where the long range of the Coulomb potential is responsible for the anomalous behaviour of $\bar{f}_{G2}$ at small Δ. We also remark from Tables 2 and 3 that neither the second Born amplitude $f_{B2} = \bar{f}_{B1} + \bar{f}_{B2}$ nor the Glauber amplitude f_G are correct through order k^{-2}. In fact, a consistent calculation of the direct scattering amplitude through that order requires the terms $\bar{f}_{B1}$, $\bar{f}_{B2}$ and Re $\bar{f}_{B3}$ (or $\bar{f}_{G3}$). Since Re $\bar{f}_{B3}$ is very difficult to evaluate, and because it is a good approximation to Re $\bar{f}_{B3}$ for large enough k, it is reasonable to use $\bar{f}_{G3}$ in place of Re $\bar{f}_{B3}$. Thus we obtain in this way the "eikonal-Born series" direct scattering amplitude

$$f_{EBS} = \bar{f}_{B1} + \bar{f}_{B2} + \bar{f}_{G3} \qquad (2.131)$$

In addition, exchange effects are taken into account by using the Ochkur amplitude g_{Och}.

A detailed account of the EBS method and its application to various electron-atom collision processes at intermediate and high energies may be found in the review article of Byron and Joachain[10]. It is apparent from the foregoing discussion that the EBS theory represents an improvement over the second Born or Glauber approximations. We recall, however, that the EBS method is a perturbative approach and it is clear from Tables 2 and 3 that the convergence of the Born series for the direct amplitude is slower at large Δ than in the small Δ region. Thus an "all-order" treatment would be clearly desirable at large Δ, and we shall now turn our attention to optical model and target eigenfunction expansion methods which can provide (approximately) such "all-order" treatments.

Optical potentials

The basic idea of the optical potential method is to analyze the elastic scattering of a particle from a complex target by replacing the complicated interactions between the projectile and the target particles by an optical potential (or pseudopotential) in which the incident particle moves[80]. Once the optical potential V_{opt} is determined, the original many-body elastic scattering problem reduces to a one-body situation. However, this reduction is in general a difficult task, and approximation methods are necessary.

At intermediate and high energies it is particularly convenient to use the multiple scattering approach[65] developed by Mittleman and Watson[81,82]. We begin by considering direct elastic scattering, for which we write the corresponding direct part V^d_{opt} of the optical potential as a multiple scattering expansion in terms of the projectile-target interaction V_d. That is

$$V^d_{opt} = V^{(1)} + V^{(2)} + V^{(3)} + \ldots \qquad (2.132)$$

Here $V^{(1)} = V_{st} = \langle o|V_d|o\rangle$ is the static potential while the second order part reads

$$V^{(2)} = \sum_{n\neq o} \frac{\langle o|V_d|n\rangle\langle n|V_d|o\rangle}{k^2/2 - K - (w_n - w_o) + i\varepsilon} \tag{2.133}$$

The static potential V_{st} is readily evaluated for simple target atoms, or when an independent particle model is used to describe the target state $|o\rangle$, which we assume here to be spherically symmetric. We note that V_{st}, which is real and of short range, does not account for polarization and absorption effects which play an important role in the energy range considered here. However, for small values of the projectile coordinate r we note that V_{st} correctly reduces to the Coulomb interaction $-Z/r$ acting between the incident electron and the target nucleus, and hence should give a good account of large angle direct elastic scattering, as we already remarked in our discussion of the Born series.

Although the second order part $V^{(2)}$ of the direct optical potential is in general a complicated non-local, complex operator, at sufficiently high energies a useful local approximation of $V^{(2)}$ may be found by introducing an average excitation energy $\bar{w}$ and using eikonal methods[83-85]. The resulting $V^{(2)}$ may be written as

$$V^{(2)} \simeq V_{pol} + i\, V_{abs} \tag{2.134}$$

where V_{pol} and V_{abs} are real and central but energy-dependent. The term V_{pol} (which falls off like r^{-4} at large r) accounts for dynamic polarization effects and $i\, V_{abs}$ for absorption effects due to loss of flux from the incident channel. Recently, the leading, local, contribution of the third order part $V^{(3)}$ has also been evaluated[74] for the case of e^--H scattering.

Having obtained a local approximation for V^d_{opt}, exchange effects may be taken into account by using a local exchange pseudopotential[82,84,85-87] V_{opt}^{ex}, so that the full optical potential is given by $V_{opt} = V_{opt}^{d} + V_{opt}^{ex}$. An "exact" (partial wave) treatment of this potential is then carried out. It is worth noting that in performing such an exact, full-wave treatment of the optical potential V_{opt}, one generates approximations to all terms of perturbation theory, a feature which is an important advantage for large angle scattering.

Target eigenfunction expansions

It is also possible to formulate the problem of fast electron-atom collisions within the framework of the target eigenfunction expansions discussed at the beginning of Section 2.3. It is clear, however, that at incident electron energies above the ionization threshold the low-energy methods described in that section - and in particular the close-coupling approximation - must be modified. This may be done by using a few well chosen pseudostates to represent the channels omitted in the close-coupling expansion. The pseudostate approach, first introduced by Damburg and Karule[50] to describe polarization effects in low-energy e^--H scattering, has been extended by Burke and Webb[88] and by Callaway and Wooten[89] to intermediate energies where the pseudostate channels are open and hence can carry away flux. Extensive calculations of elastic and inelastic differential cross sections for e^--H scattering, using the pseudostate method, have been performed by Callaway et al[90]. The method, however, presents undesirable features in the neighborhood of pseudo-resonances occuring near the pseudostate thresholds, and partly for this reason it has had only limited application thus far.

Another modification of the close coupling method, called the second order potential (SOP) method, has been applied by Bransden et al[91] to a variety of electron-atom scattering processes. The modified close-coupling equations read

$$(\nabla^2 + k_n^2)\, F_n^{\pm}(\underset{\sim}{r}) = \sum_{m=1}^{M} [U_{nm}^{\pm}(\underset{\sim}{r}) + K_{nm}^{\pm}(\underset{\sim}{r})] F_m^{\pm}(\underset{\sim}{r}) \qquad (2.135)$$

where $U_{nm}^{\pm}$ is defined by (2.92) and $K_{nm}^{\pm}(\underset{\sim}{r})$ is a potential matrix which accounts approximately for the coupling with the states $n' \geqslant M + 1$. Neglecting exchange, it is given in lowest order by $K_{nm} \simeq 2V_{nm}^{(2)}$, where

$$V_{nm}^{(2)} = \sum_{n' \geqslant M+1} \frac{\langle n|V_d|n'\rangle\langle n'|V_d|m\rangle}{k^2/2 - K - (w_{n'} - w_o) + i\varepsilon}, \quad \varepsilon \to o^+ \qquad (2.136)$$

is the expected generalization of the second order part $V^{(2)}$ of the optical potential, given by (2.133) for elastic scattering. A detailed discussion of the second order potential method and of the pseudostate approach mentionned above may be found in the review article of Bransden and McDowell[66].

Distorted waves

The basic idea of distorted wave treatments is to break the interaction in two parts, one which is treated exactly and the other which is handled by perturbation theory. Distorted wave methods are therefore conveniently discussed within the framework of the two-potential formalism . We assume that the interaction potentials in the initial and final arrangement channels may be split as

$$V_i = U_i + W_i\,, \qquad V_f = U_f + W_f \qquad (2.137)$$

We also suppose that we know the distorted waves

$$\chi_a^{(+)} = \Phi_a + (E - H_i - U_i + i\varepsilon)^{-1} U_i \Phi_a, \qquad \varepsilon \to o^+ \qquad (2.138\ a)$$

and

$$\chi_b^{(-)} = \Phi_b + (E - H_f - U_f - i\varepsilon)^{-1} U_f \Phi_b, \qquad \varepsilon \to o^+ \qquad (2.138\ b)$$

where Φ_a and Φ_b are "free waves", such that $H_i \Phi_a = E\, \Phi_a$ and

$H_f \Phi_b = E\ \Phi_b$, with $H_i = H - V_i$ and $H_f = H - V_f$. The T-matrix elements corresponding to the transition $a \to b$ are then given by

$$T_{ba} = \langle \chi_b^{(-)} | V_i - W_f | \Phi_a \rangle + \langle \chi_b^{(-)} | W_f | \Psi_a^{(+)} \rangle \qquad (2.139\text{ a})$$

or

$$T_{ba} = \langle \Phi_b | V_f - W_i | \chi_a^{(+)} \rangle + \langle \Psi_b^{(-)} | W_i | \chi_a^{(+)} \rangle \qquad (2.139\text{ b})$$

with

$$\Psi_a^{(+)} = \chi_a^{(+)} + (E - H_i - U_i + i\varepsilon)^{-1}\ W_i\ \Psi_a^{(+)}, \quad \varepsilon \to 0^+ \qquad (2.140\text{ a})$$

and

$$\Psi_b^{(-)} = \chi_b^{(-)} + (E - H_f - U_f - i\varepsilon)^{-1}\ W_f\ \Psi_b^{(-)}, \quad \varepsilon \to 0^+ \qquad (2.140\text{ b})$$

The two-potential formulae (2.139) simplify when the distorting potentials U_i and U_f cannot induce the transition $a \to b$, for example if U_i and U_f only generate elastic scattering and the transition $a \to b$ is an inelastic process or a rearrangement collision. In this case the first term on the right of the equations (2.139) vanishes, and one has

$$T_{ba} = \langle \chi_b^{(-)} | W_f | \Psi_a^{(+)} \rangle = \langle \Psi_b^{(-)} | W_i | \chi_a^{(+)} \rangle \qquad (2.141)$$

If we want to treat exactly the interactions U_i and U_f but to use perturbation theory for the interactions W_i and W_f we may solve the equations (2.140) by iteration and substitute the corresponding expansions in (2.141). In this way we generate the distorted-wave Born series, the first term of which being the distorted-wave Born approximation (DWBA)

$$T_{ba}^{DWBA} = \langle \chi_b^{(-)} | W_i | \chi_a^{(+)} \rangle = \langle \chi_b^{(-)} | W_f | \chi_a^{(+)} \rangle \qquad (2.142)$$

A variety of distorted-wave treatments have been applied to inelastic electron-atom (ion) collisions. Detailed accounts may be found in the reviews by Bransden and McDowell[66] and Henry[43]. Distorted wave approximations have also been applied to ionization process, in particular to the calculation of the triple differential cross sections[92] which are measured in (e, 2e) experiments[93,94].

Dispersion relations

To conclude this survey of electron-atom collisions, I would like to mention briefly the problem of dispersion relations for electron-atom scattering. We saw in Section 2.1 that forward dispersion relations of the form (2.70) hold in potential scattering for a large class of interaction potentials. Forward dispersion relations for electron-atom scattering were proposed some time ago by Gerjuoy and Krall[95]. However, Byron et al[96] recently pointed out that the experimental data imply that the Gerjuoy-Krall dispersion relations in fact do not hold for electron-atom scattering. Further studies[97,98] have confirmed this conclusion. Byron et al[96] also suggested that the source of the difficulty was the analytic structure of the forward exchange amplitude, g(E), as a function of the energy E of the projectile electron. This suggestion has received strong theoretical support from the work of Tip[99], Blum and Burke[100] and Byron and Joachain[101].

3. PHOTON COLLISIONS WITH ATOMS AND IONS

3.1. Semi-classical theory of the interaction of atoms with electromagnetic radiation

In this section we recall a few basic results concerning the interaction of atoms (ions) with electromagnetic radiation. In a rigorous treatment we would have to use quantum electrodynamics[102], in which the electromagnetic field is expressed in terms of its quanta - the photons. We shall proceed here by using an approximate, semi-classical model such that the radiation field is treated classically, and the atomic system by quantum mechanics[103].

The electromagnetic field

The classical electromagnetic field is described by electric and magnetic field vectors $\boldsymbol{\mathcal{E}}$ and $\boldsymbol{\mathcal{B}}$ which satisfy Maxwell's equations, and can be generated from scalar and vector potentials ϕ and $\mathbf{A}$ by

$$\boldsymbol{\mathcal{E}}(\mathbf{r},t) = -\nabla\phi - \frac{\partial}{\partial t}\mathbf{A}(\mathbf{r},t) \tag{3.1}$$

and

$$\boldsymbol{\mathcal{B}}(\mathbf{r},t) = \nabla \times \mathbf{A}(\mathbf{r},t) \tag{3.2}$$

where we are using S.I. units. The potentials are not completely defined by (3.1) and (3.2). In particular, $\boldsymbol{\mathcal{E}}$ and $\boldsymbol{\mathcal{B}}$ are unaltered by the substitution $\mathbf{A} \rightarrow \mathbf{A} + \nabla\chi$, where χ is any scalar field. This property of gauge invariance allows us to impose a further condition on $\mathbf{A}$, which we choose to be

$$\nabla \cdot \mathbf{A} = 0 \tag{3.3}$$

in which case we are said to be using the Coulomb gauge. From Maxwell's equations (without sources) it can be shown that $\mathbf{A}$ satisfies the wave equation (as do ϕ, $\boldsymbol{\mathcal{E}}$ and $\boldsymbol{\mathcal{B}}$)

$$\nabla^2\mathbf{A} - \frac{1}{c^2}\frac{\partial^2\mathbf{A}}{\partial t^2} = 0 \tag{3.4}$$

In what follows we shall set $\phi = 0$ since in empty space the most general solution of Maxwell's equations for a radiation field can be expressed in terms of potentials such that $\nabla\cdot\mathbf{A} = 0$ and $\phi = 0$.

A monochromatic plane wave solution of (3.3) and (3.4), corresponding to the angular frequency ω is given by

$$\mathbf{A}(\omega;\mathbf{r},t) = A_0(\omega)\hat{\boldsymbol{\varepsilon}}\,[e^{i(\mathbf{k}\cdot\mathbf{r}-\omega t+\delta_\omega)} + \text{c.c.}] \tag{3.5}$$

where $\hat{\boldsymbol{\varepsilon}}$ is the polarization vector, $\mathbf{k}$ the propagation vector, δ_ω is a real phase and c.c. denotes the complex conjugate. We note that (3.3) is satisfied if $\mathbf{k}\cdot\hat{\boldsymbol{\varepsilon}} = 0$, so that the wave is transverse.

Since $\hat{\varepsilon}$ lies in a plane perpendicular to $\mathbf{k}$, it can be specified by giving its components along two linearly independent vectors lying in this plane. We also remark that (3.4) is satisfied provided that $\omega = kc$.

The electric and magnetic fields associated with the vector potential (3.5) are given from (3.1) and (3.2) by

$$\boldsymbol{\mathcal{E}} = -\,2\omega\, A_0(\omega)\, \hat{\boldsymbol{\varepsilon}}\, \sin(\mathbf{k}.\mathbf{r} - \omega t + \delta_\omega) \tag{3.6}$$

and

$$\boldsymbol{\mathcal{B}} = -\,2A_0(\omega)\,(\mathbf{k} \times \hat{\boldsymbol{\varepsilon}})\, \sin(\mathbf{k}.\mathbf{r} - \omega t + \delta_\omega) \tag{3.7}$$

The expressions (3.6) and (3.7) describe a linearly polarized plane wave, having its electric field vector $\boldsymbol{\mathcal{E}}$ in the direction of the polarization vector $\hat{\boldsymbol{\varepsilon}}$. A general state of polarization for a wave propagating in the direction $\hat{\mathbf{k}}$ can be described by combining two independent linearly polarized plane waves with polarization vectors $\hat{\boldsymbol{\varepsilon}}_\lambda$ ($\lambda = 1,2$) perpendicular to $\mathbf{k}$, where the phases in the two component waves are in general different. The intensity $I(\omega)$ associated with the plane wave (3.5) is obtained by averaging over a period $(2\pi/\omega)$ the magnitude of the Poynting vector $(\boldsymbol{\mathcal{E}} \times \boldsymbol{\mathcal{B}})/\mu_o$. That is,

$$I(\omega) = 2\,\varepsilon_o\,\omega^2\, c\, A_0^2(\omega) = \rho(\omega)\, c \tag{3.8}$$

where $\rho(\omega)$ is the average energy density. In the above expressions ε_o and μ_o are respectively the permittivity and permeability of free space, with $\varepsilon_o\mu_o c^2 = 1$.

A general pulse of radiation can be described by taking $\phi = 0$ and representing $\mathbf{A}(\mathbf{r},t)$ as a superposition of the plane waves $\mathbf{A}(\omega;\mathbf{r},t)$. For example, taking each plane wave component to have the same direction of propagation $\hat{\mathbf{k}}$ and adopting a given direction of linear polarization $\hat{\boldsymbol{\varepsilon}}$, we have

$$\mathbf{A}(\mathbf{r},t) = \int_{\Delta\omega} A_0(\omega)\, \hat{\boldsymbol{\varepsilon}}[\, e^{i(\mathbf{k}.\mathbf{r}-\omega t+\delta_\omega)} + \text{c.c.}]\, d\omega \tag{3.9}$$

Atoms in an electromagnetic field

Let us consider an atomic system in an electromagnetic field. For the sake of simplicity we shall first deal with a one-electron atom or ion, of nuclear charge Ze. Neglecting small spin-dependent terms and ignoring the interaction of the radiation field with the nucleus, together with reduced mass effects, the Hamiltonian of the system reads (in S.I. units)

$$H = \frac{1}{2m} (\mathbf{p} + e\mathbf{A})^2 - \frac{Ze^2}{(4\pi\varepsilon_o)r} \tag{3.10}$$

where m is the electron mass and $\mathbf{p} = - i\hbar\nabla$ its momentum. Using the gauge condition (3.3) and treating the weak field case in which the term in $\mathbf{A}^2$ is small compared with the term linear in $\mathbf{A}$, the time-dependent Schrödinger equation to be solved is given by

$$i\hbar \frac{\partial \Psi}{\partial t} = [H_0 + H'(t)]\Psi \tag{3.11}$$

where

$$H_0 = - \frac{\hbar^2}{2m} \nabla^2 - \frac{Ze^2}{(4\pi\varepsilon_o)r} \tag{3.12}$$

is the time-independent Hamiltonian describing the one-electron atom (ion) in the absence of external fields, and

$$H'(t) = - \frac{i\hbar e}{m} \mathbf{A}(t) \cdot \nabla \tag{3.13}$$

is a perturbation which depends explicitly on the time variable through the vector potential $\mathbf{A}$.

These considerations are readily generalized to the case of many-electron atoms or ions. The Hamiltonian H_0 then corresponds to the N-electron atom (ion) in the absence of external fields, and the interaction energy between the radiation field and the electrons, having position vectors $\mathbf{r}_j$ $(j = 1,2,\ldots N)$ is given by

$$H'(t) = - \frac{i\hbar e}{m} \sum_{j=1}^{N} \mathbf{A}(\mathbf{r}_j,t)\cdot\nabla_j \tag{3.14}$$

Transition rates

The transition rate for an upward transition of an atom from a discrete state ψ_a to a state ψ_b, with the absorption of a photon, can be obtained from first order time-dependent perturbation theory. It is proportional to $|M_{ba}|^2$, where M_{ba} is the matrix element

$$M_{ba} = \langle \psi_b | \sum_{j=1}^{N} e^{i\underset{\sim}{k}\cdot\underset{\sim}{r}_j} \hat{\underset{\sim}{\varepsilon}}\cdot\underset{\sim}{\nabla}_j | \psi_a \rangle \tag{3.15}$$

The transition rate for stimulated emission may be found in a similar way. Under the same radiation field the number of transitions per second exciting the atom from the state a to the state b is the same as that de-exciting the atom from b to a; this result is consistent with the principle of detailed balancing.

In many cases of interest the quantities kr_j are small for important distances r_j of the order of an atomic unit of length (a_0). One can then replace $\exp(i\underset{\sim}{k}\cdot\underset{\sim}{r}_j)$ by unity in (3.15), which amounts to neglecting retardation across the atom. This is known as the electric dipole approximation, in which case the matrix element M_{ba} reads

$$M_{ba} = \hat{\underset{\sim}{\varepsilon}} \cdot \langle \psi_b | \sum_j \underset{\sim}{\nabla}_j | \psi_a \rangle = \hat{\underset{\sim}{\varepsilon}} \cdot \langle \psi_b | \underset{\sim}{D}_V | \psi_a \rangle \tag{3.16}$$

where $\underset{\sim}{D}_V = \sum_j \underset{\sim}{\nabla}_j$ is the dipole velocity operator. It is a simple matter to show that the matrix element (3.16) can also be written as

$$M_{ba} = -\frac{m\omega_{ba}}{\hbar} \hat{\underset{\sim}{\varepsilon}} \cdot \langle \psi_b | \sum_j \underset{\sim}{r}_j | \psi_a \rangle = -\frac{m\omega_{ba}}{\hbar} \hat{\underset{\sim}{\varepsilon}} \cdot \langle \psi_b | \underset{\sim}{D}_L | \psi_a \rangle \tag{3.17}$$

where $\omega_{ba} = (E_b - E_a)/\hbar$ and $\underset{\sim}{D}_L = \sum_j \underset{\sim}{r}_j$ is the dipole length operator.

The expressions (3.16) and (3.17) are known as the velocity and length forms of the matrix element M_{ba} (in the electric dipole approximation). A third expression known as the acceleration form is occasionally used. It is given by[104]

$$M_{ba} = - \frac{Ze}{\hbar\, \omega_{ba}} \langle \psi_b | \sum_j \frac{\mathbf{r}_j}{r_j^3} | \psi_a \rangle \qquad (3.18)$$

If exact initial and final atomic wave functions are used the three equations (3.16)-(3.18) give identical results. However, with approximate wave functions the three expressions give results which in general are different. We remark that the length formulation emphasizes the wave functions at large distances, while in the acceleration formula (3.18) involving $\mathbf{r}_j/r_j^3$ small values of r_j are important. For the matrix element (3.16) involving the gradient operators ∇_j it is found that intermediate values of $\mathbf{r}_j$ are most important.

3.2. Photoionization of atoms and ions

We shall now use the results of the preceeding section to study the photoionization of atoms and ions. The basic process which we consider is defined by

$$h\nu + A \rightarrow A^+ + e^- \qquad (3.19)$$

where A is an atom (in state i) while in the final state we have a residual ion A^+ (in state j) and a free electron. The inverse process of (3.19) is known as radiative recombination. The photoionization reaction

$$h\nu + A^- \rightarrow A + e^- \qquad (3.20)$$

involving a negative ion A^- is known as photodetachment.

In addition to their intrinsic interest in atomic physics, photoionization cross sections are also important in other fields such as astrophysics. Progress in photoionization theory have been stimulated by parallel advances in experimental techniques, the most important development being the use of synchrotron radiation sources[105-107]

Photoionization theory[108]

The photoionization cross section for a transition $i \rightarrow j$ is defined as

$$\frac{d\sigma_{ji}}{d\Omega} = \frac{W_{ji}}{F} \tag{3.21}$$

where W_{ji} is the transition rate into a group of states at energy E, and F is the incident photon flux. The quantity W_{ji} can be obtained from Fermi's Golden Rule, while the incident photon flux, of angular frequency ω, is given by the intensity (3.8), divided by $\hbar\omega$. Using the dipole approximation, it is found that the dipole velocity form of the cross section is given by

$$\frac{d\sigma_{ji}}{d\Omega} = \frac{4\pi^2 e^2 \hbar^2}{(4\pi\varepsilon_0) m^2 c\omega} |\langle \psi^-_{jE} | \hat{\varepsilon} \cdot \mathbf{D}_V | \psi_i \rangle|^2 \tag{3.22}$$

where the initial bound state is normalized according to $\langle \psi_i | \psi_i \rangle = 1$, while the final continuum state ψ^-_{jE} is normalized per unit energy interval, namely

$$\langle \psi^-_{jE} | \psi^-_{j'E'} \rangle = \delta_{jj'} \, \delta(E-E') \tag{3.23}$$

The result (3.22) corresponds to photons linearly polarized, with polarization vector $\hat{\varepsilon}$. For unpolarized photons the cross section is obtained by averaging over the two independent polarization directions. The dipole length form of the cross section (3.22) is

$$\frac{d\sigma_{ji}}{d\Omega} = \frac{4\pi^2 e^2 \omega}{(4\pi\varepsilon_0) c} |\langle \psi^-_{jE} | \hat{\varepsilon} \cdot \mathbf{D}_L | \psi_i \rangle|^2 \tag{3.24}$$

It is of interest to rewrite the results (3.22) and (3.24) in atomic units. Remembering that $a_0 = (4\pi\varepsilon_0)\hbar^2/me^2$ is the atomic unit of length while $e^2/(4\pi\varepsilon_0)a_0$ is the atomic unit of energy, we may rewrite (3.22) as

$$\frac{d\sigma_{ji}}{d\Omega} = \frac{4\pi^2 \alpha a_o^2}{E_\gamma} |\langle \psi_{jE}^- | \hat{\varepsilon} \cdot \mathbf{D}_V | \psi_i \rangle|^2 \tag{3.25}$$

while (3.24) reads

$$\frac{d\sigma_{ji}}{d\Omega} = 4\pi^2 \alpha a_o^2 E_\gamma |\langle \psi_{jE}^- | \hat{\varepsilon} \cdot \mathbf{D}_L | \psi_i \rangle|^2 \tag{3.26}$$

where $\alpha = e^2/(4\pi\varepsilon_o)\hbar c$ is the fine-structure constant and E_γ is the photon energy ($\hbar\omega$) expressed in atomic units.

Following Henry and Lipsky[109], we can expand ψ_{jE}^- in terms of the complete set of eigenchannel functions ϕ_n introduced in (2.101), corresponding to channel quantum numbers (α_n L_n S_n k_n ℓ_n s_n L S M_L M_S π). Using the Wigner-Eckart theorem, the differential cross section can then be expressed in terms of the spherical components of either $\mathbf{D}_L$ or $\mathbf{D}_V$.

The differential cross section for photoionization of an unpolarized atom (ion) by a polarized photon beam, where the spin orientations of the photoelectron and the residual ion are not observed, is obtained from (3.25) or (3.26) by summing over the final magnetic quantum numbers and taking the average over the initial magnetic quantum numbers. Thus we have for example in the dipole length form

$$\frac{d\sigma_{ji}}{d\Omega} = \frac{4\pi^2 \alpha a_o^2 E_\gamma}{(2L_i+1)(2S_i+1)} \sum_{\substack{M_{L_i}, M_{S_i} \\ M_{L_j}, M_{S_j}, m_j}} |\langle \psi_{jE}^- | \hat{\varepsilon} \cdot \mathbf{D}_L | \psi_i \rangle|^2 \tag{3.27}$$

The evaluation of (3.27) can be carried out by using Racah algebra. For an unpolarized beam of photons it is found that[108]

$$\frac{d\sigma}{d\Omega} = \frac{\sigma_{tot}}{4\pi} [1 - \frac{\beta}{2} P_2(\cos\theta)] \tag{3.28}$$

where σ_{tot} is the total photoionization cross section, θ is the angle of the photoelectron with respect to the incident photon propagation direction and the quantity β is such that $-1 \leqslant \beta \leqslant 2$. Moreover, if the atom is initially in an s-state one has $\beta = 2$ at all energies. As a result, the differential cross section (3.28) vanishes in the incident photon propagation direction.

It is only for hydrogenic atoms (ions) that photoionization cross sections can be calculated analytically. For example, the total photoionization cross section of an hydrogenic atom (ion) of nuclear charge Z from a state denoted by the principal quantum number n (averaged over ℓ substates) is given by

$$\sigma_{tot} = \frac{8\pi\,\alpha\, a_o^2}{3^{3/2}} \frac{Z^4}{n^5} E_\gamma^{-3}\, g(E_\gamma, n, Z) \qquad (3.29)$$

where the quantity g is known as the Kramers-Gaunt factor. It is a slowly varying function of the photon energy, falling off like $E_\gamma^{-1/2}$ at high photon energies[110].

For atoms (ions) with two or more electrons, approximations must be used to describe both the initial bound state wave function ψ_i and the final continuum state ψ_{jE}^-. In order of increasing sophistication, one may distinguish the single-particle approximation[111], quantum defect theory[112-114], the Hartree-Fock approximation[115], the random phase approximation[116-118], many-body perturbation theory[119-121] and configuration-interaction theory[122-124]. These approximation methods are discussed in detail in the review article of Burke[108].

Resonances in photoionization[125]

Photoionization cross sections are dominated in certain energy ranges by resonances or autoionizing states. At energies close to a resonance the photoionization process can occur either directly,

$$h\nu + A \rightarrow A^+ + e^- \qquad (3.30)$$

or via an intermediate resonant state A^*,

$$h\nu + A \rightarrow A^* \rightarrow A^+ + e^- \qquad (3.31)$$

The interference between these two mechanisms leads to the absorption line profile, which Fano and Cooper[126] have parametrized by the formula

$$\sigma(E) = \sigma_a \frac{(q + \varepsilon)^2}{1 + \varepsilon^2} + \sigma_b \qquad (3.32)$$

where σ_a and σ_b are background cross sections which vary slowly with energy, $\varepsilon = (E-E_r)/(\Gamma/2)$, E_r being the resonance energy and Γ the resonance width, and q is the line profile index which defines the line shape.

Early examples of resonances were found by Madden and Codling[127] in the photoionization of helium atoms. Using the continuous light from an electron synchrotron, they observed the resonant process

$$h\nu + He \rightarrow He^* \ (2snp \pm 2pns\ {}^1P) \rightarrow He^+ + e^- \qquad (3.33)$$

interfering with the direct process

$$h\nu + He \rightarrow He^+ + e^- \qquad (3.34)$$

This gives rise to Rydberg series of Feshbach resonances converging onto the n = 2 level of the He^+ ion. Further faint series can also be seen converging onto higher excited levels of He^+.

As another interesting example of resonances, let us consider the photodetachment of the H^- ion. Using a colliding beam method in which a laser beam is directed at variable angle across a H^- beam, Bryant et al[128] have observed a Feshbach and a shape resonance near 11 eV, which agree well with theoretical predictions[129].

Multiple photoionization

Before concluding this survey of photoionization, I would like to mention multiple ionization phenomena. These processes are of special interest since they are very sensitive probes of the details of atomic structure. Double ionization from outer shells is particularly interesting, since it depends delicately on electron correlations[130]. Several experimental investigations on double ionization in the rare gases have been carried out using both photon impact[131-133] and electron impact[134] (pseudo-photon) techniques. Theoretical calculations have been performed for helium[130,135,136], neon[137-139] and argon[140].

REFERENCES

1. B.H. Bransden, Atomic Collision Theory (Benjamin, New York, 1970).
2. C.J. Joachain, Quantum Collision Theory (North Holland Publ. Cy, Amsterdam, second printing, 2 vol., 1979).
3. P.G. Burke, Potential Scattering in Atomic Physics (Plenum Press, New York, 1977).
4. Except for the case of an infinite "wall" at r = a, where it is sufficient to require that $R_\ell(r=a) = 0$.
2. See for example ref. 2, Chapter 6.
6. R.H. Dalitz, Proc. Roy. Soc. A 206, 509 (1951); see also ref. 2, Appendix D and R.R. Lewis, Jr., Phys. Rev. 102, 537 (1956).
7. The Greek word εικοη (eikon) means "image".
8. G. Molière, Z. Naturforsch. 2A, 133 (1947).
9. R.J. Glauber, in Lectures in Theoretical Physics, vol. 1 (ed. by W.E. Brittin, Interscience, New York, 1959), p. 315.
10. F.W. Byron, Jr. and C.J. Joachain, Phys. Reports 34 C, 233 (1977).
11. F.W. Byron, Jr. and C.J. Joachain, Physica 66, 33 (1973); F.W. Byron, Jr., C.J. Joachain and E.H. Mund, Phys. Rev. D 8, 2622 (1973); D 11, 1662 (1975); C 20, 2325 (1980).
12. One can of course construct other EBS amplitudes such as $f'_{EBS} = f_E - \bar{f}_{E2} + \bar{f}_{B2}$ or $f''_{EBS} = f_E + \mathrm{Re}\,\bar{f}_{B2}$ which differ from f_{EBS} by terms of order k^{-3} (and higher).
13. L. Hulthén, Kgl. Fysiogr. Sallsk. Lund Förh. 14 (21) (1944); Arkiv. Mat. Ast. Fys. 35 A, 25 (1948).
14. I.G. Tamm, Zh. Eksper. i Teor. Fiz. 14, 21 (1944); ibid. 18, 337 (1948); ibid. 19, 74 (1949).
15. J. Schwinger, Lectures notes, Harvard Univ. (unpublished 1947), Phys. Rev. 72, 742 (1947).
16. W. Kohn, Phys. Rev. 74, 1763 (1948).
17. R.K. Nesbet, Adv. Atom. Molec. Phys. 13, 315 (1977).
18. V. de Alfaro and T. Regge, Potential Scattering (North Holland Publ. Cy, Amsterdam, 1965).

19. R.G. Newton, Scattering Theory of Waves and Particles (McGraw-Hill, New York, 1966).
20. A detailed treatment of polarized electrons may be found in J. Kessler, Polarized Electrons (Springer-Verlag, Berlin, 1976).
21. See for example ref. 2, Chapter 18. The spin-flip amplitude g(θ) defined in ref. 20 differs from that of ref. 2 by a factor of i. We shall use here the amplitude g(θ) of ref. 2.
22. For the heavy alkalis where spin-orbit interactions become important, six amplitudes are required for a complete description of the elastic electron-spin-1/2 atom scattering. See for example P.G. Burke and J.F.B. Mitchell, J. Phys. B 7, 214 (1974); P.S. Farago, J. Phys. B 7, L 28 (1974).
23. J. Macek and D.H. Jaecks, Phys. Rev. A 6, 2288 (1971).
24. J. Wykes, J. Phys. B 5, 1126 (1972).
25. U. Fano and J. Macek, Rev. Mod. Phys. 45, 553 (1973).
26. J. Macek and I.V. Hertel, J. Phys. B 7, 2173 (1974).
27. K. Blum and H. Kleinpoppen, Phys. Rev. A 9, 1902 (1974).
28. M. Eminyan, K.B. MacAdam, J. Slevin and H. Kleinpoppen, Phys. Rev. Letters 31, 576 (1973); J. Phys. B 7, 1519 (1974).
29. A. Ugbabe, P.J.O. Teubner, E. Weigold and H. Arriola, J. Phys. B 10, 71 (1977).
30. K.H. Tan, J. Fryar, P.S. Farago and J.W. McConkey, J. Phys. B 10, 1073 (1977).
31. V.C. Sutcliffe, G.N. Haddad, N.C. Steph and D.E. Golden, Phys. Rev. A 17, 100 (1978).
32. N.C. Steph and D.E. Golden, Phys. Rev. A 21, 759 (1980), A 21, 1848 (1980).
33. J. Slevin, H.Q. Porter, M. Eminyan, A. Defrance and G. Vassilev, J. Phys. B 13, L 23 (1980).
34. A.J. Dixon, S.T. Hood and E. Wiegold, Phys. Rev. Letters 40, 1262 (1978).
35. S.T. Hood, E. Weigold and A.J. Dixon, J. Phys. B 12, 631 (1979).
36. E. Weigold, L. Frost and K.J. Nygaard, Phys. Rev. A 21, 1950 (1980).
37. J. Slevin, M. Eminyan, J.M. Woolsey, G. Vassilev and H.Q. Porter, J. Phys. B 13, L 341 (1980).
38. See for example R.H.J. Jansen, F.J. de Heer, H.J. Luyken, B. van Wingerden and H.J. Blaauw, J. Phys. B 9, 185 (1976).
39. H. Kleinpoppen, in Atomic Physics, vol. 4 (ed. by G. zu Putlitz, E.W. Weber and A. Winnacker, Plenum Press, New York, 1975) p. 449.
40. I.V. Hertel and W. Stoll, J. Phys. B 7, 583 (1974); I.V. Hertel, H.W. Hermann, W. Reiland, A. Stamatović and W. Stoll, in The Physics of Electronic and Atomic Collisions (ed. by J.S. Risley and R. Geballe, Univ. of Washington Press, Seattle, 1976), p. 158.
41. I.V. Hertel and W. Stoll, Adv. Atom. Molec. Phys. 13, 113 (1977).

42. A general survey of theoretical methods for electron-atom scattering may be found in P.G. Burke and J.F. Williams, Phys. Reports 34 C, 325 (1977). See also C.J. Joachain, in Atomic and Molecular Processes in Thermonuclear Fusion (ed. by M.R.C. McDowell and A.M. Ferendeci, Plenum Press, New York, 1980) p. 147.
43. R.J.W. Henry, Physics Reports (to be published).
44. L. Castillejo, I. Percival and M.J. Seaton, Proc. Roy. Soc. A 254, 259 (1960).
45. P.G. Burke and M.J. Seaton, Methods in Comput. Phys. 10, 1 (1971).
46. A. Temkin, Phys. Rev. 107, 1004 (1957); see also R.J. Drachman and A. Temkin, in Case Studies in Atomic Collision Physics, vol. 2 (ed. by M.R.C. McDowell and E.W. McDaniel, North Holland Publ. Cy, Amsterdam, 1972), p. 399.
47. C. Schwartz, Phys. Rev. 124, 1468 (1961).
48. M. Gailitis, Soviet Physics (JETP) 20, 107 (1965).
49. P.G. Burke and A.J. Taylor, Proc. Phys. Soc. 88, 549 (1966).
50. R. Damburg and E. Karule, Proc. Phys. Soc. 90, 637 (1967).
51. Vo ky Lan, M. LeDourneuf and P.G. Burke, J. Phys. B 9, 1065 (1976).
52. See for example the recent calculations of J. Hata, L.A. Morgan and M.R.C. McDowell, J. Phys. B 13, L 347 (1980), and to be published.
53. E.P. Wigner, Phys. Rev. 70, 15, 606 (1946); E.P. Wigner and L. Eisenbud, Phys. Rev. 72, 29 (1947).
54. P.G. Burke, A. Hibbert and W.D. Robb, J. Phys. B 4, 153 (1971).
55. P.G. Burke and W.D. Robb, Adv. Atom. Molec. Phys. 11, 143 (1975).
56. A.V. Hazi and H.S. Taylor, Phys. Rev. A 1, 1109 (1970).
57. H.S. Taylor, Adv. Chem. Phys. 18, 91 (1970).
58. E.J. Heller and H.A. Yamani, Phys. Rev. A 9, 1201 (1974), A 9, 1209 (1974).
59. J.T. Broad and W.P. Reinhardt, J. Phys. B 9, 1491 (1974); Phys. Rev. A 14, 2159 (1976).
60. T.N. Rescigno and W.P. Reinhardt, Phys. Rev. A 10, 1584 (1974).
61. M.J. Seaton, Adv. Atom. Molec. Phys. 11, 83 (1975).
62. H. Feshbach, Ann. Phys. (N.Y.) 5, 357 (1958); 19, 287 (1962). See also T.F. O'Malley and S. Geltman, Phys. Rev. 137, A 1344 (1965), A.K. Bhatia and A. Temkin, Phys. Rev. A 8, 2184 (1973), A 10, 458 (1974), A 11, 2018 (1975).
63. A detailed account of the optical potential theory can be found in ref.2, Chapter 20. We shall return below to the optical potential in discussing intermediate-and high-energy electron-atom scattering.
64. Gy Csanak, H.S. Taylor and R. Yaris, Adv. Atom. Molec. Phys. 7, 287 (1971); L.D. Thomas, Gy Csanak, H.S. Taylor and B.S. Yarlagadda, J. Phys. B 7, 1719 (1974).
65. C.J. Joachain and C. Quigg, Rev. Mod. Phys. 46, 279 (1974)
66. B.H. Bransden and M.R.C. McDowell, Phys. Reports 30 C, 207 (1977); 46 C, 249 (1978).

67. We do not consider here the case of relativistic collisions.
68. See ref. 2, Chapter 19.
69. K.L. Bell and A.E. Kingston, Adv. Atom. Molec. Phys. 10, 53 (1974).
70. H.S.W. Massey and C.B.O. Mohr, Proc. Roy. Soc. A 146, 880 (1934).
71. A.R. Holt and B.L. Moiseiwitsch, J. Phys. B 1, 36 (1968); A.R. Holt, J. Hunt and B.L. Moiseiwitsch, J. Phys. B 4, 1318 (1971).
72. See for example F.W. Byron, Jr. and C.J. Joachain, J. Phys. B 10, 207 (1977); Phys. Reports 34 C, 233 (1977); D.P. Dewangan and H.R.J. Walters, J. Phys. B 10, 637 (1977).
73. A.R. Holt, J. Phys. B 5, L 6 (1972).
74. F.W. Byron, Jr. and C.J. Joachain, to be published.
75. C.J. Joachain, in Electronic and Atomic Collisions (ed. by G. Watel, North Holland Publ. Cy, Amsterdam, 1978), p. 71.
76. F.W. Byron, Jr., C.J. Joachain and B. Piraux, to be published.
77. F.W. Byron, Jr., C.J. Joachain and B. Piraux, to be published.
78. D.M. Chase, Phys. Rev. 104, 838 (1956).
79. F.W. Byron, Jr. and C.J. Joachain, Phys. Rev. A 8, 1267 (1973); A 8, 3266 (1973); J. Phys. B 7, L 212 (1974); J. Phys. B 8, L 284 (1975); J. Phys. B 10, 207 (1977), Phys. Reports 34 C, 233 (1977).
80 See ref. 2, Chapter 20.
81. M.H. Mittleman and K.M. Watson, Phys. Rev. 113, 198 (1959).
82. M.H. Mittleman and K.M. Watson, Ann. Phys. (N.Y.) 10, 268 (1960).
83. C.J. Joachain and M.H. Mittleman, Phys. Rev. A 4, 1492 (1971); F.W. Byron, Jr. and C.J. Joachain, Phys. Rev. A 9, 2559 (1974).
84. F.W. Byron, Jr. and C.J. Joachain, Phys. Letters A 49, 306 (1974); Phys. Rev. A 15, 128 (1977); C.J. Joachain, R. Vanderpoorten, K.H. Winters and F.W. Byron, Jr., J. Phys. B 10, 227 (1977).
85. R. Vanderpoorten, J. Phys. B 8, 926 (1975).
86. J.B. Furness and I.E. McCarthy, J. Phys. B 6, 2280 (1973).
87. M.E. Riley and D.G. Truhlar, J. Chem. Phys. 65, 792 (1976).
88. P.G. Burke and T.G. Webb, J. Phys. B 6, 2280 (1973). See also P.G. Burke and J.F.B. Mitchell, J. Phys. B 6, 620 (1973).
89. J. Callaway and J.W. Wooten, Phys. Letters A 45, 85 (1973); Phys. Rev. A 9, 1924 (1974); A 11, 1118 (1975).
90. J. Callaway, M.R.C. McDowell and L.A. Morgan, J. Phys. B 8, 2181 (1976); B 9, 2043 (1976).
91. B.H. Bransden and J.P. Coleman, J. Phys. B 5, 537 (1972); K.H. Winters, C.D. Clark, B.H. Bransden and J.P. Coleman, J. Phys. B 6, L 247 (1973), J. Phys. B 7, 788 (1974). See also ref. 66.
92. See for example the recent work of B.H. Bransden, J.J. Smith and K.H. Winters, J. Phys. B 11, 3095 (1978); J.J. Smith, K.H. Winters and B.H. Bransden, J. Phys. B 12, 1723 (1979).
93. H. Ehrhardt, H. Hesselbacher, K. Jung and K. Willmann, in Case Studies in Atomic Physics (ed. by M.R.C. McDowell and

E.W. McDaniel, North Holland Publ. Cy, Amsterdam, 1971), vol. 2, 159.
94. I.E. McCarthy and E. Wiegold, Phys. Reports 27 C, 275 (1976); E. Wiegold and I.E. McCarthy, Adv. Atom. Molec. Phys. 14, 127 (1978).
95. E. Gerjuoy and N.A. Krall, Phys. Rev. 119, 705 (1960); Phys. Rev. 127, 2105 (1962).
96. F.W. Byron, Jr., F.J. de Heer and C.J. Joachain, Phys. Rev. Letters 35, 1147 (1975).
97. F.J. de Heer, R.W. Wagenaar, H.J. Blaauw and A. Tip, J. Phys. B 9, L 269 (1976); F.J. de Heer, M.R.C. McDowell and R.W. Wagenaar, J. Phys. B 10, 1945 (1977).
98. P.K. Hutt, M.M. Islam, A. Rabheru and M.R.C. McDowell, J. Phys. B 9, 2447 (1976).
99. A. Tip, J. Phys. B 10, L 11 (1977); J. Phys. B 10, L 295 (1977).
100. K. Blum and P.G. Burke, Phys. Rev. A 16, 163 (1977).
101. F.W. Byron, Jr. and C.J. Joachain, Phys. Letters A 62, 217 (1977); J. Phys. B 11, 2533 (1978).
102. See for example A. Akhiezer and V.B. Bereztetski, Quantum Electrodynamics (second edition, Wiley, New York, 1963); J.D. Bjorken and S.D. Drell, Relativistic Quantum Mechanics (McGraw-Hill, New York, 1964), Relativistic Quantum Fields (McGraw-Hill, New York, 1965).
103 We follow here the treatment of B.H. Bransden and C.J. Joachain, The Physics of Atoms and Molecules (to be published).
104. H.A. Bethe and E.E. Salpeter, Quantum Mechanics of One-and Two-Electron Atoms (Springer-Verlag, Berlin, 1957).
105. K. Codling, Rep. Progr. Phys. 36, 541 (1973).
106. R.P. Madden, in The Physics of Electronic and Atomic Collisions (ed. by J.S. Risley and R. Geballe, Univ. of Washington Press, Seattle, 1976), p. 563.
107. F.J. Wuilleumier, in Electronic and Atomic Collisions (ed. by N. Oda and K. Takayanagi, North-Holland Publ. Cy, Amsterdam, 1980), p. 55.
108. For a review of the photoionization of atomic systems, see P.G. Burke, in Atomic Processes and Applications (ed. by P.G. Burke and B.L. Moiseiwitsch, North Holland Publ. Cy, Amsterdam, 1976), p. 199.
109. R.J.W. Henry and L. Lipsky, Phys. Rev. 153, 51 (1967).
110. Asymptotic expansions for the Kramers-Gaunt factor are given for example in the book by G.V. Marr, Photoionization Processes in Gases (Academic Press, New York, 1967).
111. See for example S.T. Manson and J.W. Cooper, Phys. Rev. 165, 126 (1968).
112. A. Burgess and M.J. Seaton, Mon. Not. Roy. Astr. Soc. 120, 121 (1960).
113. G. Peach, Mon. Not. Astr. Soc. 124, 371 (1962); Mem. Roy. Astr. Soc. 71, 13, 29 (1967), ibid. 73, 1 (1970).
114. J. Dubau and J. Wells, J. Phys. B 6, 1452 (1973).

115. D.J. Kennedy and S.T. Manson, Phys. Rev. A 5, 227 (1972).
116. P.L. Altick and A.E. Glassgold, Phys. Rev. 133, A 632 (1964).
117. M. Ya Amusia and N.A. Cherepkov, in Case Studies in Atomic Physics (ed. by M.R.C. McDowell and E.W. McDaniel, North-Holland Publ. Cy, Amsterdam, 1975), vol. 5, p. 47.
118. G. Wendin, J. Phys. B 4, 1080 (1971); 5, 110 (1972); 6, 42 (1973).
119. H.P. Kelly, Phys. Rev. 136, B 896 (1964).
120. E.S. Chang and M.R.C. McDowell, Phys. Rev. 176, 126 (1968).
121. J.J. Chang and H.P. Kelly, Phys. Rev. A 12, 92 (1975).
122. P.L. Altick and E.N. Moore, Phys. Rev. 147, 59 (1966); P.L. Altick, Phys. Rev. 169, 21 (1968); G.N. Bates and P.L. Altick, J. Phys. B 6, 653 (1973).
123. P.G. Burke, in Electron and Photon Interactions with Atoms (ed. by H. Kleinpoppen and M.R.C. McDowell, Plenum Press, 1976), p. 1.
124. M. Le Dourneuf, in Electronic and Atomic Collisions (ed. by G. Watel, North-Holland Publ. Cy, Amsterdam, 1978), p. 143.
125. The theory of resonances in photoionization is discussed in the reviews of P.G. Burke, Adv. Phys. 14, 521 (1965) and K. Smith, Rep. Progr. Phys. 29, 373 (1966).
126. U. Fano, Phys. Rev. 124, 1866 (1961); U. Fano and J.W. Cooper, Phys. Rev. 137, A 1364 (1965).
127. R.P. Madden and K. Codling, Phys. Rev. Letters 10, 516 (1963).
128. H.C. Bryant, B.D. Dieterle, J. Donahue, H. Sharifian, H. Tootoonchi, D.M. Wolfe, P.A.M. Gram and M.A. Yates-Williams, Phys. Rev. Letters 38, 228 (1977).
129. J. Macek, Proc. Phys. Soc. 92, 365 (1967); J.T. Broad and W.P. Reinhardt, Phys. Rev. A 14, 2159 (1976). See also J.S. Risley, in Atomic Physics, vol. 4 (ed. by G. zu Putlitz, E.W. Weber and A. Winnacker, Plenum, New York, 1974), p. 487.
130. F.W. Byron, Jr. and C.J. Joachain, Phys. Rev. 164, 1 (1967).
131. T.A. Carlson, Phys. Rev. 156, 142 (1967).
132. V. Schmidt, N. Sandner, H. Kuntzemüller, P. Dhez, F. Wuilleumier and E. Källne, Phys. Rev. A 13, 1748 (1976).
133. D.M.P. Holland, K. Codling, J.B. West and G.V. Marr, J. Phys. B 12, 2465 (1979).
134. M.J. Van der Wiel and G. Wiebes, Physica 53, 225 (1971), 54, 411 (1971); M.J. Van der Wiel and G.R. Wight, Phys. Letters A 54, 83 (1975).
135. R.L. Brown, Phys. Rev. A 1, 586 (1970).
136. M. Ya Amusia, E.G. Drukarev, V.G. Gorshov and M.P. Kazachkov, J. Phys. B 8, 1248 (1975).
137. T.N. Chang, T. Ishihara and R.T. Poe, Phys. Rev. Letters 27, 838 (1971).
138. T.N. Chang and R.T. Poe, Phys. Rev. A 12, 1432 (1975).
139. S.L. Carter and H.P. Kelly, Phys. Rev. A 16, 1525 (1977).
140. S.L. Carter and H.P. Kelly, J. Phys. B 9, L 565 (1976).

ELECTRON AND PHOTON COLLISIONS WITH MOLECULES

P.G. Burke

Queen's University, Belfast, N.Ireland
and
Daresbury Laboratory, Cheshire, England

ABSTRACT

The general theory of electron and photon collisions with molecules is presented. General expressions for angular distributions of the scattered or ejected electrons are obtained. Methods of calculation are then reviewed and are compared with illustrative results.

1. INTRODUCTION

2. THEORY OF ELECTRON-MOLECULE COLLISIONS

2.1 Fixed-Nuclei Approximation

2.2 Rotational and Vibrational Excitation

3. THEORY OF MOLECULAR PHOTOIONIZATION

4. CALCULATIONS AND ILLUSTRATIVE RESULTS

4.1 Single-Centre Expansion Method

Static Exchange Approximation

Approximate Representation of Exchange and Polarization

Photoionization

4.2 Continuum Multiple Scattering Method

4.3 L^2-Methods

Introduction

Stieltjes-Tchebycheff Moment Method

T-Matrix Method

R-Matrix Method

1. INTRODUCTION

Considerable progress has been made in the theory of electron and photon collisions with molecules in the last few years. This has been stimulated firstly by the increasing sophistication of experiments involving synchrotron radiation sources and millivolt electron energy resolution. Secondly because of the increasing need for an accurate knowledge of these processes in many applications in laboratory and astrophysics. Finally because the rapidly increasing power and availability of computers which are allowing new theoretical models to be explored.

In these lectures, because of space and time limitations, we have had to be selective in the choice of material. We have therefore decided for the most part to concentrate our attention on methods and on calculations pertaining to low incident electron and photon energies where the Born and other weak coupling theories are not appropriate. This has meant that a large and important body of work has not been covered. Nevertheless it is hoped that by making this choice a more comprehensive survey has been possible at low energies where theoretical developments have recently been most rapid, and where recent experiments have enabled a very detailed chaeck of theories to be made.

We divide these lectures into three main sections. The next section contains a detailed study of the theory of electron molecule collisions. Our main aim in this section is to derive the coupled integro-differential equations which describe the motion of the scattered electron, and then to derive equations for the cross sections in terms of the solution of these equations. We first show

how these equations are obtained in the fixed-nuclei approximation, and we then show how this assumption can be relaxed to give rotational and vibrational excitation cross sections. In the following section we carry through a similar development for molecular photoionization, placing particular emphasis on the close connection between photoionization and electron molecule scattering. Finally in the last section we summarize a number of models which have been used to enable cross sections to be calculated, and we compare illustrative results from these calculations with experiment. We consider first single-centre expansion methods. We then consider the continuum multiple scattering method. Finally we describe recent L^2 methods which appear to be particularly promising for the future.

There have been a number of important reviews on the subject in the last two years. A review of the theory of electron molecule collisions has been given by Lane (1980). A comprehensive survey of both electron-molecule and photon-molecule collisions has been edited by Rescigno, McKoy and Schneider (1979). The invited papers at a Symposium on electron-molecule collisions in Tokyo has been edited by Shimamura and Matsuzawa (1979). Finally a memorial volume for George Schulz on electron-molecule scattering has been edited by Brown (1979).

2. THEORY OF ELECTRON-MOLECULE COLLISIONS

2.1 Fixed-Nuclei Approximation

In this section we consider the approximation in which the nuclei are held fixed during the collision. The Hamiltonian then contains only electronic terms. We use a frame of reference to describe the collision which is rigidly attached to the molecule. This frame, called the molecular or body-fixed frame, is illustrated for a diatomic molecule in figure 1.

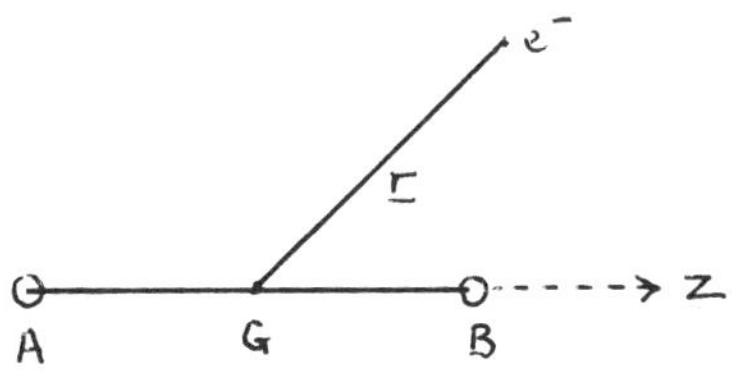

Figure 1. Molecular-Frame for a Diatomic Molecule

In this figure A and B are the two nuclei, G is their centre of gravity which is taken as the origin, and the z axis is chosen to lie along the internuclear axis. The coordinates of all electrons are referred to this frame. We will see that this approximation is appropriate when the time of the collision is short compared with typical rotation and vibration times. We will also see that it provides the first step in approximations which are valid when this criterion is not satisfied.

The Schrödinger equation describing the collision of an electron with an N electron molecule can be written using atomic units and assuming that all relativistic terms in the Hamiltonian can be neglected is

$$(H_{el} - E)\Psi = 0 \tag{1}$$

where the Hamiltonian H_{el} is given in atomic units by

$$H_{el} = -\frac{1}{2}\nabla^2_{N+1} + H_T + V \tag{2}$$

where $-\frac{1}{2}\nabla^2_{N+1}$ is the kinetic energy operator of the scattered electron H_T is the electronic Hamiltonian of the target and V is the potential interaction between the electron and the target. In the case of a diatomic molecule this is defined by

$$V(\underline{r}_1, \cdots, \underline{r}_{N+1}; R) = \sum_{i=1}^{N} \frac{1}{|\underline{r}_{N+1} - \underline{r}_i|} - \frac{Z_A}{|\underline{r}_{N+1} - \underline{r}_A|} - \frac{Z_B}{|\underline{r}_{N+1} - \underline{r}_B|} \tag{3}$$

where Z_A and Z_B are the nuclear charges located at $\underline{r}_A$ and $\underline{r}_B$, R is the internuclear distance and we have denoted the target electron coordinates, assumed to be N in number, by $\underline{r}_i$.

The total wave function Ψ in eq. (1) describing the collision can be expanded for each set of internuclear coordinates, in the form

$$\Psi_j = \mathcal{A} \sum_{i=1}^{n} \Phi_i(1, \cdots, N) F_{ij}(N+1) + \sum_{i=1}^{m} \xi_i(1, \cdots, N+1)\, a_{ij} \tag{4}$$

where the Φ_i are target electronic eigenstates, and possible pseudo-states to represent the polarizability of the target by the incident electron, the F_{ij} describes the motion of the scattered electron including its spin, $\mathcal{A}$ is the usual antisymmetrization operator, ξ_i are L^2 integrable correlation functions and finally the a_{ij} are multiplying coefficients. In practice we assume that accurate representations for the target states Φ_i are available. We also assume that a set of correlation functions ξ_i

have been chosen. Integro-differential equations can then be derived for the functions F_{ij} which are coupled to linear equations for the coefficients a_{ij}. We will consider the derivation of these equations later in this section and their solution in section 4.

We now consider the derivation of the cross section from the asymptotic form of the wave function Ψ_j. In this discussion we follow and extend the discussion of electron scattering by polyatomic molecules given by Burke et al. (1972) and by Dill and Dehmer (1974). We look for a solution of eq. (1) corresponding to an incident plane wave plus an outgoing spherical scattered wave. We introduce the laboratory frame of reference where the z' axis is along the incident beam direction as illustrated in figure 2.

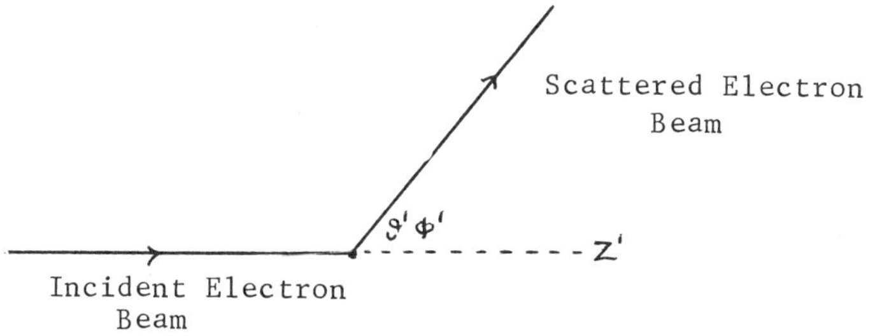

Figure 2. Laboratory-Frame for Electron Molecule Scattering

In this frame of reference, the molecular frame given in figure 1 is orientated in a direction defined by the Euler angles $\alpha\beta\gamma$. The asymptotic form of the wave function for an electron incident upon a neutral molecule is

$$\Psi_j^{(+)} \underset{r_{N+1}\to\infty}{\sim} \Phi_j(1,\cdots,N)\chi_{\frac{1}{2}m_j}(\sigma_{N+1})\, e^{ik_j z'_{N+1}}$$

$$+ \sum_i \Phi_i(1,\cdots,N)\chi_{\frac{1}{2}m_i}(\sigma_{N+1})\, f_{ij}(\vartheta'\phi')\, \frac{e^{ik_i r'_{N+1}}}{r'_{N+1}} \tag{5}$$

where as before Φ_i are the target electronic states, $\chi_{\frac{1}{2}m_i}$ are the spin functions for the scattered electron and $f_{ij}(\vartheta'\phi')$ is the scattering amplitude. The differential cross section for a transition form an initial state j to a final state i, where the scattered electron's magnetic quantum number changes from m_j to m_i is

$$\frac{d\sigma_{ij}}{d\Omega'} = \frac{k_i}{k_j}\left| f_{ij}(\vartheta'\phi')\right|^2 \tag{6}$$

where k_j and k_i are the initial and final wave numbers for the scattered electron respectively.

We now consider the asyptotic form of eq. (4). For light molecules and non-relativistic electron energies we can couple the spin of the scattered electron with the spin of the target molecule to form an eigenstate of $\underline{S}^2$ and S_z, corresponding to the quantum numbers S and M_S which are conserved in the collision. We define these eigenstates by

$$\Phi_i^{SM_S}(1,\cdots,N,\sigma_{N+1}) = \sum_{M_{S_i} m_i} (S_i M_{S_i} \tfrac{1}{2} m_i | S M_S)\, \Phi_i(1,\cdots,N)\, \chi_{\frac{1}{2} m_i}(\sigma_{N+1}) \tag{7}$$

where $(S_i M_{S_i} \frac{1}{2} m_i | S M_S)$ are Clebsch-Gordan coefficients. After re-expanding eq. (4) in terms of these eigenstates, and carrying out a partial wave analysis, we obtain for each $S M_S$ component

$$\Psi_{j \ell_j m_{\ell_j}}^{SM_S} = \mathcal{A} \sum_{i \ell_i m_{\ell_i}} \Phi_i^{SM_S}(1,\cdots,N,\sigma_{N+1})\, r_{N+1}^{-1}\, u^S_{i\ell_i m_{\ell_i}, j\ell_j m_{\ell_j}}(r_{N+1})\, Y_{\ell_i m_{\ell_i}}(\hat{\underline{r}}_{N+1}) \tag{8}$$

where for convenience of notation we have only considered the first expansion in eq. (4). Substituting this expansion for the total wave function into eq. (1), premultiplying by the channel functions $\Phi_i^{SM_S}(1,\cdots,N,\sigma_{N+1})\, r_{N+1}^{-1}\, Y_{\ell_i m_{\ell_i}}(\hat{\underline{r}}_{N+1})$ and integrating over all coordinates except the radial coordinate of the scattered electron r_{N+1} gives the following set of coupled integro-differential equations which are satisfied by the radial functions :

$$\left(\frac{d^2}{dr^2} - \frac{\ell_i(\ell_i+1)}{r^2} + \frac{2(Z-N)}{r} + k_i^2\right) u^S_{ik}(r) =$$

$$2\sum_j \left(V^S_{ij}(r) + W^S_{ij}(r)\right) u^S_{jk}(r) \tag{9}$$

where for simplicity we have written $u^S_{ij}(r) = u^S_{i\ell_i m_{\ell_i}, j\ell_j m_{\ell_j}}(r)$. In this equation Z is the sum of the charges on the nuclei so that $Z-N$ is the total charge on the molecule. V^S_{ij} is the direct potential matrix which outside the molecular charge distribution has the asymptotic form

$$V_{ij}(r) = \sum_{\lambda=1}^{\lambda_{max}} a_{ij}^{\lambda} r^{-\lambda-1} \quad , r > r_0 \tag{10}$$

where the first term in r^{-2} has a contribution arising from the dipole moment of the molecule and the second term in r^{-3} from the quadrupole moment of the molecule. Finally $W_{ij}^{S}(r)$ is the non-local exchange potential operator which has the general form

$$W_{ij}^{S} u_{jk}^{S}(r) = \int_0^{\infty} W_{ij}^{S}(r, r') u_{jk}^{S}(r') dr' \tag{11}$$

This integral term vanishes exponentially at large distances. We note that eq. (9) would have retained the same form if we had included terms in the second expansion in eq. (4), these terms just adding to the exchange potential.

We now look for a complete set of solutions of eqs. (9) satisfying the real boundary condition

$$\begin{aligned} u_{ij}^{S}(0) &= 0 \\ u_{ij}^{S}(r) &\underset{r\to\infty}{\sim} k_i^{-1/2}(\sin\theta_i \delta_{ij} + \cos\theta_i K_{ij}^{S}), \quad k_i^2 > 0 \\ &\underset{r\to\infty}{\sim} 0 \quad , k_i^2 < 0 \end{aligned} \tag{12}$$

where $\theta_i = k_i r - \frac{1}{2}\ell\pi$ for a neutral molecule. The real symmetric K-matrix defined by this equation contains all the essential information describing the collision and methods used to calculate it will be described in section 4. Here, in order to obtain an expression for the scattering amplitude, we need to introduce an alternative, but completely equivalent asymptotic form, by taking linear combinations of the solutions defined by eq. (12). We define solutions satisfying the outgoing wave boundary conditions

$$\begin{aligned} u_{ij}^{S(+)}(r) &\underset{r\to\infty}{\sim} k_i^{-1/2}(e^{-i\theta_i}\delta_{ij} - e^{i\theta_i} S_{ij}^{S}), \quad k_i^2 > 0 \\ &\underset{r\to 0}{\sim} 0 \quad , k_i^2 < 0 \end{aligned} \tag{13}$$

which defines a unitary and symmetric S-matrix which is related to the K-matrix by the matrix equation

$$\underline{S} = \frac{\underline{1} + i\underline{K}}{\underline{1} - i\underline{K}} \tag{14}$$

The S-matrix when written out explicitly in terms of all its indices is $S^S_{i\ell_i m_{\ell_i},\, j\ell_j m_{\ell_j}}$. It is diagonal in the total spin S and has non-zero matrix elements coupling the channels defined by the quantum numbers $i\ell_i m_{\ell_i}$ and $j\ell_j m_{\ell_j}$.

An expression for the scattering amplitude in terms of the S-matrix can now be obtained by expanding the plane wave term in eq. (5) as follows

$$\Phi_j(1,\cdots,N)\chi_{\frac{1}{2}m_j}(\sigma_{N+1})e^{ik_j z'_{N+1}} \underset{r'_{N+1}\to\infty}{\sim} \frac{i\pi^{1/2}}{k_j r'_{N+1}}\sum_{S\ell_j} i^{\ell_j}(2\ell_j+1)^{1/2}$$

$$\times (S_j M_{S_j}\tfrac{1}{2}m_j|SM_S)\,\Phi_j^{SM_S}(1,\cdots,N,\sigma_{N+1})\left(e^{-i\theta'_j}-e^{i\theta'_j}\right)Y_{\ell_j 0}(\hat{\underline{r}}'_{N+1}) \tag{15}$$

We then expand the solutions defined by eq. (5) in terms of the solutions defined by eqs. (8) and (13)

$$\Psi_j^{(+)} = \sum_{S\ell_j m_{\ell_j}} a^S_{\ell_j m_{\ell_j}}\,\Psi^{SM_S(+)}_{j\ell_j m_{\ell_j}} \tag{16}$$

where the coefficients $a^S_{\ell_j m_{\ell_j}}$ are obtained by equating the coefficient of the ingoing wave in eqs. (5) and (8). Remembering that

$$Y_{\ell m}(\hat{\underline{r}}') = \sum_{m'} D^{\ell *}_{m m'}(\alpha\beta\gamma)\,Y_{\ell m'}(\hat{\underline{r}}) \tag{17}$$

where we have adopted the notation of Rose (1957) for the Wigner rotation matrices $D^{\ell}_{m m'}(\alpha\beta\gamma)$ which relate the molecular and laboratory frames, we obtain

$$a^S_{\ell_j m_{\ell_j}} = i\pi^{1/2}k_j^{-1/2}\, i^{\ell_j}(2\ell_j+1)^{1/2}(S_j M_{S_j}\tfrac{1}{2}m_j|SM_S)\,D^{\ell_j *}_{0 m_{\ell_j}}(\alpha\beta\gamma) \tag{18}$$

It follows that the scattering amplitude can be expressed as

$$f_{ij}(\vartheta'\phi') = -i\left(\frac{\pi}{k_i k_j}\right)^{\frac{1}{2}} \sum_{\substack{l_i m_{l_i} \\ l_j m_{l_j} \\ S\, m''_{l_i}}} i^{\,l_j - l_i}(2l_j+1)^{1/2}(S_j M_{S_j} \tfrac{1}{2} m_j | S M_S)$$

$$\times (S_i M_{S_i} \tfrac{1}{2} m_i | S M_S)\, T^S_{i l_i m_{l_i}, j l_j m_{l_j}}\, D^{l_j *}_{0 m_{l_j}}(\alpha\beta\gamma)\, D^{l_i}_{m''_{l_i} m_{l_i}}(\alpha\beta\gamma)\, Y_{l_i m''_{l_i}}(\hat{r}') \tag{19}$$

where the T-matrix is defined in terms of the S-matrix by the matrix equation

$$\underline{T} = \underline{S} - \underline{1} \tag{20}$$

Substituting this result for the scattering amplitude into eq. (6) gives an expression for the differential cross section for electrons with a fixed component of spin m_j incident upon a molecule in state j with a fixed component of spin M_{S_j} and orientation defined by $\alpha\beta\gamma$, leading to a scattered electron with spin component m_i and a molecule in state i with spin component M_{S_i}

If the initial state is unpolarized and we do not observe the spins in the final state, then we must average over the initial spin components and sum over the final spin components. We define

$$\left(\frac{d\sigma_{ij}}{d\Omega'}\right)_{\text{Spin Av.}} = \frac{1}{2(2S_j+1)} \sum_{\substack{SS' M_{S_i} M_{S_j} \\ m_i m_j}} \frac{d\sigma_{ij}}{d\Omega'} \tag{21}$$

In addition if the electrons are scattered from a molecular gas where all molecular orientations are possible, we must average over these orientations to obtain the final observed cross section. We define therefore

$$\left(\frac{d\sigma_{ij}}{d\Omega'}\right)_{\text{Av.}} = \frac{1}{8\pi^2} \int_0^{2\pi} d\alpha \int_0^{\pi} \sin\beta\, d\beta \int_0^{2\pi} d\gamma \left(\frac{d\sigma_{ij}}{d\Omega'}\right)_{\text{Spin Av.}} \tag{22}$$

The summation over the spin magnetic quantum numbers in eq. (21) involves only the four Clebsch-Gordan coefficients, and can be carried out immediately. The integrals over the Euler angles ($\alpha\beta\gamma$) can be carried out after using the coupling rule and the orthogonality relations satisfied by the D-functions. We obtain

$$\left(\frac{d\sigma_{ij}}{d\Omega'}\right)_{Av.} = \sum_{\lambda} A_\lambda P_\lambda(\cos\vartheta') \tag{23}$$

where

$$A_\lambda = \frac{1}{8k_j^2(2S_j+1)(2\lambda+1)} \sum_{\substack{\ell_i m_{\ell_i} \ell_i' m_{\ell_i}' \\ \ell_j m_{\ell_j} \ell_j' m_{\ell_j}'}} i^{\ell_j-\ell_i-\ell_j'+\ell_i'} \left[(2\ell_i+1)(2\ell_i'+1)(2\ell_j+1)(2\ell_j'+1)\right]^{1/2}$$

$$\times (-1)^{m_{\ell_j}-m_{\ell_i}} (\ell_i 0 \ell_i' 0|\lambda 0)(\ell_j 0 \ell_j' 0|\lambda 0)(\ell_i m_{\ell_i} \ell_i' -m_{\ell_i}'|\lambda m_{\ell_i}-m_{\ell_i}')$$

$$\times (\ell_j m_{\ell_j} \ell_j' -m_{\ell_j}'|\lambda m_{\ell_j}-m_{\ell_j}') \sum_S (2S+1)\, T^S_{i\ell_i m_{\ell_i}, j\ell_j m_{\ell_j}}\, T^{S*}_{i\ell_i' m_{\ell_i}', j\ell_j' m_{\ell_j}'} \tag{24}$$

We obtain the total cross section by averaging over all scattering angles $\vartheta'\phi'$ defined in figure 2, giving

$$\sigma_{ij}^{Tot} = 2\pi \int_0^\pi \left(\frac{d\sigma_{ij}}{d\Omega'}\right)_{Av.} \sin\vartheta' d\vartheta'$$

$$= 4\pi A_0 \tag{25}$$

It follows immediately from eq. (24) that

$$\sigma_{ij}^{Tot} = \frac{\pi}{2k_j^2(2S_j+1)} \sum_{\substack{\ell_i m_{\ell_i} \ell_j m_{\ell_j} \\ S}} (2S+1) \left| T^S_{i\ell_i m_{\ell_i}, j\ell_j m_{\ell_j}} \right|^2 \tag{26}$$

Finally the momentum transfer cross section is defined by

$$\sigma_{ij}^{M} = 2\pi \int_0^\pi \left(\frac{d\sigma_{ij}}{d\Omega'}\right)_{Av.} \left(1 - \frac{k_i}{k_j}\cos\vartheta'\right) \sin\vartheta' d\vartheta'$$

$$= 4\pi \left(A_0 - \frac{1}{3}\frac{k_i}{k_j} A_1\right) \tag{27}$$

Equations (23), (25), and (27) enable the cross sections to be evaluated once the T-matrices have been calculated by solving the coupled integro-differential equations given by eq. (9).

2.2 Rotational and Vibrational Excitation

If the time of collision is short compared with typical rotation times, or what is almost equivalent, if the kinetic energy of the scattered electron is appreciably larger than the molecular rotational energy differences of interest, we can find a unitary transformation which relates the S-matrix calculated in the fixed-nuclei approximation with the S-matrix calculated in the laboratory frame in which the molecule is rotating. This transformation was first discussed by Bottcher (1969) and considered further by Burke and Sinfailam (1970) and Chang and Fano (1972).

The rotational motion of a general polyatomic molecule is described by the asymmetric top wave function

$$\Psi_{jm_j\tau}(\alpha\beta\gamma) = \left(\frac{2j+1}{8\pi^2}\right)^{\frac{1}{2}} \sum_{\lambda} a_{j\lambda\tau} D^{j*}_{m_j\lambda}(\alpha\beta\gamma) \tag{28}$$

where the coefficients $a_{j\lambda\tau}$ can be obtained by diagonalizing the rigid-rotor Hamiltonian

$$H_{rot} = \frac{J_{x'}^2}{2I_x} + \frac{J_{y'}^2}{2I_y} + \frac{J_{z'}^2}{2I_z} \tag{29}$$

in the basis of symmetric top eigen functions $\left(\frac{2j+1}{8\pi^2}\right)^{\frac{1}{2}} D^{j*}_{m_j\lambda}(\alpha\beta\gamma)$.

These coefficients are given for example by King et al. (1943, 1949). In the laboratory frame the asymmetric top wave functions are coupled with the orbital angular momentum eigen functions for the scattered electron to form eigenstates of the total angular momentum operator $\underline{J}^2$ and its z' component $J_{z'}$. These eigenstates are defined by

$$\Psi^{JM_J}_{j\tau\ell}(\alpha\beta\gamma;\hat{\underline{r}}') = \sum_{m_j m_\ell} (j\, m_j\, \ell\, m_\ell | J M_J) \Psi_{jm_j\tau}(\alpha\beta\gamma) Y_{\ell m_\ell}(\hat{\underline{r}}') \tag{30}$$

The required unitary transformation relates these eigenstates with the molecular frame orbital angular momentum eigenfunctions $Y_{\ell m_\ell}(\hat{\underline{r}})$ used in eq. (8). Using eq. (17) we find that the unitary transformation is

$$U^{JM_J}_{j\tau\ell m_\ell}(\alpha\beta\gamma) = \langle Y_{\ell m_\ell}(\hat{\underline{r}}) | \Psi^{JM_J}_{j\tau\ell}(\alpha\beta\gamma;\hat{\underline{r}}') \rangle$$

$$= \left(\frac{2j+1}{8\pi^2}\right)^{1/2} \sum_\lambda a_{j\lambda\tau}\,(j\lambda\ell m_\ell|J\lambda+m_\ell)\,D^{J*}_{M_J\lambda+m_\ell}(\alpha\beta\gamma) \tag{31}$$

The corresponding transformation of the S-matrix involving a transition from an initial electronic state i to a final electronic state i' with a corresponding change in the rotational quantum numbers is

$$S^{JS}_{i'j'\tau'\ell_i',ij\tau\ell_i} = \frac{1}{8\pi^2}\int_0^{2\pi} d\alpha \int_0^{\pi} \sin\beta\, d\beta \int_0^{2\pi} d\gamma \sum_{m_{\ell_i} m'_{\ell_i}} U^{JM_J\dagger}_{j'\tau'\ell_i' m'_{\ell_i}}(\alpha\beta\gamma)$$

$$\times\, S^{S}_{i'\ell_i' m'_{\ell_i},i\ell_i m_{\ell_i}}\; U^{JM_J}_{j\tau\ell_i m_{\ell_i}}(\alpha\beta\gamma) \tag{32}$$

where $S^{S}_{i'\ell_i' m'_{\ell_i},i\ell_i m_{\ell_i}}$ is defined by eq. (14). (Note in this and later equations we will reserve i to denote an electronic state and j to denote a rotational state). The integration over the Euler angles can be carried out using the orthogonality relation satisfied by the D-functions. We obtain

$$S^{JS}_{i'j'\tau'\ell_i',ij\tau\ell_i} = \frac{[(2j+1)(2j'+1)]^{1/2}}{2J+1} \sum_{\substack{\lambda\lambda'\\ m_{\ell_i} m'_{\ell_i}}} a^*_{j'\lambda'\tau'}\, a_{j\lambda\tau}$$

$$\times\,(j'\lambda'\ell_i' m'_{\ell_i}|J\lambda'+m'_{\ell_i})(j\lambda\ell_i m_{\ell_i}|J\lambda+m_{\ell_i})\, S^{S}_{i'\ell_i' m'_{\ell_i},i\ell_i m_{\ell_i}}\, \delta_{\lambda+m_{\ell_i},\lambda'+m'_{\ell_i}} \tag{33}$$

The scattering amplitude and cross section for a transition between states $ij\tau$ and $i'j'\tau'$ can now be obtained in a straight-forward way in terms of these S-matrix elements. We obtain the following result for the total cross section

$$\sigma^{Tot}_{i'j'\tau',ij\tau} = \frac{\pi}{2k^2_{j\tau}(2j+1)(2S_i+1)} \sum_{\ell_i \ell_i' J S} (2J+1)(2S+1) \left| T^{JS}_{i'j'\tau'\ell_i',ij\tau\ell_i} \right|^2 \tag{34}$$

where as before the T-matrix is defined by eq. (20).

Our procedure, when the rotational energy splitting can be neglected compared with the kinetic energy of the scattered electron is to first calculate S-matrix in the molecular frame defined by eq. (14). If rotational excitation cross sections are then required, the S-matrix is then transformed to the laboratory frame using eq. (33) and the summation in eq. (34) carried out.

If the kinetic energy of the scattered electron is small, or the S-matrix elements rapidly varying (e.g. near a narrow resonance) then H_{rot} can no longer be neglected in the derivation of the coupled integro-differential equations. In this case the molecular frame expansion (8) must be replaced by the laboratory frame expansion.

$$\Psi^{JM_J SM_S}_{ij\tau\ell_i} = \mathcal{A} \sum_{i'j'\tau'\ell_i'} \Phi^{SM_S}_{i'}(1,\ldots,N;\sigma_{N+1})\, r^{-1}_{N+1}\, v^{JS}_{i'j'\tau'\ell_i',ij\tau\ell_i}(r_{N+1}) \times \psi_{j'\tau'\ell_i'}(\alpha\beta\gamma;\hat{\underline{r}}'_{N+1}) \tag{35}$$

where again for convenience of notation we have not considered the correlation terms. Coupled integro-differential equations for the radial functions are then derived in a similar way to eqs. (9) by projecting the Schrödinger equation

$$(H_{el} + H_{rot} - E)\,\Psi = 0 \tag{36}$$

onto the channel functions defined in eq. (35). From the asymptotic form of these radial functions $S^{JS}_{i'j'\tau'\ell_i',ij\tau\ell_i}$ can be calculated directly. These equations and their solution were first considered in the case of a linear rotor by Arthurs and Dalgarno (1960).

The solution of the coupled equations in the laboratory frame is more difficult than the solution of the molecular frame equations because, for any given physical approximation, many more channels are coupled in the former case. However an important simplification based on a frame transformation at intermediate r was proposed by

Chang and Fano (1972). They pointed out that even when the kinetic energy of the scattered electron is small, the energy of the interaction between the electron and the molecule will still dominate H_{rot} for sufficiently small electron molecule distances. In this case the molecular frame equations neglecting H_{rot} can be used in this internal region, and the wave function on the boundary of this region transformed using eq. (31). This then provides the boundary condition for the solution of the laboratory frame equations in the outer region including H_{rot} from which $S^{JS}_{i'j'T'l_i',ijTl_i}$ is obtained. Since usually exchange and correlation effects are unimportant in this outer region these equations are simpler to solve.

We conclude this section with a brief discussion of vibrational excitation. As in the case of rotational excitation, if the time of collision is short compared with typical vibrational times, an adiabatic transformation of the S-matrix can be used to obtain vibrational excitation cross sections. This approach was first discussed by Chase (1956) in the context of the excitation of deformed nuclei.

The appropriate transformation of the S-matrix is then given by

$$S^{S}_{i'v_i'l_i'm_{l_i}',iv_il_im_{l_i}} = \langle \chi_{v_i'}(Q_1\cdots Q_n) | S^{S}_{i'l_i'm_{l_i}',il_im_{l_i}} | \chi_{v_i}(Q_1\cdots Q_n)\rangle \quad (37)$$

where an integral is carried out over all the normal coordinates $Q_1 \cdots Q_n$ of the nuclear motion corresponding to the initial and final vibrational states χ_{v_i} and $\chi_{v_i'}$. It follows that the coupled equations in the molecular frame must be solved and the S-matrix evaluated for a sufficiently large number of values of $Q_1 \cdots Q_n$ to enable the integral to be accurately evaluated. Given this transformed S-matrix the cross section formulae for vibrational excitation are easily derived. If in addition, rotational excitation cross sections are required then a further transformation using the unitary transformation given by eq. (31) can be carried out.

If the time of collision is not short compared with typical vibrational times, but is short compared with typical rotation times then a hybrid expansion suggested by Chandra and Temkin (1976a, b) can be used. In this case eq. (8) is replaced by

$$\Psi^{SM_S}_{i v_i \ell_i m_{\ell_i}} = \mathcal{A} \sum_{i' v_i' \ell_i' m_{\ell_i}'} \Phi^{SM_S}_{i'} (1, \cdots, N; \sigma_{N+1})$$

$$\times\, r^{-1}_{N+1}\, w^{S}_{i' v_i' \ell_i' m_{\ell_i}', i v_i \ell_i m_{\ell_i}} (r_{N+1})\, \chi_{v_i'} (Q_1 \cdots Q_n)\, Y_{\ell_i' m_{\ell_i}'} (\hat{\underline{r}}_{N+1}) \qquad (38)$$

Coupled integro-differential equations are then derived for the radial functions $w^{S}_{i' i}(r)$ by projecting the Schrödinger equations

$$\left(T_{Q_1 \cdots Q_n} + H_{el} - E\right) \Psi = 0 \qquad (39)$$

onto the channel function defined by eq. (38), where we have written $T_{Q_1 \cdots Q_n}$ for the nuclear kinetic energy operator. The S-matrix on the LHS of eq. (37) can then be calculated directly from the solution of these equations. In deriving these equations it is usual to make the simplifying assumption that the operation of the nuclear kinetic energy operator on Φ and on w is negligibly small. The equations resulting from this approach have had very limited use due to the very large number of vibrational states that must be retained to give convergence even for simple molecules. In practice it is more appropriate to replace eq. (38) by an expansion in terms of compound molecular resonance states, which converges much faster and is based more closely on physical intuition. This will be discussed in a later section.

Finally, if the collision time is not short compared with both vibration and rotation times an expansion in terms of vibrational and rotational eigenfunctions, combining both eqs. (35) and (38), can be used. This however gives rise to too many equations to have been solved numerically so far, and although some simplification would result from using the frame transformation approach of Chang and Fano (1972), this also has not yet been attempted.

3, THEORY OF MOLECULAR PHOTOIONIZATION

We consider in this section the differential cross section for the photoionization of a molecule which is assumed to have a fixed orientation in space. Laboratory frame axes are introduced where the z' axis is defined by the polarization direction of the incident photon as illustrated in figure 3.

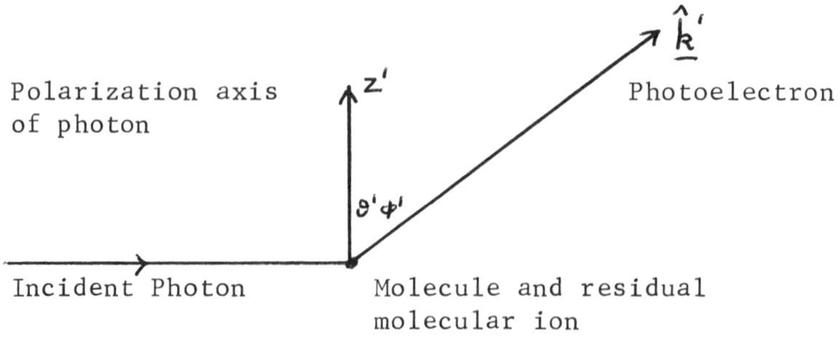

Figure 3. Laboratory-Frame for Molecular Photoionization

We assume, as we did for electron molecule scattering, that the molecular frame in which the electron-molecular ion system is described, is orientated in a direction defined by the Euler angles $\alpha\beta\gamma$. (Dill and Dehmer, 1974 and Dill, 1976).

The photoionization cross section in the dipole length approximation is given by (Bethe and Salpeter, 1957)

$$\frac{d\sigma_j^L}{d\Omega'} = 4\pi^2\alpha a_o^2 \omega \left| \langle \Psi_{jE}^{(-)} \,|\, \hat{\underline{\varepsilon}} \cdot \underline{D}^L \,|\, \Psi_o \rangle \right|^2 \tag{40}$$

and in the dipole velocity approximation is given by

$$\frac{d\sigma_j^V}{d\Omega'} = \frac{4\pi^2\alpha a_o^2}{\omega} \left| \langle \Psi_{jE}^{(-)} \,|\, \hat{\underline{\varepsilon}} \cdot \underline{D}^V \,|\, \Psi_o \rangle \right|^2 \tag{41}$$

where α is the fine structure constraint, a_o is the Bohr radius of the hydrogen atom, ω is the incident photon energy in atomic units, $\underline{D}^L$ and $\underline{D}^V$ are the dipole-length and dipole-velocity operators which in the former case has the spherical components

$$D_\mu^L = \left(\frac{4\pi}{3}\right)^{\frac{1}{2}} \sum_{i=1}^{N+1} r_i Y_{1\mu}(\hat{\underline{r}}_i') \tag{42}$$

Linear polarization along the z' axis corresponds to $\mu = 0$, while circular polarization along this axis with helicity ± 1 has $\mu = \pm 1$. As in section 2 we consider the situation where the residual molecular ion has N electrons. Finally Ψ_o and $\Psi_{jE}^{(-)}$ are the initial bound state and final scattering state of the molecule, satisfying the normalization conditions

$$\langle \Psi_0 | \Psi_0 \rangle = 1 \tag{43}$$

and

$$\langle \Psi^{(-)}_{jE} | \Psi^{(-)}_{j'E'} \rangle = \delta_{jj'} \delta(E-E') \tag{44}$$

The boundary conditions satisfied by $\Psi^{(-)}_{jE}$ can be written as

$$\Psi^{(-)}_{jE} = \Psi^{inc.}_{jE} + \Psi^{ingoing}_{jE} \tag{45}$$

where Ψ^{inc}_{jE} corresponds to a Coulomb modified plane wave in direction $\hat{k}'$ incident upon the residual molecular ion in the electronic state denoted by j and $\Psi^{ingoing}_{jE}$ corresponds to ingoing spherical waves in all open channels (Breit and Bethe, 1954). We now make a partial wave expansion of this wave function in the molecular frame. In the notation of eq. (8) we obtain, for an electron ejected with spin component m_j from a molecular ion with spin component M_{S_j},

$$\Psi^{(-)}_{jE} = \mathcal{A} \sum_{\substack{i \ell_i m_{\ell_i} \ell_j m_{\ell_j} \\ S}} (2\pi)^{-1/2} \, i^{\ell_j+1} e^{-i\sigma_{\ell_j}} Y^*_{\ell_j m_{\ell_j}}(\hat{k}) (S_j M_{S_j} \tfrac{1}{2} m_j | S M_S)$$

$$\times \Phi_i^{S M_S}(1,\ldots,N;\sigma_{N+1})\, r_{N+1}^{-1}\, u^{S(-)}_{i\ell_i m_{\ell_i}, j\ell_j m_{\ell_j}}(r_{N+1})\, Y_{\ell_i m_{\ell_i}}(\hat{r}_{N+1}) \tag{46}$$

where $u^{S(-)}_{i\ell_i m_{\ell_i}, j\ell_j m_{\ell_j}}(r)$ is a solution of the coupled integro-differential equations (9) describing the scattering of an electron by the residual molecular ion subject now to the ingoing wave boundary conditions

$$\begin{aligned} u^{S(-)}_{ij}(r) &\underset{r\to\infty}{\sim} k_i^{-1/2}\left(e^{-i\theta_i} (S^\dagger)^S_{ij} - e^{i\theta_i}\delta_{ij}\right), \quad k_i^2 > 0 \\ &\underset{r\to\infty}{\sim} 0 \quad , \quad k_i^2 < 0 \end{aligned} \tag{47}$$

Here $S^{\dagger}$ is the hermitian conjugate of the S-matrix defined in eq. (13) and

$$\theta_i = k_i r - \tfrac{1}{2} \ell_i \pi - \eta_i \ln 2k_i r + \sigma_{\ell_i} \tag{48}$$

where

$$\eta_i = -\frac{Z-N}{k_i}$$
$$\sigma_{\ell_i} = \arg \Gamma(\ell_i + 1 + i\eta_i) \tag{49}$$

and where Z is the sum of the charges on the nuclei as in eq. (9). Finally in eq. (46) $\hat{\underline{k}}$ is the direction of the ejected photoelectron referred to the molecular frame of reference.

If the molecular target is unpolarized and we do not observe the spins in the final state then we must average over the initial spin components and sum over the final spin components. We obtain

$$\left(\frac{d\sigma_j^L}{d\Omega'}\right)_{\text{Spin Av.}} = \frac{4\pi^2 \alpha a_0^2 \omega}{2S+1} \sum_{M_S M_{S_j} m_j} |\langle \Psi_{jE}^{(-)} | \hat{\underline{\varepsilon}} \cdot \underline{D}^L | \Psi_0 \rangle|^2 \tag{50}$$

in the dipole length approximation and an equivalent expression in the dipole velocity approximation. We now transform D_μ from the laboratory frame to the molecular frame using eqs. (17) and (42) giving

$$D_\mu^L = \left(\frac{4\pi}{3}\right)^{\frac{1}{2}} \sum_{i=1}^{N+1} \sum_{m_\gamma} r_i D^{1*}_{\mu m_\gamma}(\alpha\beta\gamma) Y_{1 m_\gamma}(\hat{\underline{r}}_i) \tag{51}$$

We also introduce the scattering functions

$$\Psi^{SM_S}_{j\ell_j m_{\ell_j}} = \mathcal{A} \sum_{i\ell_i m_{\ell_i}} \Phi_i^{SM_S}(1,\dots,N,\sigma_{N+1}) r_{N+1}^{-1} i(2\pi)^{-1/2} u^{S(-)}_{i\ell_i m_{\ell_i}, j\ell_j m_{\ell_j}}(r_{N+1}) \times Y_{\ell_i m_{\ell_i}}(\hat{\underline{r}}_{N+1}) \tag{52}$$

and the dipole matrix elements

$$D^{\Gamma_0(-)}_{j\ell_j m_{\ell_j} m_\gamma} = \left(\frac{4\pi}{3}\right)^{1/2} \langle \Psi^{SM_S(-)}_{j\ell_j m_{\ell_j}} \,|\, \sum_{i=1}^{N+1} r_i\, Y_{1 m_\gamma}(\hat{r}_i) \,|\, \Psi_0 \rangle \tag{53}$$

where Γ_0 specifies the quantum numbers of the initial state. We next use the coupling rule for the spherical harmonics

$$Y_{\ell_j m_{\ell_j}}(\hat{k})\, Y^*_{\ell_j' m_{\ell_j}'}(\hat{k}) = (-1)^{m_{\ell_j}} \left[\frac{(2\ell_j+1)(2\ell_j'+1)}{4\pi}\right]^{1/2} \sum_{K_e} (2K_e+1)^{-1/2}$$

$$\times (\ell_j m_{\ell_j} \ell_j' - m_{\ell_j}' | K_e M_e)(\ell_j 0 \ell_j' 0 | K_e 0)\, Y_{K_e M_e}(\hat{k}) \tag{54}$$

and for the D-functions

$$D^{1*}_{\mu m_\gamma}(\alpha\beta\gamma)\, D^{1}_{\mu m_\gamma'}(\alpha\beta\gamma) = (-1)^{\mu - m_\gamma} \sum_{K_\gamma} (1 -\mu\, 1 \mu | K_\gamma 0)$$

$$\times (1 -m_\gamma\, 1 m_\gamma' | K_\gamma M_\gamma)\, D^{K_\gamma}_{0 M_\gamma}(\alpha\beta\gamma) \tag{55}$$

Finally we transform $Y_{K_e M_e}(\hat{k})$ in eq. (54) from the molecular frame to the laboratory frame in which the observations are carried out, and combine the resultant D-function with the D-function on the RHS of eq. (55). We then find the following result for the spin averaged photoionization cross section in the dipole length approximation for a photon with polarization μ incident on a molecule with fixed orientation $\alpha\beta\gamma$.

$$\left(\frac{d\sigma_j^L}{d\Omega'}\right)_{Spin\,Av} = 4\pi^2 \alpha a_0^2 \omega \sum_{\ell_j \ell_j'} i^{\ell_j' - \ell_j} e^{i(\sigma_{\ell_j} - \sigma_{\ell_j'})} \left[\frac{(2\ell_j+1)(2\ell_j'+1)}{4\pi}\right]^{1/2}$$

$$\times \sum_{m_{\ell_j} m_{\ell_j}' m_\gamma m_\gamma'} (-1)^{\mu - m_\gamma + m_{\ell_j}'} D^{\Gamma_0(-)}_{j\ell_j m_{\ell_j} m_\gamma} D^{\Gamma_0(-)*}_{j\ell_j' m_{\ell_j}' m_\gamma'} \sum_{K_\gamma K_e K M_e'} (2K_e+1)^{-1/2}$$

$$\times (1 -\mu\, 1 \mu | K_\gamma 0)(1 -m_\gamma\, 1 m_\gamma' | K_\gamma M_\gamma)(\ell_j m_{\ell_j} \ell_j' - m_{\ell_j}' | K_e M_e)(\ell_j 0 \ell_j' 0 | K_e 0)$$

$$\times (K_e M_e' K_\gamma 0 | K M_e')(K_e M_e K_\gamma M_\gamma | K M)\, D^{K}_{M_e' M}(\alpha\beta\gamma)\, Y_{K_e M_e'}(\hat{k}') \tag{56}$$

We can write this result in the compact form

$$\left(\frac{d\sigma^L}{d\Omega'}\right)_{\text{Spin Av.}} = \sum_{K=0}^{2\ell_{max}} \sum_{M} A_{KM} Y_{KM}(\hat{k}') \tag{57}$$

where ℓ_{max} is the maximum angular momentum of the ejected electron ℓ_j or ℓ_j'.

This result simplifies considerably if we average over all orientation of the molecular axis, which is the situation in a gas. We define the angle averaged cross section by

$$\left(\frac{d\sigma_j^L}{d\Omega'}\right)_{\text{Av.}} = \frac{1}{8\pi^2}\int_0^{2\pi} d\alpha \int_0^{\pi} \sin\beta\, d\beta \int_0^{2\pi} d\gamma \left(\frac{d\sigma_j^L}{d\Omega'}\right)_{\text{Spin Av.}} \tag{58}$$

Following Fano and Dill (1972) we introduce the angular momentum transfer j_t between the incident photon and the ejected electron and we define the dipole matrix elements by

$$D^{\Gamma_0(-)}_{j\ell_j j_t m_\alpha} = \sum_{m_\gamma m_{\ell_j}} (-1)^{m_\gamma} (1 - m_\gamma \ell_j m_{\ell_j} | j_t\, m_{\ell_j} - m_\gamma)\, D^{\Gamma_0(-)}_{j\ell_j m_{\ell_j} m_\gamma} \tag{59}$$

where the summation in this equation is for fixed $m_{\ell_j} - m_\gamma = m_\alpha$.
We then find that the averaged photoionization cross section for light linearly polarized in the z' direction is

$$\left(\frac{d\sigma_j^L}{d\Omega'}\right)_{\text{Av.}} = 4\pi^2\alpha a_0^2 \omega \sum_{\ell_j \ell_j' j_t m_\alpha} (2j_t+1)^{-1} e^{i(\sigma_{\ell_j} - \sigma_{\ell_j'})}$$

$$\times D^{\Gamma_0(-)}_{j\ell_j j_t m_\alpha} D^{\Gamma_0(-)*}_{j\ell_j' j_t m_\alpha}\, \Theta(j_t \ell_j \ell_j'; \theta') \tag{60}$$

where

$$\Theta(j_t \ell_j \ell_j'; \theta') = \frac{2j_t+1}{4\pi}\left[(2\ell_j+1)(2\ell_j'+1)\right]^{1/2} (-1)^{j_t} \sum_K \begin{Bmatrix} 1 & 1 & K \\ \ell_j & \ell_j' & j_t \end{Bmatrix}$$

$$\times (1010|K0)(\ell_j 0 \ell_j' 0|K0)\, P_K(\cos\theta') \tag{61}$$

From this equation it is clear that only $K = 0$ and $K = 2$ terms can contribute to the angular distribution and so we can write

$$\left(\frac{d\sigma_j}{d\Omega'}\right)_{Av.} = \frac{\sigma_j}{4\pi}\left(1 + \beta_j\, P_2(\cos\theta')\right) \tag{62}$$

where σ_j is the total photoionization cross section and β_j is the asymmetry parameter. This result is consistent with the theorem first proved by Yang (1948), who considered the limitations imposed on the angular distribution firstly by the fact that the transition operator is an electric dipole operator, secondly that only one direction is defined by the photon polarization direction (for randomly orientated molecules) and finally that by parity conservation interference effects between even and odd parities giving rise to a $K = 1$ term do not occur.

4. CALCULATIONS AND ILLUSTRATIVE RESULTS

4.1 Single-Centre Expansion Method

In this section we describe the solution of the equations which arise when the scattered electron and target orbitals are expanded about the centre of gravity of the molecule. We will be mainly concerned with the equation which arise in the fixed-nuclei approximation. This method has had the widest application for electron scattering from simple diatomic molecules.

Static Exchange Approximation

The approximation which has been usually adopted is to retain just the ground electronic state is the basic expansion (8). This gives rise to the static-exchange approximation. The resultant coupled equations which must be solved are defined by eq. (9) which for a closed-shell diatomic molecule are

$$\left(\frac{d^2}{dr^2} - \frac{\ell(\ell+1)}{r^2} + k^2\right) u_\ell^{\Lambda S}(r) = 2\sum_{\ell'}\left[V_{\ell\ell'}^{\Lambda S} + W_{\ell\ell'}^{\Lambda S}\right] u_{\ell'}^{\Lambda S}(r) \tag{63}$$

where Λ is the z component of the orbital angular momentum of the scattered electron which is conserved in the collision, and where for convenience of notation we have omitted the second index on the radial function $u_{\ell}^{\Lambda S}(r)$.

The evaluation of the direct potential $V_{\ell\ell'}^{\Lambda S}(r)$ is straightforward provided that some suitable representation, such as an analytic SCF approximation is available for the target molecule. We define the static potential seen by the scattered electron by

$$V(\underline{r}_{N+1}) = \langle \Phi_0(1, \quad ,N) | V_{N+1} | \Phi_0(1, \quad ,N)\rangle \tag{64}$$

where Φ_0 is the ground state wave function of the target and V_{N+1} is the interaction potential defined by eq. (3). If we expand this potential in partial waves

$$V(\underline{r}) = \sum_{\lambda=0}^{\infty} V_\lambda(r)\, P_\lambda(\cos\vartheta) \tag{65}$$

then the direct potential

$$V_{\ell\ell'}^{\Lambda S}(r) = \sum_{\lambda=0}^{\infty}\left(\frac{2\ell'+1}{2\ell+1}\right)^{\frac{1}{2}}(\ell' 0 \lambda 0 | \ell 0)(\ell'\Lambda\lambda 0|\ell\Lambda)\,V_\lambda(r) \tag{66}$$

A method for calculating $V_\lambda(r)$ has been described by Faisal (1970) and programmed by Faisal and Tench (1971). More recent programs have been written by Morrison (1980) and Schmidt et al. (1980).

The exchange potential is very much more complex and it is not appropriate to write it down explicitly here, even for simple molecules. However its general form for closed-shell molecules is

$$W_{\ell\ell'}^{\Lambda S}\, u_{\ell'}^{\Lambda S}(r) = \sum_{\alpha\,\ell''\ell'''\lambda} C(\ell,\ell',\ell'',\ell''',\alpha,\lambda)\, y_\lambda(\phi_{\ell''}^{\alpha}\, u_{\ell'}^{\Lambda S}; r)\,\phi_{\ell'''}^{\alpha}(r) \tag{67}$$

where the $\phi_\ell^\alpha(r)$ are the radial functions arising from the single-centre expansion of the α th occupied target orbital

$$\phi^\alpha(\underline{r}) = \sum_\ell r^{-1}\,\phi_\ell^\alpha(r)\, Y_{\ell m_\alpha}(\hat{\underline{r}}) \tag{68}$$

the y_λ are the well-known exchange functions defined by

$$y_\lambda(u\,v\,;r) = r^{-\lambda-1}\int_0^r r'^{\lambda}u(r')v(r')\,dr' + r^{\lambda}\int_r^\infty r'^{-\lambda-1}u(r')v(r')\,dr' \qquad (69)$$

and the C's are numerical coefficients.

The main problems in the numerical solution of eqs. (62) are firstly to retain enough terms in the single-centre expansions in eqs. (8), (65) and (68) to give convergence, and secondly to accurately represent the integral terms occurring in the exchange operator eq. (67). A survey of the numerical methods adopted has recently been given by Buckley and Burke (1979). The main methods are:

(i) Non-iterative differential equations method used by Burke and Sinfailam (1971) and Raseev et al. (1978). In this approach the y_λ are represented by additional differential equations which are coupled to eqs. (63).

(ii) Integral equations method in which the coupled integro-differential equations are converted to coupled integral equations using the method introduced by Sams and Kouri (1969). The exchange potentials can be included non-iteratively in this approach using a method developed by Smith and Henry (1973). The application of this approach to electron molecule scattering has been summarized recently by Morrison (1979).

(iii) Iterative methods using either a differential equation or an integral equation algorithm. These approaches have been developed particularly by Collins et al. (1978, 1980) who have carried out the most detailed calculations to date.

Accurate static single centre exchange results have now been obtained for a number of simple diatomic molecules including H_2, N_2, CO and Li H. As an example of the convergence properties of the solutions we give in table 1 the parameters which were required by Collins et al. (1980) to obtain accurate phase shifts in their calculations on e - N_2. The potential singularities at the nuclei cause very slow convergence.

Table 1. Convergence Parameters in $e^- - N_2$ Scattering

	$^2\Sigma_g$	$^2\Sigma_u$	$^2\Pi_g$	$^2\Pi_u$	$^2\Delta_g$	$^2\Delta_u$
ℓ_m	14	15	14	15	14	13
λ_m	28	30	28	30	28	26
$\ell_m^{ex.}$	2	3	4	3	4	5
n_{1-6}^{ex}	2	2	2	2	2	2

ℓ_m is the maximum-order partial wave retained in eq. (8)

λ_m is the maximum-order partial wave retained in eq. (65)

ℓ_m^{ex} is the maximum-order partial wave retained in the expansion of the continuum orbital in the exchange potential eq. (67)

n_i^{ex} is the number of terms retained in eq. (68).

The converged phase shifts obtained by Collins et al. showed considerable dependence on the choice of target wave function as indicated for the $^2\Pi_g$ resonance state of $e^- - N_2$ in figure 4.

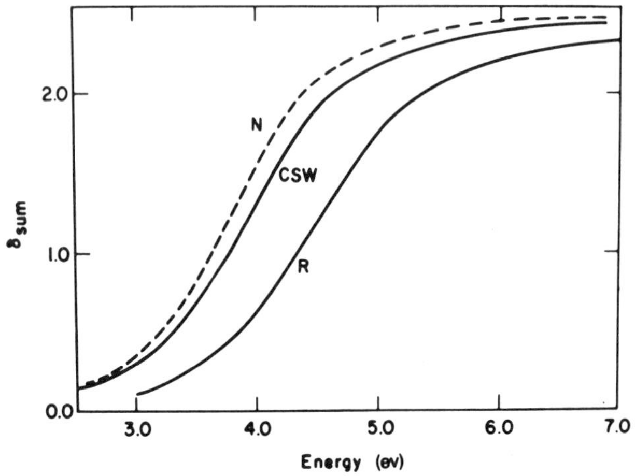

Figure 4. Comparison of eigenphase sums for $e^- - N_2$ scattering for different targets

The targets used in this work were the minimum basis set of Ransil (1960) (R), the extended-basis-set of Nesbet (1969) (N), and the near-Hartree-Fock-basis-set of Cade, Sales and Wahl (1966) (CSW).

Approximate Representation of Exchange and Polarization

The solution of eqs. (63) with the exact exchange potential becomes increasingly difficult for complex molecules. Consequently many calculations have been carried out where the exchange term is represented by an equivalent but approximate local potential. This approach is also justified since the polarization potential, which must be included to obtain agreement with experiment, is also represented by a parametrized local potential. In this way the coupled integro-differential equations are reduced to coupled differential equations which are much easier to solve.

One of the most successful and widely used methods of obtaining a local exchange potential has been the free-electron model introduced by Hara (1967). Basically the method is based on the picture introduced by Slater (1960) in which the total wave function is assumed to be composed of plane waves which are antisymmetrized in accordance with the Pauli exclusion principle and the exchange energy obtained by summing all states up to the Fermi level. The application of this approach has been discussed by Morrison et al. (1977) for $e - CO_2$ scattering and by Morrison and Collins (1978) for $e - H_2$ and $e - N_2$ scattering. The resultant form of the potential is

$$V_{ex}(\underline{r}) = -\frac{2}{\pi} k_F(\underline{r}) F(\eta) \qquad (70)$$

where $k_F(\underline{r})$ is a Fermi wave vector for the target defined as

$$k_F(\underline{r}) = \left[3\pi^2 \rho(\underline{r}) \right]^{1/3} \qquad (71)$$

$\rho(\underline{r})$ being the ground-state charge distribution. The function $F(\eta)$ is given by

$$F(\eta) = \frac{1}{2} + \frac{1-\eta^2}{4\eta} \ln\left|\frac{1+\eta}{1-\eta}\right| \qquad (72)$$

where

$$\eta = \frac{K(r)}{k_F(r)} \quad , \quad K^2 = k^2 + 2I + k_F^2(r) \tag{73}$$

I being the ionization potential of the molecule. It should be added at this point that extensive discussions of local exchange potentials for electron atom scattering have been given by Furness and McCarthy (1973) by Riley and Truhlar (1975) and by Bransden et al. (1976).

While these local exchange potentials have had considerable success they do not ensure that the continuum orbital representing the scattered electron is orthogonal to the bound orbitals which are fully occupied which is a requirement of the exact solution of the static exchange equations. The imposition of this constraint, which gives rise to an inhomogeneous term in the coupled differential equations was considered by Burke and Chandra (1972). It was shown in that paper and in subsequent papers (e.g. Chandra and Temkin (1976a, b)) that this constraint did indeed represent a substantial part of the exchange potential.

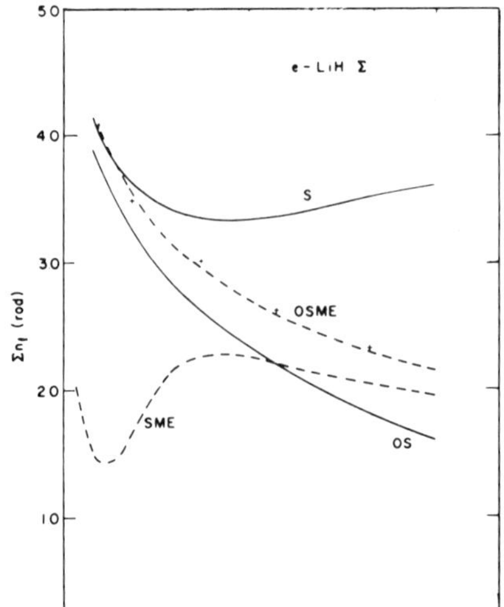

Figure 5. The $\sum$ Eigenphase Sum for e - LiH Scattering in Various Approximations

Recently Collins et al. (1979) have considered the possibility of including both a local exchange potential and imposing an orthogonality constraint. Their results for the eigenphase sum for e - LiH scattering in the Σ state are shown in figure 5. S is the static approximation neglecting exchange, OS is the static approximation with orthogonalization, SME is the static with local exchange, OSME is the static approximation with local exchange and orthogonalization and finally the crosses are the exact static exchange results. A low energy resonance which appears when only the local exchange potential is included is removed by orthogonalization and the resultant phase shifts including orthogonalization and local exchange are in excellent agreement with the exact static exchange results.

The main error in the static exchange approximation for low energy electron molecule scattering is the neglect of the long range polarization potential which behaves asymptotically as r^{-4}. This potential arises because the molecule is distorted by the field of the scattered electron. It would be represented if our original expansion eq. (8) included all excited states including continuum states coupled to the ground state by the dipole operator. However this is clearly impossible and an alternative way of including the potential must be sought. One way, suggested by Damburg and Karule (1967) for e - H scattering, is to include in expansion (8) suitably chosen pseudo-states which couple with the ground state so as to give the exact long range component of the polarizability. This approach has now been widely used for atoms, but the evaluation of the pseudo-states for molecules and the solution of the resultant scattering equations presents problems, and the only application has been to e - H_2 scattering by Schneider (1978) using the R-matrix approach discussed later in this section. The more usual approach is to represent the polarization potential by the following parametrized form

$$V_p(\underline{r}) = -\left(\frac{\alpha_0}{2r^4} + \frac{\alpha_2}{2r^4} P_2(\cos\theta)\right) C(r) \tag{74}$$

where $C(r)$ is a cut-off factor which has one of the forms

$$C_1(r) = 1 - \exp\left[-(r/r_c)^6\right] \tag{75}$$

or

$$C_2(r) = \left[1 - \exp(-r/r_c)\right]^6 \tag{76}$$

and where the parameter r_c is usually chosen to give the best agreement with experiment. This potential then augments the static potential defined by eq. (64).

The combination of a local exchange potential with a parametrized polarization potential has had wide application. For example two calculations for e - HCℓ vibrational elastic scattering are compared in Figure 6.

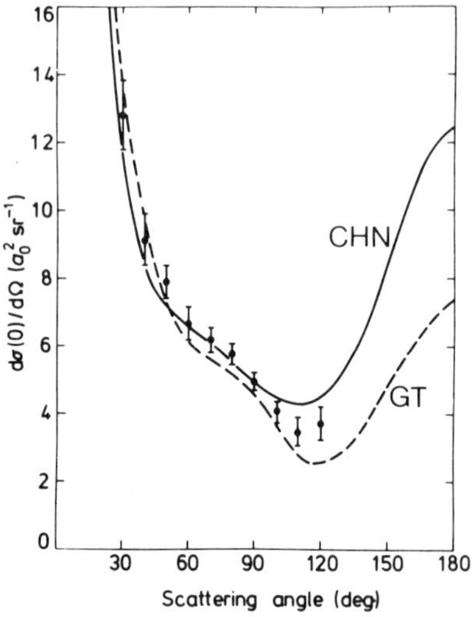

Figure 6. Elastic Differential Cross Section for e - HCℓ Scattering at 3.0 eV

The calculations of Collins, Henry and Norcross (1980) (CHN) were carried out using Hara local exchange potential and those of Gianturco and Thompson (1977) (GT) with exchange represented by the orthogonalization procedure of Burke and Chandra (1972). In both cases the cut-off in the polarization potential was used to fit a $^2\Sigma$ resonance in the 2-4 eV region. The experiments were carried out by Rohr and Linder (1975). Although the Hara local exchange model is more realistic, the flexibility allowed by the form of the polarization potential largely compensates for any difference in the two models and the two theoretical results are seen to be in equally good agreement with experiment.

The above results highlights the need to develop a more realistic treatment of polarization. At this time it seems that inclusion of polarized pseudo-states in the original expansion of eqs. (4) and (8) would probably be the best, however it involves the solution of coupled integro-differential equations representing more than an electronic state which is clearly formidable within the single-centre expansion framework.

Photoionization

The scattering wave functions obtained by one of the methods described above can be used to calculate photoionization cross sections using eqs. (60) and (61). However because of the numerical complexity, results are so far very limited, perhaps the most detailed being calculations for CO by Ritchie and Tambe (1980) and calculations for N_2 by Raseev et al. (1980).

Raseev et al. (1980) represented the final state wave function by solving the static exchange equations for an electron moving in the $X\,^2\Sigma_g^+$ state of N_2^+. They then considered three combinations of molecular orbitals for the initial $X\,^1\Sigma_g^+$ state of N_2 and the final $X\,^2\Sigma_g^+$ state of N_2^+ as indicated in table 2.

Table 2. Orbital Basis Used in Photoionization of N_2 by Raseev et al. (1980)

Calculation Number	SCF MO's used in calculation: Initial State	SCF MO's used in calculation: Final State	Quadrupole Moment of Final State of N_2^+ in a.u. at R=2.068 a_o
1	$N_2(^1\Sigma_g^+)$	$N_2(^1\Sigma_g^+)$	1.966
2	$N_2^+(^2\Sigma_g^+)$	$N_2^+(^2\Sigma_g^+)$	2.192
3	$N_2(^1\Sigma_g^+)$	$N_2^+(^2\Sigma_g^+)$	2.192

Their result for the transition

$$h\nu + N_2(X\,^1\Sigma_g^+) \rightarrow N_2^+(3\sigma_g\,\varepsilon\sigma_u\,^1\Sigma_u^+) + e^- \qquad (77)$$

is shown for each of these calculations in Figure 7.

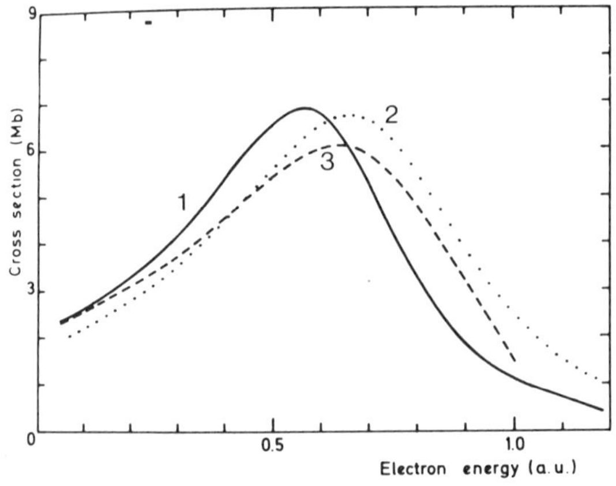

Figure 7. Photoionization Cross Section for $3\sigma_g \rightarrow k\sigma_u$ Transition in N_2

It is seen that the resonance in the final state is increased to higher energies if the relaxed orbitals of N_2^+ are used. However since the use of N_2 orbitals to describe the molecular ion, allows to some extent for polarization of the ion during the collision, it

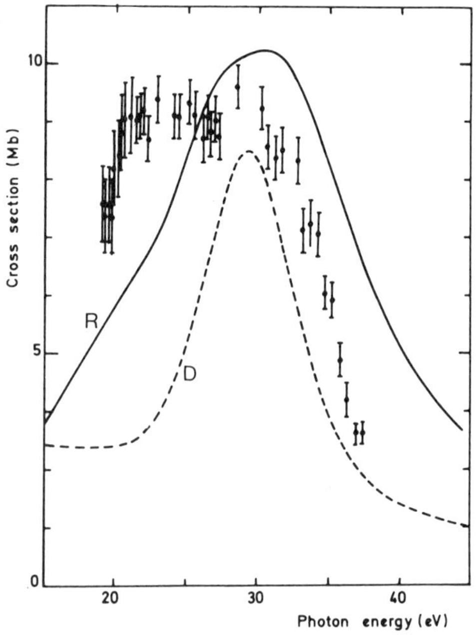

Figure 8. Vibrationally Unresolved Photoionization Cross Sections for N_2

is believed that calculation 1 is more realistic. The resultant R averaged cross section using this approximation (R), the experiments of Woodruff and Marr (1977) and the multiple scattering calculations of Dehmer et al. (1979) (D) (to be discussed in detail later) are compared in Figure 8.

The agreement between theory and experiment is good, bearing in mind that correlation effects in both the initial and final states are neglected. The inclusion of such effects would be extremely difficult in both theories, and have not been considered.

4.2 Continuum Multiple Scattering Method

This method was developed to describe electron molecule collisions by Dill and Dehmer (1974) and has been comprehensively reviewed by Dehmer and Dill (1979). This approach has also been independently considered by Ziesche and John (1976). This approach first used in bound state studies (e.g. Johnson, 1973) involves for a diatomic molecule the partition of the molecular field into three regions by spheres as illustrated in Figure 9. In region I the electron molecule potential is represented by direct and local exchange terms, in region II the potential is taken to be constant; finally in region III a polarization potential is added to the direct and local exchange potential. Several forms have been explored for the local exchange potential but that due to Hara (1967) seems to give

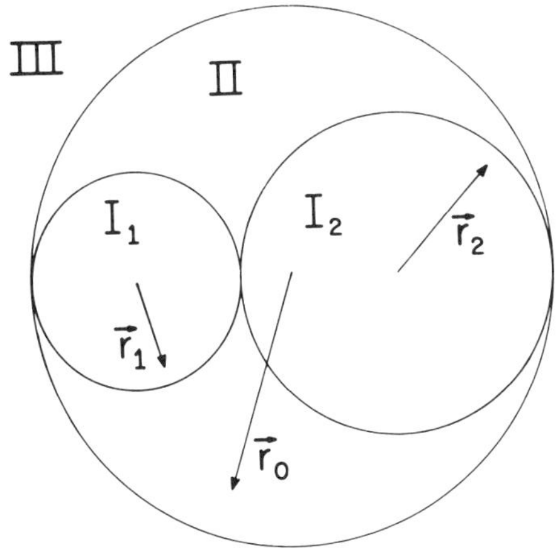

Figure 9. Partitioning of the Molecular Field in the Continuum Multiple Scattering Method

best agreement with experiment. Different expansions for the wave function for the scattered electron are now used in each region and the expansions matched the boundary. In particular, in region I, expansions centered on the respective nuclei are used, which means that problems of slow convergence, which the single-center expansion methods experience, are avoided.

The form assumed for the wave function representing the scattered electron in the three regions is as follows. In region I surrounding each nucleus

$$\psi_{I_i} = \sum_{\ell m} a_{\ell m}^{I_i} f_\ell^{I_i}(kr_i) Y_{\ell m}(\hat{r}_i) \tag{78}$$

In region II

$$\psi_{II} = \sum_{\ell m} a_{\ell m}^{II} j_\ell(\kappa r_o) Y_{\ell m}(r_o) + \sum_i \sum_{\ell m} b_{\ell m}^{II_i} n_\ell(\kappa r_i) Y_{\ell m}(\hat{r}_i) \tag{79}$$

Finally in region III

$$\psi_{III} = \sum_{\ell m} \left[a_{\ell m}^{III} f_{\ell m}^{III}(kr_o) + b_{\ell m}^{III} g_\ell^{III}(kr_o) \right] Y_{\ell m}(\hat{r}_o) \tag{80}$$

In these equations f_ℓ and g_ℓ are the regular and irregular solutions in the appropriate regions, j_ℓ and n_ℓ are the regular and irregular spherical Bessel functions κ is the local wave number in region II, and k is the continuum electron wave number $k = \sqrt{2E}$. The coefficients $a_{\ell m}$ and $b_{\ell m}$ are then determined by matching on the boundaries, and the K-matrix determined by comparing eqs. (12) and (80). Vibrational excitation cross sections are obtained by repeating the calculation for a number of internuclear separations and using the adiabatic approximation eq. (37). Photoionization cross sections are obtained by substituting the above continuum wave function into eqs. (60) and (61).

This approach has now been widely used for polyatomic as well as diatomic molecules both for electron scattering and for photoionization by Dehmer and Dill. A most interesting result which came out of these calculations was the important role which shape resonances play at intermediate energies as illustrated in the e - N_2 results shown in Figure 10.

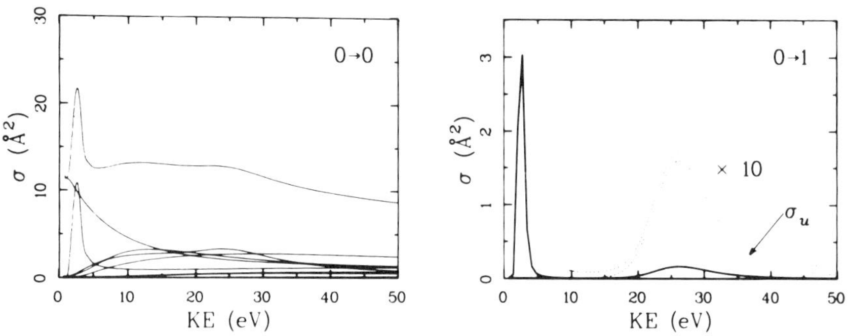

Figure 10. Vibrational Excitation for e - N_2 Scattering

The well-known low energy $^2\Pi_g$ resonance is clearly seen in both cross sections. However at higher energies a broad $^2\Sigma_u$ resonance centred at about 26 eV appears particularly in the vibrational inelastic cross section. The calculations show that it is a shape resonance with mainly f-wave symmetry. These features are also seen in the K-shell photoionization spectrum in N_2, although in this case the π_g resonance has moved into the discrete spectrum and the σ_u resonance occurs at about 10 eV above threshold due to the additional Coulomb attraction in this case.

Davenport (1976) and Dehmer and Dill have also reported calculations for photoelectron angular distributions for molecules whose orientations are fixed in space. This pertains in particular to molecules on surfaces, and the calculations enable the orientation of these molecules on the surface to be determined.

In conclusion the continuum multiple scattering method enables a rapid survey of electron scattering and photoionization of molecules to be carried out. However it is a one-electron model and so in its simple form outlined above will only treat problems in which the molecule is confined to one electronic state, and where electron correlation effects are unimportant. At present attempts are underway to remove these restrictions but no results have been reported.

4.3 L^2-Methods

Introduction

We have seen that a basic difficulty with the single-centre expansion method described in section 4.1 is its slow convergence. Although the continuum multiple scattering method overcomes this, we have seen that it has other limitations. In molecular structure calculations this problem is overcome by using multicentre configuration interaction wave functions. It is thus of interest to see if bound state techniques and computer codes can be used in collision calculations. This is the motivation behind the development of the L^2 methods, or methods based on the use of square integrable functions, which are described in this section.

The first discussions of the use of L^2 methods in electron scattering calculations were made by Temkin (1966), Hazi and Taylor (1970) and Krauss and Mies (1970). They noted that if the collision wave function was expanded as follows

$$\Psi_k = \sum_j \chi_j(1,\ldots,N+1)\, a_{kj} \tag{81}$$

where the χ_j are L^2 discrete basis functions (i.e. they are square integrable) and where the a_{kj} are determined by diagonalizing the electronic part of the Hamiltonian (using a standard molecular structure code)

$$\langle \Psi_k | H_{el} | \Psi_j \rangle = E_k \delta_{kj} \tag{82}$$

then the Ψ_k provided an accurate representation of the wave function at the energy E_k in the region where the χ_j are non-zero. Since in a typical case most of the E_k lie in the continuous spectrum of H_{el} this enables the collision wave function to be determined at these energies in an "internal region". An approximate phase shift can then be determined by projecting Ψ_k onto the appropriate channel function $\Phi_i^{SM_S} Y_{\ell_i m_{\ell_i}}$ defined by eq. (8) yielding a radial function

$$r_{N+1}^{-1} R^S_{i\ell_i m_{\ell_i} k}(r_{N+1}) = \langle \Phi_i^{SM_S} Y_{\ell_i m_{\ell_i}} | \Psi_k \rangle \tag{83}$$

where the integral is taken over all electronic coordinates except the radial coordinate of the scattered electron. The phase shift in this channel is then determined by fitting $R^{S}_{i\ell_i m_{\ell_i} k}$ to the appropriate combination of spherical Bessel functions near the boundary of the "internal region".

Stieltjes-Tchebycheff Moment Method

This method for calculating atomic and molecular photoionization cross sections was introduced by Langhoff (1973), and has recently been reviewed by Langhoff (1979). The basic idea is to replace the continuum states in the expansion for the photoionization cross section defined by eqs. (40) or (41) by an L^2 integrable basis defined as in eq. (82). Then to extract the smooth photoionization cross section from the resultant discrete spectrum.

We define the oscillator strengths

$$\tilde{f}_i = \frac{2}{3}\tilde{\varepsilon}_i \left| \langle \Psi_i | D^{L}_{\mu=0} | \Psi_0 \rangle \right|^2 \tag{84}$$

where $\tilde{\varepsilon}_i = \tilde{E}_i - E_0$ and $\tilde{E}_i$ are the eigenvalues and Ψ_i the eigenfunctions defined by eq. (82). This equation provides a discrete representation of the continuum oscillator strength, or photoionization cross section, at the energy $\tilde{E}_i$. The problem is to extract the continuum oscillator strength from this representation. Langhoff achieves this by first defining the spectral moments

$$S(-k) = \sum_{i=1}^{m} \tilde{\varepsilon}_i^{-k} \tilde{f}_i \qquad 0 \leq k \leq 2n-1 \tag{85}$$

where m is the number of basis terms retained in the original diagonalization in eq. (82). If $m \gg n$ then eq. (85) provides an accurate approximation to the correct moments. The principal representation of these moments is then constructed so as to satisfy

$$S(-k) = \sum_{i=1}^{n} \varepsilon_i^{-k} f_i \qquad 0 \leq k \leq 2n-1 \tag{86}$$

where there are 2n equations in the 2n unknowns ε_i, f_i, $i = 1, \ldots, n$. In a sense the ε_i and f_i so evaluated provide a "smoothing" of the original $\tilde{\varepsilon}_i$ and $\tilde{f}_i$ defined by equation (84). The final step is to construct the so-called Stieltjes distribution in the form

$$
\begin{aligned}
f_s^{(n)}(\varepsilon) &= 0 \,, \quad 0 < \varepsilon < \varepsilon_1 \\
f_s^{(n)}(\varepsilon) &= \sum_{i=1}^{j} f_i \,, \quad \varepsilon_j < \varepsilon < \varepsilon_{j+1} \\
f_s^{(n)}(\varepsilon) &= \sum_{i=1}^{n} f_i = S(0) \,, \quad \varepsilon_n < \varepsilon
\end{aligned}
\tag{87}
$$

where the superscript n indicates that an n-term principal pseudo-spectrum is employed in the development. The corresponding Stieltjes densities obtained from the slopes of line segments connecting succesive midpoints of the vertical portions of the Stieltjes distribution histogram defined by eq. (87)

$$
\begin{aligned}
\frac{df_s^{(n)}}{d\varepsilon} &= 0 \,, \quad 0 < \varepsilon < \varepsilon_1 \\
\frac{df_s^{(n)}}{d\varepsilon} &= \tfrac{1}{2}\left(f_{i+1} + f_i \right)/\left(\varepsilon_{i+1} - \varepsilon_i \right) \,, \quad \varepsilon_i < \varepsilon < \varepsilon_{i+1} \\
\frac{df_s^{(n)}}{d\varepsilon} &= 0 \,, \quad \varepsilon_n < \varepsilon
\end{aligned}
\tag{88}
$$

converges to the correct oscillatorstrength density in the limit of large n. Consequently a convergent histogram appropriate to the photoionization oscillator strength density is obtained in the spectral interval $(\varepsilon_t \,, \infty)$.

A useful continuum approximation to the continuum oscillator strength can be obtained by replacing the principal representation defined by eq. (86) by the representation defined by

$$
S(-k) = \sum_{i=0}^{n} \left[\varepsilon_i(\varepsilon) \right]^{k} f_i(\varepsilon) \,, \quad 0 \lesssim k \lesssim 2n \tag{89}
$$

which has one frequency fixed at a preassigned point ε in the spectrum satisfying $\varepsilon_0(\varepsilon) = \varepsilon$. The Tchebysheff distribution is then obtained from the customary derivative in the form

$$
\frac{df_t^{(n)}}{d\varepsilon} = \sum_{\varepsilon_i(\varepsilon) < \varepsilon} \frac{df_i(\varepsilon)}{d\varepsilon} + \frac{1}{2} \frac{df_0(\varepsilon)}{d\varepsilon} \tag{90}
$$

which so defined is real, non-negative and continuous on the real axis, and is convergent to the correct density in the limit of large n.

Photoionization cross sections for a large number of molecules have been determined by this method. Results obtained by Rescigno et al. (1978) for the removal of the $3\sigma_g$, $1\pi_u$ and $2\sigma_u$ valence orbitals in N_2 are shown in figure 11.

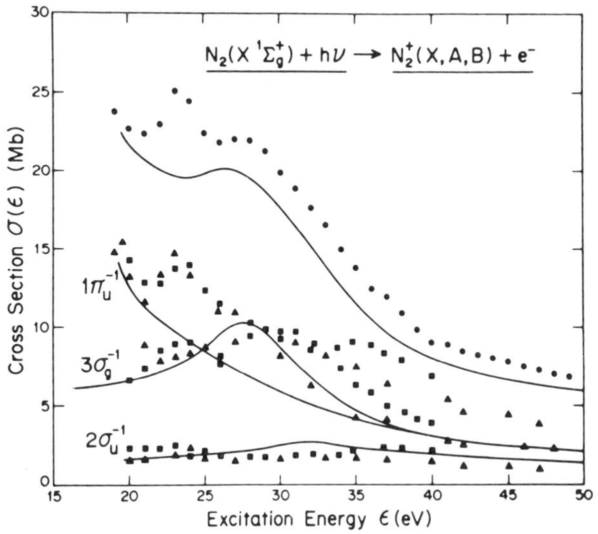

Figure 11. Partial and Total Photoionization Cross Sections in N_2

The results are in good agreement with recent experiments using line sources (Samson et al., 1979) Synchrotron radiation (Plummer et al., 1977; and Woodruff and Marr, 1977) and (e, 2e) measurements (Hamnett et al., 1976; and Wigth et al., 1976), except for the structure at about 23 eV. (A Rydberg series convergent to the $C\,^2\Sigma_u^+$ state in N_2^+ is a possible candidate for this structure). In particular the position and width of the shape resonance at 30 eV is in excellent agreement with experiment.

In conclusion, while estimates of angular distribution and autoionizing states close to threshold are difficult to obtain, the Stieltjes-Tchebycheff moment method has enabled standard bound state codes to be rapidly modified to enable a vast amount of information to be obtained on integral photoionization cross sections for molecules. We also note that recent developments of the method for calculating resonance widths have been proposed by Hazi (1979).

T-Matrix Method

Another L^2 method, this time in which the T-matrix is expanded in a discrete basis has been proposed by Rescigno et al. (1974a, b, 1975a).

The original method starts from the Lippmann-Schwinger integral equation for the transition matrix

$$T = U + U G_0^+ T \tag{91}$$

where U is twice the non-local potential interaction between the electron and the target molecule and G_0^+ is the free-particle Green's function with outgoing wave boundary conditions. We now expand U in a finite basis set of square integrable functions ϕ_α, taken in the applications to be Gaussians, giving

$$U^t(\underline{r}, \underline{r}') = \sum_{\alpha\beta} |\phi_\alpha(\underline{r})\rangle\langle\phi_\alpha(\underline{r})| \, U(\underline{r}, \underline{r}') \, |\phi_\beta(\underline{r}')\rangle\langle\phi_\beta(\underline{r}')| \tag{92}$$

Inserting this truncated potential U^t into eq. (91) gives a matrix equation for the T-matrix which can be solved yielding

$$\underline{T}^t = (\underline{1} - \underline{U}^t \underline{G}_0^+)^{-1} \underline{U}^t \tag{93}$$

where the matrix $\underline{G}_0^+$ is defined by

$$\langle\phi_\alpha | G_0^+(E) | \phi_\beta\rangle = \lim_{\varepsilon \to 0+} \int d^3k \, \frac{\langle\phi_\alpha | \underline{k}\rangle\langle \underline{k} | \phi_\beta\rangle}{q^2 - k^2 + i\varepsilon} \tag{94}$$

and where $E = \frac{1}{2}q^2$ and $|\underline{k}\rangle$ and $|\underline{k}'\rangle$ are the plane wave states

$$|\underline{k}\rangle = \frac{1}{(2\pi)^{3/2}} e^{i\underline{k}\cdot\underline{r}} \tag{95}$$

The scattering amplitude for a transition from state $\underline{k}_i$ to $\underline{k}_f$ is then given by

$$f(\underline{k}_i \to \underline{k}_f) = -2\pi^2 \sum_{\alpha\beta} \langle \underline{k}_f|\alpha\rangle\langle\alpha|T^t(E)|\beta\rangle\langle\beta|\underline{k}_i\rangle \tag{96}$$

where the matrix elements involving the plane wave states can be evaluated analytically for Gaussian basis functions (Ostlund, 1975 and Levin et al., 1980).

The theory has been used to obtain static exchange results for H_2, N_2, CO and F_2 by Rescigno et al. (1974a,b, 1975a); Fliflet et al. (1978) and Rescigno (1979). In addition the polarizability was included for H_2 by Klonover and Kaldor (1977, 1978, 1979a,b) using second-order perturbation theory. However, it has been found that errors due to truncation of the potential can only be made small, particularly at energies below 5 eV where the long range potentials are important, by the use of large basis sets and this can lead to problems of linear dependence. Although the inclusion of a variational correction using the Kohn method (Fliflet and McKoy, 1978a,b) helps to overcome this difficulty, it has proved better to use a modification of this method based on a variational principle from the outset. The method adopted has been widely used in nuclear physics (Lovelace, 1964) and involves writing the potential in the separable form

$$U^S = \sum_{\alpha\beta} U|\alpha\rangle \, (U^{-1})_{\alpha\beta} \langle\beta| U \tag{97}$$

Substituting into the Lippmann-Schwinger equation then gives

$$\langle \underline{k}_f|T^S|\underline{k}_i\rangle = \sum_{\alpha\beta} \langle \underline{k}_f|U|\alpha\rangle \left[(U - UG_0U)^{-1}\right]_{\alpha\beta} \langle\beta|U|\underline{k}_i\rangle \tag{98}$$

which can be shown to satisfy the Schwinger variational principle (Adkhikari and Sloan, 1975). An important feature of this new approach is that the expansion functions ϕ_α only appear in conjunction with the potential, which allows the use of trial functions which do not have the correct asymptotic form. This is of course in contrast to the Kohn variational method. The method involves two new types of matrix element $\langle \underline{k}|U|\alpha\rangle$ and $\langle\alpha|U G_0 U|\beta\rangle$ The first can be evaluated in closed form when Gaussians are used and the second can be approximated by closure. In practice, single centre expansions are used for evaluating the matrix elements. Results involving only 3 s - type basis functions, 1 p - type basis functions and 1d type basis functions have recently been obtained by Watson et al. (1980) for H_2 which are in good agreement with other methods and work is now underway on Li H.

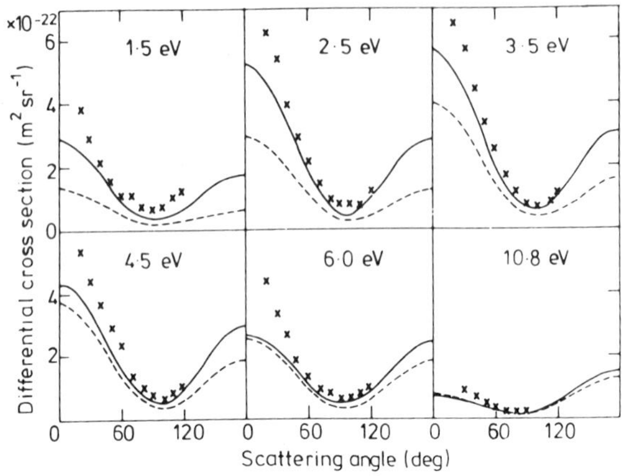

Figure 12. Differential $v = 0 \rightarrow 1$ Vibrational Excitation Cross Section for $e - H_2$ Scattering.

As an example of the application of this method we show in figure 12 the angular distribution of $v = 0 \rightarrow 1$ vibrational excitation cross section for $e - H_2$ scattering calculated by Klanover and Kaldor (1979a) compared with the measurements of Linder and Schmidt (1971). The calculations were carried out with-and without-the inclusion of the second-order polarization and the vibrational excitation cross sections were obtained using the adiabatic theory of Chase (1956). The calculation used 30 Cartesian Gaussian basis orbitals to describe the occupied H_2 orbitals and the excited orbitals which account for the polarization effects, and 61 basis orbitals to span the scattered electron orbital, some of the latter orbitals being centred on the centre of gravity. The calculations were based on eq. (93) for the T-matrix. The results show clearly the importance of the polarization effects at low energies in both the forward and backward directions.

In conclusion it is clear that the T-matrix method promises to be a powerful way of calculating electron molecule collision cross sections.

R-Matrix Method

This method was first introduced by Wigner (1946a,b) and Wigner and Eisenbud (1947) in a study of nuclear reactions, and extended to electron atom scattering by Burke and Hibbert (1969), Burke et al. (1971) and Burke and Robb (1975). Recently it has been developed further and applied extensively to electron molecule scattering by Schneider (1975a,b), Schneider and Hay (1976a,b), Morrison and Schneider (1977), Schneider (1978), Burke et al. (1977) and Buckley et al. (1979).

The method automatically takes advantage of the division of configuration space into an internal and an external region suggested by frame transformation theory. This division of space for a diatomic molecule is illustrated in Figure 13.

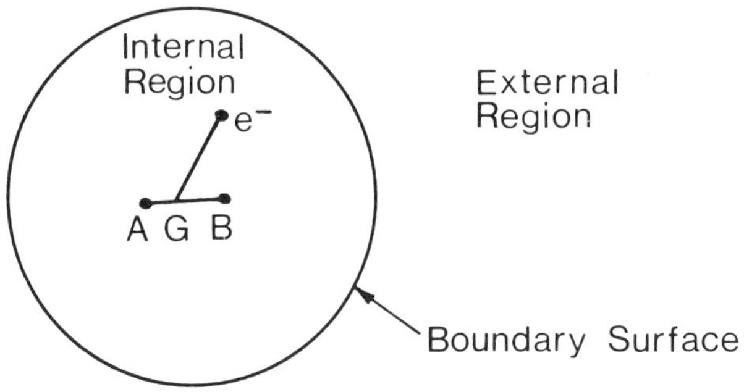

Figure 13. Division of Configuration Space in the R-Matrix Method.

In the internal region, where exchange and correlation effects are important, the wave function is expanded using the fixed-nuclei approximation in terms of a multicentre discrete basis in the molecular frame of reference. In the external region, where only the long-range components of the direct potential remain, a single-centre expansion using only a few partial waves is appropriate and either the molecular frame or the laboratory frame is used depending on the relative importance of H_{rot}. The link between these two regions is provided by the R-matrix on the surface. The basic problem of the R-matrix method is to calculate this matrix from the expansion of the wave function in the internal region.

In the following discussion we describe the R-matrix approach of Burke et al. (1977) which uses a multicentre STO basis. The work of Schneider and collaborators uses an approach based on prolate spheroidal co-ordinates which has some advantages for diatomic molecules, but which cannot be so easily extended to polyatomic molecules. We expand the wave function describing the electron molecule system in an internal region bounded by a sphere of radius $r = a$ in the form

$$\Psi_k^{SM_S} = \mathcal{A}\sum_{ij} \Phi_i^{SM_S}(1,\ldots,N;\sigma_{N+1})\,\psi_{ij}(\underline{r}_{N+1})\,c_{ijk}^{S} + \sum_i \xi_i(1,\ldots,N+1)\,c_{ik}^{S} \tag{99}$$

which replaces eqs. (4) and (8). The molecular orbitals used in the definition of the target eigenstates and pseudostates and in the correlation functions are expanded in terms of STO's centred on the nuclei which have effectively vanished by the boundary of the internal region. On the other hand the continuum molecular orbitals ψ_{ij} are expanded in terms of the same set of STO's on the nuclei but also include STO's centred on the centre of gravity G of the molecule which are of longer range and may be non-zero on the boundary. We will see that it is these STO's on G which provide the connection with the single-centre expansion used in the external region.

The coefficients c_{ijk}^{S} and c_{ik}^{S} are determined by diagonalizing the operator

$$\langle \Psi_k^{SM_S} | (H_{el} + L_b) | \Psi_{k'}^{SM_S} \rangle = E_k\,\delta_{kk'} \tag{100}$$

where the integrals in this equation are carried out over the internal region and L_b is a surface projection operator introduced by Bloch (1956) which ensures that $H_{el} + L_b$ is Hermitian. It is defined by

$$L_b = \frac{1}{2}\sum_{i\ell_i m_{\ell_i}} |\Phi_i^{SM_S} Y_{\ell_i m_{\ell_i}}\rangle\,\delta(r-a)\left(\frac{d}{dr} - \frac{b_{i\ell m}-1}{r}\right)\langle \Phi_i^{SM_S} Y_{\ell_i m_{\ell_i}}| \tag{101}$$

where the channel functions used here are those defined in eq. (8)

which involves a projection onto the angular co-ordinates of the scattered electron. We can now write the Schrödinger equation (1) as

$$(H_{el} + L_b - E)\,\Psi^{SM_S} = L_b\,\Psi^{SM_S} \tag{102}$$

which has the formal solution

$$\Psi^{SM_S} = (H_{el} + L_b - E)^{-1} L_b\,\Psi^{SM_S} \tag{103}$$

The inverse operator in this equation can be expanded in terms of the eigenfunctions defined by eq. (100) giving

$$|\Psi^{SM_S}\rangle = \sum_k \frac{|\Psi_k^{SM_S}\rangle\langle\Psi_k^{SM_S}|L_b|\Psi^{SM_S}\rangle}{E_k - E} \tag{104}$$

We now project this equation onto the channel functions $\Phi_i^{SM_S} Y_{\ell_i m_{\ell_i}}$ and evaluate it on the boundary $r = a$. If we define the radial functions

$$r_{N+1}^{-1}\, f^S_{i\ell_i m_{\ell_i}}(r_{N+1}) = \langle \Phi_i^{SM_S} Y_{\ell_i m_{\ell_i}} | \Psi^{SM_S}\rangle \tag{105}$$

where the integration in this equation is carried out over the coordinates of all $N + 1$ electrons except the radial co-ordinate of the scattered electron, then we obtain immediately

$$\underline{f}^S(a) = \underline{R}^S \left(r\frac{d\underline{f}^S}{dr} - \underline{b}\,\underline{f}^S \right)_{r=a} \tag{106}$$

where the R-matrix is given by

$$\mathcal{R}^{S}_{i\ell_i m_{\ell_i}, i'\ell'_i m_{\ell'_i}} = \sum_k \frac{\gamma^{S}_{i\ell_i m_{\ell_i} k}\, \gamma^{S}_{i'\ell'_i m_{\ell'_i} k}}{E_k - E} \tag{107}$$

The reduced width amplitudes $\gamma_{i\ell_i m_{\ell_i} k}$ are defined in terms of the value of the continuum orbitals on the boundary. Rembering that we have used a 3-centre expansion for these orbitals which can be written as

$$\psi_{ij} = \psi_{ij}(\underline{r}_A) + \psi_{ij}(\underline{r}_B) + \sum_{\ell_i m_{\ell_i}} r^{-1} v_{ij\ell_i m_{\ell_i}}(r)\, Y_{\ell_i m_{\ell_i}}(\hat{\underline{r}}) \tag{108}$$

where only the last term centred on G contributes on the boundary, then

$$\gamma^{S}_{i\ell_i m_{\ell_i} k} = \left(\frac{1}{2a}\right)^{\frac{1}{2}} \sum_j v_{ij\ell_i m_{\ell_i}}(a)\, c^{S}_{ijk} \tag{109}$$

An important point to remember is that the number of ℓ values that need to be retained on G is appropriate to the external region, and not to represent the nuclear singularities which are now represented by the orbitals centred on the nuclei. It has been found in practice that for homonuclear diatomic molecules such as N_2 at most two ℓ values and perhaps 6-8 basis orbitals on G for each value of ℓ are required to give accurate results for energies up to 10-15 eV.

The most time consuming part of this calculation is setting up and diagonalizing $H_{\mathcal{N}+1} + L_b$ which must be carried out once for each set of symmetries $S\,M_S$. This can be done by relatively small modifications of standard multicentre integral codes by subtracting the tail from these integrals involving STO's on G which extend beyond the boundary and calculating and adding on the Bloch surface terms. These modifications have recently been made to the ALCHEMY code by Kendrick and Buckley (1980). Once the R-matrix is determined then the coupled differential equations need to be solved in the external region for each energy where the cross section is required.

Results obtained for $e - H_2$ and $e - N_2$ scattering in the static exchange approximation (where just the ground electronic state is retained in expansion (99)) by Morrison and Schneider (1977) and by Buckley et al. (1979) are in good agreement. Schneider (1978) has also reported calculations on H_2 in which the polarizability was represented by pseudostates. Finally an alternative way of including the effect of polarizability has been explored by Schneider and Hay (1976a,b) for F_2. They used the F_2^- SCF orbitals rather than F_2 orbitals in representing the target, so allowing for part of the relaxation occurring during the collision.

We conclude with a discussion of the R-matrix theory of vibrational excitation. The basic approach is similar to the resonance theory of vibrational excitation and dissociative attachment introduced by Herzenberg and collaborators over the last eighteen years (Herzenberg and Mandl, 1962; Bardsley and Mandl, 1968; Herzenberg, 1978) and discussed by Bardsley elsewhere in this series of lectures. That is the total wave function rather than being expanded in terms of eigenstates of the target, which are slowly convergent, is expanded in terms of compound states of the electron molecule system. In the case of the theory described here these compound states are the R-matrix basis states defined by eq. (100).

The approach of Schneider et al. (1979a) is to adopt the following expansion for the electron diatomic molecule wave function

$$\Psi = \sum_k \Psi_k(1, \cdots, N+1)\, \xi_k(R) \tag{110}$$

where the R-matrix states Ψ_k are defined by diagonalizing $H_{el} + L_b$ for each fixed value of internuclear distance R as in eq. (100). In order to obtain an equation for the wave functions $\xi_k(R)$ describing the nuclear motion we project the Schrödinger equation

$$(T_R + H_{el} + L_b - E)\Psi = L_b \Psi \tag{111}$$

onto the R-matrix states. If we assume in accordance with the Born-Oppenheimer approximation that the nuclear kinetic energy operator acts only on the $\xi_k(R)$ we obtain the equation

$$\left(T_R + E_k(R) - E\right)\xi_k(R) = \langle \Psi_k | L_b | \Psi \rangle \tag{112}$$

In comparing this equation with that arising from the resonance theory of Herzenberg et al. we note that the potential on the left hand side is real since the R-matrix eigenvalues defined by eq. (100) are real and the surface term on the right hand side replaces the "entry amplitude".

In order to obtain the R-matrix coupling the electronic and vibrational channels we now project the total wave function defined by eq. (110) onto the appropriate channel functions and evaluate this quantity on the boundary. We obtain

$$\langle \chi_{i'v'} \Phi_{i'}^{SM_S} Y_{\ell_i' m_{\ell_i}'} | \Psi \rangle_{r=a} = \sum_k \langle \chi_{i'v'} \Phi_{i'}^{SM_S} Y_{\ell_i' m_{\ell_i}'} | \Psi_k \ \xi_k \rangle \tag{113}$$

If we now substitute for ξ_k from eq. (112) by introducing the Green's function $G_k(R, R')$ of the operator $T_R + E_k(R) - E$ we obtain

$$\langle \chi_{i'v'} \Phi_{i'}^{SM_S} Y_{\ell_i' m_{\ell_i}'} | \Psi \rangle_{r=a} = \sum_{iv\ell_i m_{\ell_i}} \mathcal{R}^S_{i'v'\ell_i' m_{\ell_i}', iv\ell_i m_{\ell_i}} \left[\left(\frac{d}{dr} - \frac{b-1}{a} \right) \langle \chi_{iv} \Phi_i^{SM_S} Y_{\ell_i m_{\ell_i}} | \Psi \rangle \right]_{r=a} \tag{114}$$

where the generalized R-matrix

$$\mathcal{R}^S_{i'v'\ell_i' m_{\ell_i}', iv\ell_i m_{\ell_i}} = \sum_k \langle \chi_{i'v'}(R') \gamma^S_{i'\ell_i' m_{\ell_i}' k}(R') G_k(R', R) \gamma^S_{i\ell_i m_{\ell_i} k}(R) \chi_{iv}(R) \rangle \tag{115}$$

and where the reduced width amplitudes are defined by eq. (109).

If dissociative attachment is possible then the Bloch operator L_b in eqs. (111) and (112) must be augmented by an additional term allowing for dissociation into $A + B^-$ type channels. Schneider et al. (1979a) show that this leads to additional terms in the generalized R-matrix coupling the electronic and vibrational channels with the dissociating channels.

We can relate eq. (115) with the result obtained using the frame transformation theory of Chang and Fano (1972). If we neglect the kinetic energy operation T_R in the internal region then the Green's function $G_k(R,R')$ can be written

$$G_k(R,R') = \frac{\delta(R-R')}{E_k(R)-E} \tag{116}$$

and then eq. (115) becomes

$$\mathcal{R}^S_{i'v'\ell_i'm_{\ell_i}',iv\ell_im_{\ell_i}} = \int_0^\infty \chi^*_{i'v'}(R)\sum_k \frac{\gamma^S_{i'\ell_i'm_{\ell_i}'k}(R)\gamma^S_{i\ell_im_{\ell_i}k}(R)}{E_k-E}\chi_{iv}(R)dR \tag{117}$$

$$= \int_0^\infty \chi^*_{i'v'}(R)\,\mathcal{R}^S_{i'\ell_i'm_{\ell_i},i\ell_im_{\ell_i}}(R)\,\chi_{iv}(R)\,dR$$

It is easy to see that while eq. (116) is a good approximation to $G_k(R,R')$ for the higher R-matrix levels where $E_k(R)-E$ is large and positive it cannot be a good approximation when $E_k(R)\approx E$ which occurs when incident energy is close to the resonance R-matrix eigenenergy. However since almost all the computational effort is expended in setting up and diagonalizing $H_{el}+L_b$ for the required set of R values and very little in determining the Green's function $G_k(R,R')$ it is just as easy and very much more accurate to use eq. (115) rather than eq. (117) to determine the R-matrix.

Once the generalized R-matrix has been determined, the final step in the calculation is to solve the collision problem in the external region. The appropriate equations in this region are the hybrid equations of Chandra and Temkin (1976a,b) without of course the exchange potential which is zero in this region. These equations are solved subject to the usual asymptotic boundary conditions where the R-matrix provides the starting condition ar r = a. In this way the K-matrix and consequently the cross section are determined.

This theory has recently been used to calculate the vibrational excitation cross section in e - N_2 scattering by Schneider et al. (1979b). They represented the target by orbitals obtained by an SCF calculation for N_2^- in its $^2\Pi_g$ state. In this way they included part of the target polarizability. They then set up the Hamiltonian matrix for the scattered electron interacting with this distorted target in the static exchange approximation. The Green's functions $G_k(R,R')$ were determined by diagonalizing $T_R+E_k(R)$

in a basis set of functions $\theta_{k\nu}(R)$ for each k. Finally the solution of the equations in the external region were obtained assuming that the electron molecule interaction was zero in this region.

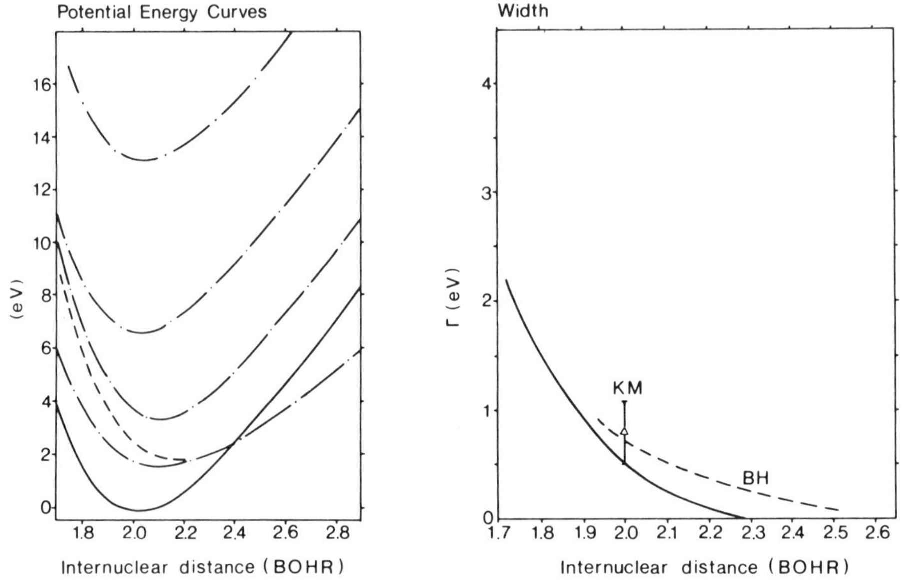

Figure 14. The Potential Energy Curves and the Resonance Width for the $^2\Pi_g$ State in e - N_2 Scattering.

We show in Figure 14 the potential energy curves and the resonance width calculated by Schneider et al. (1979b). The potential energy curves are: solid line, $^1\Sigma_g^+$ ground state of N_2; dash-dot line, R-matrix eigenenergies; dashed line, real part of resonance energy from the R-matrix calculation. The widths are: solid line, R-matrix calculation; dotted line, calculated by Birtwistle et al. (1971); triangle; calculated by Krauss and Mies (1970). We see that the real part of the resonance energy obtained by fitting the eigenphase shifts from the R-matrix calculation to the usual resonance formula is close to the second R-matrix eigenenergy at small internuclear distances and is close to the first eigenenergy at large internuclear distances. The resonance width obtained from these calculations is seen to be in reasonable agreement with earlier work. The results obtained by the R-matrix method for the vibrational excitation cross section in e - N_2 scattering is shown in Figure 15. It is seen that they are in excellent agreement with the experiment of Ehrhardt and Willmann (1967). This is the first completely ab-initio calculation for vibrational excitation for this system which shows the detailed structure of the $^2\Pi_g$ resonance and indicates that the R-matrix method can be expected to have wide applicability.

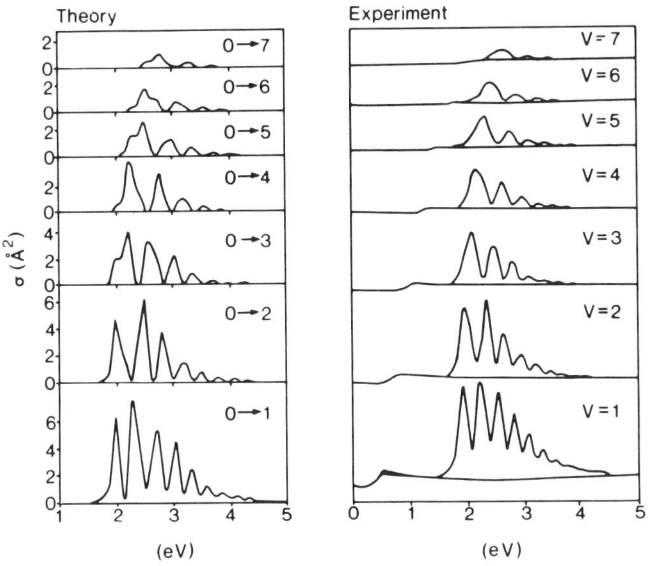

Figure 15. Total Vibrational Excitation Cross Sections for e - N_2 Scattering.

In concluding this section on the R-Matrix method we mention recent work which is underway to extend the calculations to treat coupled electronic states and to treat photoionization processes by Noble (1980). However, results have not yet been reported.

REFERENCES

Adkhikari, S.K. and Sloan, I.H., 1975, Phys. Rev. C11, 1133.

Arthurs, A.M. and Dalgarno, A., 1960, Proc. Roy. Soc. A256, 334.

Bardsley, J.N. and Mandl, F., 1968, Rep. Prog. Phys. 31, 471.

Bethe, H.A. and Salpeter, E.E., 1957, "Quantum Mechanics of One- and Two-Electron Systems" (Springer-Verlag, Berlin).

Birtwistle, D.T. and Herzenberg, A., 1971, J. Phys. B: Atom. Molec. Phys. 4, 53.

Bloch, C., 1957, Nucl. Phys. 4, 503.

Bottcher, C., 1969, Chem. Phys. Letters 4, 320.

Bransden, B.H., McDowell, M.R.C., Noble, C.J. and Scott, T., 1976, J. Phys. B: Atom. Molec. Phys. 9, L301.

Breit, G. and Bethe, H.A., 1954, Phys. Rev. 93, 888.

Brown, S.C., 1979, "Electron Molecule Scattering" (John Wiley, New York).

Buckley, B.D. and Burke, P.G., 1979, Invited paper at Symposium on Electron-Molecule Collisions, edited by I. Shimamura and M. Matsuzawa (University of Tokyo, Japan).

Buckley, B.D., Burke, P.G. and Vo Ky Lan, 1979, Comput. Phys. Commun. 17, 175.

Burke, P.G. and Chandra, N., 1972, J. Phys. B: Atom. Molec. Phys. 5, 1696.

Burke, P.G. Chandra, N. and Gianturco, F.A., 1972, J. Phys. B: Atom. Molec. Phys. 5, 2212.

Burke, P.G. and Hibbert, A., 1969, in Abstracts of VI ICPEAC (M.I.T. Cambridge, USA), 367.

Burke, P.G.,Hibbert, A. and Robb, W.D., 1971, J. Phys. B: Atom. Molec. Phys. 4, 153.

Burke, P.G.,Mackey, I. and Shimamura, I., 1977, J. Phys. B: Atom. Molec. Phys. 12, 2497.

Burke, P.G. and Robb, W.D., 1975, Adv. Atom. Molec. Phys. 11, 143.

Burke, P.G. and Sinfailam, A.L., 1970, J. Phys. B: Atom. Molec. Phys. 3, 641.

Cade, P.E., Sales, K.D. and Wahl, A.C., 1966, J. Chem. Phys. 44, 1973.

Chandra, N. and Temkin, A., 1976a, Phys. Rev. A13, 188.

Chandra, N. and Temkin, A., 1976b, Phys. Rev. A14, 507.

Chang, E.S. and Fano, U., 1972, Phys. Rev. A6, 173.

Chase, D.M., 1956, Phys. Rev. 104, 838.

Collins, L.A., Henry, R.J.W. and Norcross, D.W., 1980, J. Phys. B: Atom. Molec. Phys. 13, 2299.

Collins, L.A., Robb, W.D. and Morrison, M.A., 1978, J. Phys. B: Atom. Molec. Phys. 11, L777.

Collins, L.A., Robb, W.D. and Morrison, M.A., 1980, Phys. Rev. A21, 488.

Damburg, R. and Karule, E., 1967, Proc. Phys. Soc. 90, 677.

Davenport, J.W., 1976, Phys. Rev. Letters 36, 945.

Dehmer, J.L. and Dill, D., 1979, in "Electron-Molecule and Photon-Molecule Collisions", edited by T.N. Rescigno, V. McKoy and B.I. Schneider (Plenum Press, New York and London), 225.

Dehmer, J.L., Dill, D. and Wallace, S., 1979, Phys. Rev. Letters 43, 1005.

Dill, D. and Dehmer, J.L., 1974, J. Chem. Phys. 61, 692.

Dill, D., 1976, J. Chem. Phys. 65, 1130.

Ehrhardt, H. and Willmann, K., 1967, Z. Phys. 204, 462.

Faisal, F.H.M., 1970, J. Phys. B: Atom. Molec. Phys. 3, 636.

Faisal, F.H.M. and Tench, A.L.V., 1971, Comput. Phys. Commun. 2, 261.

Fano, V. and Dill, D., 1972, Phys. Rev. A6, 185.

Fliflet, A.W., Levin, D.A., Ma, M. and McKoy, V., 1978, Phys. Rev. A17, 160.

Fliflet, A.W. and McKoy, V., 1978a, Phys. Rev. A18, 1048.

Fliflet, A.W. and McKoy, V., 1978b, Phys. Rev. A18, 2107.

Fliflet, A.W. and McKoy, V., 1980, Phys. Rev.

Furness, J.B. and McCarthy, I.E., 1973, J. Phys. B: Atom. Molec. Phys. 6, 2280.

Gianturco, F.A. and Thompson, D.G., 1977, J. Phys. B: Atom. Molec. Phys. 10, L21.

Hamnett, A., Stoll, W. and Brian, C.E., 1976, J. Elec. Spect. Rel. Phen. 8, 367.

Hara, S., 1967, J. Phys. Soc. Japan 22, 710.

Hazi, A.U. and Taylor, H.S., 1970, Phys. Rev. A1, 1109.

Hazi, A.U., 1979, in "Electron-Molecule and Photon-Molecule Collisions", edited by T.N. Rescigno, V. McKoy and B.I. Schneider (Plenum Press, New York and London), 281.

Herzenberg, A. and Mandl, F., 1962, Proc. Roy. Soc. A270, 48.

Herzenberg, A., 1968, in "Electronic and Atomic Collisions", edited by G. Watel (North Holland Publishing Co., Amsterdam), 1.

Johnson, K.H., 1973, in "Advances in Quantum Chemistry" edited by P.O. Lowdin (Academic Press, New York), 143.

Kendrick, J. and Buckley, B.D., 1980, Daresbury Laboratory Report No. DL/SCI/TM22T.

King, G.W., Hainer, R.M. and Cross, P.C., 1943, J. Chem. Phys. 11, 27.

King, G.W., Hainer, R.M. and Cross, P.C., 1949, J. Chem. Phys. 17, 826.

Klonover, A. and Kaldor, U., 1977, Chem. Phys. Lett. 51, 321.

Klonover, A. and Kaldor, U., 1978, J. Phys. B: Atom. Molec. Phys. 11, 321.

Klonover, A. and Kaldor, U., 1979a, J. Phys. B: Atom. Molec. Phys. 12, 323.

Klonover, A. and Kaldor, U., 1979b, J. Phys. B: Atom. Molec. Phys. 12, L61.

Krauss, M. and Mies, F.H., 1970, Phys. Rev. A1, 1592.

Lane, N.F., 1980, Rev. Mod. Phys. 52, 29.

Langhoff, P.W., 1973, Chem. Phys. Letters 22, 60.

Langhoff, P.W., 1979, in "Electron-Molecule and Photon-Molecule Collisions", edited by T.N. Rescigno, V. McKoy and B.I. Schneider (Plenum Press, New York and London), 183.

Levin, D.A.,Fliflet, A.W., Ma, M. and McKoy, V., 1980, J. Comp. Phys.

Linder, F. and Schmidt, H., 1971, Z. Naturf. 26a, 1603.

Lovelace, C., 1964, Phys. Rev. 135, B1225.

Morrison, M.A., Lane, N.F. and Collins, L.A., 1977, Phys. Rev. A15, 2186.

Morrison, M.A. and Schneider, B.I., 1977, Phys. Rev. A16, 1003.

Morrison, M.A. and Collins, L.A., 1978, Phys. Rev. A17, 918.

Morrison, M.A., 1979, in "Electron-Molecule and Photon-Molecule Collisions", edited by T.N. Rescigno, V. McKoy and B.I. Schneider (Plenum Press, New York and London), 15.

Morrison, M.A., 1980, Comput. Phys. Commun. - to be published.

Nesbet, R.K., 1969, Phys. Rev. 179, 60.

Noble, C.J., 1980, private communication.

Ostlund, N.S., 1975, Chem. Phys. Letters 34, 419.

Plummer, E.W., Gustafsson, T., Gudat, W. and Eastman, D.E., 1977, Phys. Rev. A15, 2339.

Ransil, B.J., 1960, Rev. Mod. Phys. 32, 245.

Raseev, G., Guisti-Suzor, A. and Lefebvre-Brion, H., 1978, J. Phys. B: 11, 2735.

Raseev, G., Le Rouzo, H. and Lefebvre-Brion, H., 1980, to be published in J. Chem. Phys.

Rescigno, T.N., McCurdy, C.W. and McKoy, V., 1974a, Chem. Phys. Lett. 27, 401.

Rescigno, T.N., McCurdy, C.W. and McKoy, V., 1974b, Phys. Rev. A10, 2240.

Rescigno, T.H., McCurdy, C.W. and McKoy, V., 1975a, Phys. Rev. A11, 825.

Rescigno, T.N., McCurdy, C.W. and McKoy, V., 1975b, J. Phys. B: Atom. Molec. Phys. 8, L433.

Rescigno, T.N., Bender, C.F., McKoy, B.V. and Langhoff, P.W., 1978, J. Chem. Phys. 68, 970.

Rescigno, T.N., McKoy, V. and Schneider, B.I., 1979, "Electron Molecule and Photon-Molecule Collisions", edited by T.N. Rescigno, V. McKoy and B.I. Schneider (Plenum Press, New York and London).

Rescigno, T.N., 1979, private communication.

Riley, M.E. and Truhlar, D.G., 1975, J. Chem. Phys. 63, 2182.

Ritchie, B. and Tambe, B.R., 1980, J. Phys. B: Atom. Molec. Phys. 13, L225.

Rohr, K. and Linder, F., 1975, J. Phys. B: Atom. Molec. Phys. 8, L200.

Rose, M.E., 1957, in "Elementary Theory of Angular Momentum", (John Wiley, New York).

Sams, W.N. and Kouri, D.J., 1969, J. Chem. Phys. 51, 4809.

Samson, J.A.R.,Haddard, G.N. and Gardner, 1977, J. Phys. B: Atom. Molec. Phys. 10, 1749.

Schmid, G.B., Norcross, D.W. and Collins, L.A., 1980, Comput. Phys. Commun., to be published.

Schneider, B.I., 1975a, Chem. Phys. Lett. 31, 237.

Schneider, B.I., 1975b, Phys. Rev. A11, 1957.

Schneider, B.I. and Hay, P.J., 1976a, J. Phys. B: Atom. Molec. Phys. 9, L165.

Schneider, B.I. and Hay, P.J., 1976b, Phys. Rev. A13, 2049.

Schneider, B.I., 1978, in "Electronic and Atomic Collisions", ed. by G. Watel (North Holland Publishing Company, Amsterdam), 257.

Schneider, B.I., Le Dourneuf, M. and Burke, P.G., 1979a, J. Phys. B: Atom. Molec. Phys. 12, L365.

Schneider, B.I., Le Dourneuf, M. and Vo Ky Lan, 1979b, Phys. Rev. Lett. 43, 1926.

Shimamura, I. and Matsuzawa, 1979, Invited papers at Symposium on Electron Molecule Collisions (University of Tokyo, Japan).

Slater, J.C., 1960, "Quantum Theory of Atomic Structure", Vol. 2 Appendix 22 (McGraw-Hill, New York).

Smith, E.R. and Henry, R.J.W., 1973, Phys. Rev. A7, 1585.

Temkin, A., 1966, in "Autoionization", edited by A. Temkin (Mono Book Corporation, Baltimore), 55.

Watson, D.K., Lucchese, R.R., McKoy, V. and Rescigno, T.N., 1980, Phys. Rev.

Wight, G.R., Van der Wiel, M.J. and Brion, C.E., 1976, J. Phys. B: Atom. Molec. Phys. 9, 675.

Wigner, E.P., 1946a, Phys. Rev. 70, 15.

Wigner, E.P., 1946b, Phys. Rev. 70, 606.

Wigner, E.P. and Eisenbud, L., 1947, Phys. Rev. 72, 29.

Woodruff, P.R. and Marr, G.V., 1977, Proc. Roy. Soc. A358, 87.

Yang, C.N., 1948, Phys. Rev. 74, 764.

Ziesche, P. and John, W., 1976, J. Phys. B: Atom. Molec. Phys. 9, 333.

RECOMBINATION PROCESSES IN ATOMIC AND MOLECULAR PHYSICS

J.N. Bardsley

Observatoire de Meudon
92190 Meudon, France

University of Pittsburgh
Pittsburgh, Pennsylvania 15260
U.S.A.

ABSTRACT

The variety of recombination processes, leading to charge neutralization or molecule formation, is surveyed. Examples are provided of atom-atom, electron-ion and ion-ion recombination in both the two-body and many-body regimes, with species ranging from protons to ammonia clusters. The emphasis is on the theoretical development, with only brief references to experimental techniques.

1. INTRODUCTION

2. RADIATIVE RECOMBINATION

   2.1 Direct radiative recombination.
   2.2 Dielectronic recombination.
   2.3 Recombination radiation.
   2.4 $e + O^+(^4S) \rightarrow O + h$.
   2.5 $e + Mo^{32+} \rightarrow Mo^{31+} + h$.

3. DISSOCIATIVE RECOMBINATION

   3.1 Direct dissociative recombination.
   3.2 Indirect dissociative recombination.
   3.3 $e + H_2^+$.
   3.4 $e + CH^+$.
   3.5 Recombination with ion clusters.

4. MUTUAL NEUTRALIZATION

4.1 Introduction
4.2 The Landau-Zener model.
4.3 The Absorbing sphere and optical potential models.
4.4 Ion-ion neutralization in flames: NaCl and KCl.

5. RADIATIVE ASSOCIATION

6. ELECTRON-ION RECOMBINATION AT HIGH DENSITIES

6.1 Collisional-radiative recombination in an ambient electron gas.
6.2 De-excitation by collisions with neutral atoms and molecules.
6.3 Collisional-dissociative recombination.

7. HEAVY PARTICLE RECOMBINATION AT HIGH DENSITIES

7.1 Ion-ion recombination.
7.2 Ion-ion recombination in rare-gas-halide lasers.
7.3 Recombination of neutral atoms.

8. SUMMARY

1. INTRODUCTION

In these lectures we will examine the processes through which atomic particles combine to form more complex systems. In most of these processes there will be some neutralization of charge, but we will also consider reactions in which molecules are formed in collisions between neutral atoms. For example we will include radiative association

$$A + B \rightarrow AB + h\nu \tag{1.1}$$

and three-body association

$$A + B + M \rightarrow AB + M \tag{1.2}$$

However our main purpose is to study those processes which lead to the disappearance of charged particles in ionized gases.

The simplest imaginable recombination process would be one in which two particles collide and coalesce to form a compound system. However because of the laws of conservation of energy and momentum a compound system formed in this way can have only a finite lifetime. In order for a stable compound system to be created, it is essential that energy or momentum be exchanged with a third body, or emitted in the form of a photon (or neutrino). Because of the low probability of the emission of a photon in most atomic collisions, three body processes are often important even at relatively low densities.

The earliest detailed studies of recombination were motivated by a desire to understand the ionization balance in the atmosphere. For such applications we are concerned with low stages of ionization, temperatures in the range 200-2000 K, neutral particle densities up to 3×10^{19} cm^{-3} and electron or ion densities up to 10^6 cm^{-3}. Even in this environment there are many significant recombination processes involving both two-body and three-body collisions. Thus experimental studies are usually performed at sufficiently low pressures that the three-body reactions are not important.

Astrophysics gives us a greater variety of conditions. In the interstellar medium one need only consider two body processes such as radiative association and radiative recombination

$$e + A^+ \rightarrow A + h\nu \tag{1.3}$$

Here the collision energies are low and the atoms are only weakly ionized. However in stellar atmospheres the temperature is very high and the atoms are strongly ionized. Thus we must generalize this last process to

$$e + A^{n+} \rightarrow A^{(n-1)+} + h\nu \qquad (1.4)$$

and consider energies up to ∿ 10 keV. Similar conditions are found in Tokomak plasmas and a detailed understanding of the recombination processes has been an essential part of recent analyses of the effects of heavy impurity ions in hydrogen plasmas.

For most applications one needs to know the rate at which recombination occurs, given a distribution of particle velocities, rather than the cross section for a specific collision energy. For a two-body process, when the cross section for a relative velocity is $\sigma(v)$, the recombination rate coefficient is defined to be

$$\alpha = \int dv\; v\; \sigma(v)\; f(v) \qquad (1.5)$$

in which $f(v)$ gives the distribution of relative speeds in the plasma. In thermal equilibrium at a temperature T, this is

$$f(v) = \left(\frac{2\, m}{\pi k^3 T^3}\right)^{\frac{1}{2}} v^2 \exp\left(-\frac{mv^2}{2kT}\right) \qquad (1.6)$$

where m is the reduced mass of the colliding particles. For reaction (1), for example the number of recombination events per unit volume per unit time is then given by $\alpha\, N_A\, N_B$, where N_A and N_B are the densities of atoms A and B. The usual unit for α is $cm^3\, s^{-1}$. For three-body processes, such as (2) one often defines an effective two-body recombination, α, for the production of AB, which is dependent on the density of the third species M. Alternatively, at low pressures, it is sometimes convenient to introduce a three-body rate coefficient, K, so that the number of recombination events, in unit volume and time, is $K\, N_A\, N_B\, N_M$.

In electron-ion collisions at thermal energies the energy dependence of two-body recombination cross sections is often well represented by the Wigner threshold law for exothermic processes, $\sigma \propto E^{-1}$. In such cases the temperature dependence of the recombination coefficient should be close to $T^{-0.5}$. There are two useful formulas for calculating the recombination rate in these situations. First let us suppose that the cross section has been calculated using atomic units, giving

$$\sigma(E) = \frac{C_1}{E} \qquad (1.7)$$

with σ in a_o^2 and E in Hartree. The recombination rate coefficient is then

$$\alpha = 3.2\ C_1 \left(\frac{300}{T}\right)^{0.5} \times 10^{-7} cm^3\ s^{-1} \quad (1.8)$$

Alternatively if $\sigma(E)$ is expressed in $Å^2$ and E in eV, so that

$$\sigma(E) = \frac{C_2}{E}\ Å^2 (eV)^{-1} \quad (1.9)$$

then $$\alpha = 4.2\ C_2 \left(\frac{300}{T}\right)^{0.5} \times 10^{-8} cm^3\ s^{-1} \quad (1.10)$$

The main purpose of these lectures is to provide an introduction to the wide variety of recombination processes and to discuss their relation to other collision processes, such as excitation and ionization. To avoid a completely superficial treatment, while staying within the allotted space and time, we will not treat all the aspects uniformly, but will give more details in some areas and refer to recent papers and reviews for additional information on other topics.

We will begin by studying two-body processes. In section 2 we will consider radiative recombination of electrons with atomic ions, including both direct and dielectronic recombination. In section 3 we will discuss dissociative recombination

$$e + AB^+ \rightarrow A + B \quad (1.11)$$

which is usually the dominant mode of electron-ion recombination whenever molecular ions are present. Turning to heavy particle collisions we will treat mutual neutralization

$$A^+ + B^- \rightarrow A + B \quad (1.12)$$

in section 4, and radiative association (1) in section 5. Three-body reactions will then be described, with electron processes such as

$$e + A^+ + e \rightarrow A + e \quad (1.13)$$

and $$e + A^+ + M \rightarrow A + M\ , \quad (1.14)$$

in section 6, and ion-ion recombination

$$A^+ + B^- + M \rightarrow AB + M \quad (1.15)$$

in section 7. The final section will include comments about the significance of recombination processes, and the present status of the theory.

2. RADIATIVE RECOMBINATION

2.1 Direct Radiative Recombination

In principle the radiative recombination process (1.3) is just the inverse of photoionization. The cross section, for a given initial and final state, can be obtained from that of the reverse reaction through the relation (in atomic units)

$$\sigma_{ia}^{RR}(E) = \frac{(h\nu)^2}{E}\left(\frac{1}{137}\right)^2 \frac{g_a}{2g_i}\, \sigma_{ai}^{PI}(h\nu) \tag{2.1}$$

In this expression E and $h\nu$ represent the energies of the electron and photon, and g_a and g_i are the degeneracies of the atomic and ionic states involved. At energies very close to threshold, the photoionization cross section σ^{PI} approaches a constant limit, and so the recombination cross section σ^{RR} shows the expected E^{-1} behavior.

In order to calculate the total recombination cross section one must include transitions from a given ionic state to every state of the neutral atom. Hence a calculation of the photoionization cross section for the ground state of an atom does not permit the deduction of the total cross section for radiative recombination. Nevertheless, if accurate calculations of photoionization cross sections are available for each of the states with the same configuration as the ground state, then the contributions associated with the higher states can be obtained from quantum defect theory and the total recombination can be deduced with high precision.

In the calculation of recombination rates at moderate temperatures, one further working rule is useful. Let us suppose that the threshold value of a relevant photoionization cross section is $C_o \times 10^{-18}$ cm^2 [this unit, sometimes called the Megabarn, is often used for such cross sections]. The rate coefficient for recombination through this same transition is then given by

$$\alpha = 1.5 \times 10^{-13} \times C_o \frac{g_a}{2g_i}\left(\frac{I}{I_H}\right)^2\left(\frac{300}{T}\right)^{\frac{1}{2}} \text{cm}^3\ \text{s}^{-1} \tag{2.2}$$

in which I/I_H is the ratio of the relevant ionization energy to that of H(1s) (13.6 eV). This formula can be used only for those temperatures such that the photoionization cross section does not deviate significantly from its threshold value within the main body of the electron energy distribution.

For recombination of non-relativistic electrons with H^+ or other fully stripped ions, the cross sections can be obtained analytically (Burgess, 1964). The hydrogenic model, when adapted to take account of the quantum defects, has been used extensively to describe recombination with highly ionized atoms and excited states of weakly-ionized systems. To obtain accurate results for atoms with high Z, we must take into account relativistic effects (Lee and Pratt, 1975), but Hahn and Rule (1977) have suggested that, following simple rules to allow for screening, one can obtain cross sections accurate to 20% with a hydrogenic model for all highly ionized atoms with $Z \lesssim 50$.

If one examines carefully the photoionization cross section for any atom or ion with more than one electron, one finds a very large number of resonances, arising from the formation of autoionizing states of the initial system. These are quasi-stationary states that are usually formed by the excitation of an inner-shell electron or the simultaneous excitation of two outer electrons. These states can be very important in recombination at high temperatures. Although they can be formed only in collisions of electrons with specific energies, the relatively long duration of a resonant collision leads to a significant increase in the probability of photon emission. Recombination that proceeds through these autoionizing states is usually called dielectronic recombination.

The intermediate states in dielectronic recombination belong to Rydberg series terminating on excited states of the initial ion. For example, a typical reaction in e-He^+ recombination would be

$$e + He^+ (1s) \rightarrow He(2pn\ell) \rightarrow He(1sn\ell) + h\nu \qquad (2.3)$$

The need for high temperatures is clear since the lowest such resonance appears near 35 eV. However in other systems the autoionizing states appear at much lower energies.

In many problems the autoionizing states can be regarded as non-overlapping resonances, and the interference with non-resonant scattering can be neglected. The recombination cross section is then given by the standard Breit-Wigner formula

$$\sigma(E) = \frac{\pi}{2E}\,\omega\,\frac{\Gamma_r\Gamma_a}{(E-E_q)^2 + \tfrac{1}{4}\Gamma_t^2} \tag{2.4}$$

In this expression, ω is the ratio of degeneracies

$$\omega = \frac{g_a}{2g_i} \tag{2.5}$$

Γ_a and Γ_r are the widths corresponding to decay by auto-ionization leaving the ion in its initial state, and by radiative decay. The total width Γ_t is equal to $\Gamma_a + \Gamma_r$ when there are no other open autoionization channels, and E_q denotes the resonance energy, with q used as a collective index.

The widths of these resonant states are so small that their effect on the recombination rate is almost the same as that of a delta function

$$\sigma(E) = \frac{\pi^2}{E}\,\omega\,\frac{\Gamma_r\Gamma_a}{\Gamma_t}\,\delta(E-E_q) \tag{2.6}$$

Usually $\Gamma_r \ll \Gamma_a$, and if there is only one open autoionization channel we find,

$$\sigma(E) \approx \frac{\pi^2}{E}\,\omega\,\Gamma_r\,\delta(E-E_q) \tag{2.7}$$

The contribution of this state to the recombination rate is then

$$\alpha_q(T) \approx \left(\frac{2\pi}{kT}\right)^{3/2}\Gamma_r\omega\,\exp\left[-\frac{E_q}{kT}\right] \tag{2.8}$$

Let us now return to the specific example of the process (2.3), and compare the roles of states of different n in the same Rydberg series. In the radiative transition that stabilizes the intermediate state the active electron is the inner one, so that Γ_r depends only very weakly upon n. The autoionization width Γ_a depends strongly on n, varying approximately as n^{-3}. However, as can be seen from eq. (2.7), the contribution to the recombination rate is insensitive to Γ_a, provided that $\Gamma_a \gg \Gamma_r$. For very large n this condition fails and the contribution then decreases with n, but while this condition is satisfied the n-dependence of the contribution arises mainly from the changing position of the resonance. Hence for low stages of ionization at high temperature many resonant states can be important, and Rydberg states with very high n are formed. This leads to very large deviations in the populations of excited states from those expected in Saha equilibrium.

The significance of dielectronic recombination was clearly demonstrated by Burgess (1964, 1965), who also gave a simple formula for estimating its effect on the recombination rate for an arbitrary ion. The rule seems to have been remarkably successful (Seaton and Storey, 1976) and is usually accurate to within $\pm$ 40% (Burgess and Tworkowski, 1976). However for some ions the recombination may be reduced significantly due to the presence of additional autoionization channels in which an electron is emitted and the ion is left in an excited state. For example in neon-like ions the incident electron may be captured by exciting a 2p electron into a 3d orbit, creating a doubly excited state with the outer configuration $(3d, n\ell)$. Since the 2p orbital is still almost completely full, autoionization leaving the tenth electron in the 3p rather than the 2p orbital may be dominant. Jacobs et al. (1977a,b) have studied the magnitude of this effect for several ions of Si and Fe and find some significant deviations from the values predicted by the simple Burgess formula.

The importance of stabilization of the intermediate state by radiative transitions of the outer $(n\ell)$ electron has been examined recently by several authors. Raymond (1978) shows that for Fe^{23+} ions such transitions add between 10 and 20% to the recombination rate, which is well within the error expected in the Burgess formula. Beigman & Chichkov (1980) have suggested for ions, such as $O^+(^4S)$, with low-lying metastable states processes such as

$$e + O^+(^4S) \rightarrow O^*(^2D, n\ell) \rightarrow O^*(^2D, n'\ell') + h\nu \qquad (2.9)$$

may be important for a small range of temperatures, say between 4000 K and 10,000 K.

For multiply-charged ions of high Z the competition between autoionization and radiative decay must be reexamined. As one follows a given resonance along an isoelectric sequence one finds approximately that $\Gamma_r \propto Z^4$ whereas Γ_a depends only weakly on Z. Thus the condition $\Gamma_a \gg \Gamma_r$ becomes invalid at smaller n and the importance of states with high values of n is diminished. This is apparent in the calculations by Gau et al. (1980) that will be described below. One must also consider the effects of the breakdown of L-S coupling on the relative magnitudes of the autoionization and radiative decay widths.

Recent calculations by Bely-Dubau et al. (1979) on e + Fe^{23+} show the rate of dielectronic recombination to peak at 5 x 10^7 K with a maximum value of $\sim 5 \times 10^{-13} cm^3 s^{-1}$. Although this maximum value is very close to that predicted by the Burgess formula, the peak occurs at a lower temperature and recombination rate calculated at temperatures around 10^7 K is much higher than that obtained with the simple general formula.

2.3 Recombination Radiation

Both direct radiative recombination and dielectronic recombination lead to characteristic radiation that is a valuable diagnostic tool both for natural and artificial plasmas. For example the study of radio-recombination lines, involving transitions between states with very high n, has become an integral part of radio astronomy. More recently Seaton (1978) has calculated the intensities of the Balmer ($n_f = 2$) and Paschen ($n_f = 3$) emission lines in He II that should result from the recombination of He^{++} ions at temperatures between 5000 and 20000 K to facilitate UV observations of this recombination.

A good example of the study of recombination radiation in the atmosphere is provided by the satellite observations of the OI lines at 1304 Å (3s $^3S \rightarrow$ 2p ^3P) and 1356 Å (3s $^5S \rightarrow$ 2p ^3P) in the tropic nightglow. The history of these measurements and their analysis, as chronicled by Bates (1979), shows that the interpretation of such data is often not simple. We will not delve into the aeronomy here, but will discuss the basic atomic physics below.

In tokomak plasmas the intensity of the "satellite lines" that result from dielectronic recombination has given useful checks of the electron temperature. Returning to process (2.3), if n is not too large the radiative transition can be separated from the parent line, 2p → 1s in He^+, and the electron temperature can be deduced from the ratio of the intensities of the satellite lines and their parent. This technique, which is also used in astronomy, has recently been improved to allow for cascading (Dubau et al., 1980) and has been reviewed thoroughly by Dubau and Volonte (1980).

2.4 $e + O^+(^4S) \rightarrow O + h\nu$

For our first example, let us examine the recombination of low-energy electrons with singly-ionized oxygen in its ground state. Of the three states of O(^3P, ^1D and ^1S) belonging to the configuration $(2p)^3$, only the ground ^3P state can be formed through an electric dipole transition. Thus one can hope to obtain accurate results by calculating the rate for recombination to the ground state from ab initio calculations of the ground-state photoionization cross sections and by treating recombination to excited states by the quantum defect method.

R-matrix calculations reported by Berrington and Crees (1979) give a cross section which varies little between 0 and 1 eV with a value of 4×10^{-18} cm^2. This leads to a rate coefficient, for recombination to the ground state, of

$$\alpha^{G}(T) = 6.6 \times 10^{-13}\left(\frac{300}{T}\right)^{0.5} \text{cm}^3 \text{ s}^{-1} \qquad (2.10)$$

Quantum defect calculations of recombination to excited states have been made bv Davis and Lewis (1978) and by Julienne et al. (1974) but their main interest is in the intensity of UV emission and they do not give a complete set of data on the recombination rates. Tinsley et al. (1973) quote unpublished calculations by R. C. Kirkpatrick which indicate that the total recombination cross section at low temperatures is given by

$$\alpha^{T}(T) = 3.9 \times 10^{-12}\left(\frac{300}{T}\right)^{0.69} \text{cm}^3 \text{ s}^{-1} \qquad (2.11)$$

However the temperature dependence implied by this formula may not be very accurate because the cross sections for photoionization from excited states of O seem to vary significantly within 1 eV of threshold.

The lowest autoionizing states that contribute significantly to dielectronic recombination are those with configurations like $2p^3$ (^2D) 4s and $2p^3(^2D)$ 3d which can decay by allowed transitions to the ground $(2p)^4$ ^3P state of O. Beigman and Chichkov (1980) estimate that at temperatures between 4000 K and 15000 K the contribution from all of the autoionizing states below the ^2P metastable level of O^+ is slightly greater than that of the direct process. Beigman and Chichkov suggest that dielectronic recombination through higher resonant states, with configurations $(2p)^3$ 3ℓ $n\ell'$, is dominant for temperatures above 10^4 K.

Gould (1978) has reported calculations of radiative recombination coefficients for the first four stages of ionization of the elements C, N, O, Ne, Mg, Si, S and Ar. For O^+ he finds a rate of 75×10^{-13} cm^3 s^{-1} at 100 K and 3.3×10^{-13} cm^3 s^{-1} at 10^4 K. The calculations discussed above suggest that his value at 100 K may be too high and that at 10^4 K too low.

2.5 $e + Mo^{32+} \rightarrow Mo^{31+} + h\nu$

For our second example let us take a case involving a highly ionized atom for which relativistic effects might be important. Lee and Pratt (1975, 1976) have developed a relativistic theory in which they examine carefully the cross sections for direct recombination to the lowest states of each symmetry from numerical solutions of the Dirac equation with a Hartree-Slater potential. Then they compute the cross sections for bremsstrahlung in the tip region in which a low energy electron is produced and use quantum defect theory to derive the cross sections for recombination to the more highly excited states of the final ion. For neon-like Mo^{32+} the total recombination cross section decreases from 2.2×10^{-21} cm^2 at 1 keV to 2.7×10^{-24} cm^2 at 50 keV.

Hahn and Rule (1977) have examined this same system using a non-relativistic model with a very simple procedure for treating the screening effects. For an ion with nuclear charge Z that is n times ionized they take a hydrogenic model with an effective charge given by

$$Z_{eff} = \tfrac{1}{2}(Z + n) \qquad (2.12)$$

The total recombination cross section calculated in this way is 2.6×10^{-21} cm^2 at 1 keV and 2.4×10^{-24} cm^2 at 50 keV. They claim that this technique should be good to 20% accuracy, or better, whenever

$$Z \lesssim 50, \qquad 0.4 \lesssim \frac{n}{Z}, \qquad \text{and } 0.3 \text{ keV} \lesssim E \lesssim 100 \text{ keV}.$$

The effects of dielectronic recombination, which were ignored in these calculations, have been studied more recently by Gau, Hahn and Retter (1980). They find that the usual schemes for averaging over states of the same configuration but different angular momenta lead to considerable error. At a temperature of 10^7 K (for which $kT \approx 1$ keV) their estimate of the total rate of dielectronic recombination is $\sim 3 \times 10^{-11}$ cm^3 s^{-1}, which is 10 times larger than the contribution from direct recombination.

Results of similar calculations on Ne-like Fe and Ar, with a discussion of the scaling properties along this isoelectronic sequence, have been reported by Hahn et al. (1980).

3. DISSOCIATIVE RECOMBINATION

3.1 Direct Dissociative Recombination

At this point the recombination saga increases in pace by six orders of magnitude and we begin to examine processes that can compete with ion-molecule reactions at standard gas pressures. The greater efficiency of two-body electron-ion recombination that is found when the ions are in molecular form arises because most of the compound states that are formed in the electron-ion collisions can be stabilized through dissociation. Hence one can have two particles in the final state, without the emission of a photon, and the conservation of momentum and energy can be satisfied simultaneously. At 300 K the recombination rate is usually between 2 x 10^{-7} cm^3 s^{-1} and 5 x 10^{-6} cm^3 s^{-1}.

The dissociative recombination process (1.11) usually leaves at least one of the fragments in an excited state and the present challenge to the theorist is to calculate the total rate for the reaction and the branching ratios for the production of specific final states.

Since one can view dissociative recombination as the formation and dissociation of a quasistationary molecular state,

$$e + AB^+ \rightarrow AB^{**} \rightarrow A + B^* \qquad (3.1)$$

it is natural to apply resonant scattering theory to the process. In terms of their electronic structure the intermediate states AB^{**} are very similar to the autoionizing states that are present in dielectronic recombination, involving inner-shell excitations or double excitations from the ground state AB. However the potential energy curve associated with AB^* is usually repulsive, and in terms of the nuclear motion there is usually no sense in which the state AB^{**} is a discrete state embedded in the continuum of the A + B^* scattering states. The two modes of decay of the state AB^*, autoionization and predissociation, may then have to be treated in different ways. Although autoionization can be treated as a spontaneous process, which may occur at any instant with a known probability, it is not at all clear that dissociation can be regarded in the same manner.

This problem has been solved through the introduction of the Born-Oppenheimer separation of electronic and nuclear motion. Let us assume that during the lifetime of the intermediate state AB^* the wave function can be written as a product of an electronic wave function $\phi_d(r, R)$ and a nuclear function $\xi(R)$. The function $\phi_d(r, R)$ describes an autoionizing state of a molecule in which the nuclei are held fixed and the nuclear coordinate R acts only as a parameter in this electronic wave function. The expectation value of the electronic Hamiltonian $H^{el}(r, R)$,

$$E_d(R) = \langle \phi_d(r, R) | H^{el}(r, R) | \phi_d(r, R) \rangle \qquad (3.2)$$

then provides the potential for the nuclear motion.

Resonant scattering theory for electron-molecule collisions

can be developed following the configuration interaction approach of Fano (1961), as shown by Bardsley (1968a, 1978) and Fiquet-Fayard (1974). If one suppresses the rotational motion of the molecule, the complete wave function for the scattering of an electron of energy E by an ion in the vibrational state v can be written in the form

$$\Phi_{vE}(r, R) = \phi_d(r, R)\, \xi_d(R) + \sum_{v'} \int dE'\, b_{v'}(E')\, \Psi_{v'E'}(r, R) \tag{3.3}$$

Here, as above, r is being used to denote all of the electronic coordinates. The functions $\Psi_{v'E}(r,R)$ represent the non-resonant scattering of electrons by ions in the vibrational levels v'. It is usually assumed that no vibrational excitation or dissociation would occur in the absence of the resonance. The main aim of the resonance theory is to study the couplings that are induced among the $\Psi_{v'E'}$ and the loss of flux into the dissociative capture channel that result from the formation of the autoionizing state that is represented by the first term in eq. (3.3).

In the standard theory of dissociative recombination it is assumed that as the nuclei separate the autoionizing state $\phi_d(r, R)$ becomes stable against electron emission and that the cross section for dissociative recombination can be obtained from the asymptotic behavior of $\xi_d(R)$. Specifically, if the scattering functions are normalized to delta functions in energy

$$\sigma_{DR} = \frac{\pi^2}{E} \frac{g_o}{2g_+} \frac{K}{M} \lim_{R\to\infty} |\xi_d(R)|^2 \tag{3.4}$$

in which g_o and g_+ are the statistical weights of the resonant state AB^{**o} and the initial ionic state AB^+, respectively, and K and M are the wave number and reduced mass of the dissociating fragments. However the path to dissociation involves an infinite number of crossings, and avoided crossings, of potential curves. Although the assumption that the nuclei dissociate smoothly along some kind of adiabatic potential curve probably leads to no serious error in the calculation of the total recombination rate, a more thorough analysis is needed if the states of the fragments are to be identified with confidence.

To calculate the nuclear function $\xi_d(R)$ we diagonalize the total Hamiltonion with the space spanned by the functions $\phi_d(r, R)$ and $\Psi_{v'E'}(r, R)$. This leads to an integro-differential equal for $\xi_d(R)$. Let us suppose that the non-resonant scattering functions can be written as products of electronic and nuclear functions

$$\Psi_{vE}(r, R) = \psi_E(r, R)\ \zeta_v(R) \tag{3.5}$$

in which the $\zeta_v(R)$ represent the vibrational states of the initial ion. The coupling between the resonant and non-resonant electronic functions can then be described by the matrix element

$$V_E(R) = <\ \phi_d(r, R)|H_{el}(r, R)\ |\psi_E(r, R)\ > \tag{3.6}$$

Neglecting the shift induced in the potential curve $E_d(R)$ by this coupling, one obtains the equation

$$\left[-\frac{\hbar^2}{2M}\nabla_R{}^2 + E_d(R) - E_v - E\right]\xi_d(R) = V_E\ (R)\ \zeta_v(R)$$

$$+\ \pi i \sum_{v'} V_E(R)\quad \zeta_{v'}(R) \int dR'\ \zeta^*_{v'}(R')\ V^*{}_E(R')\ \xi_d(R') \tag{3.7}$$

The first term on the R.H.S. of this equation is a source term which leads to the formation of the autoionizating state; the index v refers to the initial ionic state and E_v is the corresponding energy of the ion. The second term describes the decay of the intermediate state through autoionization. The summation should be carried out over all open channels.

The second term on the RHS of eq. (3.7) can be regarded as the imaginary part of a non-local optical potential. It can be approximated by a local potential if there are enough open channels that the sum over v' can be performed using the completeness relation

$$\sum_v \zeta_v(R)\ \zeta_v^*(R') = \delta(R-R') \tag{3.8}$$

The equation then becomes

$$\{-\frac{\hbar^2}{2M}\nabla_R^2 + E_d(R) - \frac{i}{2}\Gamma(R) - E_v - E\}\ \xi_d(R) = V_E(R)\zeta_v(R) \tag{3.9}$$

$$\text{with } \Gamma(R) = 2\pi\ |\ V_E(R)|^2 \tag{3.10}$$

In either form the nuclear wave equation can be solved through the computation of the associated Green function, and iteration is not needed for eq. (3.7). If the non-local optical potential is retained the formalism properly treats the competition between dissociative recombination and vibrational excitation. The recombination cross section should then fall suddenly at each vibrational excitation threshold.

The calculation of dissociative recombination cross sections involves three steps. First the relevant autoionizing states AB^* must be identified, their potential curves must be calculated and the strength of their coupling with the continuum determined. The calculation of potential curves can be accomplished by the stabilization method or through the introduction of projection operators. The coupling matrix elements $V_E(R)$ can then be obtained from an approximate representation of the non-resonant electron-scattering continuum functions $\psi_E(r, R)$ or by the Stieltjes technique (Hazi, 1978). Both of these procedures give the energy dependence of these matrix elements and so permit the calculation of the shift and width of the potential curve for AB^{**}. Alternatively both the position and width of the autoionizing state potential curve can be deduced from ab initio calculations of the scattering of electrons from an ion with fixed nuclei, provided that the scattering wave function includes a good representation of the resonant state. At least three separate methods are being developed currently, the close coupling method, the R-matrix method, and the variational approach using prolate spheroidal coordinates.

The second stage of the calculation involves the solution of the nuclear wave equation (3.7) or (3.9) to determine the effects of the competition between autoionization and dissociation. Finally one must follow the dissociating atoms through the maze of potential curves which must be traversed before dissociation is complete. Very little work has been done on this third stage and further development of the theory of diabatic states would be helpful in this regard.

It is often convenient to think of the cross section for recombination as a product

$$\sigma_{DR}(E) = \sigma_{cap}(E)\ S(E)$$

Here $\sigma_{cap}(E)$ represents the probability that an electron of energy E can be captured to form the state AB^{**}, and S(E) is the survival factor which gives the probability that this state dissociates without emitting an electron.

The capture cross section will be large if there are autoionizing states AB^{**} with potential curves so that the transition from AB^+ to AB^{**} can be accomplished without significant change in the nuclear motion. This is possible if the separation R_c, for which the difference in these potential energy curves is equal to the incident energy E, is within the range of nuclear vibrations in the initial ionic state. One also requires that the coupling matrix element $V_E(R)$, or autoionization width $\Gamma(R)$, be not too small.

For the capture of thermal electrons, at moderate temperatures, one looks for potential curves that cross those of the ion near its equilibrium nuclear separation. With such curves the energy dependence of the capture cross section should be close to E^{-1}.

The survival factor depends on the size of $V_E(R)$, the number of open vibrational channels through which autoionization can occur, and the time needed for the AB^{**} state to become stable through the separation of the nuclei. This time depends on the slope of the potential curve and the distance between the point of capture and the stabilization point R_c, at which the curves for AB^{**} and AB^+ intersect. It has usually been assumed that for energies $\lesssim$ 0.1 eV this survival factor is close to 1 so that the overall recombination cross section should also vary as E^{-1}. Most of the recombination studies performed in afterglows give rates with a temperature dependence close to $T^{-0.5}$, which is consistent with this energy variation. The cross sections obtained in beam experiments and ion-traps show that the situation is more complicated at higher energies, as is expected from the theory.

With many of the experimental techniques that have been used to study dissociative recombination the ions occupy a mixture of vibrational states. Some of the discrepancies between the results obtained by different groups may be due to variations in the vibrational state populations. From the study of dissociative attachment to neutral molecules one knows that the attachment cross sections for excited vibrational states can exceed that of the ground state by many orders of magnitude. A similar enhancement may be found in the recombination of low energy electrons with H_2^+ and He_2^+. On the other hand, O'Malley et al. (1972) have suggested that for Ne_2^+ and Ar_2^+ the cross sections for recombination to vibrationally excited ions may be much less than those of ground state ions. Although this could be the case if, for each of the significant autoionizing states, the stabilization point is close to the equilibrium separation in the ion, it seems unlikely that there would be no important states which cross at larger distances.

3.2 Indirect Dissociative Recombination

If one examines the photon absorption cross sections for molecules just above the ionization threshold, one finds that the spectrum is often dominated by many discrete lines. Indeed, for H_2 the lines are so strong that it is hard to see the underlying photoionization continuum. Further examination of these peaks reveals that they are due to the formation of autoionizing states of the molecule and that many of these states are also subject to predissociation. Very close to threshold these lines represent excited rotational or vibrational levels of high Rydberg states. For example for H_2 the excited vibrational levels of states with n $\gtrsim$ 8 can autoionize, as can excited rotational levels of states with n $\gtrsim$ 30.

It was pointed out by Bardsley (1968b) that these states could play a role in dissociative recombination, if they decay both by autoionization and dissociation. An example of this indirect dissociate recombination is

$$e + AB^+(v = 0) \to AB^R(n = 8,\ v = 1)$$

$$\to AB^{**} \to A + B^*$$

In the first step of inverse autoionization, the electron is captured through the exchange of energy with the nuclear motion. It drops into a bound Rydberg orbital and excites the vibrational motion of the nuclei. This Rydberg state then predissociates through a doubly excited state AB^{**} of the kind that were introduced in sec. (3.1).

In contrast to direct dissociative recombination, which can occur for any energy of the incident electron, the indirect process can only be important when the total energy is close to that of the one of the Rydberg levels. It therefore leads to structure in the cross section, which should be most obvious below the threshold for vibrational excitation. In simple situations it can be described by a simple Breit-Wigner formula. As in the case of dielectronic recombination the contribution of a given level of AB^R may be insensitive to its autoionization rate, provided that this is greater than the rate of predissociation. Furthermore, since the predissociation rate for AB^R and the proability of direct capture into the state AB^{**} depend on the same electronic coupling, there is a close connection between the rate of direct and indirect recombination. Indeed Bottcher (1976) has suggested that the rates should be equal, but his argument is not convincing.

The indirect process should be important for recombination with ground state ions in cases, like H_2^+ and He_2^+, where there are not favorable curve crossings for the direct process. However, for most diatomic molecules its contribution to the recombination rate is probably small. It may be more important for polyatomic molecules and cluster ions for which the large number of modes of vibrational motion may lead to more efficient electron trapping.

3.3 $e + H_2^+$

At low energies the dominant intermediate state is believed to be the $(1\sigma_u)^2\ {}^1\Sigma_g^+$ state of H_2, whose potential curve crosses that of H_2^+ near 2.8 a_o, well outside the range of ground-state vibrations in H_2^+. The other doubly-excited states cross at even larger separations. Although the potential curve for this state is well known (see e.g. Bottcher 1976) the width $\Gamma(R)$ has not yet been established. Unpublished calculations by Robb give a width of $\sim$ 1.3 eV (5 x 10^{-2} a.u.) at R = 2 a_o which increases with increasing R. The values at large R appear to be greater than those derived from an analysis of the double minima in the excited $^1\Sigma_g^+$ potential curves by Bardsley et al. (1980), and further study is warranted. Calculations incorporating Robb's widths show the recombination cross section to be large at thermal energies only for v $\geq$ 2. Discontinuities are observed at the vibrational excitation thresholds, and the indirect process leads to significant enhancement of the ground state recombination cross sections at the resonant energies. When averaged over several vibrational levels the cross sections below 0.3 eV are close to the values measured by Auerbach et al. (1977), but above 0.3 eV the calculated cross sections are somewhat smaller.

The state $(1\sigma_u)(2\sigma_g)\ ^1\Sigma_u^+$ has been examined by several groups, and again there is little disagreement concerning the real part of the potential curve. There is encouraging qualitative agreement for the width also. Kirby et al. (1979) use the Fermi golden rule with a two-center Coulomb function to represent the continuum function and find the width to be 1.4 x 10^{-2} a.u. at R = 2.0. By the Stieltjes technique, in which this continuum is replaced by a set of discrete states, Hazi (1978) finds a corresponding value of 2.7 x 10^{-2} a.u., and the Kohn variational calculation by Takagi and Nakamura (1980) gives 3.4 x 10^{-2} a.u. In unpublished two state close-coupling calculations Robb obtained a value of 2.5 x 10^{-2} a.u. For this state also the width increases with R and approaches a finite value at the stabilization point, R_S.

For the lowest two $^1\Sigma$ states the widths are large and there is severe competition between autoionization and dissociation. Indeed most of the molecules that are formed in these states will decay by electron emission. In this situation, as pointed out by Zhdanov and Chibisov (1978) the higher members of the same Rydberg sequence $(1\sigma_u)(n\ell)$ may be important in the recombination of electrons with energy of several eV, or for slower electrons colliding with ions in highly excited vibrational states. One would then expect to find significant fractions of highly excited atoms among the dissociating H atoms, although whether these states can easily dissociate to the limit H + $H^*(n)$ (rather than to H^{-*} + H^+) has been questioned (M. Barat, private communication).

The triplet states, such as $(1\sigma_u)(2\sigma_g)\ ^3\Sigma_u$ appear to have smaller autoionization widths. Thus although fewer H_2 molecules are formed in this state there is a much greater probability that dissociation will result.

3.4 $e + CH^+$

There has been considerable discussion over the last ten years about the magnitude of the rate of recombination of slow electrons with CH^+ ions in the ground state. The controversy arose out of the models of interstellar clouds (see e.g. Solomon and Klemperer, 1972), in which it seemed that the observed densities of CH^+ molecules could be explained only if the rate of dissociative recombination is less than 10^{-9} cm^3 s^{-1}, at temperatures of around 100 K. This rate is at least two orders of magnitude smaller than the typical rate for diatomic molecules. There appear to be at least two other molecules, H_2^+ and He_2^+, for which the recombination rate under the same conditions is very small, but calculations by Krauss and Julienne (1973) and by Bardsley and Junker (1973) indicated that there is an autoionizing state with a favorable curve crossing and that the rate should not be anomalously low for CH^+. Giusti-Suzor and Lefebvre-Brion (1977) have criticized these calculations, with a very plausible argument, and conclude that the rate may indeed be small. However Mitchell and McGowan (1978) have measured the cross section in a merged beam experiment. Their result leads to a rate constant of 3×10^{-7} cm^3 s^{-1} at 120 K. Meanwhile the astrophysicists have searched for more efficient sources of CH^+ (see e.g. Elitzur and Watson 1978), but the problem has not yet been clearly resolved.

3.5 Recombination with Ion Clusters

Through microwave studies of afterglows, Huang et al. (1976) have measured the recombination coefficients for ion clusters of ammonia, $NH_4^+ \cdot (NH_3)_n$ with $0 \leq n \leq 4$. Previously Leu et al. (1973) had studied the recombination coefficients for water clusters, $H_3O^+ \cdot (H_2O)_n$ with $1 \leq n \leq 5$, over a region of temperatures between 300 K and 6000 K. At room temperature the rates are very large, in the range $1 - 5 \times 10^{-6}$ cm^3 s^{-1}. This is not unexpected since the large size of the clusters facilitates electron trapping and the indirect process may be enhanced, as discussed above. However the remarkable feature of their results is that they find a very weak temperature variation of the recombination coefficient in the clusters with $n \geq 1$, for $300 \lesssim T \lesssim 6000$ K. In order for such high recombination rates to persist up to temperatures around 5000 K, one requires cross sections of the order of 10^{-13} cm^2 at energies of 0.5 eV. This is an enormous cross section for electrons. It could be achieved, for example, if all partial waves with $\ell \leq 6$ were completely absorbed by the ion.

In spite of studies of the indirect process by Smirnov (1977) and Bottcher (1978), and the extended discussion of the direct process by Bates (1979), no quantitative explanation of this phenomena has yet emerged.

4. MUTUAL NEUTRALIZATION

4.1 Introduction

Let us now consider the simplest ion-recombination process, in which during the collision of a positive and negative ion an electron is exchanged without the emission of radiation

$$A^+ + B^- \to A + B \tag{4.1}$$

This process is a form of charge transfer and the general techniques for ab initio calculations of charge transfer cross sections have been discussed in other lectures. However the application of these methods is very difficult. In this lecture we will point out some of the problems that are encountered when one attempts a direct solution, and discuss some of the simple models that have been devised to provide estimates of the cross sections. We will confine our attention to low energies ($E \ll 1$ keV) since our major interest is in thermal collisions.

For low energy collisions it is useful to consider the states of the composite molecule AB. If we imagine a set of diabatic molecular states, which retain their ionic or covalent character at large R, then one finds that the potential curves corresponding to $A^+ \cdot B^-$ are strongly attractive, whereas those corresponding to the states $A^* \cdot B$ are only weakly attractive at long range. Let us suppose that for a particular pair of final states, say $A^* + B$, the reaction (4.1) is exothermic with an energy defect of ΔE. Then one expects to find a curve crossing at an internuclear distance R_x very close to $1/\Delta E$.

As an example let us consider

$$H^+ + H^- \to H^*(n\ell) + H \tag{4.2}$$

This reaction is exothermic when $n \leq 4$, and the crossings occur near 10 a_0 ($n = 2$), 35 a_0 ($n = 3$) and 300 a_0 ($n = 4$). However the positions of these crossings are very sensitive to the energy assumed for the H^- ion. Hence even for this system the performance of a completely ab initio calculation is difficult, and it may be necessary to incorporate a model potential for the H^- ion in order to obtain the correct asymptotic forms for the potential curves involved.

It seems plausible that the crossing distance for the states within $n = 4$ is so large that they can be neglected. Perhaps then, in our expansion of the total wave function over molecular states, one need only retain six Σ, three Π and one Δ state. Unpublished calculations by Browne and Victor have suggested that rotational coupling may be important and so all 10 states may have to be included simultaneously. In the higher molecular states the outer electron is moving extremely slowly and momentum transfer factors may have to be introduced even for relatively low collision velocities. Finally the choice of molecular states and potential curves becomes difficult at small R because of the strong interactions between the Rydberg and non-Rydberg states of H_2 and the possibility of autoionization.

In view of the problems that are anticipated for the simplest reaction of this kind, examination of simple models is clearly worthwhile. However one should be suspicious about the results of all calculations of these processes, and should never regard agreement between theory and experiment on a small number of reactions as a proof of the validity of a model. Instead one should only make relative judgements about the apparent reliability of different models and try to assess the range of validity of each.

4.2 The Landau-Zener Model

The earliest analyses of many charge exchange reactions at low energy were based on the assumption that the electron transfer occurs only at separations close to the crossing point R_x of the initial and final molecular potential curves. It should be stressed that the electronic matrix element $H_{12}(R)$ that describes the transition does not peak at the crossing point, indeed it usually decreases monotonically with R, but that an electronic transition can occur near R_x with minimal disturbance to the nuclear motion. The approximation is therefore within the spirit of the Born-Oppenheimer approximation.

Let us imagine that we have two diabatic states and that the diagonal elements of the electronic Hamiltonian H^{el} for these states are $H_{11}(R)$ and $H_{22}(R)$. If we now look for adiabatic states, by diagonalizing H^{el} with respect to these two states one obtains the adiabatic potential curves

$$E(R) = E_o(R) \pm \sqrt{\Delta^2(R) + H_{12}{}^2(R)}$$

$$\text{with } E_o(R) = \tfrac{1}{2}\,(H_{11}(R) + H_{22}(R)) \qquad (4.3)$$

$$\text{and } \Delta(R) = H_{11}(R) - H_{22}(R)$$

In the usual Landau-Zener theory one assumes that $\Delta(R)$ is a linear function of R

$$\Delta(R) = (R-R_x)\,\Delta'(R_x) \qquad (4.4)$$

and that $H_{12}(R)$ can be replaced by $H_{12}(R_x)$. The minimum separation between the two adiabatic curves is then 2 $H_{12}(R_x)$ and occurs at $R=R_x$.

According to the Landau-Zener theory, the probability of a transition from one adiabatic state to another is given by

$$p = \exp\left(- \frac{2\pi \; H_{12}{}^{2}(R_x)}{v_x \; \Delta'(R_x)} \right) \qquad (4.5)$$

in which v_x is the relative radial velocity of the nuclei at R_x. For a collision with a given impact parameter b the probability of charge exchange is then given by

$$P_{ex}(b) = 2p(1 - p) \qquad (4.6)$$

and the total cross section is

$$Q_{ex}(E) = 2\pi \int_{o}^{b_x} b \; P_{ex}(b) \; db \qquad (4.7)$$

The upper limit arises because there is a maximum impact parameter for which the crossing point is traversed that is given by

$$b_x = R_x \left[1 + \frac{\Delta E}{E} \right]^{\frac{1}{2}} = R_x \left[1 + \frac{1}{R_x E} \right]^{\frac{1}{2}} \qquad (4.8)$$

in which E is the relative kinetic energy and ΔE is the energy defect.

At large R the matrix element $H_{12}(R)$, that describes the probability of an electron jump from one center to another, can be calculated by the theory of asymptotic interactions (see e.g. Smirnov, 1965, and Janev, 1976). However, Olson, Smith and Bauer (1971) have suggested the use of an empirical formula

$$H_{12}(R_x) = \frac{\alpha\gamma}{2} \beta R_x \exp(- 0.86 \; \beta R_x) \qquad (4.9)$$

in which $\frac{1}{2}\alpha^2$ is the electron affinity of B (in a.u.), $\frac{1}{2}\gamma^2$ is the ionization energy of A, and $\beta = \frac{1}{2}(\alpha + \gamma)$. In comparing this formula with the values derived from calculations and from fitting to cross section data they found that "83% of the data were within a factor of 3 to this line". Although this result is undoubtedly of value in estimating cross sections, it should be stressed that this matrix element is squared, divided by a small number and then exponentiated in the calculation of the probability of charge transfer. An error of a factor of 3 in H_{12} could therefore be serious.

Before we turn to more complicated problems let us make a brief empirical analysis of the reaction (4.2) at low energies. For a relative energy of 1 eV the cross section, as measured by Moseley, Aberth and Peterson (1970), is $\sim 600\ \pi\ a_o^2$. Theory suggests that at such a low energy the n = 3 crossing at $R = 36\ a_o$ is most important. The maximum impact parameter at which this crossing can be reached is $\sim$ 48 a.u., and the average probability of charge transfer is thus $\sim \frac{1}{4}$. Using the asymptotic formula of Smirnov, Olson et al. (1970) were able to get very good agreement with experiment for this case for all energies below 10 eV. Above 10 eV transitions at smaller R seem to be important.

The review by Moseley, Olson and Peterson (1975) contains a comparison of the results of Landau-Zener calculations with experiment for several systems involving atomic ions. Later we will discuss briefly some of the reactions not considered in that review.

At thermal energies, the radial velocity of the nuclei near the crossing point is determined mainly by the amount of acceleration arising from the Coulomb interaction and is often insensitive to the initial velocity. The probability of charge exchange then varies only slightly with b and

$$Q_{ex}(E) \approx \pi\ R_x^{\ 2} \left(1 + \frac{\Delta E}{E}\right)\ P_{ex}(o) \tag{4.10}$$

A much more careful analysis of this kind is outlined by Moseley et al. (1975). Again the behavior at zero energy is consistent with the 1/E threshold law.

4.3 The Absorbing Sphere and Optical Potential Models

For collisions involving molecular ions there are many possible final states and crossing points and a statistical treatment is desirable. Olson (1972) has developed a very simple model in which he assumes that the reaction proceeds with unit probability if the nuclear separation becomes less than a critical value R_c. This is a reasonable assumption if for $R \lesssim R_c$ the electron can easily jump from one atom to the other and if there is a very large ratio of the density of states of covalent configurations to that of ionic configurations. The neutralization cross section is then given (in a.u.) by

$$\sigma(E) = \frac{\pi R_c}{E} + \pi R_c^{\ 2} \tag{4.11}$$

Using a somewhat arbitrary modification of two-state Landau-Zener theory, Olson suggested that R_c should be chosen so that

$$M^{0.25}\ R_c^{1.25}\ H_{12}(R_c) = 0.184 \tag{4.12}$$

in which M is the reduced mass of the ion pair in a.u. The critical radius and cross section can then be evaluated for any system using eq. (4.9)

Hickman (1979) has extended this model by assuming a complex optical potential to describe the motion of the nuclei while the electrons are in their original ionic configuration. The imaginary part of this potential is

$$\Gamma(R) = 2\pi\rho\ |H_{12}(R)|^2 \tag{4.13}$$

in which ρ is the density of final states. For atomic ions ρ is given by

$$\rho = [2(\frac{1}{R} + E_B)]^{-3/2} \tag{4.14}$$

where E_B is the electron affinity of B. Hickman allows for exchange of energy with the internal rotational and vibrational modes using semi-classical perturbation theory.

The basic assumption in this model is still that, once the extra electron jumps from the negative ion to the positive ion, it does not return. However the probability of such a jump is no longer represented by a simple step function. From an analysis of his results, Hickmann suggests the following universal formula for neutralization involving complex clusters.

$$\alpha = 2.28 \times 10^{-5}\ (\frac{T}{300})^{-0.5}\ M^{-0.5}\ (E_B)^{-0.4}\ cm^3\ s^{-1} \tag{4.15}$$

when E_B is in eV. The experimentalists have joined the trail, and Miller (1980) has endorsed this kind of formula, but suggests the use of slightly different powers.

4.4 Ion-ion Neutralization in Flames: Na Cℓ and KCℓ.

The current interest in combustion problems has led to several investigations of ion-ion recombination in alkali-halide systems at temperatures around 2000 K. Burdett and Hayhurst (1979) found that most of their data on two-body recombination rates shows a strong correlation between the reaction rate and the position of the crossing point, as expected from the Landau-Zener theory. There are several conspicuous exceptions, one being KCℓ. In contrast NaCℓ behaves normally, and so we will compare recent calculations on KCℓ and NaCℓ. The measured recombination rates at 1800 K are 1.6×10^{-10} $cm^3\ s^{-1}$ and 6.0×10^{-10} $cm^3\ s^{-1}$, respectively.

In determining the matrix elements $H_{12}(R)$, ab initio calculations will be very poor unless the energy defect for the reaction is predicted accurately. More accurate results might be obtained by using asymptotic interaction theory (Janev and Radulovic 1977) or a pseudo-potential model which guarantees the correct dissociation limits (Grice and Herschbach 1974, Adelman and Herschbach 1977). These latter authors have discussed the merits of allowing for the polarization of the negative ion. If one includes an unscreened polarization term ($-\alpha/2R^4$) in the ion-ion potential curve, one can make serious errors in estimating the position of crossings that occur at small R. However, the polarizabilities of negative ions are so big that complete neglect of the effects of polarization may lead to significant errors also.

Neglecting polarization, Janev and Radulovic (1977) find the crossing points to be at 17.8 a.u. for NaCℓ and 37.4 a.u. for KCℓ. The splittings $\Delta(R_x)$ are then very different, 1.01×10^{-4} a.u. for NaCℓ and 8.87×10^{-9} a.u. for KCℓ. The crude pseudopotential model of Grice and Herschbach gives splittings of 1.8×10^{-4} and 1.4×10^{-8} respectively. Using the Landau-Zener model, Janev and Radulovic calculate the recombination rate for NaCℓ formation to be 4.5×10^{-10} cm^3 s^{-1}, in good agreement with experiment, but their value of 5.0×10^{-17} cm^3 s^{-1} for KCℓ at the same temperature is over 6 orders of magnitude smaller than the experimental value. Olson (1977) has performed two-state close-coupling calculations with coupling matrix elements $H_{12}(R)$ obtained from parametric fits to the results of Grice and Herschbach (1974). His calculated rate, 8.8×10^{-11}cm^3 s^{-1} is approximately half the measured rate. These results, if confirmed, suggest that the error arising from the Landau-Zener model may be very much larger than has been anticipated for transitions with crossings at large R. Further experimental and theoretical work would be worthwhile.

5. RADIATIVE ASSOCIATION

The radiative association process (1.1) occurs most readily when, in the collision between atoms A and B, an electronically excited state of AB is formed that can decay to a lower state by a dipole-allowed transition. An important example of this process is

$$C^+(^2P_{1/2}) + H(^2S) \rightarrow CH^+(A^1\Pi)$$

$$\rightarrow CH^+(X^1\Sigma) + h\nu \qquad (5.1)$$

Bates (1951) calculated the rate of this reaction by a semi-classical technique in which the motion of the nuclei was described classically. This approach was developed further by Giusti-Suzor, Roueff and van Regemorter (1976). The probability for photon emission during a collision with relative energy E and impact parameter b is given by

$$P(E,b) = 2\,g(E,b) \int_{R_T}^{\infty} A(R) \left|\frac{dt}{dR}\right| dR \qquad (5.2)$$

in which A(R) is the Einstein coefficient. This is calculated from

$$A(R) = \frac{4}{3\,\hbar^4 c^3} \,|D(R)|^2\, |\Delta V(R)|^3 , \qquad (5.3)$$

where D(R) is the electric-dipole matrix element and $\Delta V(R)$ is the difference between the potential curves for the initial and final molecular states. g(E,b) gives the probability that the molecule is in the radiating state during the collision and R_T is the classical turning point. Integration over b and E leads to reaction rates of the order of 10^{-17} cm^3 s^{-1}.

A fully quantal theory was developed by Smith et al. (1973) and Abgrall et al. (1976). The cross section is written as

$$\sigma(E) = \frac{8}{3h^3c^3}\,\frac{\pi}{k^2}$$

$$\times \sum_{v\ell} g_\ell(E)(2\ell + 1)(\Delta E_v)^3 \left| \int P_{v\ell}(R)D(R)\,F_{E\ell}(R)dR\right|^2 \qquad (5.4)$$

in which k is the wave-number for the relative motion, $F_{E\ell}(R)$ is the initial continuum wave function for the final molecular state and ΔE_v is the energy carried away by the emitted photon.

The quantal approach clearly takes account of the discrete vibrational spectrum in the molecule being formed. More significantly, it also includes a better treatment of the effects of the centrifugal barrier, since reflection can occur at energies just above the barrier and tunnelling is allowed for energies just below the barrier height. This tunnelling leads to the existence of predissociating states of the molecule which act as resonances in the association processes. The role of these resonances has been discussed by Bain and Bardsley (1972) and by Carrington (1972). Study of these resonances is especially important in CH formation at low energies since there appears to be a potential barrier hindering the approach of C and H along the upper potential curve even for $\ell = 0$.

The most difficult aspect of this calculation for CH^+ is the determination of $g_\ell(E)$, the probability that the A $^1\Pi$ state is formed during the collision. If the C^+ ion is initially in the ground $^2P_{1/2}$ state and the collision is adiabatic then the ground state of $^2CH^+$ must be formed and $g_\ell(E) = 0$. However, according to close coupling calculations by Weisheit and Lane (1971), non-adiabatic effects are important even at thermal velocities. Solomon and Klemperer (1972) have pointed out that the $A^1\Pi$ state of CH^+ may support predissociating levels between the $C^+(^2P_{1/2})$ and $C^+(^2P_{3/2})$ dissociation limits and that these levels will also act as resonances in C^+ + H scattering. These states have been observed in photodissociation by Cosby et al. (1980), but their effect on the association rate has not yet been determined accurately.

Ramaker and Peek (1976) have studied a simpler system

$$H^+ + H(1s) \rightarrow H_2^+\,(^2\Sigma_u) \rightarrow H_2^+(^2\Sigma_g) + h\nu \qquad (5.5)$$

for which $g_\ell(E) = 0.5$ at low and intermediate energies. Since the $^2\Sigma_u$ potential is repulsive and the probability of photon emission decreases rapidly with increasing R, the effects of tunnelling are particularly important and quantal calculations give larger association rates than are obtained by the semi-classical method.

Frommhold and Pickett (1978) have studied the formation of HeH^+ in H^+ + He collisions. Since there is only one molecular state connected to this dissociation limit, the photon emission arises from an infra-red transition within the ground electronic state. The rate is several orders of magnitude smaller than that for CH^+ formation. Frommhold and Pickett also examined the formation of HD^+ in H + D^+ and H^+ + D collisions. Their treatment of the non-adiabatic effects at large R was very simple and this problem deserves further study.

There have been no direct measurements of the rate for radiative association, although studies of the inverse process of photodissociation provide checks on these calculations. Also the analogous process of radiative charge transfer

$$He^{++} + He \rightarrow He^+ + He^+ + h\nu$$

has been studied by Cohen and Bardsley (1978), using the complex potential method. The results are in very good agreement with the measurements of Johnsen and Biondi (1978).

6. ELECTRON-ION RECOMBINATION AT HIGH DENSITIES

6.1 Collisional-Radiative Recombination in an Ambient Electron Gas

The rate of radiative transitions involving highly excited states is so low that excitation and deexcitation through electron impact competes even at moderate densities. The process of electron-ion recombination through the combined effects of radiative emission and absorption and collisions with ambient electrons is called collisional-radiative recombination. A single recombination event may involve many steps and can almost be regarded as a random walk through the excited states of the atom that is being formed.

In full thermodynamic equilibrium the ratio of the number density n(p) of atoms with principal quantum number p to the number density of free electrons is given by the Saha equation

$$\frac{n(p)}{n(e)} = \frac{n_s\ (p)}{n(e)} \equiv p^2\left(\frac{2\pi\hbar^2}{mkT}\right)^{3/2} \exp\left(\frac{me^4}{2\hbar^2p^2kT}\right) n(e) \qquad (6.1)$$

During recombination the excited state populations n(p) vary slowly in comparison with the recombination time and one often writes

$$n(p) = b(p)\ n_s(p) \qquad (6.2)$$

in which b(p) is called the Saha decrement. By equating the rates of formation and destruction of atoms in level p one can obtain a set of linear equations for n(p). There are often many equations in this set, but the number can be made finite because $b(p) \to 1$ as $p \to \infty$. The major uncertainty in setting up the equations lies in the choice of the coefficients K(p, q) for collisions leading to a transition from state p to q.

Byron, Stabler and Bortz (1962) have suggested that the rate of recombination is effectively controlled by the number of downward transitions across a defile, or bottleneck. This arises because the deexcitation rate has a minimum near $p = p^*$. Using cross sections obtained from Monte-Carlo trajectory calculations, Mansback and Keck (1969) estimate that

$$p^* \simeq 10 \left(\frac{410}{T}\right)^{0.5} \qquad (6.3)$$

This feature suggests ways of avoiding the full solution of the large number of coupled equations when one needs only approximate results, but also indicates which values of p should be kept in mind when one decides upon a method for evaluating the collision rates.

In the first comprehensive calculations of the rate of collisional-radiative recombination, Bates, Kingston and McWhirter (1962a, b) used collision rates obtained by the classical impulse approximation. Later Stevefelt, Boulmer and Delpech (1975) performed further calculations using the Mansbeck-Keck cross sections, whereas semi-empirical cross sections were adopted by Johnson and Hinnov (1973), and by Drawin and Emard (1978). Most of these authors have performed calculations for several different optical conditions, since the recombination rate is reduced if emitted radiation is reabsorbed.

A thorough discussion of the construction and solution of the equations relating the excited state populations has been given by Burgess and Summers (1977) and by Summers (1978). These authors consider the effects of external radiation fields and examine the extent to which one needs to distinguish between states of the same principal quantum number but different angular momentum.

We have seen in section 2 that dielectronic recombination in tenuous plasmas leads to very large Saha decrements. Thus collisions with ambient electrons could have a large effect on these populations and thus on the recombination rate. These effects have been reviewed by Seaton and Storey (1976). For example, for recombination of C^{2+} at 10^5 K the recombination rate is reduced from 10^{-10} cm^3 s^{-1} for $N_e = 0$ to 10^{-11} cm^3 s^{-1} at $N_e = 10^{10}$ cm^{-3} and 10^{-12} cm^3 s^{-1} at $N_e = 10^{14}$ cm^{-3}. More recently Jacobs and Davis (1978) have studied the recombination of multiply-charged Fe ions for N_e up to 10^{16} cm^{-3}. Weisheit (1975) has discussed some of the additional effects that occur in plasmas produced by laser irradiation, such as continuum lowering.

6.2 Deexcitation by Collisions with Neutral Atoms and Molecules

For low electron densities the neutral atoms or molecules may be important in absorbing energy from the recombining particles. Although the Thomson theory of ion-ion recombination had been adopted to treat electron-ion recombination, the earliest theory specially developed for this problem was by Pitaevskii (1962). Since an electron usually suffers a large change in momentum but a small change in energy when it collides with a neutral atom, the time required for the electron energy distribution to reach equilibrium is much longer than for the relaxation of other variables. This led Pitaevskii to regard the recombination process in terms of diffusion through energy space and to apply the Fokker-Planck equation. The relationship between the Thomson and Pitaevskii approaches has recently been investigated thoroughly by Bates (1980b) and so neither will be described in detail here.

In 1965, Bates and Khare adapted the theory of collisional-radiative recombination to allow for stabilization through collisions with neutral atoms. They used the bottleneck concept, as developed by Bates and Kingston (1964), and calculated the deexcitation cross sections by the binary encounter method. For alkali ions at 250 K in helium, they found three body recombination rates of $\sim 10^{-27}$ cm^6 s^{-1}. The rate decreases with electron temperature and the rate of $\sim 4 \times 10^{-29}$ cm^6 s^{-1} recently measured for Cs^+ in He at 625 K by Gousset, Sayer and Berlande (1977) seems consistent with their theory. The Pitaevskii theory predicts a rate of 4×10^{-28} cm^6 s^{-1}. In contrast, the measured rate for He^+ and He_2^+ at 300 K, in a gas of He, is larger than the values calculated by both theories. It is possible that the excitation of vibrational motion in He_2^+ may be responsible for the enhanced recombination there, and that the long-range resonance interactions may be important for He^+, but further theoretical work on the helium systems is needed.

Bates, Malaviya and Young (1971) have demonstrated that molecules can be much more efficient absorbers of energy, through rotational or vibrational excitation. Water vapour is particularly potent, and three-body rates over 10^{-27} cm^6 s^{-1} are found in flames containing water vapour, even at 2000 K. For N_2 at 250 K the recombination coefficient is 1.1×10^{-27} cm^6 s^{-1}.

6.3 Collisional-Dissociative Recombination

Collins (1965) has considered recombination in molecules which are not subject to dissociative recombination, because of the absence of suitable autoionizing states that can act as intermediate states in the reactions. This situation may pertain for alkali-inert-gas systems, for heteronuclear inert-gas systems like $XeNe^+$, and possibly for He_2^+ in the ground vibrational state. In this case the recombination would proceed through the standard collisional-radiative process until a state is reached with a repulsive potential curve that leads to dissociation. The dissociation should have a relatively small effect on the overall rate of collisional-radiative recombination, since it will usually take place for levels below the bottleneck, but the determination of the atomic states produced in this mode of recombination presents an interesting problem.

Bates (1980a) has recently examined dissociative recombination at high electron densities. For an ion such as Ar_2^+, he finds that the collisional effects become important for $N_e \gtrsim 10^{12}$ cm^{-3} at 250 K and for $N_e \gtrsim 10^{15}$ cm^{-3} at 2000 K.

7. HEAVY PARTICLE RECOMBINATION AT HIGH DENSITIES

7.1 Ion-Ion Recombination

Thomson (1924) presented a classical model for ion-ion recombination at low pressures that has survived for many years despite the development of quantum mechanics. As a particular pair of oppositely charged ions approach one another their kinetic energy is increased by the Coulomb force. If one of the ions collides with a neutral atom or molecule it will, on average, lose energy and the pair may be left with insufficient kinetic energy to separate. If the collision occurs at an ion separation R, the average kinetic energy of relative motion before the collision is

$$K_i = \frac{3}{2} kT + \frac{e^2}{R} \quad . \tag{7.1}$$

Let us express the average relative kinetic energy of the ion-pair after collision in the form

$$K_f = \frac{3}{2} kT + \frac{e^2}{R} \quad (1-F) \tag{7.2}$$

Thomson assumed that F = 1, which seems to lead to good results but has not been justified in terms of collision dynamics. Wadehra and Bardsley (1978) have calculated F, with the assumptions that the ion-neutral scattering is elastic and isotropic and the probability that the ions undergo collisions at a separation R is independent of their velocities. When all three particles have equal mass one finds F = 3/8. If K_f is less than e^2/R then the average ion-pair will not be able to separate (unless they pick up energy from a second collision). They will become trapped if

$$R \ < R_T = \frac{2}{3} \frac{e^2}{kT} F \tag{7.3}$$

Thus if a collision occurs while the ion-pair separation is less than the trapping radius, the ion-pair becomes temporarily bound, and in the Thomson model it is assumed that they will recombine. For low densities, with F taken as one, the recombination rate then becomes

$$\alpha_T = \frac{32}{27} \frac{\pi e^6}{(kT)^{5/2} (3M)^{1/2}} \left[\frac{1}{\ell^+} + \frac{1}{\ell^-}\right] \tag{7.4}$$

in which ℓ^+ and ℓ^- are the mean free paths for ion-neutral collisions and M is the reduced mass of the ion pair. This two-body recombination rate is proportional to the neutral gas density.

We have already demonstrated one way in which the Thomson model can be "improved". Many similar modifications have been suggested, most intended to extend the range of validity of the model to higher densities. Many of these papers are cited in the reviews of Flannery (1976) and Bates (1979). However, since the model is based on the properties of the "average" ion-pair and the effects of deviations from the average behavior are not considered, it seems unwise to increase the complexity of the model.

Bates and Moffett (1966) carried out a more rigorous analysis for the systems

$$X^+ + X^- + X \rightarrow X_2 + X \tag{7.5}$$

in which they incorporate the collision dynamics appropriate to heavy particle collisions into the theory of electron-ion collisional recombination and calculate the quasi-static energy distribution for the ion pairs. They allow for symmetric charge transfer which provides an efficient mechanism for energy exchange between the ions and neutral atoms. Bates and Flannery (1968) applied the same method to the general case, with three different atomic species, using an ion-atom interaction with a long-range polarization attraction and a hard core at a short distance. Their results confirmed the predictions of the simple Thomson theory to within a factor of 2.

At large pressures one cannot assume that the ions move in freely to a separation such as R_T and must consider the effects of multiple collisions. The approach of two ions becomes limited by the ionic mobility and once two ions get close together (say $R < 200\ a_0$) they rarely escape. Langevin (1902) showed that the high pressure limit of the effective two-body recombination rate is

$$\alpha_L = 4\pi e\ (K^+ + K^-) \tag{7.6}$$

in which K^+ and K^- are the low-field mobilities of the positive and negative ions in the ambient gas. Since these are inversely proportional to gas density, so is the recombination rate and there must be a maximum in the rate at some intermediate pressure.

Although Natanson (1959) and others have developed analytic models applicable to the whole range of gas densities, their validity remains in doubt and numerical calculations appear to be necessary. Two programs have been developed for Monte Carlo simulation of ion-ion recombination. This approach is very powerful since, in principle, one can incorporate any description of the ion-neutral collision dynamics and allow for inelastic collisions, gas mixtures, and plasma screening effects. However the initial calculations have assumed isotropic elastic collisions with simple cross sections.

In the code developed by Bates and Mendas (1977) an attempt is made to follow each ion-pair from the first time that they approach one another in order to calculate the probability that they will recombine. A large separation R_0 is chosen such that the motion of ions with separations greater than R_0 is almost exactly thermal, with uniform ion density and a Maxwellian distribution of velocities. At the beginning of each computer run, an ion-pair is moving inward with separation R_0 and it is assumed that their separation has not previously been less than R_0. The computer follows this pair through many collisions and records whether the pair recombine (i.e. attain a large negative energy) or separate to infinity. Since collisions are relatively frequent, the program must allow for subsequent approaches of pairs that separate beyond R_0. In contrast, Bardsley and Wadehra (1980) study all ion pairs whose separation becomes less than R_0 and determine the probability that each such pair recombines before the separation next exceeds R_0. The advantages of this latter approach are that one does not need to worry about subsequent approaches of separating pairs and that the velocity distribution of all ion-pairs at R_0 is much closer to Maxwellian than that of ions that are on first approach. The results obtained by the two methods cannot be compared directly, since they have been applied to different systems with different collision cross sections.

In the Monte-Carlo method it is possible to incorporate mutual-neutralization, which is essentially a quantum process that was not considered by Thomson. Bates (1979b) has pointed out that the mutual-neutralization process may be enhanced by the presence of an ambient gas, since any temporary loss of energy of the ion-pair to the neutral atoms reduces the velocity of the ions in the region where the potential curves cross and often leads to many traversals of this region. Thus as the pressure is increased one expects to find a smooth transition from the rate for the true two-body process to the many-body ion-ion recombination rate. This effect may be particularly important in flames, since at high temperatures the two-body and many-body processes may have similar rates at atmospheric pressure (Bates 1980c).

7.2 Ion-Ion Recombination in Rare-Gas-Halide Lasers

Ion-ion recombination in KrF and XeCℓ often leads directly to the radiating states in these powerful gas lasers. Knowledge of the rate of this process also is important in calculations of the charge density in laser plasmas. Analytic models have been applied to these systems by Flannery and Yang (1978a, b) and by Wadehra and Bardsley (1978), and the Monte Carlo method has been applied by Bardsley and Wadehra (1980). At 300 K the two-body recombination rate peaks at $\sim 2 \times 10^{-6}$ cm^3 s^{-1} at a pressure of around 2 atm. However in plasmas with a high density of charged particles ($N_+ \gtrsim 10^{13}$ cm^{-3}) these rates may be reduced because of the screening of the Coulomb interactions. A discussion of the atomic collision processes pertinent to rare-gas-halide lasers has been given by Flannery (1979), who has also studied ion-ion recombination in mercury-halide lasers (Flannery 1978).

7.3 Recombination of Neutral Atoms

To most chemists the most important recombination process is that in which neutral atoms recombine to form molecules, e.g.

$$H + H + M \rightarrow H_2 + M \ . \tag{7.7}$$

Classical statistical theories of this process (e.g. Shui and Appleton 1971) give results in qualitative agreement with experiment. However, as in radiative association, the presence of centrifugal barriers leads to resonances. Although the resonances are quantum phenomena, their formation and decay can be treated semi-classically (Whitlock, Muckerman and Roberts 1974). The importance of these resonances has not yet been determined. The theoretical arguments in favour of a resonance model for recombination seem strong, but the experimental evidence on the ortho-para ratio among the molecules created in this way supports the statistical theory (Walkauskas and Kaufman (1976)).

Atom-atom recombination may proceed through an alternative "bound complex" mechanism involving two atom-transfer reactions. For recombination of atoms A and B in a gas of molecules M this process can be written

$$A + M + M \rightarrow AM + M \tag{7.8}$$

$$AM + B \rightarrow AB + M$$

The end result of these two steps is the same as in reaction (1.2). A resonant scattering model for this process was suggested by Pack et al. (1972). Snider (1979) has recently examined the relative magnitude of the processes (1.2) and (7.8) for Br and I atoms in the rare gases. He finds that the former process always dominates at higher temperature, but that the bound complex mechanism may be important in Xe at low temperatures ($\lesssim$ 300 K).

8. Summary

The recombination processes discussed in these lectures are among the most important reactions in the physics and chemisty of gases. They are mostly complex processes that present a formidable challenge to the theorist. However the success of simple models, such as those of Langevin (1902) and Thomson (1924) demonstrates that theory can make useful contributions to our understanding of these processes. In some areas, such as dissociative recombination and mutual neutralization, there are reliable experimental data that can be used to check one's calculations, but for other processes, such as dielectronic recombination and ion-ion recombination in the many-body regime, there is little experimental information and thus there is a greater need for accurate calculations.

Excellent reviews of this field have been given recently by Bates (1979) and Flannery (1976). These articles include discussion of many aspects of the subject that have been omitted in these lectures.

Acknowledgements: The author is very grateful to the Observatoire de Meudon, for their hospitality during a period when most of these lectures were written, and to Ms. Noramae Herald, who prepared the manuscript with the assistance of Dr. J. M. Wadehra.

References:

H. Abgrall, A. Guisti-Suzor and E. Roueff, 1976, Quantum effects in the formation of CH^+ by radiative association, Astrophys. J., 207:L69.

S. Adelman and D. R. Herschbach, 1977, On the long range configuration-interaction between ionic and covalent states, Mol. Phys., 33:793.

D. Auerbach, R. Cacak, R. Caudano, T. D. Gaily, C. J. Keyser, J. W. McGowan, J.B.A. Mitchell and S.F.J. Wilk, 1977, Merged electron-ion beam experiments I. Method of measurements of e-H_2^+ and e-H_3^+ dissociative recombination cross sections, J. Phys. B, 10:3797.

R. A. Bain and J. N. Bardsley, 1972, Shape resonances in atom-atom collisions I. Radiative association, J. Phys. B, 5:277.

J. N. Bardsley, 1968a, Configuration intereaction in the continuum states of molecules, J. Phys. B, 1:349.

J. N. Bardsley, 1968b, The theory of dissociative recombination, J. Phys. B, 1:365.

J. N. Bardsley, 1979, Molecular resonance phenomena, in "Electron-molecule and photon-molecule collisions", T. Rescigno, V. McKoy, B. Schneider, eds., Plenum, New York.

J. N. Bardsley and B. R. Junker, 1973, Dissociative recombination of CH^+ ions, Astrophys. J., 183:L135.

J. N. Bardsley and J. M. Wadehra, 1980, Monte Carlo simulation of three-body ion-ion recombination, Chem. Phys. Lett. (in press).

J. N. Bardsley, C. Derkits and J. M. Wadehra, 1980, Dissociative attachment in H_2, F_2 and $C\ell_2$ and dissociative recombination in H_2^+, to be published.

D. R. Bates, 1951, Rate of formation of molecules by radiative association, Mon. Not. Roy. Astron. Soc., 111:303.

D. R. Bates, 1979a, Aspects of recombination, Adv. Atom. Molec. Phys. 15:235.

D. R. Bates, 1979b, Dependence of the binary ionic recombination coefficient on the density of ambient gas at the low density limit, J. Phys. B, 12:L505.

D. R. Bates, 1980a, Dissociative recombination in an ambient electron gas, J. Phys. B, 13:L51.

D. R. Bates, 1980b, Classical theory of electron-ion recombination in an ambient gas, J. Phys. B, (in press).

D. R. Bates, 1980c, Mutual neutralization of ions in flames. J. Phys. B, 13:205.

D. R. Bates and M. R. Flannery, 1968, Three-body recombination of positive and negative ions II. General third body. Proc. Roy. Soc. Lond. 302:367.

D. R. Bates and S. P. Khare, 1965, Recombination of positive ions and electrons in a dense neutral gas, Proc. Phys. Soc. Lond., 85:231.

D. R. Bates and A. E. Kingston, 1964, Collisional-radiative recombination at low temperatures and densities, Proc. Phys. Soc. Lond., 83:43.
D. R. Bates and I. Mendas, 1978, Ionic recombination in an ambient gas, II. Computer experiment with specific allowance for binary recombination, Proc. Roy. Soc. Lond., A 359:287.
D. R. Bates and R. J. Moffett, 1965, Three-body recombination of positive and negative ions I. Ions recombining in their parent gas, Proc. Roy. Soc. Lond., 291:1
D. R. Bates, A. E. Kingston and R.W.P. McWhirter, 1962a, Recombination between electrons and atomic ions I. Optically thin plasmas, Proc. Roy. Soc. Lond., 267:297.
D. R. Bates, A. E. Kingston and R.W.P. McWhirter, 1962b, Recombination between electrons and atomic ions II. Optically thick plasmas, Proc. Roy. Soc. Lond., A 270:155.
D. R. Bates, V. Malaviya and N. A. Young, 1971, Electron-ion recombination in a dense molecular gas, Proc. Roy. Soc., A 320:437.
I. L. Beigman and B. N. Chichkov, 1980, Dielectronic recombination through the forbidden levels, J. Phys. B, 13:565.
F. Bely-Dubau, A. H. Gabriel and S. Volonte, 1979, Dielectronic satellite spectra for highly charged helium-like ions V. Effect of total satellite contribution on the solar flare iron spectra, Mon. Not. Roy. Astron. Soc., 189:801.
K. Berrington and M. Crees, 1979, R-matrix calculations for electron-molecule scattering, Comp. Phys. Commun., 17:181.
C. Bottcher, 1976, Dissociative recombination of the hydrogen molecular ion, J. Phys. B, 9:2899.
C. Bottcher, 1978, Dissociative recombination of large molecular ions, J. Phys. B, 11:3887.
N. A. Burdett and A. N. Hayhurst, 1979, Determination of the rate coefficients of ion-pair production and ion-ion recombination in flames, Phil. Trans. Roy. Soc. Lond., 290:300.
A. Burgess, 1964, Hydrogenic photoionization cross sections and recombination coefficients, Mem. Roy. Astron. Soc., 69:1.
A. Burgess, 1964, Dielectronic recombination and the temperature of the solar corona, Astrophys. J., 139:776.
A. Burgess, 1965, A general formula for the estimation of dielectronic recombination coefficients in low-density plasmas, Astrophys. J., 141:1588.
A. Burgess and H. P. Summers, 1976, The recombination and level populations of ions I. Hydrogen and hydrogenic ions, Mon. Not. Roy. Astron. Soc., 174:345.
A. Burgess and A. S. Tworkowski, 1976, Dielectronic recombination to form helium-like ions, Astrophys. J., 205:L105.
S. Byron, R. C. Stabler and P. I. Bortz, Electron-ion recombination by collisional and radiative processes, 1962, Phys. Rev. Lett., 8:376.
T. Carrington, 1972, Radiative recombination of atoms as a resonance scattering process, J. Chem. Phys., 57:2033.

J. S. Cohen and J. N. Bardsley, 1978, Calculation of radiative single-charge-transfer cross sections for collisions of He^{++} with He at low energy, Phys. Rev. A, 18:1004.
C. B. Collins, 1965, Collisional-dissociative recombination of electrons with molecular ions, Phys. Rev. A, 140:1850.
P. C. Cosby, H. Helm and J. T. Moseley, 1980, Observation of predissociated levels of $CH^+(A\ ^1\Pi)$ Astrophys. J., 235:52.
J. Davis and J. Lewis, 1973, Continuum emission from recombining oxygen and nitrogen plasmas, J. Quant. Spectrosc. Radiat. Transfer 13:871.
H. W. Drawin and F. Emard, 1978, Ground state populations of hydrogenic atoms in nonthermal plasmas, and collisional-radiative recombination and ionization coefficients, Physica, 94C:134.
J. Dubau and S. Volonte, 1980, Dielectronic recombination and its applications in astronomy, Rep. Prog. Phys., 43:199.
J. Dubau, M. Loulergue and L. Steenman-Clark, 1980, Calculations of the atomic parameters for the dielectronic satellite lines of the Mg X11 resonance line, Mon. Not. Roy. Astron. Soc. 190:125.
M. Elitzur and W. D. Watson, 1978, Formation of molecular CH^+ in interstellar shocks, Astrophys. J., 222:L141.
U. Fano, 1961, Effects of configuration interaction on intensities and phase shifts, Phys. Rev., 124:1866.
F. Fiquet-Fayard, 1974, Theoretical problems in the interpretation of dissociative attachment cross sections, Vacuum, 24:533.
M. R. Flannery, 1976, Ionic Recombination, in "Atomic processes and applications", P. G. Burke and B. L. Moiseiwitsch, eds., North-Holland, Amsterdam.
M. R. Flannery, 1978, Three-body ion-ion recombination in mercury-halide gas lasers, Chem. Phys. Lett., 56:143.
M. R. Flannery, 1979, Atomic and molecular collision processes in rare-gas-halide lasers and rare-gas excimer lasers, Int. J. Quant. Chem. Symp., 13:501.
M. R. Flannery and T. P. Yang, 1978a, Ionic recombination of rare-gas atomic ions X^+ with F^- in a dense gas X, Appl. Phys. Lett., 32:327.
M. R. Flannery and T. P. Yang, 1978b, Ionic recombination of Kr^+ and Kr_2^+ with F^- in dense buffer rare gases, Appl. Phys. Lett., 33:574.
L. Frommhold and H. M. Pickett, 1978, Rates of radiative recombination to form HD^+ and HeH^+, Chem. Phys., 28:441.
J. N. Gau, Y. Hahn and J. A. Retter, 1980, Dielectronic recombination of positive ions III. Rate coefficients for Mo^{32+}, J. Quant. Spectrosc. Radiat. Transfer, 23:147.
A. Giusti-Suzor and H. Lefebvre-Brion, 1977, The dissociative recombination of CH^+ ions, Astrophys. J. 214:L101.
A. Giusti-Suzor, E. Roueff and H. van Regemorter, 1976, Comparison between semi-classical and quantal calculations of the radiative association rate for CH^+, J. Phys. B, 9:1021.

R. J. Gould, 1978, Radiative recombination of complex ions, Astrophys. J., 219:250.
G. Gousset, B. Sayer and J. Berlande, 1977, Electron-Cs^+-ion recombination in the presence of neutral helium atoms, Phys. Rev., A 16:1070.
R. Grice and D. R. Herschbach, 1974, Long range configuration interaction of ionic and covalent states, Mol. Phys., 27:159.
Y. Hahn and D. W. Rule, 1977, Direct radiative capture of high-energy electrons by atomic ions, J. Phys. B, 10:2689.
Y. Hahn, J. N. Gau, R. Luddy and J. A. Retter, 1980, Scaling properties of the dielectronic recombination rate for the Ne-sequence, J. Quant. Spectrosc. Radiat. Transfer, 23:65.
A. U. Hazi, 1978, A purely L^2 method for calculating resonance widths, J. Phys. B., 11:L259.
A. P. Hickman, 1979, Approximate scaling formula for ion-ion mutual neutralization rates, J. Chem. Phys., 70:4872.
C. M. Huang, M. A. Biondi and R. Johnsen, 1976, Recombination of electrons with $NH_4^+\cdot(NH_3)_n$ series ions, Phys. Rev. A 14:984.
V. L. Jacobs and J. Davis, 1978, Effects of collisions on level populations and dielectronic recombination rates of multiply charged ions, Phys. Rev. A, 18:697.
V. L. Jacobs, J. Davis, P. C. Kepple and M. Blaha, 1977a, The influence of autoionization accompanied by excitation on dielectronic recombination and ionization equilibrium, Astrophys. J., 211:605.
V. L. Jacobs, J. Davis, P. C. Kepple and M. Blaha, 1977b, Dielectronic recombination and the ionization equilibrium of silicon ions, Astrophys. J., 215:690.
R. K. Janev, 1976, On the long-range configuration interaction between ionic and covalent states, J. Chem. Phys. 64:1891.
R. K. Janev and Z. M. Radulovic, 1977, Recombination of alkali positive and halogen negative ions, J. Chem. Phys., 67:2856.
P. S. Julienne, J. Davis and E. Oran, 1974, Oxygen recombination in the tropical nightglow, J. Geophys. Res., 75:3862.
K. Kirby, S. Guberman and A. Dalgarno, 1979, Resonant dissociative photoionization of H_2, J. Chem. Phys., 70:4635.
M. Krauss and P. S. Julienne, 1973, Dissociative recombination of $e + CH^+(X\ ^1\Sigma^+)$, Astrophys. J., 183:L139.
P. Langevin, 1902, Recherches sur les gaz ionisés, C.R. Acad. Sci., 134:414.
C. M. Lee and R. H. Pratt, 1975, Radiative capture of high energy electrons, Phys. Rev. A, 12:1825.
C. M. Lee and R. H. Pratt, 1976, Radiative capture of electrons by Mo ions, Phys. Rev. A., 14:990.
M. T. Leu, M. A. Biondi and R. Johnsen, 1973, Measurements of the recombination of electrons with $H_3O^+\cdot(H_2O)_n$ - series ions, Phys. Rev. A, 7:292.
R. Johnsen and M. A. Biondi, 1978, Measurements of radiative-charge transfer reactions of doubly and singly charged rare-gas ions with rare gas atoms at thermal energies, Phys. Rev. A, 18:996.

L. C. Johnson and E. Hinnov, 1973, Ionization, recombination, and population of excited levels in hydrogen plasmas, J. Quant. Spectrosc. Radiat. Transfer, 13:333.

P. Mansbach and J. Keck, 1969, Monte Carlo trajectory calculations of atomic excitation and ionization by thermal electrons, Phys. Rev., 181:275.

T. M. Miller, 1980, Parametrization of ion-ion mutual neutralization rate coefficients, J. Chem. Phys., 72:4659.

J.B.A. Mitchell and J. W. McGowan, 1978, The dissociative recombination of CH^+ X $^1\Sigma^+$ (v = 0), Astrophys. J., 222:L77.

J. T. Moseley, W. Aberth and J. R. Peterson, 1970, The H^+-H^- mutual neutralization cross section obtained with superimposed beams, Phys. Rev. Lett. 24:435.

J. T. Moseley, R. E. Olson and J. R. Peterson, 1975, Ion-ion mutual neutralization, Case Studies in Atomic Physics, 5:1.

G. L. Natanson, 1959, The theory of volume recombination of ions, Sov. Phys.-Tech. Phys., 4:1263.

W. L. Nighan, 1977, Influence of recombination and ion chemistry on the stability of externally sustained molecular discharges, Phys. Rev., A 16:1209.

R. E. Olson, 1972, Absorbing sphere model for calculating ion-ion recombination total cross sections, J. Chem. Phys.,56:2979.

R. E. Olson, 1977, Theoretical chemi-ionization and recombination cross sections for reactions of Na and K with Cℓ, Comb. Flame, 30:243.

R. E. Olson, J. R. Peterson and J. T. Moseley, 1970, Ion-ion recombination total cross sections - atomic species, J. Chem. Phys., 53:3391.

R. E. Olson, F. T. Smith and E. Bauer, 1971, Estimation of the coupling matrix elements for one-electron transfer systems, Appl. Optics, 10:1848.

T. F. O'Malley, A. J. Cunningham and R. M. Hobson, 1972, Dissociative recombination at elevated temperatures II. Comparison between theory and experiment in neon and argon afterglow, J. Phys. B, 5:2126.

R. T. Pack, R. L. Snow and W. D. Smith, 1972, On the mechanism of low-temperature termolecular atomic recombination, J. Chem. Phys., 56:926.

L. P. Pitaevskii, 1962, Electron recombination in a monatomic gas, Soviet Phys. JETP, 15:919.

D. E. Ramaker and J. M. Peek, 1976, Molecule formation in tenuous media:Quantum effects in spontaneous radiative association, Phys. Rev. A, 13:58.

J. C. Raymond, 1978, On dielectronic recombination and resonances in excitation cross sections, Astrophys. J., 222:1114.

M. J. Seaton, 1978, Calculated intensities of He II recombination lines in the ultraviolet, Mon. Not. Roy. Astron. Soc., 185:5p.

M. J. Seaton and P. J. Storey, 1976, Dielectronic recombination, in "Atomic Processes and Applications", P. G. Burke and

B. L. Moiseiwitsch, eds., North Holland, Amsterdam, p. 133.

V. H. Shui and J. P. Appleton, 1971, Gas phase recombination of hydrogen. A comparison between theory and experiment, J. Chem. Phys., 55:3126.

B. M. Smirnov, 1965, Formation and decay of negative ions, Sov. Phys.-Dokl., 10:218.

B. M. Smirnov, 1977, Recombination of an electron and a complex ion, Sov. Phys. JETP, 45:731.

W. H. Smith, H. S. Liszt and B. L. Lutz, 1973, Diatomic processes leading to CH and CH^+ formation in the interstellar medium, Astrophys. J., 153:69.

N. Snider, 1978, Bound complex and triple collision mechanisms for diatom dissociation and recombination, Can. J. Chem., 57:1167.

P. M. Solomon and W. Klemperer, 1972, The formation of diatomic molecules in interstellar clouds, Astrophys. J., 178:389.

J. Stevefelt, J. Boulmer and J. F. Delpech, 1975, Collisional-radiative recombination in cold plasmas, Phys. Rev. A, 12:1246.

H. P. Summers, 1977, The recombination and level populations of ions II. Resolution of angular momentum states, Mon. Not. Roy. Astron. Soc., 178:101.

H. Takagi and H. Nakamura, 1980, Elastic scattering of electrons from H_2^+:Phaseshifts, quantum defects and two-electron excited states, J. Phys. B (in press).

J. J. Thomson, 1924, Recombination of gaseous ions, the chemical combination of gases and monomolecular reactions, Phil. Mag. Ser 6, 47:337.

B. A. Tinsley, A. B. Christensen, J. Bittencourt, H. Gouvela, P. D. Andreji and H. Takahashi, 1973, Excitation of oxygen permitted line emissions in the tropical nightglow, J. Geophys. Res. 78:1174.

J. M. Wadehra and J. N. Bardsley, 1978, Calculations of ion-ion recombination rates at high pressures, Appl. Phys. Lett., 32:76.

L. P. Walkauskas and F. Kaufman, 1976, Measurement of ortho-para ratio in gas phase hydrogen atom recombination, J. Chem. Phys., 64:3885.

J. C. Weisheit, 1975, Recombination in dense plasmas, J. Phys. B, 8:2556.

J. C. Weisheit and N. F. Lane, 1971, Low energy elastic and fine-structure excitation scattering of ground state C^+ ions by hydrogen atoms, Phys. Rev. A, 4:171.

P. A. Whitlock, J. T. Muckerman and R. E. Roberts, 1974, Classical mechanics of recombination via the resonance mechanism: $H + H + M \rightarrow H_2 + M$, J. Chem. Phys., 60:3658.

V. P. Zhdanov and M. I. Chibisov, 1978, Dissociative recombination of electrons with molecular ions involving the formation of excited states, Sov. Phys. JETP, 47:38.

CHARGE EXCHANGE AND IONIZATION IN

ION-ATOM COLLISIONS

Ronald McCarroll

Laboratoire d'Astrophysique
Université de Bordeaux I
33405 Talence, France

ABSTRACT

The general theory of charge exchange in atom-atom and ion-atom collisions is examined at the various levels of approximation that are made possible by comparing collision energies and typical electron binding energies.

The impact parameter approximation and the high energy methods are examined and the corresponding merits of several possible functional expansion methods are discussed.

The molecular state expansion is seen both within the impact parameter and the quantum mechanical formulations, while diabatic representations are also analyzed.

A few important illustrative examples are reported in the last section.

1. INTRODUCTION

2. GENERAL THEORY - BASIC RESULTS

3. IMPACT PARAMETER APPROXIMATION

4. HIGH ENERGY METHODS

5. SOLUTION OF THE EIKONAL EQUATION

   5.1. Atomic eigen-function expansion method
   5.2. Sturmian function basis set

5.3. Pseudo-state expansion
5.3. Scaled hydrogenic basis set - ionization to the continuum

6. MOLECULAR STATE EXPANSION (IMPACT PARAMETER FORMULATION)

6.1. Perturbed stationary state method
6.2. Translation factors
6.3. Radial and rotational coupling terms

7. MOLECULAR STATE EXPANSION (QUANTUM MECHANICAL FORMULATION)

7.1. General considerations
7.2. Partial wave decomposition
7.3. Diabatic transformation

8. DIABATIC REPRESENTATION

9. ILLUSTRATIONS

9.1. Resonant charge transfer
9.2. Electron capture by multicharged ions
9.3. Inner shell excitation and vacancy sharing

1. INTRODUCTION

Charge exchange has been and continues to be a widely studied atomic collision process. Although qualitatively a well understood process, it has proved difficult to predict the cross sections with reliability and the theory of charge exchange still remains an interesting challenge.

The practical importance of charge exchange in many laboratory and astrophysical plasmas has been recognized only quite recently. For example, in the interstellar medium and in gaseous nebulae, the charge exchange rate for multiply charged ions may exceed the recombination rate by several orders of magnitude. Collisions between multiply charged ions and atomic hydrogen are important in determining the radiation losses and neutral beam heating efficiencies in fusion plasmas.

The phenomena of charge exchange and ionization in atom-atom collisions are typical of reactive scattering events since they involve a rearrangement of the electrons of the colliding particles. Although the application of the formal theory of reactive scattering to charge exchange has received considerable attention

(Bransden[1,2], Coleman and McDowell[3]), few of the practical approaches to the calculation of collision cross sections are rigorously based on formal theory. In this course of lectures, emphasis will be therefore directed more towards the development of reliable theoretical models of charge exchange under specific conditions rather than to the general theory. Of course some results of formal theory will serve as a useful guide for the classification and validity of the various approximate methods. For a recent review of the theoretical problem see Basu et al[4], Belkić et al[5], Briggs and Taulbjerg[6], Shakeshaft and Spruch[7].

Certain aspects of charge exchange processes are common to all types of atom (ion)-atom collisions. When the collision energy E greatly exceeds the typical electron binding energy, it is legitimate to adopt the impact parameter (eikonal) method, in which the electronic motion is described quantum mechanically, while the relative motion of the colliding systems is described classically. The impact parameter method is particularly convenient when the total wave-function is expanded on a finite basis set. It may be remarked however, that for calculations based on a Born type approximation the impact parameter method offers little real advantage over a quantal calculation.

At low energies (thermal - several eV), orbiting and resonance effects occur. In general these are best treated by a quantum mechanical approach, although in many particular cases semi classical methods may be adequate. The choice of method is a practical one depending on the precision required and on the energy range under investigation.

Atomic units will be used throughout except where stated otherwise.

2. GENERAL THEORY - BASIC RESULTS

It is useful to recall briefly the essential features of the quantum-mechanical theory of the charge exchange process. For illustration purpose, we shall consider a typical reaction of the form

$$B + (A + e) \rightarrow (B + e) + A \qquad 2.1$$

where A and B represent singly or multiple charged ionic cores.

Let $\underline{R}$ be the position vector of B relative to A and let $\underline{r}_a$, $\underline{r}_b$ and $\underline{r}_g$ be the position vectors of the electron relative respectively to A, B and the centre of mass of A and B. It is convenient to introduce $\underline{R}_A$ the position vector of B relative to the centre of mass of (A + e) and $\underline{R}_b$ the position vector of A relative to the centre of mass of (B + e).

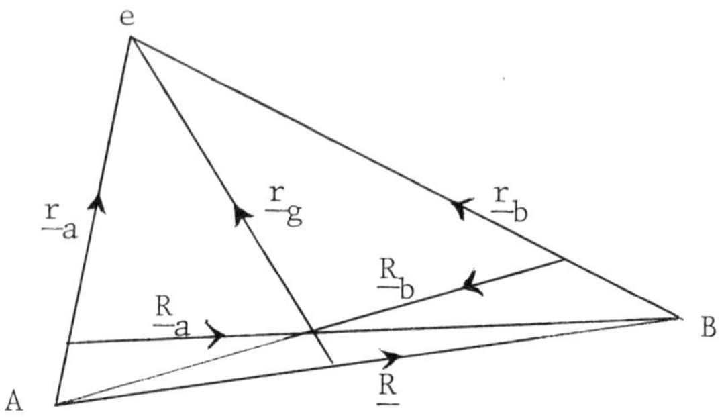

The vectors $\underline{R}_a$, $\underline{R}_b$ and $\underline{r}_g$ intersect at the centre of mass of the colliding system. There are three equivalent sets of independent centre of mass variables $(\underline{r}_g, \underline{R})$, $(\underline{r}_a, \underline{R}_a)$, $(\underline{r}_b, \underline{R}_b)$. The centre of mass kinetic energy operator T can be expressed in terms of these variables as follows

$$T = \frac{P^2}{2\mu} + \frac{p_g^2}{2m} = \frac{P_a^2}{2\mu_a} + \frac{p_a^2}{2m_a} = \frac{P_b^2}{2\mu_b} + \frac{p_b^2}{2m_b} \qquad 2.2$$

where $\underline{P}$, $\underline{P}_a$, $\underline{P}_b$, $\underline{p}_g$, $\underline{p}_a$, $\underline{p}_b$ are respectively the momenta conjugate to $\underline{R}$, $\underline{R}_a$, $\underline{R}_b$; $\underline{r}_g$, $\underline{r}_a$, $\underline{r}_b$ and M_a, M_b, m_e the masses of A, B and e

$$\frac{1}{\mu} = \frac{M_a+M_b}{M_a M_b} \qquad \frac{1}{\mu_a} = \frac{M_a+M_b+m_e}{M_b(M_a+m_e)} \qquad \frac{1}{\mu_b} = \frac{M_a+M_b+m_e}{M_a(M_b+m_e)} \qquad 2.2$$

$$\frac{1}{m} = \frac{M_a+M_b+m_e}{m_e(M_a+M_b)} \qquad \frac{1}{m_a} = \frac{m_e+M_a}{m_e M_a} \qquad \frac{1}{m_b} = \frac{m_e+M_b}{m_e M_b} \qquad 2.3$$

The variables $(\underline{r}_a, \underline{R}_a)$ are appropriate to describe the entry channel, the variables $(\underline{r}_b, \underline{R}_b)$ are appropriate for the rearrangement (charge exhange) channel. The adiabatic coordinates $(\underline{r}_g, \underline{R})$ are the most convenient for the calculation of the molecular states of the system (A + B + e).

The potential energy operator V may be written in the form

$$V = V_{ae}(\underline{r}_a) + V_i = V_{be}(\underline{r}_b) + V_f \qquad 2.4$$

where

$$V_i = V_{be}(\underline{r}_b) + V_{ab}(\underline{R}) \qquad V_f = V_{ae}(\underline{r}_a) + V_{ab}(\underline{R}) \qquad 2.5$$

The asymptotic entry channel states are then described by the eigen function $\Phi^a_n(\underline{r}_a,\underline{R}_a)$ defined by

$$(T + V_{ae})\, \Phi^a_n(\underline{r}_a,\underline{R}_a) = E\, \Phi^a_n(\underline{r}_a,\underline{R}_a) \qquad 2.6$$

The solution of (2.6) corresponding to (A + e) in the ith state and for relative momentum $\hbar\underline{k}_i$ of B with respect to (A + e) is

$$\Phi^a_i(\underline{r}_a,\underline{R}_a) = \exp(i\underline{k}_i \cdot \underline{R}_a)\, \phi^a_i(\underline{r}_a) \qquad 2.7$$

where

$$\left(\frac{p_a^2}{2m_a} + V_{ae}\right) \phi^a_i(\underline{r}_a) = \varepsilon^a_i\, \phi^a_i(\underline{r}_a) \qquad 2.8$$

and

$$k_i^2 = 2\mu_a (E - \varepsilon^a_i) \qquad 2.9$$

Similarly the asymptotic exit channel states are described by

$$(T + V_{be})\, \Phi^b_m(\underline{r}_b,\underline{R}_b) = E\, \Phi^b_m(\underline{r}_b,\underline{R}_b) \qquad 2.10$$

The solution of (2.10) corresponding to (B + e) in the state f with $\hbar\underline{k}_f$, the momentum of (B + e) relative to A is

$$\Phi^b_f(\underline{r}_b,\underline{R}_b) = \exp(-i\underline{k}_f \cdot \underline{R}_b)\, \phi^b_f(\underline{r}_b) \qquad 2.11$$

where

$$\left(\frac{p_b^2}{2m_b} + V_{be}\right) \phi^b_f(\underline{r}_b) = \varepsilon^b_f\, \phi^b_f(r_b) \qquad 2.12$$

and

$$k_f^2 = 2\mu_b \ (E - \varepsilon_f^b) \qquad 2.13$$

The collision process is described by the eigen-function Ψ_i^+, a solution of the Schrödinger equation

$$H \ \Psi_i^+ = E \ \Psi_i^+ \qquad 2.14$$

which behaves asymptotically as

$$\Psi_i^+ \underset{R_a \to \infty}{\sim} \phi_n^a(\underline{r}_a) \left[\exp\ (i\underline{k}_i \cdot \underline{R}_a) \ \delta(n,i) + f_{in}^a(\Omega_a) \frac{1}{R_a} \exp(ik_n R_a)\right]$$

2.15

$$\underset{R_b \to \infty}{\sim} \phi_m^b(\underline{r}_b) \ f_{im}^b(\Omega_b) \frac{1}{R_b} \exp\ (i\ k_m R_b) \qquad 2.16$$

The cross-section for charge exchange in state f of (B + e) is obtained directly from (2.16). For example the differential scattering cross section for the scattering of B in the direction Ω_b is given by

$$\frac{d\sigma_{i\to f}}{d\ \Omega_b} = \frac{\mu_a}{\mu_b} \frac{k_f}{k_i} \left| f_{if}^b(\Omega_b) \right|^2 \qquad 2.17$$

From scattering theory it is recalled that the scattering amplitude may be expressed in the following integral representation

$$f_{if}^b(\Omega_b) = - \frac{M_b}{2\pi} <\Phi_f^b \left|V_f\right| \Psi_i^+ > \qquad 2.18$$

It may be remarked that the scattering amplitude may also be derived from the eigen-function Ψ_f^-, defined an analogous way to Ψ_i^+ with the outgoing waves of (2.18) replaced by ingoing waves.

In a quantum mechanical formulation there are two types of methods commonly used to determine the scattering amplitudes. At very low energies, a partial wave decomposition of Ψ_i^+ may be adopted using a basis set of atomic or molecular type orbitals. The problem is thereby reduced to one of solving a set coupled second-order differential equations, from which the scattering S-matrix can then be extracted by conventional methods. At higher energies, where the partial wave decomposition is unpractical the impact parameter approach (see next section) is more appropriate. However if the energy is sufficiently high, so that Ψ_i^+ is weakly perturbed by the collision, some approximate form of Ψ_i^+ (based on a Born or distorted wave type approximation) may be used to calculate the scattering amplitude directly using the integral representation (2.18).

3. IMPACT PARAMETER APPROXIMATION

Whenever, the collision energy is much greater than the typical electron binding energies, much simplification can be obtained by the adoption of the impact parameter approach. This approach avoids the use of a partial wave expansion and is particularly adopted to the finite expansion method at energies of the order of ∿50 eV or greater.

The impact parameter method makes explicit use of the fact that at these energies, scattering takes place primarily in the forward direction (usually in a cone of the order of a few degrees).

This suggests that it is useful to write Ψ_i^+ in the form

$$\Psi_i^+ = \exp\,(i\,\underline{k}\cdot\underline{R}_a)\ \psi\,(\underline{r}_a, \underline{R}) \qquad 3.1$$

where $\underline{k}$ is a vector usually taken to be in the direction of $\underline{k}_i$ and whose magnitude is given by

$$k^2 = 2\,E \qquad 3.2$$

Substitution of (3.1) in the Schödinger equation (2.14) yields

$$-\frac{1}{2\mu_a}\left[\nabla^2_{\underline{R}_a} + 2\,i\underline{k}\cdot\nabla_{\underline{R}_a}\right]\psi + H_{el}\,\psi = 0 \qquad 3.3$$

where

$$H_{el} = \frac{p_a^2}{2m} + V \qquad 3.4$$

When the collision energy is sufficiently large, the function $\psi(\underline{r}_a,\underline{R})$ will vary slowly with R as compared with $\exp\,(i\underline{k}\cdot\underline{R}_a)$. It is then legitimate to assume that

$$|\nabla^2_{\underline{R}_a}\,\psi| \ll |\underline{k}\cdot\nabla_{\underline{R}_a}\,\psi| \qquad 3.5$$

Setting $\underline{v} = \underline{k}/\mu_a$, then equation (3.3) may be written as

$$\left[H_{el} - i\, \underline{v} \cdot \nabla_{\underline{R}_a} \right] \psi = 0 \qquad 3.6$$

This equation (3.6) may be transformed into a more recognizable form by introducing the time variable t, defined by the relation

$$\underline{R}_a = \underline{b} + \underline{v}t \qquad 3.7$$

where $\underline{b}$, a vector in the plane perpendicular to $\underline{v}$, may be identified as the impact parameter. Then equation (3.6) becomes the well-known impact parameter or eikonal equation

$$\{H_{el} - i \left.\frac{\partial}{\partial t}\right]_a\} \psi = 0 \qquad 3.8$$

where the symbol $\Big]_a$ signifies that the derivate will respect to t be carried out for $\underline{r}_a$ constant. It may be remarked that the particular time derivative in (3.8) is a direct consequence of the eikonal approximation. More generally, if instead of (3.1), we had written

$$\Psi_i^+ = \exp\,(i\, \underline{k} \cdot \underline{R}')\, \psi' \qquad 3.9$$

where $\underline{R}' = p\, \underline{R}_a + (1 - p)\, \underline{R}_b$ and p is an arbitrary constant $o \lesssim p \lesssim 1$, the impact parameter equations becomes

$$\{H_{el} - i \left.\frac{\partial}{\partial t}\right]_{\underline{r}'}\} \psi = 0 \qquad 3.10$$

where $\underline{r}'$ is the position vector of the electron with respect to a point P on the internuclear axis, such that $\underline{AP} = p\underline{R}$. Provided the condition (3.6) is satisfied the eikonal approximation should be independent of small variations of $\underline{R}'$. This is equivalent to the requirement of Galilean invariance of the impact parameter method in the case of rectilinear trajectories. However, one should note that the condition (3.6) will be satisfied only if a "reasonable" choice of $\underline{R}'$ is adopted.

As we shall see later, a suitable choice of $\underline{R}'$ can in some particular cases lead to further simplications.

The scattering amplitude may be calculated from ψ using the integral representation (2.18).This is necessary since (3.1) is a good approximation for Ψ_i^+ only in the interaction region. Then

$$f_{if}(\Omega_b) = -\frac{M_b}{2\pi}\iint d\underline{r}_a\, d\underline{R}_a \exp\left[i\underline{k}\cdot\underline{R}_a + i\underline{k}_f\cdot\underline{R}_b\right] \phi_f^b V_f \psi \qquad 3.11$$

The coordinates of $\underline{R}_a$ (X, Y, Z) are chosen as follows: the Z-axis is taken in the direction of $\underline{k}$ and the X-axis in the plane containing $\underline{k}$ and $\underline{k}_f$.

Let $\underline{\eta}$ be a vector directed along the X-axis and of magnitude equal to the projection $\underline{k}_f$ on the axis. Then to the order of (m/μ)

$$\underline{k}\cdot\underline{R}_a + \underline{k}_f\cdot\underline{R}_b = \frac{\varepsilon_f Z}{v} - \underline{\eta}\cdot\underline{b} + \underline{v}\cdot\underline{r}_a \qquad 3.12$$

so that

$$f_{if}(\Omega_b) = -\frac{M_b}{2\pi}\int_{-\infty}^{\infty} dz \exp\left(i\frac{\varepsilon_f Z}{v}\right)\int d\phi\, db\, b \exp(-i\eta\, b\cos\phi)$$

$$\langle \phi_f^b | V_f e^{i\,\underline{v}\cdot\underline{r}_a} | \psi \rangle \qquad 3.13$$

The term $\langle \phi_f^b | V_f\ e^{i\,\underline{v}\cdot\underline{r}_a} | \psi \rangle$ is directly related to the transition amplitude $P_{if}(b)$ obtained from the eikonal equation (3.8)

$$P_{if}(\underline{b}) = \lim_{t\to\infty} \langle \phi_f^b\, e^{-i\varepsilon_f t}\, | e^{i\underline{v}\cdot\underline{r}_a} | \psi \rangle \qquad 3.14$$

In other words $P_{if}(\underline{b})$ is the projection of $|\psi\rangle$ on the final state $|\phi_f^b\rangle$ in the limit as $t\to\infty$. Note the presence of the term $e^{i\underline{v}\cdot\underline{r}_a}$, which arises from the necessity to describe the final state will respect to origin A. Using the eikonal equation it is easy to show that

$$P_{if}(\underline{b}) = -\frac{i}{v}\int_{-\infty}^{\infty} dz\ e^{i\varepsilon_f z/v} \langle \phi_f^b | V_f\ e^{i\underline{v}\cdot\underline{r}_a} | \psi \rangle \qquad 3.15$$

and we may thus write

$$f_{if}(\Omega_b) = \frac{i\, M_b\, v}{2\pi}\int d\underline{b}\ e^{-i\,\underline{\eta}\cdot\underline{b}}\, P_{if}(\underline{b}) \qquad 3.16$$

The quantum mechanical scattering amplitude is simply the Fourier transform of the transition probability $P_{if}(\underline{b})$. Using the fact that

$$\eta \, d\eta \approx \mu \, V \sin\theta \, d\theta \qquad 3.17$$

it is straightforward to establish that the total cross section is given by

$$Q_{if} = \int d\underline{b} \, |P_{if}(\underline{b})|^2 \qquad 3.18$$

which is of course the standard expression obtained by the impact parameter method. However, the derivation given here, not only clarifies the nature and limitations of the eikonal approximation, but also provides a direct way of calculating the differential scattering cross section. See McCarroll and Salin[8], Gaussorgues et al[9], Taulbjerg[90].

4. HIGH ENERGY METHODS

Although the treatment of charge exchange at high energies is outside the scope of these lectures, it is useful to recall some of the basic problems associated with the use of the Born approximation.

In the quantal formulation, the first Born approximation is obtained by replacing Ψ_i^+ in the expression (2.18) for the scattering amplitude by Φ_i^a; in the impact parameter treatment, it suffices to replace ψ in expression (3.15) by ϕ_i^a.

Oppenheimer[10] and Brinkman and Kramers[11] assumed that in calculating (2.18) or (3.15), it is legitimate to neglect the contribution of the internuclear potential in calculating the total cross section. It should be remarked however that the internuclear potential must be included when calculating the differential scattering cross section. The reason for this can be deduced easily from the expression for the scattering amplitude (3.15) in the eikonal approximation. The internuclear potential only has an influence on the phase of $P_{if}(\underline{b})$. As a result, the differential cross section is sensitive to the internuclear potential whereas the total cross section is not (see Belkić et al[5]).

The Oppenheimer-Brinkman-Kramers (OBK) cross sections are in general easy to calculate. For example the cross section for capture into the ground state in H^+ - H collision is

$$Q_{OBK} = \frac{2^8}{5(1+v^2/4)^5 \, v^2} \pi a_o^2 \qquad 4.1$$

The cross sections for capture into excited states are also straightforward to determine (see Bates and McCarroll[12]).

Comparison with experimental data reveals that the OBK cross section is several times too large even at very high energies, where the Born approximation is excellent for direct excitation and ionization. Bates and Dalgarno[13] and Jackson and Schiff[14] argued in favour of retaining the internuclear potential, but while this improves the agreement with experiment in some cases, it has proved unsatisfactory in general (see Belkić et al[5]). In the case of capture into the ground state in H^+ - H collisions, the Jackson-Schiff cross section, in the high energy limit is given by

$$Q_{JS} = 0.661\ Q_{OBK}$$

This ambiguity with respect to the first Born approximation is a strong indication that charge exchange cannot be described by a first order approximation. The Born series for charge exchange was first investigated by Drisko[87]; he showed that including all terms up to second other, the cross section for capture into the ground state in H^+ - H collisions has the asymptotic form

$$Q_{B2} = (0.295 + \frac{5\pi v}{2^{12}})\ Q_{OBK} \qquad 4.2$$

The contribution of the internuclear potential vanishes exactly in the second Born approximation. However this cancellation of the internuclear potential appears to be accidental for ground state capture. Belkić et al[5] indicate that this is not the case for capture into excited states.

The leading term in Q_{B2} exhibits a v^{-11} behaviour in contrast with the v^{-12} dependence of Q_{OBK}. Physically the origin of the v^{-11} term arises from the double scattering of the electron, via the continuum states of the target and projectile nuclei. Such a picture also appears in the classical binary encounter model of Thomas[15]. Firstly, the bound electron is scattered by the projectile in the direction of the target nucleus, where it undergoes a second collision with its parent nucleus. Thomas assumes that capture takes place if, as a result of the second collision, the electron emerges with the same velocity as the projectile. The classical value of the v^{-11} coefficient is however different from that of the second Born approximation. The correspondence between the Thomas model and the second Born approximation has been established by Shakeshaft[16]. See also the review of Shakeshaft and Spruch[7];

A number of approximate methods have been proposed to take into account the effect of coupling with the intermediate continuum channels; for example the impulse approximation (I.A), continuum distorted wave (CDW) and continuum intermediate state (CIS) approximations. The asymptotic form of the corresponding cross sections for capture into the ground state in H^+ - H collisions are closely related to the second Born form

$$Q_{CDW} = Q_{B2} \qquad 4.3$$

Of these the CDW and CIS methods have proved to be the most successful, giving results in excellent agreement with experiment for a wide variety of systems at high energies. A detailed review of these high energy methods is given by Belkić et al[5].

5. SOLUTION OF THE EIKONAL EQUATION

Most of the difficulties connected with charge exchange can be illustrated most simply in the framework of the eikonal approximation. Let us write the eikonal equation

$$\left[H_{el} - i\frac{\partial}{\partial t}\right]\psi = 0 \qquad 5.1$$

where H_{el} depends on t through the relation

$$\underline{R} \cdot \underline{b} + \underline{v}t \qquad \underline{b} \cdot \underline{v} = 0 \qquad 5.2$$

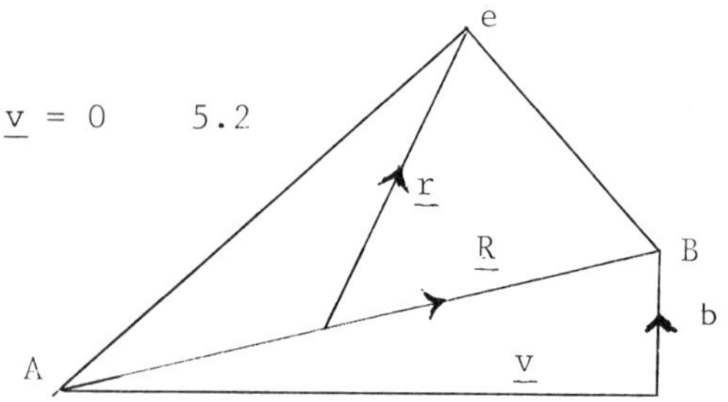

From the discussion of section 3, it is implicitly understood that the choice of independent variables ($\underline{r}$,t) is to some extent arbitrary. Within the limits of validity of the eikonal approximation, the scattering amplitude should be insensitive to the position of the origin 0. It is therefore convenient to define 0 by a parameter p such that $\vec{AO} = p\,\vec{AB}$ and to limit variations of p to the range

$0 \leq p \leq 1$. A basic requirement is then that the scattering amplitude be independent of p.

In the energy range, where the collision velocity is less than or of the some order as the classical velocity of an electron in a Bohr orbit, the common way of solving (5.1) is to expand ψ on a suitable basis set. In the case of collisions where $v \gtrsim 4v_e$, the high energy methods of section 4 are appropriate. For velocities $v \sim v_e$, corresponding to the energy range 5 - 100 keV per nucleon, an atomic type basis set has proved to be quite successful. For lower energies a molecular basis set is to be preferred.

5.1. Atomic eigen-function expansion method

To describe the asymptotic states in a charge exchange collision, it is obvious that a two-centre atomic expansion is required. Such a method was originally proposed by Bates[17], who recognized the necessity to take account of the translational velocity of B with respect to A. Thus we may expand ψ in the form

$$\psi(\underline{r},t) = \sum_n a_n(t)\,\Phi_n^a(\underline{r},t) + \sum_m b_m(t)\,\Phi_m^b(\underline{r},t) \qquad 5.3$$

where

$$\Phi_n^a(\underline{r},t) = \phi_n^a(\underline{r}_a)\exp-i\left[\varepsilon_n^a t + p\,\underline{v}\cdot\underline{r} + \frac{1}{2}p^2v^2t\right] \qquad 5.4$$

$$\Phi_m^b(\underline{r},t) = \phi_m^b(\underline{r}_b)\exp-i\left[\varepsilon_m^b t - (1-p)\underline{v}\cdot\underline{r} + \frac{1}{2}(1-p)^2v^2t\right] \qquad 5.5$$

The velocity-dependent phase-factor $\exp-(ip\,\underline{v}\cdot\underline{r})$ and $\exp i(1-p)\underline{v}\cdot\underline{r}$ are required to preserve translational invariance. The phase factors depending on t are included to simplify the form of the differential equations for a_n and b_n.

Using the relations

$$\left.\frac{\partial}{\partial t}\right]_{\underline{r}} = \left.\frac{\partial}{\partial t}\right]_{\underline{r}_a} + p\,\underline{v}\cdot\nabla_{\underline{r}} = \left.\frac{\partial}{\partial t}\right]_{\underline{r}_b} - (1-p)\,\underline{v}\cdot\nabla_{\underline{r}} \qquad 5.6$$

$$\nabla_{\underline{r}} = \nabla_{\underline{r}_a} = \nabla_{\underline{r}_b} \qquad 5.7$$

we may easily deduce that

$$H_{el}\,\Phi_n^a(\underline{r},t) = \left[(\varepsilon_n^a + V_{eb} + V_{ab} + \frac{1}{2}p^2v^2)\,{}_n^a + p\,\underline{v}\cdot\underline{\nabla}_{\underline{r}}\,\phi_n^a\right]\cdot$$

$$\cdot\ \exp{-i}\left[\varepsilon_n^a t + p\,\underline{v}\cdot\underline{r} + \frac{1}{2}p^2v^2t\right] \qquad 5.8$$

and

$$\{H_{el} - i\frac{\partial}{\partial t}\Big]_{\underline{r}}\}\,\Phi_n^a\,(\vec{r},t) = (V_{eb} + V_{ab})\,\Phi_n^a(\underline{r},t) \qquad 5.9$$

Similarly

$$\{H_{el} - i\frac{\partial}{\partial t}\Big]_{\underline{r}}\}\,\Phi_m^b\,(\underline{r},t) = (V_{ea} + V_{ab})\,\Phi_m^b(\underline{r},t) \qquad 5.10$$

The coupled equations for the functions $a_n(t)$ and $b_n(t)$ may then be obtained from the relations

$$<\Phi_n^a\,|H_{el} - i\frac{\partial}{\partial t}|\,\psi> = 0 \qquad 5.11$$

$$<\Phi_m^b\,|H_{el} - i\frac{\partial}{\partial t}|\,\psi> = 0 \qquad 5.12$$

which yield using (5.10)

$$i\,\dot{a}_n(t) + i\sum_k \dot{b}_k(t)S_{nk}^{ab} = \sum_j a_j(H_{nj}^{aa} + \delta_{nj}V_{ab}) + \sum_k b_k(H_{nk}^{ab} + V_{ab}S_{nk}^{ab}) \qquad 5.13$$

$$i\,\dot{b}_m(t) + i\sum_j \dot{a}_j\,S_{mj}^{ba} = \sum_k b_k(H_{mk}^{bb} + \delta_{mk}V_{ab}) + \sum_j a_j(H_{mj}^{ba} + V_{ab}S_{mj}^{ba}) \qquad 5.14$$

$$S_{jk}^{ab} = <\phi_j^a(\underline{r}_a)\,|e^{i\underline{v}\cdot\underline{r}_m}|\,\phi_k^b(\underline{r}_b)>\exp i(\varepsilon_j^a - \varepsilon_k^b)t \qquad 5.15$$

$$S_{jk}^{ba} = <\phi_j^b(\underline{r}_b)\,|e^{-i\underline{v}\cdot\underline{r}_m}|\,\phi_k^a(\underline{r}_a)>\exp i(\varepsilon_j^b - \varepsilon_k^a)t = S_{kj}^{ab} \qquad 5.16$$

$$H_{jk}^{aa} = <\phi_j^a(\underline{r}_a)\,|V_{eb}|\,\phi_k^a(\underline{r}_a)>\ \exp i(\varepsilon_j^a - \varepsilon_k^a)\,t \qquad 5.17$$

$$H_{jk}^{bb} = <\phi_j^b(\underline{r}_b)\,|V_{ea}|\,\phi_k^b(\underline{r}_b)>\exp i(\varepsilon_j^b - \varepsilon_k^b)\,t \qquad 5.18$$

$$H_{jk}^{ab} = \langle \phi_j^a(\underline{r}_a) \, | V_{eb} \, e^{i \, \underline{v} \cdot \underline{r}_m} | \, \phi_k^b(\underline{r}_b) \rangle \exp i(\varepsilon_j^a - \varepsilon_k^b) \, t \qquad 5.19$$

$$H_{jk}^{ba} = \langle \phi_j^b(\underline{r}_b) \, | V_{ea} \, e^{-i \, \underline{v} \cdot \underline{r}_m} | \, \phi_k^a(\underline{r}_a) \rangle \exp i(\varepsilon_j^b - \varepsilon_k^a) \, t \qquad 5.20$$

where $\underline{r}_m = (\underline{r}_a + \underline{r}_b)/2$.

The initial conditions are such that

$$a_n(-\infty) = \delta_{ni} \qquad b_n(-\infty) = 0 \qquad 5.21$$

We may note that the internuclear potential V_{ab} intervenes only in the phase of the transition amplitudes V_{ab}. To illustrate this let us substitute

$$a_n(t) = \exp \left[-i \, \phi(t) \right] \, a_n'(t) \qquad 5.22$$

$$b_n(t) = \exp \left[-i \, \phi(t) \right] \, b_n'(t) \qquad 5.23$$

where $\phi(t) = \int^t V_{ab} \, dt$.

Then equations (5.13), (5.14) become

$$i \, \dot{a}_n'(t) + i \sum_k \dot{b}_k'(t) \, S_{nk}^{ab} = \sum_j a_j'(t) \, H_{nj}^{aa} + \sum_k b_k'(t) \, H_{nk}^{ab} \qquad 5.24$$

$$i \, \dot{b}_m'(t) + i \sum_j \dot{a}_j'(t) \, S_{mj}^{ba} = \sum_k b_k'(t) \, H_{mk}^{bb} + \sum_j a_j'(t) \, H_{mj}^{ba} \qquad 5.25$$

Since $|a_n|^2 = |a_n'|^2$ and $|b_n|^2 = |b_n'|^2$ it follows that the total cross sections are independent of V_{ab} in the eikonal approximation. On the other hand, the differential scattering cross sections will depend on V_{ab}. In this respect, it may be remarked that approximate solutions of (5.13) and (5.14) may exhibit a spurious dependence on V_{ab}. For example, in the first Born approximation for charge exchange in state f, the system of coupled equations is reduced to

$$i \, \dot{b}_f = H_{fi}^{ba} + V_{ab} \, S_{fi}^{ba} \qquad 5.26$$

and the corresponding cross section will therefore exhibit a dependence on V_{ab}, even in the high energy range. This non physical result illustrates the inability of first order methods to describe the charge exchange process.

The coupled equations (5.13, 5.14) have been solved with basis sets of different sizes for various simple systems. The simplest approximation is a two state expansion involving only the initial and final states. Although the two-state approximation is expected to be unsatisfactory both at low velocities where adiabatic distortion effects are significant and at high velocities where coupling with many intermediate states is non-negligible, it often gives reasonably accurate results in the intermediate energy range, where the cross section is maximum.

It is instructive to apply the two state method to the case of resonant charge transfer, for example electron capture in the ground state in proton hydrogen collisions McCarroll[18]. The coupled equations then reduce to

$$i\,\dot{a}_1 + i\,\dot{b}_1\,S = a_1\,c + b_1\,h \qquad 5.27$$

$$i\,\dot{b}_1 + i\,\dot{a}_1\,S = b_1\,c + a_1\,h \qquad 5.28$$

where for symmetry reasons

$$H_{11}^{ab} = H_{11}^{ba} = h \qquad 5.29$$

$$S_{11}^{ab} = S_{11}^{ab} = S \qquad 5.30$$

$$H_{11}^{aa} = H_{11}^{bb} = c \qquad 5.31$$

Adding and subtracting 5.27 and 5.28 we obtain the decoupled equation

$$i(1+S)\ (\dot{a}_1 + \dot{b}_1) = (a_1 + b_1)\ (c+h) \qquad 5.32$$

$$i(1-S)\ (\dot{a}_1 - \dot{b}_1) = (a_1 - b_1)\ (c-h) \qquad 5.33$$

which are easily solved to yield for the charge exchange transition amplitude

$$|b_1(\infty)|^2 = \sin^2 F \qquad 5.34$$

where

$$F = \int_{-\infty}^{\infty} \frac{h-sc}{1-S^2}\, dt \qquad 5.35$$

It is obvious from the form of (5.34) that the two state approximation for resonant charge transfer is much less sensitive to errors when F is large. Formula (5.34) also allows us to assess the importance of momentum transfer in a simple way. For example, we may compare the result with that obtained by neglecting the phase factors exp $(i\underline{v}\cdot\underline{r}_m)$ in the calculation of the matrix elements $\hbar$ and S. In the case of proton-hydrogen collisions, momentum transfer effects become important at energies above 5 keV.

The two state approximation has proved particularly successful for K - K electron transfer in heavy ion collisions. Electron transfer plays an important rôle in the production of inner shell vacancies, especially in near symmetrical collisions. Comparison with experimental data (Tawara et al[19] on F^{9+} - Si collisions) is very satisfactory (Lin et al[20]).

In collisions involving multicharged ions, electron capture from the outer shells takes place mainly into excited states. However, since the transition occurs in fairly distant collisions, the atomic state expansion can be adequate provided account is taken of all the interacting states. Ryufuku and Watanabe[21] have proposed a simple unitarized distorted ware approximation, closely similar to the method of Bates which has been used successfully to calculate charge transfer cross sections involving highly charged ions with atomic hydrogen. Comparison with experiment is satisfactory.

However, except in these cases such as the above-mentioned examples where coupling with a finite number of states dominates, the atomic state expansion is often unsatisfactory. This is especially true at high energies where coupling with continuum states becomes of increasing importance. The most extensive calculations have been carried out for proton-hydrogen collisions, which while not exactly a typical case does illustrate the main defects of the atomic state method. Calculations, based on a two centre expansion including the 1s, 2s and 2p states have been carried out by Wilets and Gallaher[22], Cheshire et al[23], Rapp and Dinwiddie[24].

It is found, that the cross section for electron capture is

scarcely changed by inclusion of the 2s and 2p states at energies below 20 keV. For energies above 20 keV, the influence of coupling with the 2s and 2p states is of the order of 10%. The calculations of Rapp and Dinwiddie[24] who also included the 3s and 3p states, indicate that coupling with these states has not an appreciable effect. Even the cross section for capture in the 2s or 2p states is not significantly changed by coupling will the 3s and 3p states, particularly in the high energy range.

Unfortunately this convergence is only apparent in the high energy range. Agreement with experiment is poor at high energies. Asymptotically the two state approximation behaves as the Brinkman-Kramers approximation, which suggests that at high energies the cross section for ground state capture is considerably overestimated. The obvious conclusion is that account must be taken of continuum states.

Various approaches to include continuum states have been proposed. Amongst these, the most significant are the basis set of Sturmian functions first introduced by Gallaher and Wilets[25], the pseudo-state expansion first proposed by Cheshire et al[23], and the scaled hydrogenic basis set of Shakeshaft[26].

5.2. Sturmian function basis set

Tha atomic wave-functions ϕ_n occurring in (5.3) are replaced by function of the type

$$\hat{\phi}_n(\underline{r}) = \frac{1}{r} S_{nl}(r) \; Y_{lm}(\theta,\phi) \qquad 5.36$$

where $S_{nl}(r)$ is a radial function satisfying the equation

$$\left(-\frac{1}{2}\frac{d^2}{dr^2} + \frac{\ell(\ell+1)}{2r^2} - \frac{\alpha_{n\ell}}{r}\right) S_{nl}(r) = E_1 \, S_{nl}(r) \qquad 5.37$$

This equation, known as the Sturmian equation is similar to the Schrödinger equation, except that the energy E_1 is taken as a fixed parameter, while the effective charge α_{nl} acts as the eigenvalue. For example Gallaher and Wilets choose $E_1 = -1/\left[2(1+1)^2\right]$. As boundary conditions, it is supposed that $S_{nl}(r)$ be zero at the origin and decays exponentially at infinity. The $\hat{\phi}_{nl}(r)$ form an infinite discrete and complete set of states and unlike the hydrogenic functions there is no continuum.

The Sturmian functions are explicity given by scaled hydrogenic functions

$$S_{nl}(r) = \alpha_{nl}^{1/2} \; R_{nl}(\alpha_{nl} r) \qquad 5.38$$

Where $R_{nl}(r)$ is the usual radial hydrogenic function. The normalisation is chosen so that

$$< \hat{\phi}_k \,|\, \hat{\phi}_k > = 1 \qquad 5.39$$

It should be remarked that the Sturmian functions are not orthogonal. However, they are orthogonal with a weighting factor of r^{-1}. Choosing E_1 as in Gallaher and Wilets, we have

$$\alpha_{nl} = n/(l+1) \qquad 5.40$$

yielding the mean energy of

$$\hat{E}_k = < \hat{\phi}_k \,|\, H_o \,|\, \hat{\phi}_k > = - \frac{1}{n(l+1)} + \frac{1}{2(l+1)^2} \qquad 5.41$$

It is to be remarked that with this definition of basis set, the 1s, 2p, 3d Sturmian states coincide with the hydrogenic ϕ_{nl}.

However as the Sturmian functions with one or more nodes are not equivalent to the hydrogenic states, difficulties arise in defining the transition amplitudes. For example the 2s state is poorly represented by a 2s Sturmian function. As a consequence the corresponding transition probabilities contain oscillatory components which do not vanish as $t \to \infty$. Shakeshaft[26] has shown that this defect can be removed if a large basis set of Sturmian function is used. The most extensive calculations are those of Shakeshaft who included the 1s to 6s and 2p to 4p Sturmian states around each centre.

5.3. Pseudo-state expansion

The pseudo-state method attempts to preserve certain interesting features of the atomic state expansion while introducing pseudo-state functions to deal with the problem of the continuum states. A finite number of atomic states are introduced explicitly. The pseudo-state functions are constructed so as to represent the continuum in the most efficient manner. To avoid problems in the asymptotic region, the pseudo-state function are chosen so as to be orthogonal to all other functions included in the basis.

In this respect it is instructive to examine the pseudo-state functions used by Cheshire et al[23] in the $H^+ - H$ problem. In this work the 1s, 2s, 2p states are included explicitly in the expansion. This enables one to treat electron capture into these states without the projection difficulties encountered in the Sturmian calculation and the long range coupling between the 2s and 2p states can be fully represented. To simplify the calculation of the exchange matrix elements, the number of exponents are kept to a minimum. The simplest pseudo-state functions consistent with these constraints

$$\bar{\phi}_{3s} = \sqrt{\frac{150}{17\pi}} \exp(-r) \left[1 - \frac{16}{15} r + \frac{1}{5} r^2\right] \qquad 5.42$$

$$\bar{\phi}_{3p} = \frac{1}{2}\sqrt{\frac{5}{6}} \exp(-r/2) \left[r(1 - \frac{r}{5})\right] Y_{1,m}(\theta,\phi) \qquad 5.43$$

where the subscripts 3s, 3p indicate the character of the radial function.

The sums of the squares of the overlap integrals of $\bar{\phi}_{3s}$ and $\bar{\phi}_{3p}$ with the complete set of hydrogenic bound states have the values of 0.039 and 0.833 respectively. Thus $\bar{\phi}_{3s}$ is almost entirely on the hydrogenic continuum. The nodes of the 3s pseudo-state functions occur near the first two maxima of the ns atomic wave function, which accounts for the near orthogonality. In addition the pseudo-state functions have a considerable overlap with the lower bound states of He^+. So it may be expected that the pseudo-state expansion should also be reasonable at low energies where it is crucial to have a good representation of molecular states correlated to the 2p state of He^+. Using a 7 state expansion (4 atomic state + 3 pseudo states) Cheshire et al[23] have obtained a very impressive agreement with experiment. However, a more thorough investigation of the pseudo-state expansion method is required before a definitive conclusion can be drawn.

Another type of pseudo-state expansion has been introduced by Dose and Semini[27], who used a basis set of Gaussian functions. The advantage of Gaussian functions has in the fact that the two centre integrals involving momentum transfer terms (integrals of the type H^{BA} or S^{BA}) can be calculated analytically (cf. Errea and Riera[28]).

(All of the methods devised for calculating those integrals in a Slatertype or atomic basis set, McCarroll[18], Cheshire[29], Shakeshaft[26] usually involve a numerical integration at some stage of the calculation. The calculations of Dose and Semini are less ambitious than those of Cheshire in that only the ground is state is explicitly included. Their results are in good agreement with Cheshire et al[23] for capture into the ground state and for the angular variation of the charge exchange probability. Once again more work is required to investigate the practical usefulness of a Gaussian pseudo-state basis set.

5.4. Scaled Hydrogenic basis set - Ionization to the Continuum

As we have already remarked, the Sturmian and pseudo-state methods allow for coupling with the continuum states. They can therefore in principle be used to compute ionization cross sections.

The most extensive calculations, exploiting this property, are those of Shakeshaft[30] using a scaled hydrogenic basis set. This basis set is very similar to the Sturmian set, but with the scaling factors determined by the requirement that when the atomic Hamiltonian is diagonalized, the resulting energy values almost coincide with the 1s, 2s, 2p, 3s, 3p, 3d states and overlap the low energy part of the continuous spectrum. The basis functions $\hat{\phi}_{nlm}(\underline{r})$ satisfy the equation

$$\left[-\frac{1}{2}\nabla_r^2 - \frac{\lambda_{nl}}{r} + \frac{1}{2}\frac{\lambda_{nl}^2}{n^2}\right]\hat{\phi}_{nlm}(\underline{r}) = 0 \qquad 5.44$$

The expansion used by Shakeshaft included 35 such functions centred about each proton with $0 \leq l \leq 2$, $0 \leq m \leq l$, $1 \leq n \leq N_l$. The scale factors λ_{nl} were chosen as follows: $\lambda_{no} = 0.75n$, $\lambda_{n1} = 0.7n$, $\lambda_{n2} =$ $= 0.6n$. The eigenvalues in atomic units of the diagonalized Hamiltonian are shown in the table (5.1) below.

Table 5.1

n \ l	0	1	2
1	- 0.500		
2	- 0.125	- 0.125	
3	- 0.0552	- 0.0552	- 0.0551
4	- 0.0163	- 0.0158	- 0.0119
5	0.0541	0.0595	0.0967
6	0.199	0.227	0.507

(continued)

Table 5.1 (Continued)

n \ l	0	1	2
7	0.532	0.678	
8	1.552	2.612	
9	7.640		

It is seen that this basis set gives a good representation of the first few excited states and the low energy part of the continuum oof the hydrogen atom.

The total wave-function of the colliding system is then expanded in the eigen-vectors that diagonalize the atomic hamiltonian.

A selection of the results obtained are represented in the Table 5.2 below. Rows (a) refer to excitation, rows (b) to charge transfer. The ionization cross section includes both the contribution from direct ionization and from charge transfer to the continuum.

Table 5.2

Energy keV		1s	2S	Final State 2p	3s	3p	3d	Continuum
15	a		0.88	2.46	0.18	0.47	0.20	3.1
	b	58.3	3.41	3.10	0.33	0.47	0.20	
25	a		1.56	4.85	0.44	0.86	0.34	8.5
	b	30.4	3.98	1.74	0.93	0.52	0.06	
40	a		2.10	6.89	0.37	1.17	0.44	14.6
	b	11.9	2.33	0.72	0.67	0.24	0.026	
75	a		1.19	7.97	0.28	1.21	0.23	15.0
	b	2.10	0.42	0.086	0.14	0.030	0.004	
145	a		0.80	6.53	0.13	1.20	0.152	9.4
	b	0.19	0.040	0.0077	0.012	0.0025	0.0001	
200	a		0.49	5.55	0.11	0.99	0.065	6.5
	b	0.047	0.009	0.0014	0.0030	0.0005	0.00006	

The ionization cross section was obtained from the overlap matrix element of the eigen-vector with the continuum of the hydrogen atom about whose nucleus the eigen-vector is centered. It was thus natural to divide the ionization probability into two separate contributions, one for the probability of direct ionization obtained by summing over those eigen-vectors centred on the target nucleus the other for the probability for "charge transfer to the continuum" (CTTC) obtained by summing over the eigen-vectors centred on the projectile. The sum of these two contributions is given in table 5.2. It is noteworthy that the ionization cross section is the dominating inelastic process except at low energies.

The distinction between direct ionization and CTTC is to some extent arbitrary, since there is no way of identifying in an experiment the origin of the ejected electrons. However, the notion of CTTC has proved to be very useful in the interpretation of the ejected electron spectra in ion-atom collisions. The energy spectrum of the ejected electrons emitted in the forward direction can be identified as CTTC.

The cross sections for direct ionization and CTTC are presented in figure 5.1 and the total ionization cross section compared with the experimental results of Park et al[31].

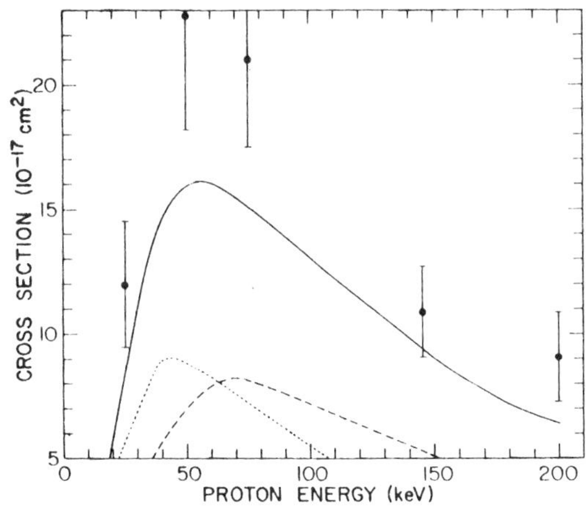

Fig. 5.1 - Total cross section for ionization,— theoretical results, . experimental results of Park et al.; --- theoretical contribution from direct ionization, ...theoretical contribution from charge transfer to the continuum. The solid curve is the sum of the two contributions. Note that the vertical scale begins at 5×10^{-17} cm^2. (Shakeshaft[30]).

The theoretical and experimental data for the total cross section are in reasonably good agreement. At energies below about 60 keV, the CTTC cross section is larger than the cross section for direct ionization. This is a remarkable (and unsuspected) result and illustrates the necessity of taking account of CTTC in ion atom collisions. The position of the peak in the CTTC cross section at 40 keV can be understood from simple classical physical arguments (Shakeshaft[30]). The probability for CTTC is maximum when the velocity of the ejected electron is equal to the velocity of the projectile, that is to say when the internal energy of the projectile-electron sub-system is zero. Let $\underline{q}$ and ε denote the initial momentum and energy of the electron in the laboratory from of reference. Let $\underline{v}$ be the initial velocity of the projectile and $\underline{u}$ the final velocity of the electron-projectile sub-system. Neglecting any momentum transferred to the target nucleus during the collision, it follows from momentum and energy conservation that if the electron-projectile sub-system has zero internal energy

$$M\,\underline{v} + \underline{q} = (M+1)\,\underline{u} \tag{5.45}$$

$$\frac{1}{2} M v^2 + \varepsilon = \frac{1}{2} (M+1)\, u^2 \tag{5.46}$$

Eliminating $\underline{u}$ from equations (5.45) and (5.46) and neglecting corrections of order 1/M, we obtain

$$v^2 - 2\,\underline{v}\cdot\underline{q} + 2\varepsilon = 0 \tag{5.47}$$

Setting $\underline{v}\cdot\underline{q} = vq\cos\beta$, where β is the angle between $\underline{u}$ and $\underline{q}$, and using the fact that $\varepsilon = -1/2$ (in atomic units), it follows that

$$v^2 = 2q^2\cos^2\beta + 1 + 2q\cos\beta\,(q^2\cos^2\beta + 1)^{1/2} \tag{5.48}$$

The square of the momentum transform of the initial is electron is (in atomic units)

$$\frac{8}{\pi^2}\frac{1}{(1+q^2)^4} \tag{5.49}$$

Hence the average value of v^2

$$\langle v^2 \rangle = \int \alpha^3 \underline{q}\, \frac{8}{\pi(1+q^2)^4}\left[2q^2\cos^2\beta + 1 + 2q\cos\beta\,(q^2\cos^2\beta + 1)^{1/2}\right]$$

$$= \frac{5}{3} \tag{5.50}$$

The energy at which the CTTC cross section is maximum is

$$\frac{1}{2} M \langle v^2 \rangle \cong 42 \text{ keV}$$

a prediction which is in excellent agreement with the detailed calculations.

6. MOLECULAR STATE EXPANSION (Impact parameter formulation)

When the relative velocity of the collision is slow compared with the classical velocities of a bound electron in a Bohr orbit, the electronic motion is well described by the adiabatic approximation. An inelastic transition, involving either charge exchange or excitation can take place only at those internuclear distances where there is near degeneracy of the adiabatic electronic states. In these circumstances, the collision is most simply described by a molecular state expansion. This method is often termed the perturbed stationary state (PSS) method.

Unfortunately the PSS method has a number of serious defects arising from the inability of a finite molecular basis set to represent correctly the asymptotic conditions. To illustrate the problem, let us obtain the conventional PSS equations.

6.1. Perturbed stationary state method

The wave function $\psi(\underline{r},t)$ is expanded on a basis set of adiabatic molecular wave function $\chi_n(\underline{r},\underline{R})$ defined by

$$H_{el}\ \chi_n(\underline{r};\ \underline{R}) = \varepsilon_n(R)\ \chi_n(\underline{r};\ \underline{R}) \qquad 6.1$$

We note that the functions χ_n are also eigen-functions of L_z, the projection of the electronic angular momentum on the internuclear axis

$$L_z\ \chi_n\ (\underline{r};\ \underline{R}) = \Lambda_n\ \chi_n\ (\underline{r};\ \underline{R}) \qquad 6.2$$

In the case of a non symmetric system $\chi_n(\underline{r},\ \underline{R})$ will tend in the dissociation limit $(R\to\infty)$ to an atomic wave function centred either on A or B

$$\chi_n(\underline{r};\ \underline{R}) \underset{R\to\infty}{\sim} \phi_n\ (\underline{r}_a) \qquad 6.3$$

$$\chi_m(\underline{r};\ \underline{R}) \underset{R\to\infty}{\sim} \phi_m(\underline{r}_b) \qquad 6.4)$$

For a symmetric system

$$\chi_n(\underline{r};\ \underline{R}) \underset{R\to\infty}{\sim} \phi_n(\underline{r}_a) \pm \phi_n(\underline{r}_b) \qquad 6.5$$

depending on whether the state n is of gerade or ungerade symmetry. Then expanding $\psi(\underline{r},\ t)$ in the form

$$\psi(\underline{r},t) = \sum c_n(t)\ \chi_n(\underline{r};\underline{R}) \exp\left[-i\int^t \varepsilon_n(R)dt'\right] \qquad 6.6$$

one obtains the set of coupled differential equations for the coefficients $c_n(t)$

$$i\ \dot{c}_n = \sum_{n'\neq n} a_{n'}\ V_{nn'}\ \exp\left[-i\int^t (\varepsilon_n - \varepsilon_{n'})dt'\right] \qquad 6.7$$

where

$$V_{nn'} = \langle \chi_n \left|\frac{\partial}{\partial t}\right|_{\underline{r}} |\chi_{n'} \rangle \qquad 6.8$$

Unfortunately the matrix elements $V_{nn'}$ exhibit a linear dependence on the origin of the reference system $\underline{r}$ and certain of them do not even vanish asymptotically as $t \to \infty$. For example, consider the case of the matrix elements between two states j and k which dissociate both to the centre A

$$\chi_j(\underline{r};\ \underline{R}) \sim \phi_i(\underline{r}_a) \qquad 6.9$$

$$\chi_k(\underline{r};\ \underline{R}) \sim \phi_k(\underline{r}_a) \qquad 6.10$$

Then using relation (5.6)

$$\langle \chi_j \left|\frac{\partial}{\partial t}\right]_{\underline{r}} | \chi_k \rangle = \langle \chi_j \left|\frac{\partial}{\partial t}\right]_{\underline{r}_a} + p\ \underline{v}\cdot\nabla_{\underline{r}_a} | \chi_k \rangle \qquad 6.11$$

$$\langle \chi_j \left|\frac{\partial}{\partial t}\right]_{\underline{r}} | \chi_k \rangle \underset{k\to\infty}{\sim} \langle \phi_i |p\ \underline{v}\cdot\nabla_{\underline{r}_a} |\phi_k \rangle \qquad 6.12$$

which is non-zero if j and k are connected by a dipole transition. The equation (6.7) will not then yield an unambiguous probability amplitude.

Must we conclude that the molecular expansion is unsatisfactory? No general reply can be given since it all depends on how sensitive are the results to the lack of Galilean invariance.

In the following section we shall discuss a number of modifications of the expansion (6.6) which have been proposed to ensure translational invariance of the scattering equations even when a finite basis set is used.

6.2. Translation factors

The asymptotic conditions require that the basis function at large separation approach the separated atom functions multiplied by an appropriate translation factor of the type already used in the atomic state method. Thus would suggest expanding ψ in the form

$$\psi(\underline{r},t) = \sum c_n(t)\ \chi_n(\underline{r};R)\ \exp\left[i f_n(\underline{r},\underline{R})\underline{v}\cdot\underline{r} - i\int^t \varepsilon_n\, dt'\right] \qquad 6.13$$

Each basis function will then have the correct asymptotic form if $f_n \to p$ when $R\to\infty$ with r_a finite and $f_n \to (1-p)$ when $R\to\infty$ with r_b finite. The form of the translation factor $\exp\left[i\ f_n\ \underline{v}\cdot\underline{r}\right]$ for finite R is in principle arbitrary.

The first type of translation factor is the plane wave type proposed by Bates and McCarroll[32]. They choose

$$\begin{aligned} f_n &= p \quad \text{if } \chi_n \text{ dissociates to centre A} \\ &= (1-p) \quad \text{if } \chi_n \text{ dissociates to centre B} \end{aligned} \qquad 6.14$$

To get the scattering equations, it is convenient to separate the direct and rearrangement channels and rewrite (6.13) as

$$\psi(\underline{r},t) = \sum\left[a_n(t)\ \psi_n^a(\underline{r},t) + b_n\ \psi_n^b\ (\underline{r},t)\right] \qquad 6.15$$

where

$$\psi_n^a = \chi_n^a \exp\left[-i\ p\ \underline{v}\cdot\underline{r} - i\int^t (\varepsilon_n^a + \frac{1}{2}\ p^2 v^2)dt'\right] \qquad 6.16$$

$$\psi_n^b = \chi_b^b \exp\left[i(1-p)\underline{v}\cdot\underline{r} - i\int^t\{\varepsilon_n^b + \frac{1}{2}(1-p)^2\ v^2\}dt'\right] \qquad 6.17$$

It is straightforward to show that

$$\{H_{el} - i \frac{\partial}{\partial t}\Big]_{r}\}\, \psi_n^a = -i \frac{\partial \chi_n^a}{\partial t}\Big]_{\underline{r}_a} \exp\left[-i\, p\, \underline{v}\cdot\underline{r} - i\int(\varepsilon_n^a + \frac{1}{2} p^2 v^2)\, dt'\right] \qquad 6.18$$

$$\{H_{el} - i \frac{\partial}{\partial t}\Big]_{\underline{r}}\}\, \psi_n^b = -i \frac{\partial \chi_n^b}{\partial t}\Big]_{\underline{r}_b} \exp\left[i(1-p)\underline{v}\cdot\underline{r} - i\int^t\{\varepsilon_n^b + \frac{1}{2}(1-p)^2 v^2\}\, dt'\right] \qquad 6.19$$

Substitution of (6.15) in the eikonal equation and using 6.18, 6.19, yields the system of coupled equations, with $\underline{r}_m = (\underline{r}_a + \underline{r}_b)/2$.

$$\dot{a}_n + \sum_m \dot{b}_m S_{nm}^{ab} = -\sum_m a_m V_{nm}^{aa} - \sum_m b_m W_{nm}^{AB} \qquad 6.20$$

$$\dot{b}_n + \sum_m \dot{a}_m S_{nm}^{ba} = -\sum_m b_m V_{nm}^{bb} - \sum_m a_m W_{nm}^{ba} \qquad 6.21$$

where

$$S_{nm}^{ab} = \langle\psi_n^a|\psi_m^b\rangle = \langle\chi_n^a|\, e^{i\underline{v}\cdot\underline{r}_m}\, |\chi_m^b\rangle \exp\left[i\int(\varepsilon_n^a - \varepsilon_m^b)\, dt'\right] \qquad 6.22$$

$$S_{nm}^{ba} = \langle\psi_n^b|\psi_m^a\rangle = \langle\chi_n^b|\, e^{-i\underline{v}\cdot\underline{r}_m}\, |\chi_m^a\rangle \exp\left[i\int(\varepsilon_n^b - \varepsilon_m^a)\, dt'\right] = S_{mn}^{ab} \qquad 6.23$$

$$V_{nm}^{aa} = -i \langle\chi_n^a|\, \frac{\partial}{\partial t}\Big]_{\underline{r}_a} |\chi_m^a\rangle \exp\left[i\int^t(\varepsilon_n^a - \varepsilon_m^a)\, dt'\right] \qquad 6.24$$

$$V_{nm}^{bb} = -i \langle\chi_n^b|\, \frac{\partial}{\partial t}\Big]_{\underline{r}_b} |\chi_m^a\rangle \exp i\int(\varepsilon_n^b - \varepsilon_m^b)\, dt' \qquad 6.25$$

$$W_{nm}^{ab} = -i \langle\chi_n^a|\, e^{i\underline{v}\cdot\underline{r}_m} \frac{\partial}{\partial t}\Big]_{\underline{r}_b} |\chi_m^b\rangle \exp\left[i\int^t(\varepsilon_n^a - \varepsilon_m^b)\, dt'\right] \qquad 6.26$$

$$W_{nm}^{ba} = -i \langle\chi_n^b|\, e^{-i\underline{v}\cdot\underline{r}_m} \frac{\partial}{\partial t}\Big]_{\underline{r}_a} |\chi_m^a\rangle \exp\left[i\int^t(\varepsilon_n^b - \varepsilon_m^a)\, dt'\right] \qquad 6.27$$

It may be remarked that since H_{el} is an Hermitian operator

$$W_{mn}^{ba} - W_{nm}^{ab} = i\,\dot{S}_{nm}^{ab} \qquad 6.28$$

$$V_{jk}^{bb} = V_{kj}^{bb} \qquad V_{jk}^{aa} = V_{kj}^{aa} \qquad 6.29$$

These conditions guarantee that in equations 6.20, 6.21, the probability is conserved.

We thus see that with the introduction of plane wave translation factors, the difficulties of the PSS method are apparently removed. All the coupling matrix elements vanish for large R and there is no dependence on the reference frame. (However, it should be borne in mind that the improvement may be purely formal, since the choice of translation factor is arbitrary in the range of finite R).

Unfortunately the exchange matrix elements W_{jk}^{AB} are lengthy to evaluate, especially since they must be calculated separately for each velocity. It is not surprising that few applications of the method have been carried out. The most extensive application is the recent work on He^{+2} - H collisions reported by Hatton et al[33] and Winter and Hatton[34], Winter et al[91].

Because of the complexity of the exchange integrals various approximate forms of the matrix elements have been suggested. For example at low velocities it might appear legitimate to expand (exp $(i\underline{v}\cdot\underline{r}_m)$ in powers of $\underline{v}$ and retain only leading terms. To a first approximation the coupled equations become

$$\dot{a}_n = - \sum a_m V_{nm}^{aa} - \sum b_m W_{nm}^{ab} \qquad 6.30$$

$$\dot{b}_n = - \sum b_m V_{nm}^{bb} - \sum a_m W_{nm}^{ba} \qquad 6.31$$

where

$$W_{nm}^{ab} = \langle \chi_n^a \left| \frac{\partial}{\partial t} \right]_{\underline{r}_b} \left| \chi_m^b \right\rangle \exp\left[i\int^t (\varepsilon_n^a - \varepsilon_m^b)\, dt' \right] \qquad 6.32$$

$$W_{nm}^{ba} = \langle \chi_n^b \left| \frac{\partial}{\partial t} \right|_{\underline{r}_a} \left| \chi_m^a \right\rangle \exp\left[i\int^t (\varepsilon_m^a - \varepsilon_n^b)\, dt' \right] \qquad 6.33$$

Unfortunately the relation (6.28) is then no longer verified and the solutions of 6.32, 6.33 will not satisfy unitarity. But since, preservation of unitary is a desirable feature in a numerical calculation several procedures have been proposed to force unitarity. Briggs and Taulbjerg[35] replace 6.32 by exchange elements evaluated at a common origin, chosen in their case to be the centre of mass. Bates and Williams[36] replace 6.32 and 6.33 by their mean value. Although no formal justification for these ad-hoc symmetrization procedures can be given (Green[37]) it may be argued that the approach is a reasonable one if the exchange matrix elements are insensitive to choice of origin. See also Riera and Salin[38], Ponce[39], Schimd[40], Crothers and Hughes[41].

Apart from the complicated exchange and overlap integrals characteristic of the Bates-McCarroll method, there are other difficulties associated with the plane wave translation factors. In the interaction region of a slow encounter, an electron belongs to neither centre A nor centre B. For example an electron in state n, described by χ_n^a is not localized on centre A except at large internuclear distances. It is therefore unreasonable to take f_n as in (6.14), particularly if χ_n^a is subject to an avoided crossing with a state χ_m^b. For this and other reasons, Schneiderman and Russek[42] argued that a more flexible form of f_n should be chosen. Instead of (6.14), we might take by analogy with Schneiderman and Russek[42].

$$f_n(\underline{r}, \underline{R}) = \left[-p - \frac{1}{2}(1 - \underline{\hat{r}}\cdot\underline{\hat{R}})\right] / (1 + d/R^2) \qquad 6.34$$

where d is to be regarded as a relatively small distance below which the electron essentially cases to be localized on either nucleus. For R large, we see that for an electron in state χ_n^a, $\underline{\hat{r}}\cdot\underline{\hat{R}} \to -1$, while for an electron in state χ_n^b, $\underline{\hat{r}}\cdot\underline{\hat{R}} \to 1$. Asymptotically the form (6.35) will yield the plane wave form. Although the functional form of (6.35) is relatively complex, it does have one important simplifying feature in that it is independent of n. The basis states will then be orthogonal. The overlap matrix reduces to the unit matrix, the $\partial/\partial t$ matrix elements will not involve operators containing exp $(i\underline{v}\cdot\underline{r})$. There is no necessity to separate the direct and rearrangement channels as with the plane wave factors. We may thus expand $\psi(\underline{r},t)$ as

$$\psi(\underline{r},t) = \sum c_n(t)\ \psi_n(\underline{r},t) \qquad 6.35$$

where

$$\psi_n(\underline{r},t) = \chi_n(\underline{r};\ R)\ \exp\ (i\, f\, \underline{v}\cdot\underline{r}) \qquad 6.36$$

Using the identity

$$[g, H_e] = \underline{\nabla}\, g \cdot \underline{\nabla} + \frac{1}{2} \nabla^2 g \qquad 6.37$$

It is easily shown that

$$\langle \psi_n | H_{el} | \psi_m \rangle =$$

$$= \delta_{nm}\, \varepsilon_n + i(\varepsilon_n - \varepsilon_m) \langle \chi_n | f\, \underline{v} \cdot \underline{r} | \chi_m \rangle + \frac{1}{2} \langle \chi_n | [\underline{\nabla} f\, \underline{v} \cdot \underline{r}]^2\, \chi_m \rangle \qquad 6.38$$

The resulting coupled equations then become

$$i\, \dot{c}_n = \sum_m c_m V_{nm} \exp i\int^t (\varepsilon_n - \varepsilon_m)\, dt' \qquad 6.39$$

where

$$V_{nm} = -i(\varepsilon_n - \varepsilon_n) \langle \chi_n | f\, \underline{v} \cdot \underline{r} | \chi_m \rangle + \frac{1}{2} \langle \chi_n | [\underline{\nabla}(f\, \underline{v} \cdot \underline{r})]^2 | \chi_m \rangle$$

$$-i \langle \chi_n | \frac{\partial}{\partial t}\Big]_{r} | \chi_m \rangle + \langle \chi_n | \dot{f}\, \underline{v} \cdot \underline{r} | \chi_m \rangle \qquad 6.40$$

It is easily verified from the asymptotic form of f ($\underline{r}$, $\underline{R}$) that for large R all the non diagonal coupling matrix elements vanish. Provided some reasonable choice of switching function is adopted, the calculation of V_{nm} involves standard molecular integrals.

Unfortunately there is no guarantee that any particular choice of f is optimal or even reasonable. The elimination of the formal difficulties of the PSS method has been achieved only at the expense of introducing a new problem - the sensitivity of the results to the form of switching function. In the absence of any well defined criteria, it is legitimate to ask if the explicit introduction of translation factors presents any real advantage over the ad-hoc methods frequently used to solve the PSS equations. Indeed many ad-hoc procedures simply correspond to a particular type of switching function. For example, taking f to be a constant f_o in the interaction region $R < R_o$, the matrix element V_{nm} reduces to

$$V_{nm} = -i(\varepsilon_n - \varepsilon_m) f_o \langle \chi_n | \underline{v} \cdot \underline{r} | \chi_m \rangle - i \langle \chi_n | \frac{\partial}{\partial t}\Big]_{\underline{r}} | \chi_m \rangle =$$

$$= -i \langle \chi_n | \frac{\partial}{\partial t}\Big]_{\underline{r}'} | \chi_m \rangle \qquad 6.41$$

where $\underline{r}' = \underline{r} + f_0 \underline{R}$. In other words a constant f corresponds to a simple change of origin for the $\partial/\partial t$ matrix elements. In this case the optimal choice of f_o will be that for which the convergence of the basis set is most rapid.

Ideally, a variational principle may be used to determine the switching function. Riley and Green[43], using the Euler Lagrange method have derived general equations for f. However, these equations are exceedingly complicated and they have only been applied to the case where f is independent or r. (See also Crothers and Hughes[41]). The prospects of finding a practical procedure for optimizing the switching function do not therefore seem promising. However, there have been so few detailed calculations on the influence of translation factors that it is premature to draw a definite conclusion. Most of the applications have been restricted to one electron systems such as $H^+ - H$, $He^{2+} - H$ collisions. Unfortunately, although these systems offer the advantage that the matrix elements can be calculated with precision, they are not typical of most atom-ion collisions. Because of the symmetries associated with a pure two-centre Coulomb potential, curve crossings are allowed which in more complex systems are avoided. As a consequence, the dynamic matrix elements become important only at relatively high velocities, where translation effects are large. On the other hand for complex systems, the $\partial/\partial t$ matrix elements are often not known with precision and spurious translation effects may arise. When approximate wave functions are used, an incorrect dependence on the origin could result, which might lead to non physical translation effects.

6.3. Radial and Rotational coupling terms

In the calculation of the matrix element

$$< \chi_n \left|\frac{\partial}{\partial t}\right| \chi_m > \qquad 6.42$$

it should be remarked the time derivative must be taken with respect to a fixed space axis. In general the molecular wave functions are specified with respect to a body-fixed axis with the internuclear axis taken as the axis of quantization. Account must then be taken of the rotation of the internuclear axis during the collision. Let α be the angle between $\underline{R}$ and $\underline{v}$

$$\sin \alpha = b/R \qquad \cos \alpha = vt/R \qquad 6.43$$

If (x', y', z') are the components of $\underline{r}$ with respect to the rotating axis and (x, y, z) the components with respect to a space fixed axis, with z taken in the incident direction, then choosing the (x, z) plane to be the plane of rotation, we have

$$x' = x \cos\alpha + z \sin\alpha \qquad 6.44$$

$$y' = y \qquad 6.45$$

$$z' = z \cos\alpha - x \sin\alpha \qquad 6.46$$

It is easily established that

$$\frac{\partial}{\partial t} = \frac{\partial'}{\partial t} + \frac{d\alpha}{dt}\left(z' \frac{\partial}{\partial x'} - x' \frac{\partial}{\partial z'}\right) \qquad 6.47$$

where $\partial'/\partial t$ designates differentiation with respect to x', y', z' fixed. Since

$$\frac{d\alpha}{dt} = -\frac{bv}{R^2} \qquad 6.48$$

we have

$$\frac{\partial}{\partial t} = \frac{\partial'}{\partial t} + i\,\frac{bv}{R^2}\,Ly' \qquad 6.49$$

where Ly' is the y' component of the angular momentum operator. Conventionally, the first term of (6.49) $\partial'/\partial t$, which represents the time variation with respect to the body-fixed axis is termed radial coupling. The second term, which represents the Coriolis force arising from the rotation of the coordinate system used to define the molecular wave function, is termed rotational coupling.

The two terms of (6.49) are subject to different selection rules. For the radial coupling term, the axial component of the electron angular momentum is conserved, while for rotational coupling $\Lambda_n = \Lambda_m \pm 1$. If in addition as is the case in the quasi diabatic representation, the wave-functions χ_n are described by a single configuration of molecular orbitals, the matrix elements are non-vanishing only if χ_m and χ_n differ by one molecular orbital. As a general rule, radial coupling is the dominant contribution in the vicinity of an avoided crossing between states of the same symmetry. However in other situations, the radial and rotational contributions can be of comparable importance. Rotational coupling can be very important at small internuclear distances, as the Σ, π, Δ states become degenerate in the united atom limit.

7. MOLECULAR STATE EXPANSION (quantum mechanical formulation)

When the energy transfer is comparable with the collision energy, the eikonal approximation is unsatisfactory and a quantum mechanical approach is called for. Since the relative velocity is then low, we need only consider the formulation of the problem in a molecular state (PSS) expansion.

Just as in the eikonal approximation, the existence of spurious non physical coupling terms in the asymptotic region may render the standard PSS method formally unacceptable and in principle the introduction of appropriate translation factors may become necessary.

However, in practice, at the low collision energies where a quantum mechanical treatment is required, the problem is usually not critical since inelastic transitions occur only in the vicinity of a pseudo crossing of the molecular potential energy surfaces. The explicit introduction of translation factors may then be avoided by the use of some ad-hoc procedure with no serious loss of precision. A new method of incorporating translation factors has been proposed by Delos and Thorson[44], but a detailed discussion at this stage is perhaps premature. In the following section we shall limit ourselves to a consideration of some ad-hoc methods.

7.1. General considerations

It is instructive to rewrite the kinetic energy operator T in terms of the coordinates $\underline{R}$ and $\underline{r}$ used in section 5. We have

$$T = -\frac{1}{2\mu}\nabla_R^2 - \frac{1}{2m_r}\nabla_r^2 + \underline{\nabla}_R\cdot\underline{\nabla}_r\left(-\frac{p}{\mu}+\frac{1}{M_a}\right) \qquad 7.1$$

where μ is as defined in 2.2 and 2.3 and

$$\frac{1}{m_r} = \frac{1}{m_e} + \frac{(1-p)^2}{M_a} + \frac{p^2}{M_b} \qquad 7.2$$

Of course, when $p = M_b/(M_a + M_b)$, $\underline{r}$ becomes $\underline{r}_g$ and the cross term in (7.1) vanishes yielding

$$T = -\frac{1}{2\mu}\nabla_R^2 - \frac{1}{2m}\nabla_{r_g}^2 \qquad 7.3$$

Expanding the total wave-function $\Psi(\underline{r}, \underline{R})$ on a molecular basis set

$$\Psi(\underline{r}, \underline{R}) = \sum_n F_n(\underline{R})\,\chi_n(\underline{r}, \underline{R}) \qquad 7.4$$

with $\underline{r}$ and $\underline{R}$ considered as independent variables, applying the condition

$$< \chi_n \mid H - E \mid \Psi > = 0 \tag{7.5}$$

and neglecting certain small isotopic terms, we obtain the coupled equations

$$\nabla_R^2 F_n + \sum_n \{2\underline{A}_{nm} \cdot \underline{\nabla}_R F_m + F_m B_{nm}\} + 2 \ (E - \varepsilon_n) F_n = 0 \tag{7.6}$$

where

$$\underline{A}_{nm} = < \chi_n \mid (\underline{\nabla}_R)_{\underline{r}_g} \mid \chi_m > \tag{7.7}$$

$$B_{nm} = < \chi_n \mid (\nabla_R^2)_{\underline{r}_g} \mid \chi_m > \tag{7.8}$$

It should be remarked that the gradient operator in 7.7 and 7.8 is defined with respect to $\underline{r}_g$ fixed, irrespective of the initial choice of p. The matrix elements $\underline{A}_{nm}$ do not vanish asymptotically when m and n are connected by a dipole transition in the separated atoms limit, with the result that it is not strictly possible to extract the scattering amplitudes. In collisions where excitation and rearrangement processes are not strongly coupled, various ad-hoc solutions to the problem have been suggested. For example, in the process of electron capture by a multicharged ion from a neutral atom, Butler et al.[68] have proposed the simple replacement of 7.7 and 7.8 by

$$\underline{A}_{nm} = <\chi_n \mid (\underline{\nabla}_R)_{\underline{r}_b} \mid \chi_m > \tag{7.9}$$

$$B_{nm} = <\chi_n \mid (\nabla_R^2)_{\underline{r}_b} \mid \chi_m > \tag{7.10}$$

where the gradient operator is defined with respect to $\underline{r}_b$ fixed rather than $\underline{r}_g$. (It is assumed that the nucleus of the multicharged ion is located at B.) The spurious coupling terms are thereby eliminated. Unfortunately, the procedure has no formal justification; indeed it is strictly equivalent to neglecting the cross term $\underline{\nabla}_R \cdot \underline{\nabla}_r$ in the kinetic energy operator 7.1. In the case where $M_b >> M_a$, the centre of mass is close to B, the use of 7.9 and 7.10 may be acceptable. However, calculations by Gargaud et al.[66] would indicate that the adoption of such a procedure may lead to more serious inconsistencies than those we are seeking to eliminate.

In consequence, when confronted with a problem in which

spurious asymptotic couplings arise, the safest approach in the present state of the art is to introduce a simple cut-off procedure. For example, one might introduce the notion of an interaction sphere of radius R_0, an arbitrary internuclear distance beyond which the eigen functions χ_n are purely atomic in character. The spurious couplings are then taken to be zero for R greater than R_0. The sensitivity of the scattering amplitudes to R_0 and the convergence of the basis set provide a measure of the consistency of the calculations. This procedure may appear to be a crude one, but it does have the virtue of yielding a quantitative measure of the defects of the molecular model.

It is worth remarking that in actual practice, it is easy to confuse errors arising from approximations used in the molecular structure calculations of ε_n and $\underline{A}_{mn}$ with the defects of the PSS method referred to in the preceding paragraphs. For, only if the accuracy of the dynamic coupling matrix elements can be guaranteed to high precision, is it possible to analyse the PSS method with any degree of confidence.

7.2. Partial wave decomposition

Let us now return to equation (7. 6). Since the molecular wave functions are quantized with respect to the internuclear axis it is convenient to introduce a rotating system $(\bar{X}, \bar{Y}, \bar{Z})$ obtained by a rotation of the (X', Y', Z') reference frame through the Euler angles (η, ξ, 0), where (ξ, η) are the azimuthal coordinates of the internuclear axis in the space-fixed system. With this choice, the $\bar{Z}$ axis is in the direction of the internuclear axis and the $\bar{Y}$ axis is in the X' Y' plane. The electronic coordinates in the rotating axis are $\underline{r}$ $(\bar{x}, \bar{y}, \bar{z})$.

The kinetic energy operator in the $\underline{R}'$ system transforms to the rotating system as

$$-\frac{1}{2\mu}\nabla^2_{\underline{R}'} = -\frac{1}{2\mu R^2}\frac{\partial}{\partial R'}\left(R^2\frac{\partial}{\partial R'}\right) + T_r + T_c \qquad 7.11$$

where

$$T_r = -\frac{1}{2\mu R^2 \sin^2\xi}\left[\left(\sin\xi\frac{\partial}{\partial\xi}\right)^2 + \left(\frac{\partial}{\partial\eta} - i\cos\xi\, L_{\bar{z}}\right)^2\right] \qquad 7.12$$

$$T_c = \frac{1}{2\mu R^2}\left[L_{\bar{x}}^2 + L_{\bar{y}}^2 + 2i\, L_{\bar{y}}\frac{\partial}{\partial\xi} - \frac{2i}{\sin\xi}L_{\bar{x}}\frac{\partial}{\partial\eta} - 2\cot\xi L_{\bar{x}} L_{\bar{z}}\right]$$

7.13

and $L_{\bar{x}}$, $L_{\bar{y}}$, $L_{\bar{z}}$ are the $\bar{x}$, $\bar{y}$, $\bar{z}$ components of the electronic angular momentum operator.

Then adopting a partial wave expansion for $F_n(\underline{R})$ of the form

$$F_n(\underline{R}) = \frac{1}{R} \sum_K F_n^K(R) \; Y_{K_{\Lambda n}}(\xi, \eta) \qquad 7.14$$

where Λ_n is the quantum number associated with state χ_n it can be shown that equation (7.23) reduces to the following set of coupled differential equations

$$\frac{d^2}{dR^2} F_m^K + \left[2\mu \left[E - \varepsilon_m(R) \right] - \frac{K(K+1) - \Lambda_m^2}{R^2} \right] F_m^K$$

$$= \sum_n F_n^K \left< Y_{K\,\Lambda_m} \chi_m \left| 2\mu \, T_c - \frac{\partial^2}{\partial R^2} \right| \chi_n \, Y_{K\,\Lambda_n} \right> \qquad 7.15$$

$$- 2 \sum_n \frac{d F_n^K}{dR} \left< Y_{K\,\Lambda_m} \chi_m \left| \frac{\partial}{\partial R'} \right| \chi_n \, Y_{K,\Lambda_n} \right>$$

where in the Dirac brackets, it is understood that integration is over the electron coordinates $\underline{r}$ and the angular variables ξ, η. To a good approximation (Van Vleck[45]), the right hand side of (7.15) can be simplified in most cases of physical interest to yield

$$\frac{d^2 F_m^K}{dR^2} + \left[2\mu \left[E - \varepsilon_m(R) \right] - \frac{K(K+1) - \Lambda_m^2}{R^2} \right] F_m^K$$

$$= \pm \frac{i}{R^2} \sum_n F_n^K \left< \chi_m \left| L_{\bar{y}} \right| \chi_n \right> \delta(\Lambda_m, \Lambda_n \pm 1)$$

$$\times \left[(K \mp \Lambda_m)(K \pm \Lambda_m + 1) \right]^{1/2}$$

$$-2 \sum_n \frac{dF_n^K}{dR} \left< \chi_m \left| \frac{\partial}{\partial R'} \right| \chi_n \right> \delta(\Lambda_m, \Lambda_n)$$

$$- \sum_n F_n^K \left< \chi_m \left| \frac{\partial^2}{\partial R^2} \right| \chi_n \right> \delta(\Lambda_m, \Lambda_n) \qquad 7.16$$

The last term involving $<\chi_m|\partial^2/\partial R'^2|\chi_n>$ is also considered to be of minor importance but Zimmermann and George[46] in a study of atom molecule collisions that neglect of such a term can lead to instability problems in solving the differential equations. In any case, as we shall see in the following section, the inclusion of this term is straightforward.

From equation (7.16), we thus see that as in the impact parameter approximation two types of dynamic coupling are responsible for inelastic transitions - radial coupling between molecular states of the same symmetry and rotational coupling between states whose value of Λ_n differ by ± 1.

7.3. Diabatic transformations

The presence of a first order derivative in (7.16) is inconvenient for the application of many standard numerical methods for the solution of the coupled differential equations. It is thus common in practice to transform equation (7.16) to one in which first order derivatives are absent. Such a transformation is equivalent to the diabatic representation as introduced by Smith[47]. See for example Baer[48], Heil and Dalgarno[49].

To illustrate the problem, let us rewrite equation (7.16) in matrix form. Since the coefficients of the first order derivates of F_n do not involve the rotational coupling matrix elements, the transformation will concern only those states coupled by the radial coupling terms. Limiting consideration to the radial coupling terms, we then have

$$\frac{d^2}{dR^2}\underline{F}^K + 2\,\underline{A}\,\frac{d\underline{F}^K}{dR} + \underline{B}\,\underline{F}^K - 2\mu\,\underline{\varepsilon}\,\underline{F}^K =$$

$$= \left[2\mu\,E - \frac{K(L+1)}{R^2} + \frac{\Lambda^2}{R^2}\right]\underline{F}^K \qquad 7.17$$

where

$$A_{ij} = <\chi_i|\frac{\partial}{\partial R}|\chi_j> \qquad 7.18$$

$$B_{ij} = <\chi_i|\frac{\partial^2}{\partial R^2}|\chi_j> \qquad 7.19$$

and $\underline{\varepsilon}$, the adiabatic energy matrix, is diagonal.

Let us now write

$$\underline{F}^{K}(R) = \underline{C}(R)\ \underline{G}^{K}(R) \qquad 7.20$$

where $\underline{C}(R)$ is the solution of the equation

$$\frac{d}{dR}\underline{C}(R) + \underline{A}(R)\ \underline{C}(R) = 0 \qquad 7.21$$

which satisfies the condition

$$\underline{C}(R) \underset{R\to\infty}{\sim} \underline{I} \qquad 7.22$$

Making use of the identify

$$\underline{B} = \underline{A}^{2} + \frac{d}{dR}\underline{A} \qquad 7.23$$

it is easily established that $\underline{G}^{K}(R)$ is solution of the equation

$$\frac{d^{2}}{dR^{2}}\underline{G}^{K}(R) - \underline{V}^{d}\ \underline{G}^{K}(R) = \left[2\mu E - \frac{K(K+1)}{R^{2}} + \frac{\Lambda^{2}}{R^{2}}\right]\underline{G}^{K}(R) \qquad 7.24$$

where

$$\underline{V}^{d} = \underline{C}^{-1}(R)\ \underline{\varepsilon}(R)\ \underline{C}(R) \qquad 7.25$$

The matrix $\underline{V}^{d}$ may now be considered as the diabatic potential matrix. Although $\underline{V}^{d}$ is by construction identical to the adiabatic potential matrix in the asymptotic limit, it will differ in general for finite R. The difference will be particularly striking in the region of an avoided crossing of the adiabatic potential energies, where certain elements of the $\underline{A}$ matrix may be large. The diagonal elements of $\underline{V}^{d}$ may then cross.

Let us now consider the diabatic molecular eigen functions χ_{n}^{d} as defined by

$$\chi_{j}^{d} = \sum_{k} \chi_{k}\, C_{kj} \qquad 7.26$$

The matrix of the $\partial/\partial R$ operator in the diabatic representation then becomes

$$<\chi_i^d |\frac{\partial}{\partial R}| \chi_j^d> = \sum_k \sum_{k'} c_{ik} <\chi_k |\frac{\partial}{\partial R}| \chi_{k'}> c_{k'j} \qquad 7.27$$

$$= \sum_k c_{ik} \sum_{k'} \{A_{kk'} c_{k'j} + \frac{d}{dR} c_{k'j} \delta_{kk'}\}$$

$$= \underline{C} \{\underline{A}\, \underline{C} + \frac{d}{dR} \underline{C}\}_{ij} = 0 \qquad 7.28$$

We thus see that the diabatic representation as defined by (7.20) and (7.21) is identical to that of Smith[47].

A simple illustration of the problem is provided by the two-state expansion, where an analytic solution can be found. The matrix $\underline{A}$ is then

$$\underline{A} = \begin{pmatrix} 0 & A_{12} \\ -A_{12} & 0 \end{pmatrix} \qquad 7.29$$

where

$$A_{12} = <\chi_1 |\frac{\partial}{\partial R}| \chi_2 > \qquad 7.30$$

It is then trivial to establish that the solution of (7.21) is

$$\underline{C}(R) = \begin{pmatrix} \cos\zeta(R) & \sin\zeta(R) \\ -\sin\zeta(R) & \cos\zeta(R) \end{pmatrix} \qquad 7.31$$

where

$$(R) = \int_R^{\infty} A_{12}(R')\, dR' \qquad 7.32$$

and the diabatic potential matrix $\underline{V}^d$ is

$$\underline{V}^d = \begin{pmatrix} \varepsilon_1 \cos^2\zeta + \varepsilon_2 \sin^2\zeta & \frac{1}{2}(\varepsilon_1-\varepsilon_2)\sin 2\zeta \\ \frac{1}{2}(\varepsilon_1-\varepsilon_2)\sin 2\zeta & \varepsilon_1 \sin^2\zeta + \varepsilon_2 \cos^2\zeta \end{pmatrix} \qquad 7.33$$

The diagonal terms of $\underline{V}^d$ will therefore cross whenever ζ passes through $(2n + 1)\,\pi/4$ for integer n.

The form of $\underline{V}^d$ would appear to give a certain justification for the Landau-Zener approach to treat transitions occuring at avoided crossings. However, there are some fundamental differences which should be mentioned. In the usual version of the Landau-Zener treatment (Nikitin[50]) the adiabatic representation is taken as the starting point; a suitable linear combination of the two adiabatic states is then constructed so that the diagonal terms of the hamiltonian matrix vary smoothly through the avoided crossing. This procedure therefore assumes implicitly that

i) A_{12} is only of importance in the vicinity of the crossing point $R = R_x$

ii) $\zeta \to \pm\,\pi/2$ for $R \ll R_x$

Neither condition is necessarily satisfied in reality, with the result that no simple physical interpretation can be given to the diabatic representation as defined by (7.21). Of course, it remains a simple and elegant way of reducing the scattering equations to a form suitable for a numerical computations.

A striking illustration is furnished by the charge exchange process

$$N^{+3} + H \to N^{+2} + H^{+} \qquad 7.34$$

The molecular system $(NH)^{3+}$ has been the object of several independent investigations (Christensen et al[51], McCarroll and Valiron[52]; Butler[53]). The calculated potential energy curves, shown in fig. (7.2) exhibit only one effective avoided crossing involving the σ_{1sH} orbital correlated to the entry channel and the σ_{3sN} orbital correlated to the exit channel with N^{+2} in the state $N^{2+}\ (1s^2\ 2s^2\ 3s)\,^2S$. Detailed calculations of the radial coupling matrix have been calculated by Hanssen[54] as a function of R for several different origins of the electron coordinates. In the vicinity of the avoided crossing $R \cong R_x$, the radial coupling matrix is large and its sensitivity to the choice of origin is feeble. However, for $R < R_x$, as the radial coupling matrix weakens, its sensitivity to the choice of origin increases. This behaviour is easily understood from (6.11), the sensitivity being proportional to the matrix element

$$(\varepsilon_i - \varepsilon_f) < \chi_i | \underline{r} | \chi_f >$$

Because of this, $\zeta(R)$ may differ considerably from $\pm\pi/2$ for $R \ll R_x$. An illustration of this is given in figure (7.2) where the elements of the diabatic matrix V^{-d} are plotted for an origin centred on the N nucleus and an origin centred on the H nucleus. In neither case do the diabatic curves become identical to the Landau-Zener model. As a consequence, it would be unwise to attach too much physical importance to V^{-d} when the diagonal elements differ from the adiabatic energies at distancrs when the radial coupling is small. For example, at the inner crossing of the diabatic states observed when the origin is on the H nucleus, the non-diagonal elements are so large, that no simple interpretation is possible. This is confirmed by the fact that the calculated cross sections do not depend on the origin at low collision energies ($\lesssim$ 100 eV).

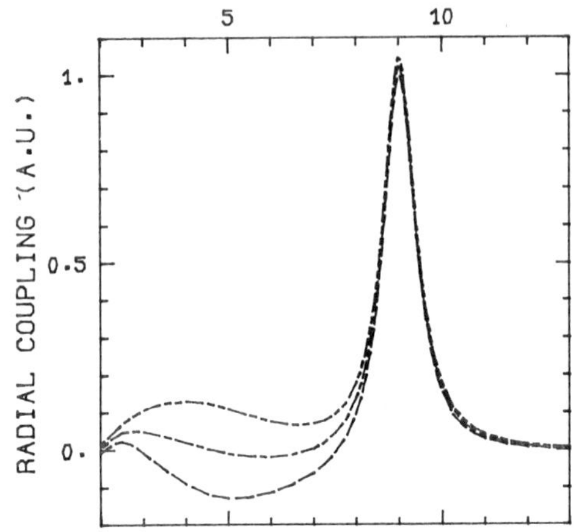

Fig. 7.1 - Radial coupling matrix element between the $4\ ^2\Sigma^+$ and $5^2\Sigma^+$ states of NH^{+3} (see fig. 9.7). Origin on N nucleus - - -; origin on mid-point of axis -.-. ; origin on H nucleus -..-

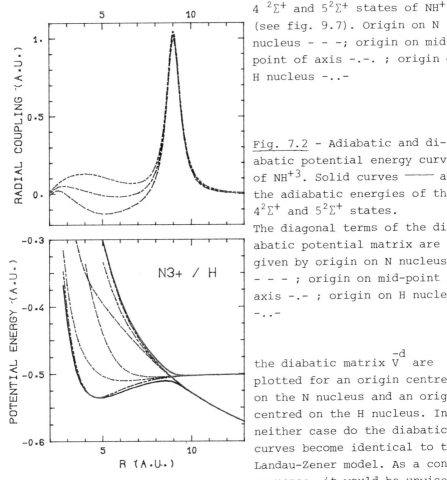

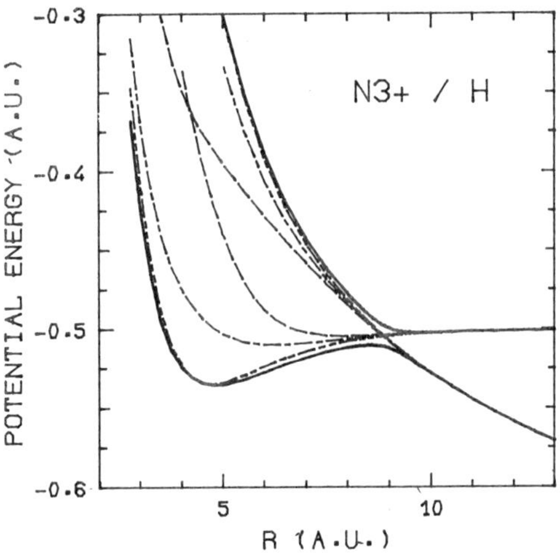

Fig. 7.2 - Adiabatic and diabatic potential energy curves of NH^{+3}. Solid curves —— are the adiabatic energies of the $4^2\Sigma^+$ and $5^2\Sigma^+$ states.
The diagonal terms of the diabatic potential matrix are given by origin on N nucleus - - - ; origin on mid-point of axis -.- ; origin on H nucleus -..-

8. DIABATIC REPRESENTATIONS

In section 7.c, we have seen how to obtain a diabatic representation by diagonalizing the radial coupling matrix of the adiabatic states. As a practical procedure, such a diabatic representation is limited to those systems where the number of adiabatic states effectively coupled in the collision process remains small. But under these conditions, the transformation to a diabatic representation is only necessary in very low energy collisions when quantal methods must be employed. So wherein lies the interest to diabatic representations in collision theory?

The basic problem is that in many systems, the number of accessible avoided crossings in the adiabatic representation may be very great. To illustrate the point let us consider the simple system He_2^+. The collision of He^+ and He in their ground state will give rise adiabatically to two molecular states, ${}^2\Sigma_u^+$, ${}^2\Sigma_g^+$. Adopting the molecular orbital approach to describe the electronic molecular states, we may attribute the configuration $(1s\sigma_g^2)\ 2p\sigma_u$ to the ${}^2\Sigma_u^+$ state and the configuration $1s\sigma_g(2p\sigma_u^2)$ to the ${}^2\Sigma_g^+$ state. Since the $1s\sigma_g$ orbital is attractive and the $2p\sigma_u$ orbital repulsive, it is trivial to establish that the ${}^2\Sigma_u^+$ state is the ground state and possesses a potential minimum, while the ${}^2\Sigma_g^+$ state is repulsive. According to the molecular orbital correlation diagram for symmetric molecules (Herzberg[56]), the ${}^2\Sigma_u^+$ state is correlated in the united atom limit to Be^+ $(1s^2\ 2p)$; the ${}^2\Sigma_g^+$ state is correlated to Be^+ $(1s\ 2p^2)$, all states of which lie above the first ionization threshold of Be^+. As a consequence in the independent particle model, the ${}^2\Sigma_g^+$ state will cross state of the configuration $(1s\sigma_g)^2 \cdot$ $\cdot (nl\sigma_g)$, which are correlated in the united atom limit to $Be^+(1s^2nl)$. When account is taken of electron correlation effects these crossings become avoided crossings. At each avoided crossing, the adiabatic states are strongly coupled, so that an infinite number of states are required to describe the collision process. A schematic diagram is given in figure (8.2). Sidis has termed such crossing Diabatic II.

In such cases the use of an adiabatic representation is obviously out of question. An approach based on the independent particle model seems more appropriate. Let us then define an approximate wave function χ_n^d, based for example on the Hartree-Fock procedure or on some other judicious choice of one-electron molecular orbitals. The function χ_n^d will then be an antisymmetrized

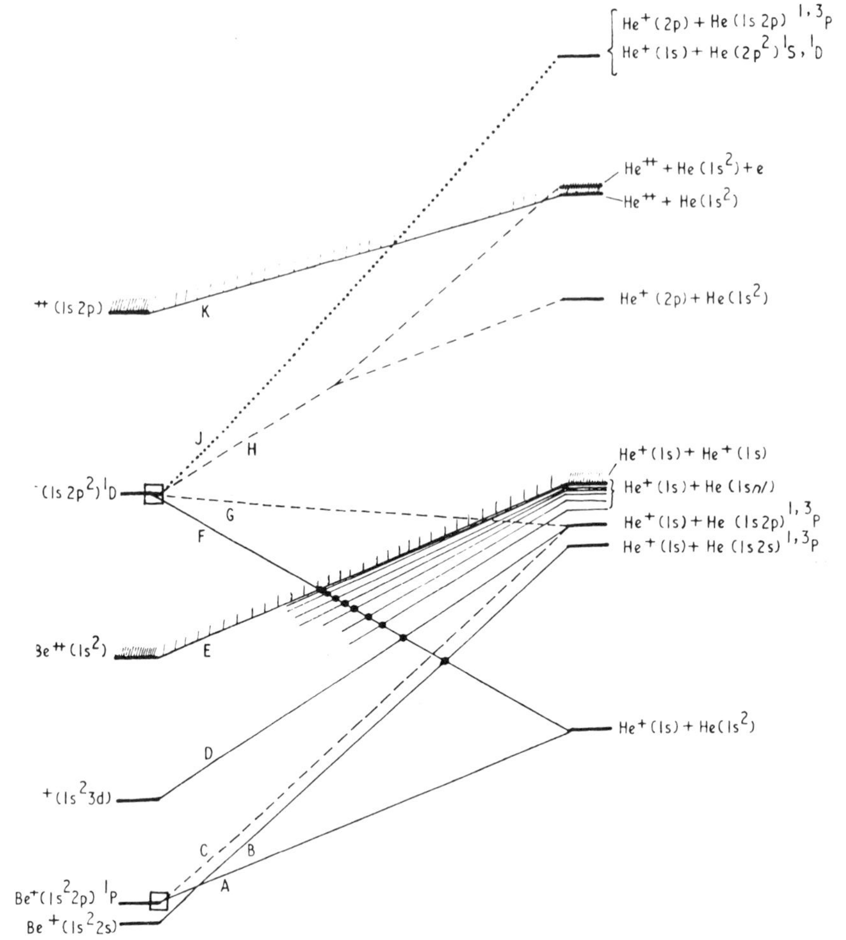

Fig. 8.1 - Schematic correlation diagram for the He^+ - He system. A, $\{(1s\sigma_g)^2\ (2p\sigma_u)\}\ ^2\Sigma_u^+$; B, $\{(1s\sigma_g)^2\ (2s\sigma_g)\}\ ^2\Sigma_g$; C, $\{(1s\sigma_g)^2 (2p\pi_u)\}\ ^2\Pi_u$; D, $\{(1s\sigma_g)^2\ (3d\sigma_g)\}\ ^2\Sigma_g$; E, $\{(1s\sigma_g)^2\}\ ^1\Sigma_g^+$ (of He_2^{2+}); F, $\{(1s\sigma_g)\ (2p\sigma_u)^2\}\ ^2\Sigma_g^+$; G, $\{((1s\sigma_g)\ (2p\sigma_u)\ ^3\Sigma_u)\ (2p\pi_u)\}\ ^2\Pi_g$; H, $((1s\sigma_g)\ (2p\sigma_u)\ ^1\Sigma_u\ (2p\pi_u)\ ^2\Pi_g$; J, $\{(1s\sigma_g)\ (2p\pi_u)^2\}\ ^2\Delta_g$; K, $\{(1s\sigma_g)\ (2p\sigma_u)\}\ ^1\Sigma_u$ (of He_2^{2+}). Σ-states (solid line); Π-states (dashed line); Δ-states (dotted line). Curve crossings where radial coupling induced inelastic transitions are marked by full circles; those where rotational coupling occurs are marked by squares (Barat et al[55]).

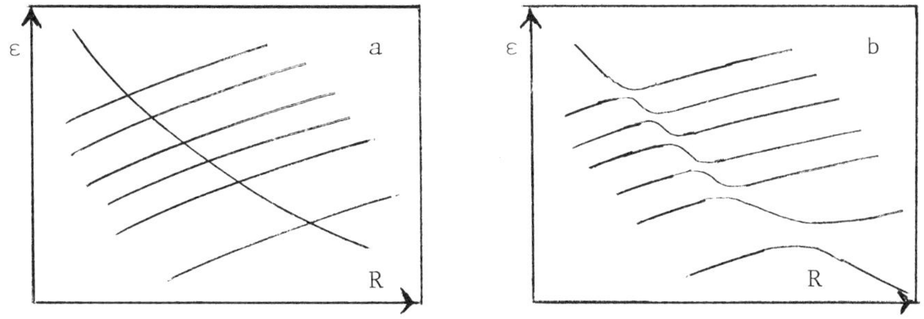

Fig. 8.2 - Schematic diagram of (a) diabatic representation and (b) adiabatic representation of potential energies.

product of one electron orbitals. It may then be remarked that the radial coupling matrix between two determinants χ_n^d, $\chi_{n'}^d$, that differ by at least two orbitals is zero.

In the case of He_2^+, the $^2\Sigma_g^+$ crossings involve states which differ by two orbitals; consequently the $\partial/\partial R$ matrix elements vanish. Of course the χ_n^d functions are not now eigenfunctions of H_e. Non-diagonal matrix elements will then arise from $<\chi_n^d|H_{el}|\chi_{n'}^d>$. These electrostatic matrix elements are of importance only for the first few excited configurations. A basis set constructed using the function has been termed a quasi-diabatic basis by Sidis[57]. It has been successfully used to treat the collision problem.

Another type of diabatic crossings may occur when the many electron configuration states differ only by these orbitals which, because of their symmetry, may cross. Examples of these crossings, termed diabatic I, occur frequently in neutral symmetric rare gas systems, He_2, Ne_2 etc.

However, the identification of diabatic state with a single configuration suffers from one serious drawback. The molecular orbitals, from which the states are constructed, tend at large R to interacting mixtures of separated atomic states. As a result, the correct description of the proper asymptotic conditions may be poorly represented by a basis set of diabatic states, which is satisfactory to describe the excitation mechanisms at small and intermediate internuclear distances. This problem can in principle be avoided by using a larger basis set but then one loses many of the practical advantages of a diabatic representation. One possible solution is to change from a basis set of simple configuration states to a decoupled basis set of appropriate atomic states at some sufficiently large R.

9. ILLUSTRATIONS

From the preceding analysis of the molecular model, it is evident that since the positions of the avoided (and real) curve crossings may vary widely from one system to another, a general classification of charge exchange reactions is well nigh impossible. It is however instructive to consider in detail a number of particular cases as representative of the major types of charge exchange reactions.

9.1. Resonant charge transfer

This is the simplest type of charge transfer reaction involving a collision of an ion with its parent atom

$$A^+ + A \rightarrow A + A^+ \qquad 9.1$$

In the case where the ground state of the ion and the atom involve only S states, for example $H^+ - H$, $He^+ - He$, $Li^+ - Li$, $Na^+ - Na$ etc. the treatment is straightforward. The resonant entry and exit channels are correlated either to a Σ_g or a Σ_u state. If the origin of electronic coordinates is taken to be the centre of symmetry of the system the radial coupling between the u and g states vanishes. Let us now assume that these Σ states are not subject to avoided crossings with higher states. (This assumption is often not verified for the upper Σ state, which may interact with excited Σ or Π states at small internuclear distances. However, since the upper Σ state is strongly repulsive, the avoided crossings are usually not classically accessible in low (eV) energy collisions).

The total wave function then becomes simply

$$\Psi(\underline{r}, \underline{R}) = F_g(\underline{R})\ \chi_g(\underline{r};R) + F_u(\underline{R})\ \chi_u(\underline{r};R) \qquad 9.2$$

where F_g, F_u satisfy the uncoupled equations

$$-\frac{1}{2\mu}\Delta^2_{\underline{R}} F_g + (E-\varepsilon_g) F_g = 0 \qquad 9.3$$

$$-\frac{1}{2\mu}\Delta^2_{\underline{R}} F_u + (E-\varepsilon_u) F_u = 0 \qquad 9.4$$

ε_g, ε_u are respectively the potential energies of the Σ_g and Σ_u states of the molecular ion. Equations (9.3) (9.4) are none other then the equations for elastic scattering. Let us define the corresponding scattering amplitude by $f_g(\theta)$, $f_u(\theta)$

$$F_{g,u}(\underline{R}) \underset{R\to\infty}{\sim} \left[\exp(i\,\underline{k}\cdot\underline{R}) + \frac{1}{R}\exp(ikR)\, f_{g,u}(\theta)\right] \qquad 9.5$$

where θ is the scattering angle. In the limit as $R\to\infty$ the initial and final atomic states are obtained from suitable linear combinations of χ_g, and χ_u

$$\frac{1}{\sqrt{2}}(\chi_g + \chi_u) \underset{R\to\infty}{\sim} \phi_o(\underline{r}_b)$$

$$\frac{1}{\sqrt{2}}(\chi_g - \chi_u) \underset{R\to\infty}{\sim} \phi_o(\underline{r}_a) \qquad 9.6$$

where $\phi_o(\underline{r})$ is the wave function of the ground state of A. The corresponding eigen-energy ε_o is given by

$$\varepsilon_o = \lim_{R\to\infty} \varepsilon_g(R) = \lim_{R\to\infty} \varepsilon_u(R) \qquad 9.7$$

It is then easily deduced that the differential cross section for charge transfer is given by

$$I_{ex} = \frac{1}{4}\left|f_g(\theta) - f_u(\theta)\right|^2 \qquad 9.8$$

The total cross section is then obtained from the relation

$$Q = \frac{\pi}{2}\int \left|f_g(\theta) - f_u(\theta)\right|^2 \sin\theta\, d\theta \qquad 9.9$$

The scattering amplitudes $f_{g,u}(\theta)$ may be calculated using the partial wave decomposition

$$f_{g,u} = \frac{1}{2ik}\sum_{K=0} (2K+1)\left[\exp(2i\eta_{g,u} - 1)\right] P_K(\cos\theta) \qquad 9.10$$

where the phase shifts $\eta_{g,u}$ are obtained from the solutions of

$$\left[\frac{d^2}{dR^2} - \frac{K(K+1)}{R^2} + k^2 - 2\mu\ \varepsilon_{g,u}\right] F^K_{g,u}(R) = 0 \qquad 9.11$$

which vanish at $R = 0$ and have the asymptotic form

$$F^K_{g,u} \sim \frac{1}{k} \sin\left(kR - \frac{1}{2} K\Pi + \eta_{g,u}\right) \qquad 9.12$$

Using (9.9), the total cross section may be expressed in terms of the phase shifts as

$$Q = \frac{\Pi}{k^2} \sum_{K=0}^{\infty} (2K+1) \sin^2 (\eta_g - \eta_u) \qquad 9.13$$

At low energies this form is the most convenient. At high energies ($\gtrsim$ 100 eV) the wave number k becomes large and the functions $F_{g,u}$ oscillate rapidly as a function of R. The numerical integration of (9.11) can then be quite time consuming. Besides as k increases, the number of partial waves which contribute effectively to the cross section also increase. Under these conditions the impact parameter approach is more appropriate. Using the theory developed in section (3) it is easily deduced that in the impact parameter treatment, the total cross section may be expressed as

$$Q = 2\pi \int_0^{\infty} b \sin^2 P(b)\, db \qquad 9.14$$

where

$$P(b) = \frac{1}{v} \int_b^{\infty} (\varepsilon_g - \varepsilon_u) \left[1 - \frac{b^2}{R^2}\right]^{-1/2} dR \qquad 9.15$$

As an example let us consider the system $Na^+ - Na$. The potential energy curves for the Na_2^+ molecular ion have been calculated by Bottcher et al[58] using a model potential method (see figure 9.1). Subsequently they have used (9.15) to determine the total cross section in the energy range 0.6 - 30 keV. The results presented in figure 9.2 are in excellent agreement with experiment. Of special interest in this case is the oscillatory nature of the total cross section as a function of energy. Such a pronounced structure, peculiar to alkali ion-alkali atom collisions, is due to the existence of a minimum in the energy difference $(\varepsilon_g - \varepsilon_u)$ near $4a_o$.

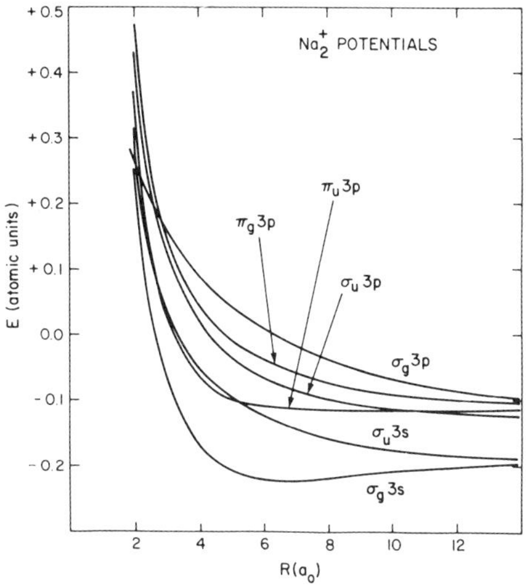

Fig. 9.1 - Potential energy curves of Na_2^+ (Bottcher et al[58])

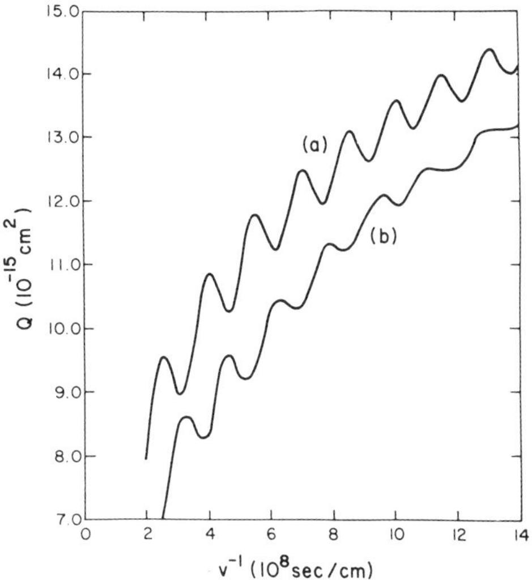

Fig. 9.2 - Total cross section for resonant charge transfer in Na^+-Na collisions, (a) theory (b) experiment. (Bottcher et al[58])

In general, the simple theoretical model for resonant charge transfer provides very reliable results for total cross sections for all systems. However, it is not a very sensitive test of the theory since the main contribution to the total cross section arises from small angle scattering (corresponding to large K or large impact parameters). Differential scattering cross sections provide a much more sensitive probe of the theoretical model. In particular large angle scattering, (corrsponding to small K or small impact parameters) allows one to investigate the result of close collisions. As we have already remarked the upper Σ state may be subject to dynamical coupling with other excited states.

The classic example is the simple H^+ - H system where the state of H_2^+ is subject to relational coupling with the first $^2\Pi_u$ state at close internuclear separations, Bates and Williams[36], McCarroll and Piacentini[59]. See figure 9.3.

The matrix element

$$< \chi_{2_\Sigma} \left| \frac{b}{R^2} \; L_y \right| \chi_{2_{\Pi_u}} > \qquad 9.16$$

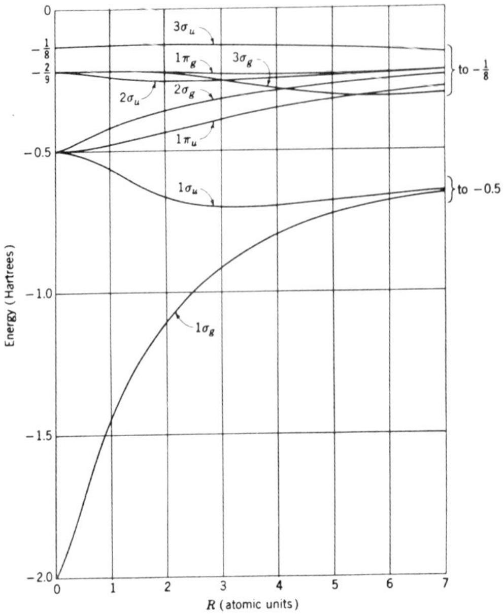

Fig. 9.3
Lowest energy levels of H_2^+ as a function of internuclear distance. Internuclear repulsive energy is not included (Slater[60]).

is important only for small values of R where the $^2\Sigma_u$ and $^2\Pi_u$ states are quasi degenerate. For small R this matrix element is equal to $-i\ b/R^2$. The effect of rotational coupling is then important only in close collisions. Physically this is due to the fact that in general the electron follows adiabatically the rotation of the molecular axis; however for almost head on collisions, the rotation of the molecular axis can become large in the vicinity of small R and the electron is no longer able to follow the molecular axis.

Since the energies and the coupling matrix element of the H_2^+ system can be calculated exactly, the $H^+ - H$ system provides a crucial lest for the theoretical model. A representative selection of the results (McCarroll and Piacentini[59]) for charge exchange probabilities as a function of scattering angle and energy and for the differential scattering cross section of the various processes ar given in figure 9.3, 9.7. The agreement with the experimental results of Helbing and Everhart[61] and Houver et al[62] is excellent. For a detailed investigation of translation factors in $H^+ - H$ collisions see Crothers and Hughes[63].

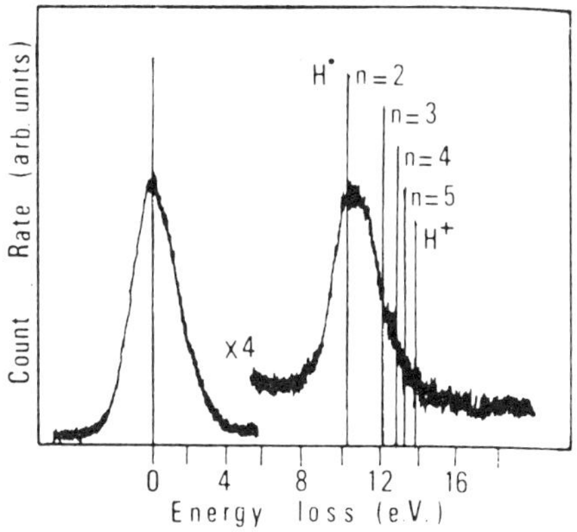

Fig. 9.4 - Energy loss spectrum of H^+ on H at an incident energy of 1 keV and scattering angle of 1°. The widths of the two peaks (elastic and inelastic) are almost equal which indicates that the n = 2 level is the only one substantially excited. The energies of the excited levels of H atoms are indicated. (Houver et al[62]).

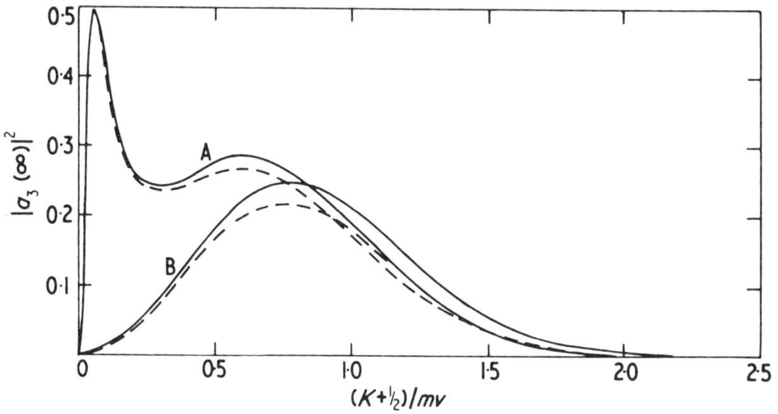

Fig. 9.5 - Excitation probability plotted against impact parameter for a laboratory energy of 500 eV. A, classical trajectory taking account of short-range coulomb repulsion of nuclei, B, straight line trajectory. Full curve: calculations using the exact value of the rotational coupling matrix element; broken curve, calculations using the united atom limit (Gaussorgues et al[64]).

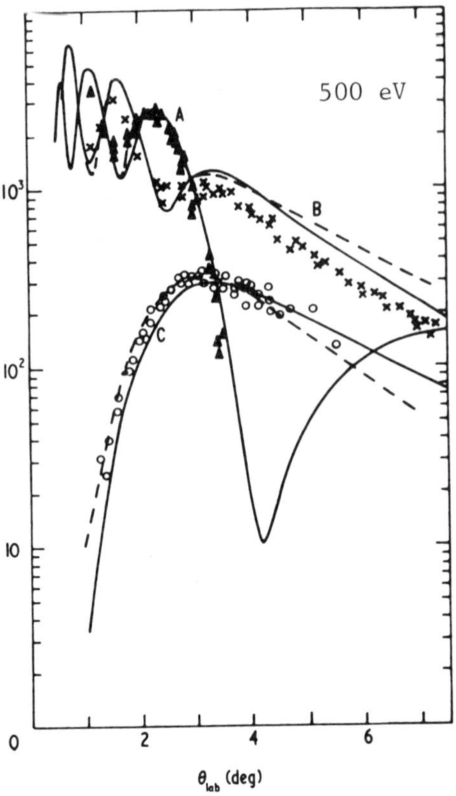

Fig. 9.6 - Reduced differential cross sections $\theta^2_{lab}\, d\sigma/d\Omega$ plotted as a function of laboratory scattering angle θ_{lab}. A, direct elastic scattering cross section, B, total exchange cross section, C, direct excitation cross section. Full curve: classical trajectory; broken curve, straight-line approximation. Experimental results are from Houver et al[62]: direct elastic scattering, x total exchange scattering, o direct excitation scattering. (Gaussorgues et al[64], McCarroll and Piacentini[59]).

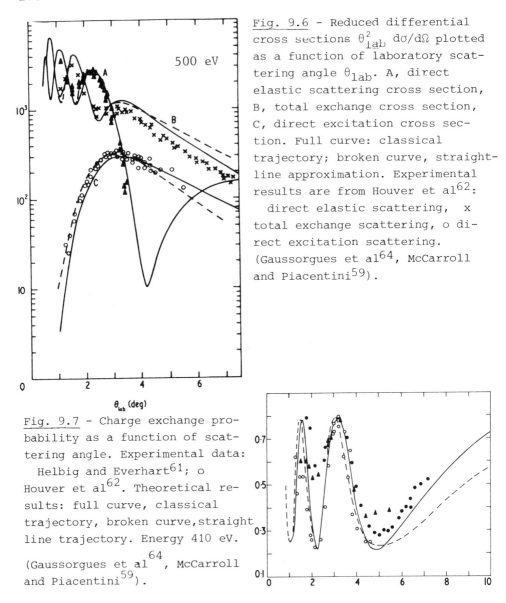

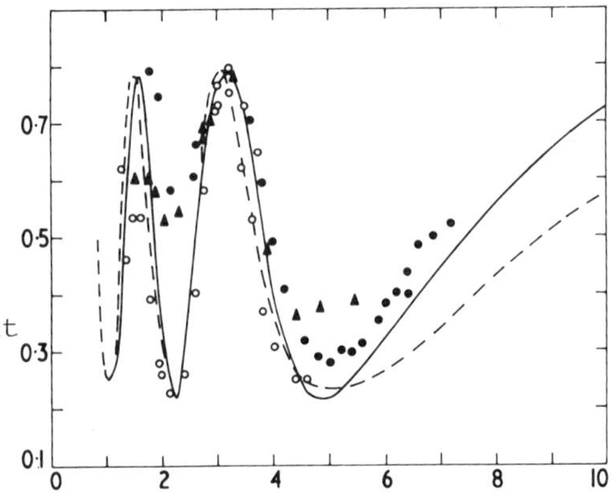

Fig. 9.7 - Charge exchange probability as a function of scattering angle. Experimental data: Helbig and Everhart[61]; o Houver et al[62]. Theoretical results: full curve, classical trajectory, broken curve, straight line trajectory. Energy 410 eV. (Gaussorgues et al[64], McCarroll and Piacentini[59]).

9.2. Electron capture by multicharged ions

In many respects the process of electron capture by multicharged ions is simpler to investigate theoretically than capture by singly charged ions. However, it is only recently that the problem has been seriously studied. The reasons for this were mainly practical ones since efficient sources for the production of low velocity multicharged ion beams have become available only since around 1974. Besides, the recognition of charge exchange with multicharged ions as an important recombination mechanism in many fusion and astrophysical plasmas, dates from around 1976. Since the process is usually exothermic, the reaction rate at thermal energies can be large ($\sim 10^{-9}$ cm^3 s^{-1}); in consequence the ionization equilibrium of certain impurity ions in the plasma can be greatly modified. For example, in the interstellar medium the predicted intensity of radiation expected from many ions was too large by several orders of magnitude because of the neglect of charge exchange (Christensen et al[51]); estimates of the temperature in stellar chromosphere and corona based on a commonly usded indicator like Si^{+2} can be seriously in error if charge exchange is omitted (Baliunas and Butler[65]).

Let us confine our attention in this illustration to the problem of charge exchange of multicharged ions with atomic hydrogen or helium, for example,

$$A^{+q} + H \rightarrow A^{+q-1} + H^{+} + \Delta\varepsilon_n \qquad 9.17$$

Although the number of avoided crossings can in principle be quite large for $q \leq 10$ the number of effective crossings is rarely excessive. The position R_x of the avoided crossing is determined approximately by the relation

$$R_x \simeq q / \Delta\varepsilon_n \qquad 9.18$$

If R_x is small ($\leq 4\ a_o$), the exchange interaction is important and the energy separation at the avoided crossing is large, whereas if R_x is large ($\geq 15\ a_o$) the exchange interaction is very weak and the energy separation at the crossing very small. In the first case, the collision is adiabatic while in the second case diabatic conditions prevail. These conditions limit the effective number of exit channels quite severely in practice. For example, the reaction

$$N^{+3} + H \rightarrow N^{+2} + H^{+} \qquad 9.19$$

takes place primarily via the $(1s^2\ 2s^2\ 3s)\ ^2S$ state of N^{+2}. The calculated potential energy curves of Butler[53] exhibit several avoided crossing (see figure 9.8), but all except the crossing at $9a_o$ involve more than one orbital change. For example the crossing at 3.1a_o leads to formation of N^{+2} in the ($1s^2$ 2s $2p^2$) state. Reactions leading to rearrangement of the core configurations are in general weaker than those where no change of core configuration occurs. Other crossings involving only one orbital change do occur with the (σ3p) and (σ3d) orbitals, but at such large distances that they do not contribute (at least in the eV region) to the total cross section.

Another typical example is the case of C^{+4}-H shown in figure 9.9. Three well defined avoided crossings occur involving the $(\sigma 3s_c)^2\Sigma$, $(\sigma 3p_c)^2\Sigma$ and $(\sigma 3d_c)^2\Sigma$ states. The dominant reaction at low energies will be capture via the 3d state of C^{3+}, but capture will also occur via the 3p and 3s states. We may note that in the case of C^{+4} - H, rotational coupling via the π3p, π3d, δ3d states may also be of importance.

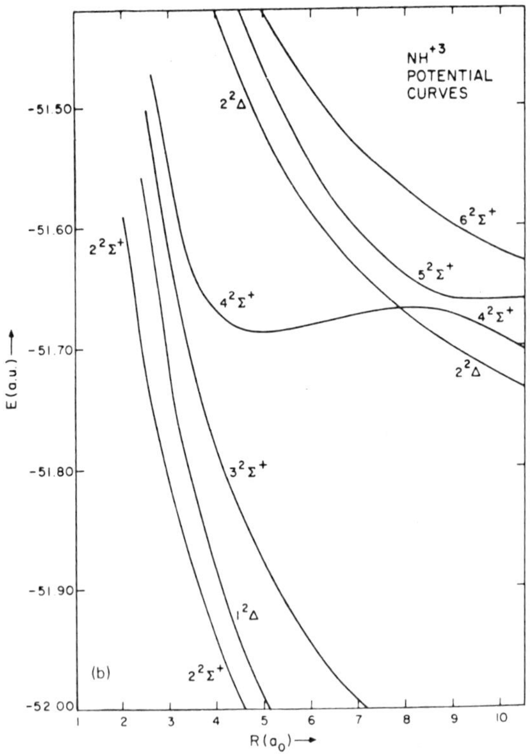

Fig. 9.8

Potential energy curves of NH^{+3} as a function of internuclear distances (Butler[53]).

For reactions of this type, which are important in many astrophysical applications, the determination of the cross section proceeds along the lines outlined in section 7 - calculation of the non diagonal radial coupling matrix elements followed by

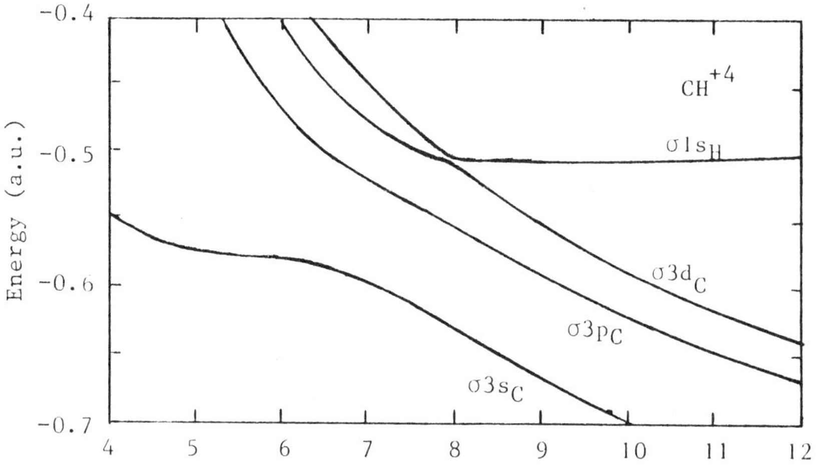

Fig. 9.9 - Potential energy curves of CH^{+4} as a function of internuclear distance. (Gargaud et al[66]).

the subsequent construction of the diabatic potential matrix. The scattering equations are then solved to obtain the scattering matrix, usually by the log-derivative method of Johnson[67]. For details of a typical calculation see McCarroll and Valiron[52] or Butler, Heil and Dalgarno[68].

Table 9.1 - Charge exchange cross section in N^{+3} - H collisions. The columns a, b, c refer to different origins for the calculation of the radial matrix elements: a - origin on N nucleus, b - origin at midpoint of internuclear axis, c - origin on H-nucleus

Energy (eV)	Cross-section (units of a_o^2) a	b	c
0.0082	70.39	70.54	70.48
0.082	27.50	27.60	27.67
0.82	65.30	65.52	65.67
8.2	127.6	128.0	128.4
27.2	129.2	128.6	128.6
136	103.5	97.0	93.0

As we have already remarked in section 7 the radial coupling matrix and the diabatic potential matrix are origin dependent (see figure 7.2). Butler et al[68] have chosen simply to place the origin on the heavy nucleus thereby eliminating spurious long range couplings. However, they do not give any estimate of the possible sensitivity to the choice of origin. Preliminary results by Gargaud et al[66], indicate that in the case of N^{+3} - H, where only radial coupling is involved the scattering matrix is indeed independent of the origin at best for energies up to ∿ 100 eV, see table 9.1. But much more work is necessary before the problem is clarified.

Unfortunately no detailed experimental data is available at such low energies. In the keV range, there is now a considerable amount of data, which can serve as a test for our theoretical model. One interesting case is the Ar^{+6} - He system, where capture is expected to take place primarily via capture into the 4s, 4p or 3d states of Ar^{+5} (see figure 9.10 for the potential energy curves of Ar He^{+6}). The measurements of Panov et al[69] indeed reveal that only these three states are populated by the charge transfer process. The calculated cross sections (McCarroll et al[70]) are in good qualitative agreement with the measured cross sections (figure 9.11).

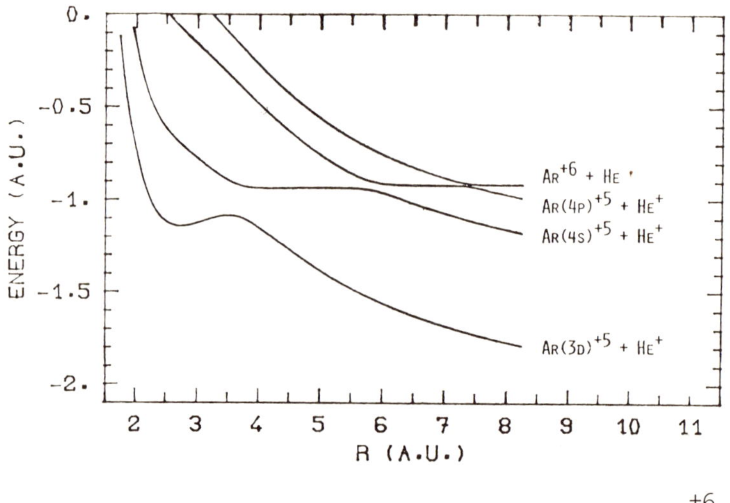

Fig. 9.10 - Potential energy curves of the system Ar^{+6} - He (Opradolce et al[71])

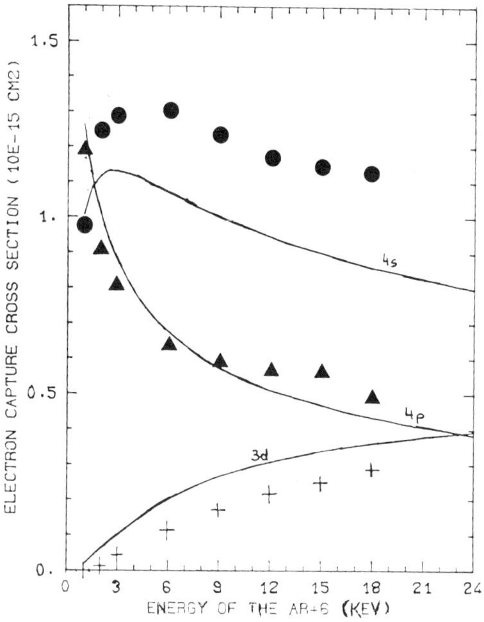

Fig. 9.11 - Collision cross section for capture into Ar^{+5} ($3s^2$ nl) in Ar^{+6} - He collisions. Impact paramet calculations: ● 4s, ▲ 4p, + 3d. Landau-Zener calculations, full line curves.

However the quantitative agreement is not so satisfactory as would expect from the theoretical model in this energy range (6 - 50 keV). From the experiments of El-Sherbini et al[72], it is observed that at higher energies (∿ 200 keV), the 4d state also become populated.

9.3. Inner shell excitation and vacancy sharing

The quasi-diabatic molecular model based on single configuration of molecular orbitals has proved successful in describing many inelastic processes involving inner shell electrons. The limitation of the molecular model is that the collision velocity be smaller than the velocity associated with the inner shell electrons. Thus, for example in collisions between ions heavier than Ne, we may expect the molecular model to be valid in the MeV energy range for K-shell transions (Fano and Lichten[73], Barat and Lichten[74]).

According to the molecular orbital correlation diagram, two successive mechanisms are responsible for K-shell vacancy production in slow ion-atom collisions. In the incoming part of the collision, vacancies in the outer shells may be transferred into

$2p\pi$ molecular orbital. It is supposed that the $2p\pi$ vacancy occupation number is N_π. At small internuclear distances, as in the H^+-H collision, the $2p\pi$ orbital is subject to rotational coupling with the $2p\sigma$ orbital. As a result $2p\pi$ vacancies may be transferred to the 2p orbital. This process is characterized by σ_{rot}, the rotation coupling cross section. It is then assumed that the corresponding cross section for K-shell excitation σ_K is given by $N_\pi\ \sigma_{rot}$.

Calculation of σ_{rot} have been made by Briggs[75] for the system Ne^+ - Ne using the impact parameter method with screened Coulomb trajectories. At low collision velocities, the probability of the $2p\pi$ - $2p\sigma$ transition is greatly reduced by the Coulomb deflection. At higher collision velocities, the transition probability a sharp peak occurs at small impact parameters, see figures 9.12 (a) and (b) (see also fig. 9.5). This peak is situated at 90° in the CM system and is caused by the sudden rotation of the internuclear axis. However this peak contributes little to the total

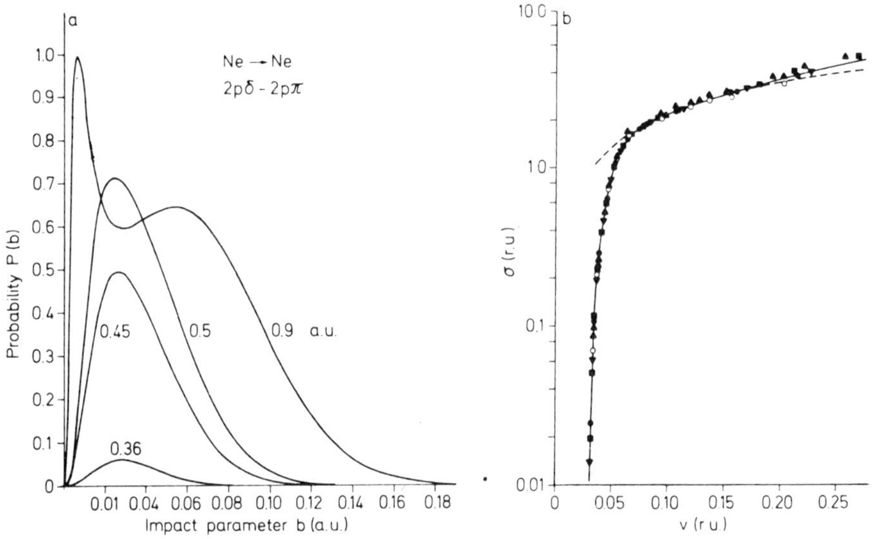

Fig. 9.12 (a) and (b) - Theoretical calculation for creation of K-shell vacancies in Ne - Ne collisions $2p\sigma$ - $2p\pi$ rotational coupling (a) Probability as a function of impact parameter for different collision velocities (b) Total cross section. The dashed curve is for straight line paths. (Briggs and Taulbjerg[35], Taulbjerg et al[76]).

cross section. The total cross sections figure exhibits a threshold behaviour due to the Coulomb deflection. Scaling laws apply to both the differential and total cross sections (Taulbjerg et al[77]).

For example

$$\sigma_{rot}(v;q) \sim q^{-2}\, \sigma_{rot}\left(\frac{v}{q};\, 1\right) \qquad 9.20$$

In the incoming part of the collision there are two processes which produces vacancies in the 2pπ orbital. Firstly, available 2p vacancies are statistically destributed among the molecular orbitals correlated with the 2p level, which produces a static contribution N_o to the 2pπ vacancy occupation number. Macek and Briggs[78] estimate N_o to be n/6 for symmetric systems and to be n/3 for non symmetric systems, when n is the number of vacancies in the lower-lying 2p level prior to the collision. Secondly, Stolterhoht et al[79], have shown a velocity dependent contribution N_v, arising from coupling with higher molecular orbitals. For example, in the Ne^+ - Ne collision, N_o = 1/6. Neglecting N_v, the calculated cross section (Briggs[75], Briggs and Macek[77]) are compared with the experimental results of Stolterfoht et al[79] in figure (9.13). The discrepancy between experimental and theory at high velocities presumably arises from the neglect of N_v.

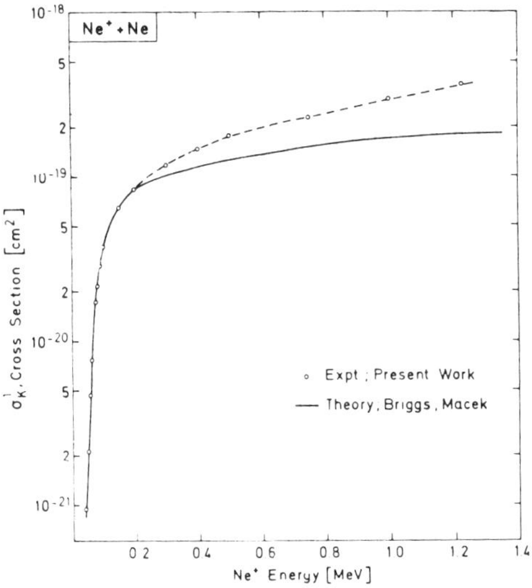

Fig. 9.13 - Total cross section for K excitation in Ne^+ - Ne collisions as a function of the projectile energy. Experimental data are from Stolterfoht et al[79]. Theoretical results are from Briggs[75] and Briggs and Macek[77].

More detailed information on the $2p\pi$ vacancy can be obtained from variation of the incident charge state. Figure 9.14 shows the ratios of cross sections σ_K^q for K excitation in Ne^{+q} + Ne collisions measured by Stolterfoht et al[79] for q = 0, 1, 2. In the low energy limit, $N_\pi = N_o$ and it is expected that $\sigma_K^2/\sigma_K^1 \approx 2$ and that $\sigma_K^1/\sigma_K^0 \to \infty$. These limits are in excellent agreement with experiment. As the energy increases, the relative importance of N_π increases and σ_K^2, σ_K^1, σ_K^0 approach each other in the high energy limit.

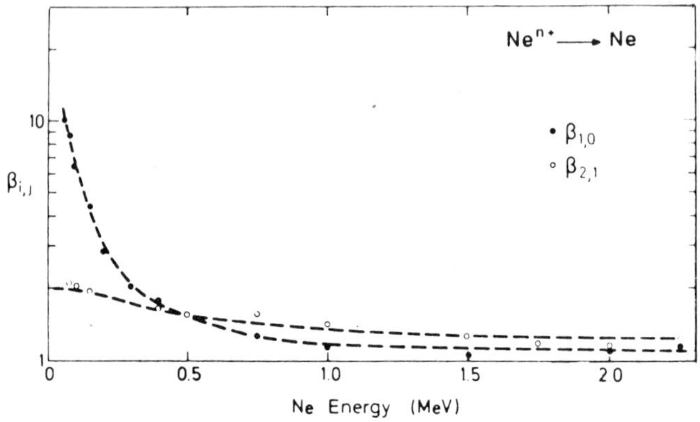

Fig. 9.14 - Cross section ratio $\beta_{1,0} = \sigma_K^1/\sigma_K^0$ and $\beta_{2,1} = \sigma_K^2/\sigma_K^1$ as a function of incident particle energy. σ_K^q denotes the K excitation cross sections in Ne^{+q} - Ne collisions. (Stoltherfoht et al[79]).

In non-symmetric collisions, the $2p\sigma$ orbital is correlated with the upper 1s level. It is thus to be expected that K-shell vacancies are preferentially produced in the lighter species. However, there is a finite probability for vacancy transfer to the K-shell of the heavier partner in the outgoing part of the collision (Meyerhof[80]). Non symmetric systems do not have g - u symmetry and there may be an important radial coupling between the $2p\sigma$ and $1s\sigma$ orbitals as their energies approach one another at intermediate internuclear distances. A detailed calculation, using the Hartree-Fock method to construct the relevant molecular orbitals has been carried out by Briggs and Taulbjerg[35] for the Ne - O system. The theoretical results obtained using $N_\pi = 1/3$ presented in figure 9.15 are in excellent agreement with the experiments of Stolterfoht

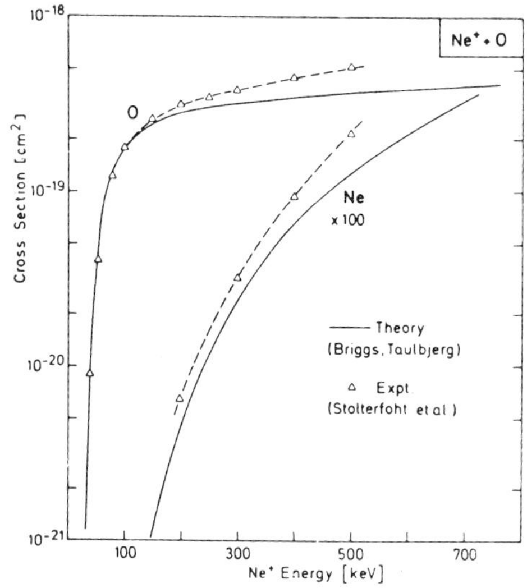

Fig. 9.15 -

Experimental cross sections for K excitation of oxygen and neon in Ne^+ - O_2 collisions compared with theoretical data Briggs and Taulbjerg[35], Stolterfoht et al[79].

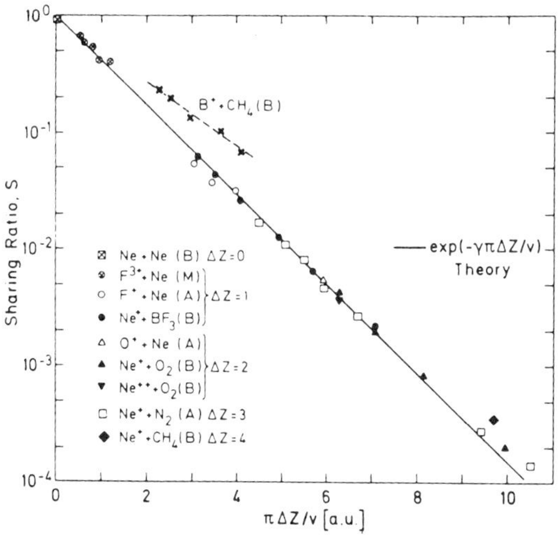

Fig. 9.16 -

Ratio for K-vacancy sharing as a function of $\pi\Delta Z/v$, where v is the projectile velocity,and ΔZ is the difference of the atomic numbers of the colliding species. Labels A, B, M refer to measurements of Fastrup et al[84], Stolterfoht et al[85] and Woods et al[86]. The theoretical curve with $\nu = 0.87$ is obtained using the formula of Meyerhof[80].

et al[79] on Auger electrons produced in Ne^+ - O_2 collisions.

Since the $2p\pi$ - $2p\sigma$ rotational coupling mechanism takes place near the distance of closest approach, while the $2p\sigma$ - $1s\sigma$ radial coupling is important only at intermediate distances, the two processes are independent of each other. In consequence, the distribution of the K-shell vacancies between the two collisions partners, should be independent of N_π. Uncertainties arising from the value of N_π can then be avoided by considering cross section ratios. By this means the vacancy sharing process may be examined independently.

A simple model to describe the K-vacancy sharing process has been proposed by Meyerhof[80], based on atomic two-state model of Demkov[81] and Nikitin[82] for charge exchange, when the coupling matrix is approximated by an exponential dependence on the internuclear distance. The sharing ratio is then given by

$$S = \exp(-\alpha / v) \qquad 9.21$$

with

$$\alpha = \pi(\sqrt{2 I_H} - \sqrt{2 I_L}) \qquad 9.22$$

where I_H and I_L are the K-shell bonding energies of the heavier and lighter particles respectively. Since the K-shell binding energy scales as Z^2, we may write

$$S = \exp\left(-\gamma\frac{\pi\Delta q}{v}\right) \qquad 9.23$$

where $\alpha = \gamma\pi\Delta q$ and Δq is the difference of the atomic numbers of the collision partners. It turns out that γ is virtually independent of the collision system. Using the experimental K-shell bonding energies (Lederer et al[83]), $\gamma = 0.87$ to within $\pm 2\%$ for collision particles from q = 5 for 11.

The theoretical results, when presented in a semi-logarithmic plot, give a straight line with slope γ, which represents a universal curve for the sharing ratios. In figure 9.16, experimental data for several collision systems (Fastrup et al[84], Stolterfoht[85], Woods[86]) are compared with the theoretical curve. The agreement is excellent.

For a review of inner shell transitions in ion atom collisions, see Stolterfoht[88], Mokler and Folkmann[89].

References

1 Bransden BH, 1965 Adv. Atom. Molec. Phys. 1, 85

2 Bransden BH, 1972 Rep. Prog. Phys. 35, 949

3 McDowell MRC and Coleman JP, 1969 Introduction to the theory of Ion-Atom Collisions (North-Holland, Amsterdam)

4 Basu D, Mukherjee SC and Sural DP, 1978 Phys. Reports 42, 145

5 Belkić Dz, Gayet R and Salin A, 1979 Phys. Reports 56, 279

6 Briggs JS and Taulbjerg K, 1978 Structure and Collisions of Ions and Atoms : edited by I.A. Sellin (Springer-Verlag, Berlin, Heidelberg), 106

7 Shakeshaft R and Spruch L, 1979 Rev. Mod. Phys. 51, 369

8 McCarroll R and Salin A, 1968 J. Phys. B : Atom. Molec. Phys. 1, 163

9 Gaussorgues C, Le Sech C, Masnou-Seeuws F, McCarroll R, Riera A, 1975 J. Phys. B : Atom. Molec. Phys. 8, 239

10 Oppenheimer JR, 1928 Phys. Rev. 31, 66

11 Brinkman HC and Kramers HA, 1930 Proc. Acad. Sci. Amsterdam 33, 973

12 Bates DR and McCarroll R, 1962 Adv. Phys. 11, 39

13 Bates DR and Dalgarno A, 1952 Proc. Phys. Soc. A 65, 919

14 Jackson J.D and Schiff H, 1953 Phys. Rev. 89, 359

15 Thomas LH, 1927 Proc. Roy Soc. 114, 561

16 Shakeshaft R, 1974 J. Phys. B : Atom. Molec. Phys. 7, 1059

17 Bates DR, 1958 Proc. Roy. Soc. A 247, 294

18 McCarroll R, 1961 Proc. Roy. Soc. A 264, 547

19 Tawara H, Richard P, Gray T, Newcomb J, Jamison KA, Schmiedekamp C and Hall J , 1978 Phys. Rev. A 18, 1373

20 Lin CD and Tunnell LN, 1979 J. Phys. B : Atom. Molec. Phys. 12, L 485

21 Ryufuku H and Watanabe T, 1978 Phys. Rev. A 18, 2005

22 Wilets L and Gallaher DF, 1966 Phys. Rev. 147, 13

23 Cheshire IM, Gallaher DF, Taylor AJ, 1970 J. Phys. B : Atom Molec. Phys. 3, 813

24 Rapp D and Dinwiddie D, 1972 J. Chem. Phys. 57, 4919
25 Gallaher DF and Wilets L, 1968 Phys. Rev. 169, 139
26 Shakeshaft R, 1976 Phys. Rev. A 14, 1626
27 Dose V and Semini C, 1974 Helv. Phys. Acta 47, 609
28 Errea LF, Méndez L and Riera A, 1979 J. Phys. B : Atom. Molec. Phys. 12, 69
29 Cheshire IM, 1967 Proc. Phys. Soc. 92, 862
30 Shakeshaft R, 1978 Phys. Rev. A, 18, 1930
31 Park JT, Aldag JE, George JM, Peacher JL and McGuire JH, 1977 Phys. Rev. A 15, 508
32 Bates DR and McCarroll R, 1958 Proc. Roy. Soc. A 245, 175
33 Hatton GJ, Lane NF and Winter TG, 1979 J. Phys. B : Atom. Molec. Phys. 12, L 571
34 Winter TG and Hatton GJ, 1980 Phys. Rev. A 21, 793
35 Briggs JS and Taulbjerg K, 1975 J. Phys. B : Atom. Molec. Phys. 8, 1909
36 Bates DR and Williams DA, 1964 Proc. Phys. Soc. 83, 425
37 Green TA, 1965 Proc. Phys. Soc. 86, 1017
38 Riera A and Salin A, 1976 J. Phys. B : Atom. Molec. Phys. 9, 2877
39 Ponce VH, 1979 J. Phys. B : Atom. Molec. Phys. 12, 3731
40 Schmid GB, 1979 J. Phys. B : Atom. Molec. Phys. 12, 3909
41 Crothers DSF and Hughes JG, 1978 Proc. Roy. Soc. A 359, 345
42 Schneiderman SB and Russek A, 1969 Phys. Rev. A 247, 194
43 Riley ME and Green TA, 1971 Phys. Rev. A 4, 619
44 Thorson WR and Delos JB, 1078 Phys. Rev. A 18, 117
45 Van Vleck JH, 1928 Phys. Rev. 33, 467
46 Zimmerman IH and George TF, 1975 Chem. Phys. 7, 323
47 Smith FT, 1969 Phys. Rev. 179, 111
48 Baer M, 1975 Chem. Phys. Lett. 35, 112
49 Heil TG and Dalgarno A, 1979 J. Phys. B : Atom. Molec. Phys. 12, L 557

50 Nikitin EE, 1968 Chemische Elementar prozesse ed. Hartmann H, Springer-Verlag Berlin Heidelberg, 43

51 Christensen RB, Watson WD and Blint RJ, 1977 Astrophys. J. 213, 712

52 McCarroll R and Valiron P, 1979 Astron. Astrophys. 78, 177

53 Butler SE, 19 Phys. Rev. A 20, 2317

54 Hanssen J, 1980 Thèse de 3e cycle, Université de Bordeaux I

55 Barat M, Dhuicq D, François R, McCarroll R, Piacentini RD and Salin A, 1972 J. Phys. B : Atom. Molec. Phys. 5, 1343

56 Herzberg G, 1950 Spectra of Diatomic Molecules - Van Nostrand

57 Sidis V, 1975 Thèse de Docteur ès-Sciences, Université de Paris VI

58 Bottcher C, Allison AC and Dalgarno A, 1971 Chem. Phys. Lett. 11, 307

59 McCarroll R and Piacentini RD, 1970 J. Phys. B : Atom. Molec. Phys. 7, 1358

60 Slater JC, 1968 Quantum theory of matter (McGraw-Hill, New York)

61 Helbig HF and Everhart E, 1965 Phys. Rev. 140, 1715

62 Houver JC, Fayeton J, Abignoli M, Barat M, 1972 Phys. Rev. Lett. 28, 1433

63 Crothers DSF and Hughes JG, 1979 Phil. Trans. Roy. Soc. A 292, 539

64 Gaussorgues C, Le Sech C, Masnou-Seeuws F, McCarroll R and Riera A, 1975 J. Phys. B : Atom. Molec. Phys. 8, 253

65 Baliunas SL and Butler SE, 1980 Astrophys. J. (Letters) to appear

66 Gargaud M, Hanssen J, McCarroll R and Valiron P, 1980 VIII Colloque sur les Collisions Atomiques et Electroniques, Louvain; 1981 J. Phys. B: Atom. Molec. Phys. to appear.

67 Johnson BR, 1973 J. Comp. Phys. 13, 445

68 Butler SE, Heil TG and Dalgarno A, 1980 Astrophysical J. 221, 442

69 Panov MN, 1980 Invited Lectures and progress reports, Electronic and Atomic Collisions, edited by Oda N and Takayanagi K (North-Holland, Amsterdam)

70 McCarroll R, Opradolce L, Valiron P, 1980 Second Technical Committe Meeting on Atomic and Molecular Data for Fusion

71 Opradolce L, McCarroll R and Valiron P, 1979 Abstracts of Contributed Papers : XI ICPEAC, Kyoto 1979, 564

72 El-Sherbini TM, Salop A, Bloemen E and De Heer FJ, 1979 J. Phys. B : Atom. Molec. Phys. 12, L 579

73 Fano U and Lichten W, 1965 Phys. Rev. Lett. 14, 627

74 Barat M and Lichten W, 1972 Phys. Rev. A 6, 211

75 Briggs JS, 1976 Rep. Prog. Phys. 39, 217

76 Taulbjerg K and Briggs JS, 1975 J. Phys. B : Atom. Molec. Phys. 8, 1895

77 Taulbjerg K, Briggs JS and Vaaben J, 1976 J. Phys. B : Atom. Molec. Phys. 9, 1351

78 Macek JH and Briggs JS, 1973 J. Phys. B : Atom. Molec. Phys. 6, 841

79 Stolterfoht N, Schneider D, Burch D, Aagaard B, Boving E and Fastrup B, 1975 Phys. Rev. A 12, 1313

80 Meyerhof WE, 1973 Phys. Rev. Lett. 31, 1341

81 Demkov YN, 1963 Soviet Phys. JETP, 18, 138

82 Nikitin EE, 1970 Advances in Quantum Chemistry, 5, 135

83 Lederer CM, Hollander JM and Perlman J, 1967 Table of Isotopes (Wiley, New York)

84 Fastrup B, Aagaard B, Boving E, Schneider D, Ziem P, Stolterfoht N, 1975 Proc. of IX ICPEAC, Abstracts of contributed papers, (Univ. of Washington Press, Seattle), 1058

85 Stolterfoht N, Schneider D and Ridder D, 1975 Proc. IX ICPEAC, Abstracts of Contributed papers, (Univ. of Washington, Seattle), 1060

86 Woods CW, Kauffman RL, Stolterfoht N and Richard P, 1976 Phys. Rev. A 13, 1358

87 Drisko RM, 1955 Ph. D. Thesis, Carnegie Institute of Technology

88 Stolterfoht N, 1978 Structure and Collisions of Ions and Atoms : edited by Sellin AE (Springer-Verlag, Berlin, Heidelberg)

89 Mokler PH and Folkmann F, 1978 Structure and Collisions of Ions and Atoms : edited by Sellin AE (Springer-Verlag, Berlin, Heidelberg)

90 Taulbjerg K, 1977 J. Phys. B : Atom. Molec. Phys. 10, L 341

91 Winter TG, Hatton GJ and Lane NF, 1980 Phys. Rev. A 22, 930

ATOMS IN ASTROPHYSICAL PLASMAS

Henri Van Regemorter

Observatoire de Paris

92190 Meudon

ABSTRACT

In low densities astrophysical plasmas, like planetary nebulae or the solar corona, photoionization, radiative and dielectronic recombination, autoionization, electron impact excitation and ionization rates must be known for interpreting the spectrum and finding spectroscopic diagnostics of the temperature and of the density. Emphasis is given on the rôle of electron-ion collisions and on the importance of accurate and approximate theories of the ion-electron system for the calculation of all the different processes which are reviewed.

CONTENTS

I. ATOMS IN LOW DENSITY PLASMAS

I.1. Thermodynamical equilibrium.

I.2. The competition of collision and radiative processes.

I.3. Temperature and density diagnostics.

I.4. Low density astrophysical ionized plasmas.

II. ATOMIC PROCESSES IN PLANETARY NEBULAE

II.1. The line spectrum.

II.2. Recombination theory.

II.3. The rôle of collisions.

II.4. Radio and non hydrogenic recombination lines.

II.5. Forbidden lines intensities.

II.6. The (X^+ + e) system.

III. ATOMIC PROCESSES IN THE SOLAR CORONA

III.1. The line spectrum.

III.2. Ionization equilibrium in the solar corona.

III.3. Level population mechanisms.

III.4. The satellite lines.

III.5. Electron-ion collisions.

I. ATOMS IN LOW DENSITY PLASMAS

I.1. Thermodynamical equilibrium

Most of the information on the structure, physical conditions and chemical composition of astrophysical objects in the universe are deduced from the study of continuous spectra and of intensities of spectral lines. The intensity of a given spectral line is proportional to the abundance of the chemical element concerned, to the population of the upper level of the transition and to the transition probability.

The rôle of collisions in astrophysical plasmas, like in laboratory plasmas where one cannot assume thermodynamical equilibrium, is crucial to determine the state of equilibrium of the atom or of the molecules, i.e., the population of the levels, the state of ionization or of dissociation.

If we consider a star or a nebulae as a mass of gas surrounding an energy source, these objects cannot be in a perfect thermodynamical equilibrium essentially because the radiation can more or less freely escape into space. The continuous spectra of these objects are not spectra of black bodies.

In thermodynamical equilibrium, the radiation intensity is given by the Planck law

$$B_\nu(T) = \frac{2h\nu^3}{c^2}\left[e^{\frac{h\nu}{kT}} - 1\right]^{-1} \qquad (1)$$

The distribution of velocities is given by the Maxwell law

$$f(v) = 4\pi\left(\frac{m}{2\pi kT}\right)^{3/2} e^{-\frac{mv^2}{2kT}}\, dv \quad \text{with} \int_0^\infty f(v)\, dv = 1 \qquad (2)$$

The ratio of the level population is given by the Boltzmann law

$$\frac{N_i}{N_j} = \frac{\omega_i}{\omega_j}\, e^{\frac{-(E_i - E_j)}{kT}} \qquad (3)$$

where N_i is the number of atoms per cm^3, with energy E_i and statistical weight ω_i

Taking account of all possible states of ionization , if $N_i(X^{+q})$ is the population of level i of ion X^{+q}, $N(X^{+q+1})$ the number of ion X^{+q+1} in their ground state a with statistical weight ω_a^{+q+1}, I_i^a the ionization of level i, of statistical weight ω_i, the Bolzmann-Saha law.

$$\frac{N_i(X^{+q})}{N_a(X^{+q+1})\,N_e} = \frac{\omega_i}{2\,\omega_a^{+q+1}}\left[\frac{h^2}{2\pi m k T}\right]^{3/2} e^{\frac{I_i^a}{kT}} \tag{4}$$

gives the level population in thermodynamical equilibrium.

Complete thermodynamical equilibrium is never realized, but sometimes one can assume "local" thermodynamical equilibrium, which means that the temperature is not uniform and that the radiation intensity in each region of the plasma is given by the Planck equation (1) at the local temperature T. The other three Maxwell-Boltzmann-Saha laws are still valid.

This assumption of "Local Thermodynamical Equilibrium", has been extensively used in the study of stellar atmospheres in which temperature increases with depth : inner parts of the atmospheres are "protected" by outer parts, local thermodynamical equilibrium occurs long before the light quantum can escape in space.

When the local thermodynamical equilibrium (L.T.E.) is justified, the level population and the state of ionization are simply given by equation (3) and (4). They depend only on the temperature, on the electron density and on the chemical abundance of the element. They are independent of the excitation and desexcitation processes as well as the ionization and recombination processes of the atoms.

The spectral line intensity in this case, which is proportional to the upper level population of the transition, will give straight away information on T, N_e and the chemical abundance.

I.2. The competition of collision and radiative processes.

In most of the astrophysical conditions, i.e., the external layers of a star (chromosphere, corona) the planetary nebulae (cloud surrounding a hot star), the different molecular clouds in the interstellar medium, the local thermodynamical equilibrium is not valid, the level population are not given by the Boltzmann and Saha equations, all microscopic excitation and deexcitation processes have to be known in order to interpret the spectral observations.

Let us consider a two level atom and suppose $E_2 > E_1$ the ground state energy.

If ρ_ν is the radiation density, A_{21} the spontaneous transition probability, B_{21} and B_{12} the Einstein induced probabilities,

$\alpha_{12} N_e$ and $\alpha_{21} N_e$ the excitation and desexcitation collision probabilities (supposing that the electron collisions are the more effective, N_e being the electron density) one can write

$$N_1 \left[B_{12} \rho_\nu + \alpha_{12} N_e \right] = N_2 \left[A_{21} + B_{21} \rho_\nu + \alpha_{21} N_e \right] \tag{5}$$

corresponding to a steady state of equilibrium of the excitation and deexcitation processes. The collision rates must be known if we want to calculate the observed line intensity $N_2 \, A_{21} \, h\nu_{21}$.

In many astrophysical conditions, the radiation field ρ_ν is very weak and one can neglect absorption and induced emission of radiation $B_{12} \rho_\nu$ and $B_{21} \rho_\nu$: the medium is optically thin in the line. In the opposite case of an optically thick plasma, every photon is immediately reabsorbed by the atom before the atom experiences a collision. In the intermediate case, when radiative processes and collisional processes compete, multilevel equations of type (5) giving the level populations and radiative transfer equations must be solved simultaneously. This is a major problem in theoretical astrophysics which will not be considered here.

Let us assume in the following that the medium is optically thin, a reasonable assumption in many astrophysical conditions.

The collisional transition probability $\alpha_{21} N_e$ is given in terms of the cross section Q_{21}

$$\alpha_{21} = \int_0^\infty f(v) \; Q_{21} \; v \, dv \tag{6}$$

v being the initial electron velocity. The average is over the Maxwellian electron velocity distribution f (v), which is satisfied in a very good approximation because of the effectiveness of electronic collisions in redistributing kinetic energy.

From the detailed balance principle, which states that for collision processes, the number of $i \rightarrow j$ transitions is equal to the number of inverse $j \rightarrow i$ transitions one can obtain the relation between excitation and deexcitation rates

$$\alpha_{12} = \frac{\omega_2}{\omega_1} \, \alpha_{21} \, e^{-\frac{E_{21}}{kT}} \tag{7}$$

This can also be deduced from the symmetry of the collision process, i.e., from the symmetry of the S matrix, as we shall see later on.

Let us define the ratio ε, which characterizes the competition between collision processes and spontaneous decay by

$$\varepsilon = \frac{N_e \alpha_{21}}{A_{21}} \tag{8}$$

ε is the probability that the atom experiences electron collision during its lifetime.

In the high density limit, ε is much bigger than 1 and from (7) and (5), neglecting the radiative induced processes, one finds the Boltzmann relation (2). Thermodynamical equilibrium is due to the effectiveness of the collision processes. Each excitation is followed by collisional deexcitation.

In the low density case, very frequent in the study of low density astrophysical plasmas, on the contrary $\varepsilon \ll 1$ and

$$N_2 A_{21} = N_1 \alpha_{12} N_e \tag{9}$$

Each excitation is followed by quantum emission. In order to interpret the observed line intensity $N_2 A_{21} h\nu_{21}$, one must know the collision excitation rate α_{12}. The radiative transition probability does not enter the expression of the line intensity.

One can understand in this way why, in low density conditions, forbidden lines can be as strong or stronger than permitted lines. The strongest lines in the spectra of most nebulae are the forbidden transitions within the ground configuration of O^{+2} (OIII):

$$2p^2\ ^1D \rightarrow 2p^2\ ^3P_J + h\nu \tag{10}$$

The transitions probabilities of these forbidden electric quadrupole or magnetic dipole transitions are very small (A of the order of 1 sec^{-1} or 10^{-5} sec^{-1}, compared to A $\sim 10^8$ sec^{-1} for permitted lines), but the density N_e is so low (10^4 cm^{-3}) that the ratio ε is much smaller than 1 and thus equation (9) is valid. The intensity therefore depends only on the rate at which inelastic collisions produce excited atoms, and this rate does not differ greatly from those for permitted lines.

This is the simplest way to understand the paradoxal importance of collisions processes when collisions are rare, in a very large emitting volume as in an astrophysical object.

When collisional excitation is followed by quantum emission, part of kinetic energy of the colliding electrons (or other particles) is given to the radiation field and the medium is cooled. The interstellar matter is cooled in this way by the collisional excitation of fine structure transitions or other forbidden transitions within the ground configuration by electrons (or eventually by protons) in the ionized region, or by H_2 in the interstellar low temperature clouds.

I.3. Temperature and density diagnostics.

From the simple equations above one can guess how to define spectroscopic diagnostics of temperature or density of the emitting medium. When equation (9) is valid for 2 levels (2 and 3, 1 being the ground) the ratio of the two line intensities 2 - 1 and 3 - 1 is independent of N_e and from (7) depends of T as exp ΔE_{23} / kT This ratio may be a good temperature diagnostic.

On the contrary, when one excited level is related to the ground by a forbidden transition and also is collisionally deexcited, the ratio of the two lines is a good diagnostic of the electron density. Ratio of forbidden lines of different kinds can also be used in this way.

Figure 1 shows the energy diagram of the lowest levels of helium like ions. The radiative transition probabilities are given to show the importance of the ionic charge Z - 1 for choosing good diagnostics.

When these transitions are collisionally excited, as it is the case in the solar corona, for densities $N_e \sim 10^9$, one can use for density criterion :

- the ratio $\frac{\text{Forbidden line}}{\text{Intercombination line}}$ at low densities $N_e < 10^{10}$ for Z=5

which decreases with N_e (because of the $2\ ^3S$ - $2\ ^3P$ collisions)

- the ratio $\frac{\text{Intercombination line}}{\text{Resonance line}}$ at higher densities $10^9 < N_e < 10^{15}$ for Z=5

which decreases with N_e (because of collision transfer between singulet and triplet terms).

For very high density the intercombination line becomes too weak to be measurable. The limit beyond which L.T.E. conditions apply can be easily determined.

On the other hand, the two level atom model is sufficient to understand why electron collisions insure a thermodynamic equilibrium of the high atomic levels, a situation which is compatible with the Maxwellian distribution of velocities of the free electrons.

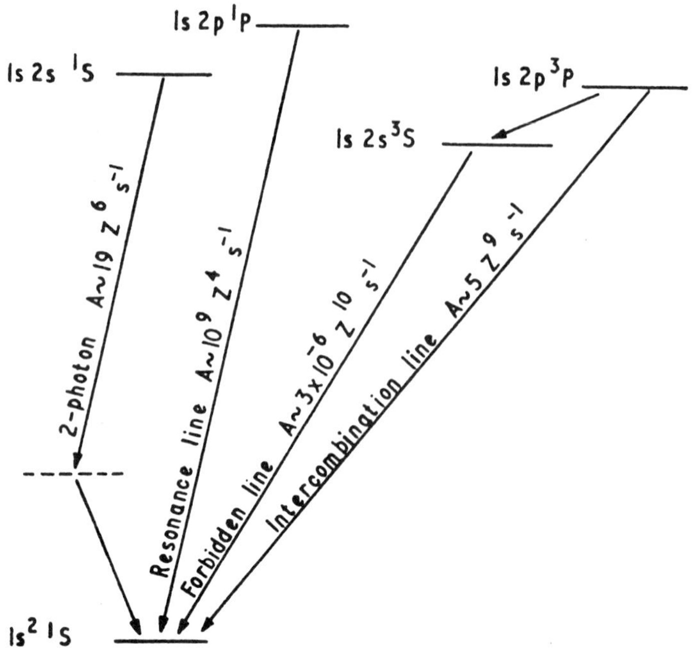

Fig. 1. Energy level diagram of helium-like ions

For transitions between high excited states $n \rightarrow n' = n + \alpha$, α being small, the ratio

$$\varepsilon = \alpha_{n'n} N_e / A_{n'n}$$

increases with n and there is always a critical value n_c beyond which ε is bigger than 1 : collisional excitation is followed by collisional deexcitation. The Boltzmann distribution of the different sublevels $n\ell$ for the same n according to their statistical weights $2\ell + 1$ is also due to collisions with electrons or protons (the last are the more effective when the energy defect ΔE ($n\ell$, $n\ell'$) is very small). For a given electron (or proton) density there is a critical value of n below which this Boltzmann distribution is not valid.

In the case of a realistic multilevel atom, a system of linear differential equations has to be solved to determine the level populations and the state of ionization, taking account of all the important excitation and deexcitation processes together with all ionization and recombination processes.

The relaxation time corresponding to all these processes being very short compared to the time scale of evolution of the astrophysical plasmas, steady state is realized. Exceptionally some plasmas can be transient : the evolution of the eruptive solar flare is so rapid that the ionization equilibrium is not realized and dN^i/dt, the number of ions in state i varies with time as it is confirmed by the variation with time of the observed spectra.

Important simplifications of the level population equations can be made according to the physical conditions. We shall examine briefly the cases of nebulae and of the solar corona.

I.4. Low density astrophysical ionized plasmas

We shall concentrate on three examples of low density astrophysical ionized plasmas : diffuse nebulae, planetary nebulae and solar corona.

Diffuse nebulae are ionized regions of the interstellar matter in the vicinity of hot stars (O and B stars of temperature of the order of 10^4 - 10^5 °K) which are sources of strong ultraviolet radiation. All photons shorter than 912 Å, i.e. more energetic than 13,60 eV, contribute to ionize hydrogen, by far the most abundant element (90%) : this is why diffuse nebulae are called H II regions to distinguish them from the neutral gas of the Galaxy. Temperatures are of the order of 10^3 - 10^4 and densities vary from 10^2 to 10^4 cm^{-3}.

Planetary nebulae are masses of gas ejected by stars of low mass (smaller than 4 Solar Masses) at a late stage of their evolution. This gas surrounding the hot star, whose temperature can be greater that of O or B star, is ionized by the UV radiation from the star. Their study is important to understand how stellar matter is ejected from the star to refill the interstellar medium.

Both the diffuse nebulae and the planetary nebulae are photon ionized gases and their physics are in fact very similar, the thermodynamic structure being determined by the interplay of heating by photoionization and cooling by inelastic collisions. In the following we shall be concerned with planetary nebulae.

The electron temperatures are of the order of T = 10^4. The density is approximatively 10^4 - 10^5 cm^{-3} (N $\doteq$ N (protons) $\cong$ N (electrons))

The Solar Corona is also a low density ionized plasma (N = 10^9) and very hot (T = 10^6 K). In this case the mechanism of heating is not entirely elucidated : acoustic and magnetohydrodynamic waves play an important rôle to increase the temperature from 4.300°K of the solar atmosphere to 10^6°K of the Corona. Thanks to space observations in the extreme UV, it is not only possible to have good information on the solar corona but to have the first direct observations of other stellar coronae. This promises to be one of the most exciting fieldsduring the Space Laboratory era. Of peculiar interest are the study of active regions, in particular of the solar flares where temperatures can reach 2 10^6 - 5 10^7 °K and the particle densities never exceed 10^{13}.

II. ATOMIC PROCESSES IN PLANETARY NEBULAE

II.1. The line spectrum

In the nebulae, energy is given by photoionization of hydroen (neutral and ionized helium have also to be taken into account). The photon flux coming from the central star (s) provides the medium with kinetic energy input.

Three major energy loss mechanisms balance this gain :

- Electron recombination $H^+ + e \rightarrow H + h\nu$
- Free Free transitions $H^+ + e \rightarrow H^+ + h\nu + e$
- Electron Excitation of forbidden lines of the most abundant elements (C,N,O ions).

The first two give rise to a weak measurable continuum emission and to recombination lines in all wavelengths of the spectrum (from the UV to the radio), which can provide information on the temperature and the density of the gas. Despite the low abundance of complex ions compared to hydrogen, the last excitation process is the most important energy loss process which gives rise to strong emission lines (collisional excitation being followed by spontaneous decay).

Let us first recall the cosmic abundance of the most abundant elements compared to hydrogen. These values are given here to see which atoms are of interest. The correct determination of the abundance of elements is a major problem in astrophysics which will not be discussed here (2.6^{-5} means 2.6 10^{-5}).

Element	H	He	C	N	O	Ne
Abundance	1	$8,5^{-2}$	$3,3^{-4}$	$9,1^{-5}$	$6,6^{-4}$	$8,3^{-5}$
Element	Mg	Si	S	Ca	Fe	Ni
Abundance	$2,6^{-5}$	$3,3^{-5}$	$1,6^{-5}$	$2,6^{-6}$	$4,0^{-5}$	$2,0^{-6}$

Looking at the spectra of the planetary nebulae, in addition to the hydrogen and the He recombination lines, one observes a great number of lines, allowed or forbidden. It is usual to characterize a forbidden line with two brackets [ ], for example the [O III] 4363 line is the $2p^2$ (^1S - ^1D) transition in O^{+2}, and to use one bracket to characterize an intercombination line, for example the C III] 1909 line is the $2s2p\ ^3P$ - $2s^2\ ^1S$ transition in C^{+2}. Lines with no bracket are just ordinary permitted lines.

TABLE 1 - Important lines in nebulae

H	H 6560 - H 4861 - H 4341.
He	He I 4471, 5876 - He II 4686, 1640.
C	[C I] 9849, 8727 - C II] 2326 - C III] 1908, 1909 - C IV 1549
N	[N I] 5200, 10400, 3466 - [N II] 6584, 6548, 5755 - N IV] 1488
O	[O I] 6300, 6363, 5577 - [O II] 2470, 3727, 7323 - O III] 1664 [O III] 4959, 5007, 4363 - O IV] 1403.
Ne	[Ne III] 3869, 3967, 3342 - [Ne IV] 2422, 2424, 2431, 4724 - [Ne V] 3426, 3346, 2972.
Mg	Mg I 2852 - Mg II 2798 - [Mg V] 2750 - [Mg VI] 3488.
S	[S I] 7725 - [S II] 6716, 6731, 10320, 4070 - [S III] 9069, 9532, 6312, 3722.

Wavelengths are given in Å. This list does not include the many strong fine-structure transitions of ions with ground state configuration $2p^q$ and $3p^q$, which appear in the far infrared (from 10 μ - 200 μ).

Table 1 gives a sample of observed emission lines. Many of them, in particular, intercombination and permitted recombination lines are in the UV spectrum. Some of them have been observed recently. With the International Explorer (I.U.E.) satellite alone, many dozens of planetary nebulae have been observed in the last few years.

From this table, it is also possible to see for which elements and for which stages of ionization cross sections are needed. We must now review the important radiative and collisional processes which are involved.

II.2. Recombination theory.

In nebulae, photoionization is balanced by radiative recombination. At T = 10^4 the mean kinetic energy of the electron is of 1 eV, approximatively half of the excitation energy of the forbidden lines within the ground state, which are the main transitions excited by electron collisions. Most of the other lines are recombination lines.

The equation of ionization equilibrium is

$$N_1(X^{+q}) \int F_\nu \, a_1(\nu) \, d\nu = N_e \, N_1(X^{+q+1}) \sum_{j=1}^{\infty} \alpha_j \qquad (11)$$

where $a_1(\nu)$ is the cross section for photoionization from the ground state N_1 (X^{+q}) of ion X^{+q}. $F_\nu \, d\nu$ (ergs cm^{-2} sec^{-1}) is the photon flux coming from the central star. α_j is the radiative recombination rate on the level j of ion X^{+q}. Recombination occurs to all levels j.

This recombination rate to one specific level j is given in terms of the capture cross section Q_j^c (v) by the average over the electron velocities.

$$\alpha_j = \int Q_j^c(v) \, f(v) \, v \, dv \qquad (12)$$

Capture cross section to a given level j is related to the photoionization cross section from this level $a_j(\nu)$,

$$X_j^{+q} + h\nu \rightleftarrows X_1^{+q+1} + e(v) \qquad (13)$$

the inverse process , by

$$Q_j^c(v) = \frac{h^2 \nu^2}{c^2 m^2 v^2} \, \frac{\omega_j}{\omega^{+q+1}} \, a_j(\nu) \qquad (14)$$

Here $h\nu = \frac{1}{2} mv^2 + I_j$ where all quantities have been defined above (see (4)).

No sum appears on the left hand side of equation (11) because only photoionization from the ground state N_1 (X^{+q}) is important (sometimes it is necessary to take account of a few metastable levels). Under interstellar conditions, when the low levels are populated by recombination and cascades, only the ground level is usually populated and is very much overpopulated compared to L.T.E.

For $T > 10^4$ dielectronic recombination start to play a rôle but has a small influence on equation (11) for most nebulae.

The system of linear equation giving the level population is of the form

$$N_e N(X^+)[\alpha_i + \gamma_i N_e] + \sum_{i'} N_{i'} [\alpha_{i'i} N_e + A_{i'i}] = N_i \left[\sum_{i'} (A_{ii'} + \alpha_{ii'} N_e) + \alpha_{ic} N_e \right] \tag{15}$$

For a steady state, the number of atoms leaving state i is equal to the number of atoms entering state i. To simplify the writing we put q = 0 (neutral case). N^+ is the number of ions X^+ .

In addition to recombination in state i, we have included all collisional processes :
$\alpha_{ii'}$ are the electron collision excitation and desexcitation rate, (see (6)), α_{ic} is the ionization collisional rate for the process

$$X + e \longrightarrow X^+ + e + e \tag{16}$$

and γ_i is the rate of the inverse processes of (16), the three body capture.

At $T = 10^4$ only the very low sublevels of the ground configuration are excited by collisions. Excitation is followed by radiative decay because the density is very low. Electron collisions may play a rôle in populating the intermediate excited states. Many of the important lines are recombination lines. But electron collisions start again to be important for redistribution energy between very excited states.

Let us take the case of hydrogen and characterize the level by $i = n\ell$. If $N_{n\ell}$ is the number of atoms in state $n\ell$, it is usual to write

$$N_{n\ell} = b_{n\ell} N_{n\ell}(LTE) \tag{17}$$

where the $b_{n\ell}$ characterizes the divergence from local thermodynamical equilibrium that one can expect at very low density.

From the paragraphs I.2. and 3., it is clear that collisional processes have to be included in equation (15). Pure recombination theory would imply population by radiative capture and cascade, but collisional processes become always dominant for high levels. One must take account of the redistribution in energy due to electron impact and to redistribution in angular momenta due to charged particles (the last processes is not included in the equation above only to simplify the writing).

In writing equations (15), the plasma is supposed to be optically "thin" in all the lines, in the sense of discussion in I.2. All radiative induced processes have been omitted. The UV radiation from the central star can play a rôle in excitation of the Lyman lines as it plays a rôle in photoionization from the ground state N_1 (see equation (11)).

In the pioneering work of Baker and Menzel [1938], two different cases are considered : case A in which the nebulae is optically thin in all the lines, and case B in which the Lyman lines are supposed to be optically thick (absorption followed by radiative emission).
Leaving out this last crucial point, which is a problem related to the radiative transfer of radiation in the nebulae, we shall concentrate on the rôle of collisions in recombination theory.

II.3. The rôle of collisions

Electron-electron collisions insure the Maxwellian distribution of the free electron velocities. In the limit of large n electron inelastic collisions will be important to insure thermodynamical equilibrium : $b_n \rightarrow 1$. For high n inelastic collisions between high levels, collisional ionization (the term $\alpha_{ic} N_e$ in equation (15)), as well as the three body capture ($N_e^2 N^+ (X^+) \gamma_i$) will tend to give a Boltzmann-Saha distribution for these levels.

The steady state equations (15) have been solved with and without the collision terms included. Results are given in figure 2, giving the b_n for recombining H atoms. The dotted line is based on electron-ion radiative recombination followed by radiative decay cascading. The dashed curve (a) includes collisional ionization and three body recombination. The dashed curve (b) includes instead the collisional transition $n \rightarrow n + 1$. The solid curve includes all these processes (see M.J. Seaton, 1964).

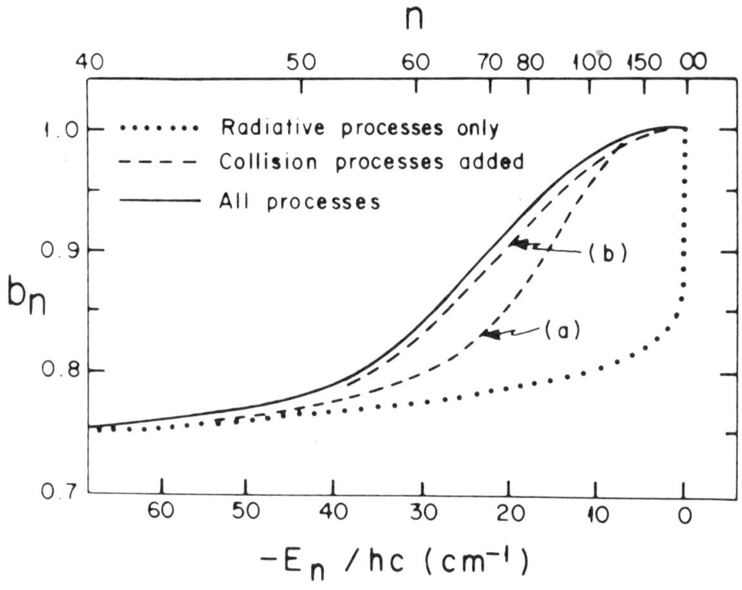

Fig. 2. Values of b_n for recombining H atoms.

The collision problems involving highly excited states are a very difficult three body problem which cannot be solved by accurate quantum mechanical techniques (long range interaction between two electrons at large distances from the nucleus). Classically, the problem can be solved numerically using Monte Carlo methods involving random samplings of initial conditions.

For electron excitation between highly excited states, different approximations (quantal, semi-classical, classical) have their own defined domain of validity. Typical errors in the recommended cross sections are of the order of 10%. Further information can be found in the lectures of I.C. Percival (see also Percival and Richards 1975).

For n not very large each $b_{n\ell}$ for the different values of ℓ must be calculated. The redistribution of angular momentum is due to collisions with protons. In the case of H, the different $n\ell$ levels for the same n are degenerate. In this case, the collision rate of excitation by a given particle p is proportional to v_r^{-1} or to the square root of its mass m_p and protons are much more effective than electrons

$$H(n\ell) + H^+ \longrightarrow H(n\ell') + H^+ \tag{18}$$

For non hydrogenic ions electron excitation or proton excitation can dominate according to the spacing of the level ΔE ($n\ell$, $n\ell'$). For all collisions excitation of allowed transitions between closed levels, large impact parameters are dominant and good results are obtained using the semi-classical approach .

When the rate of reaction (18) is bigger than the rate of radiative decay ($\varepsilon \gg 1$ in the sense of relation (8)), the quantum sublevels $n\ell$ are populated according to their statistical weights $2\ell+1$

$$N(n\ell) = \frac{2\ell+1}{n^2} N(n) \tag{19}$$

For typical densities in the nebulae ($N(H^+) \sim N_e \sim 10^4 - 10^5$) this Boltzmann distribution starts to be valid for n bigger than 30 - 40.
Calculation of b_n for high levels and of $b_{n\ell}$ levels for low levels have been done in this way by Brocklehurst [1970, 1971] .

The ground state is overpopulated ($b_1 \gg 1$) and absorption of the ambient radiation can play an important rôle in populating the two levels ($n \leq 3$). Collisional processes are important for the higher levels. Table 2 gives the ratio of the Balmer lines compared to observed values in the nebula NGC 7027, for which the temperature and the electronic density have been derived in this way in agreement with the values deduced from the study of forbidden lines.

II.4. Radio recombination lines and lines of non hydrogenic ions

A very exciting subject is the study of radio recombination lines, i.e., of the transitions between very high Rydberg states $n \rightarrow n+1$ (α lines) $n \rightarrow n+2$ (β lines) for n of the order of 100. The H 109_α line , and some others have been observed in nebulae.

In the ordinary conditions in a laboratory,these very high levels are destroyed by collisions or by wall effects. In the interstellar conditions of H II regions or nebulae, these large atoms of

TABLE 2. Intensities of Balmer lines relative to H_β in NGC 7027

Line	Observations NGC 7027	Baker & Menzel 1938	Penguelly 1964	Brocklehurst 1971
H_α	281	254	281	279
H_β	100	100	100	100
H_γ	47,3	50,5	47,0	47,2
H_δ	25,0	30,5	25,9	26,2
H_{15}	1,61	2,3	1,55	1,68
H_{16}	1,40		1,27	1,41
H_{17}	1,22		1,06	1,20
H_{18}	1,11		0,88	1,03
H_{19}	0,94		0,76	0,90

the size $n^2 a_o = 6{,}4\ 10^{-5}$ cm for n = 109 (the size of a biological microorganism), can live without much trouble. At $N_+ = N_e = 10^4$ the mean distance between this atom and the closest charged particle is of the order of $4\ 10^{-2}$, a thousand times the size of the orbit radius.

As we have just seen, electron collisions between these high levels have to be treated by classical methods. These collisions give rise to a collision broadening of the $n \rightarrow n' = n + 1$ transition. The half width due to this broadening is given by

$$\gamma = 2\, N_e \langle v\, (Q'_n + Q'_{n'}) \rangle \tag{20}$$

where $Q'_n \sim Q_{n'} = \sum_{n''} Q(n' \rightarrow n'')$ is the sum of all the collision cross sections between level n' and all the adjacent levels.

Calculations of the line profile of radio recombination lines have been done in this way by Brocklehurst and Seaton [1972]. Unfortunately the observations available are very few, observed lines are weak and it is difficult to distinguish the collisional broadening in the line wings from the continuous background. There is little chance that the collisional physics implied in these calculations be in error. Better observations are needed to confirm the other assumptions made in modeling the nebulae. Experimental studies of collision effects involving high excited states up to $n \sim 80$ are now possible using dye laser selective excitation of alkali atoms and are carried out in different laboratories (Bell Laboratory, Rice University, Orsay University). Measurements of electron induced transfer agree with classical results.

As shown in Table 1, many lines of non-hydrogenic ions or or atoms are observed in the visible and UV spectrum. Thanks to the recent satellite observations a considerable number of studies are carried out in order to use appropriate ratio of these lines to determine the physical conditions in the nebulae and also on the abundance of the principal elements like C, N and O relative to H.

There are many specific cases. The helium recombination spectra has been carefully studied. Explicit allowance of individual angular momentum must be taken into account because level of the same n are not degenerate. Recombination and cascade tend to overpopulate the $2\,^3S$ metastable level and line issued from this level can be optically thick, but this level is in fact depopulated by photoionization by the stellar continuum, by collisional transfer to $2\,^1S$ and $2\,^1P$ and by one photon decay $2\,^3S - 1\,^1S$.

Recent works have been concerned with visible and UV lines in CII, CIII, CIV, OIII, O IV, in conjunction with forbidden lines in the ground states of the same ions to deduce new information on temperature and on densities. Recombination theory must be applied with care : resonance absorption of stellar radiation, accidental fluorescence mechanism, electronic excitation by energetic electrons produced by photo-ionization of hydrogen by Lyman photons of He^+ of the inner zone of the nebula can play a role. Results are quite sensitive to the accuracy of photoionization, recombination and excitation rates concerning these non hydrogenic ions. Most of the time recombination rates on excited states are estimated in using hydrogenic approximations. Ion C III is one of the best studied from the atomic physics standpoint.

II.5. Forbidden line intensities.

At low densities, forbidden lines within the ground configuration of atoms or ions with outer electrons $2p^q$ and $3p^q$ with q = 2, 3, 4, are observed. These configurations have three spectral terms denoted by n = 1, 2, 3 in order of increasing excitation energy and are represented in figure 3 in the case of O III, O II, and N II.

Lines of the same type in other ions have been observed in nebulae as one can see in Table 1, in particular O I, C I, N I, S II

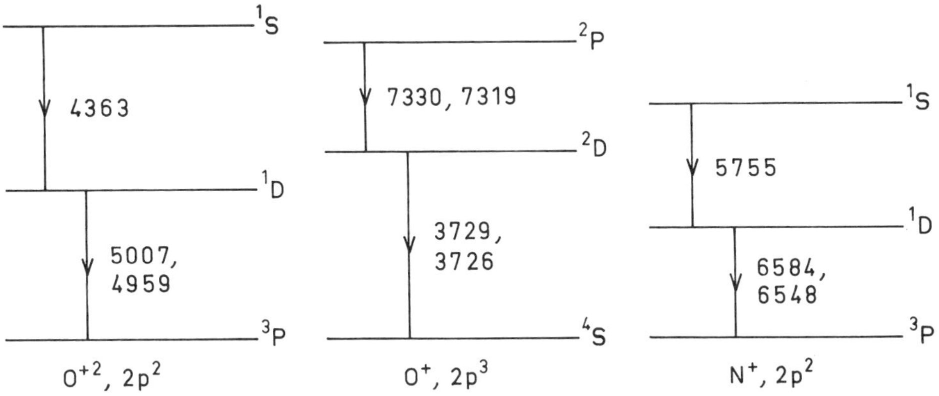

Fig. 3. Energy Level Diagrams For Forbidden Lines Observed in Nebulae

We have already explained before why the ratio of some lines can be good temperature diagnostic, some ratio of other lines of the electron density.

For example, the ratio of the strongest line in most nebulae, the 2 - 1 line (λ = 5007), to the 3 - 2 line (λ = 4363) is given at low density, when formula (9) applies, i.e., when for both lines collisional excitation is followed by quantum emission, by

$$\frac{I_{21}}{I_{32}} = \frac{N_1 N_e \alpha_{12} h\nu_{21}}{N_1 N_e \alpha_{13} h\nu_{32}} = \frac{\omega_2}{\omega_3} \frac{\alpha_{21}}{\alpha_{31}} e^{\frac{h\nu_{32}}{kT}} \frac{h\nu_{21}}{h\nu_{32}} \qquad (21)$$

using the detailed balance relation (7). The ratio of the collision deexcitation rates α_{21} and α_{31} do not depend much on the temperature. The ratio of the two lines is independent of N_e and is a good temperature diagnostic.

On the contrary, ratio of other lines, for example the I_{32}/I_{21} ratio of the λ 5755 line to the sum of the two $\lambda = 6584 + 6548$ lines in N^+ is a good diagnostic of electron density.

All these different ratios can be combined to determine the T and N_e in the nebulae. Each ratio gives one curve and the different curves would cross at the same point for an ideal nebulae with uniform temperature and density. In addition these values of T and N_e should be compatible with other determination of T and N_e by observation of the continuum spectrum and of the ratio of different Balmer lines.

Figure 4 shows the results obtained for different ratios of forbidden lines in O I, O II, O III, N I, N II, for the planetary nebulae NGC 7027. O I, O II, N II observations agree and S II (not shown in the graph) observations are exactly compatible for conditions $N_e \sim 8\ 10^4\ cm^{-3}$ and $T \sim 10^4$ K, but the ratios lines in O III and N I would tend to give a high temperature and a lower density (see Pequignot and Aldrovandi, 1977).

Leaving here the astrophysical problem of the modeling of the nebulae (the temperature is not constant throughout the entire volume, density fluctuations have been assumed), it is necessary to consider the accuracy of the atomic data used in the computation.

Results are very sensitive to the transitions probabilities which are well known for quadrupole and magnetic dipole transitions and also to the excitation rates α by electron collisions. At $T = 10^4$, the electron energy is of the order of the excitation energy of these transitions, the collision cross sections, as we shall see later, are very difficult to calculate accurately. At low energies, they are very sensitive to resonance effects and to configuration interaction in the target. Table 3 gives the comparison of excitation rates obtained for C I, N I, O I, with different cross sections calculations. N I is the most sensitive case.

Accurate calculations of collision rates for $2s^2\,2p^q$ and $3s^2\,3p^q$ have been made recently. To take only one example, in the case of S II, they differ from previous ones by a factor 2, which affects the deduced electron density at least by a factor 2 at $N_e \sim 10^4$.

Detailed calculations of the ratios of the population of fine structure levels of some of these ions (C II, N II, N III, O I, C I, Si II) have also been calculated and used to determine the temperature at the density of the interstellar gas. Special mention must be made for a systematic study of the intensity ratio

$$I\,[{}^2D_{5/2} \to {}^4S] \,/\, I\,[{}^2D_{3/2} \to {}^4S]$$

in O II, and S II, which is very sensitive to the electron density and for which the magnetic dipole transition probabilities must be calculated with extreme care.

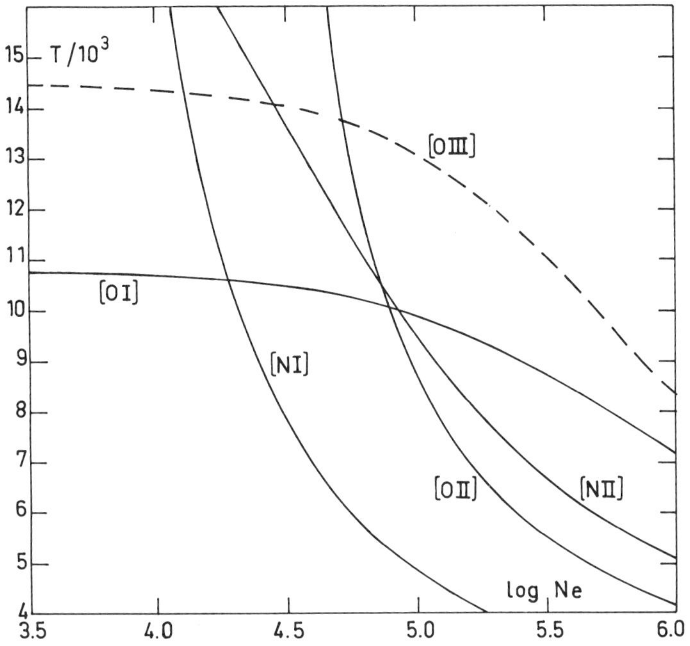

Figure 4

TABLE 3. Comparison of different excitation rate calculations of forbidden transitions in the ground configuration of O I N I

T (K)	Nitrogen I					
	$^2D - ^2P$		$^4S - ^2P$		$^4S - ^2D$	
	S/PA	HW/PA	S/PA	HW/PA	S/PA	HW/PA
$5x10^2$	0.083	0.565	0.795	1.28	0.715	2.75
10^3	0.137	0.704	0.844	1.47	0.806	2.75
$5x10^3$	0.380	0.828	1.12	1.51	1.16	2.40
10^4	0.578	0.866	1.18	1.37	1.15	2.06
$2x10^4$		0.896		1.30		1.78

TABLE 3. (continuation)

T	Oxygen I					
	$^1S - {}^1D$		$^3P - {}^1S$		$^3P - {}^1D$	
	S/PA	HW/PA	S/PA	HW/PA	S/PA	HW/PA
5×10^2	0.041	0.388	1.04	0.605	0.867	1.11
10^3	0.063	0.544	1.11	1.19	1.09	1.15
5×10^3	0.207	0.836	1.44	0.802	1.52	1.18
10^4	0.326	0.915	1.46	0.844	1.45	1.16
2×10^4		0.988		0.861		1.14

S refers to calculations by Seaton (1956) assuming an exact resonance approximation, HW to one configuration calculations by Henry and Williams (1968), PA refers to rate calculations by Pequignot and Aldrovandi (1976) from multiconfiguration calculation of excitation cross section by Le Dourneuf et al [1975].

In nebulae, as in all astrophysical objects, many different observations must be compatible. When some discrepancies remain, new considerations implying atomic physics can be taken into account. Assuming for example different kinds of astrophysical models for the nebulae, it is found that it is impossible to conciliate all the different observations if one assumes that the ionization equilibrium is given by the balance between photoionization and recombination. It has been found that charge transfer with hydrogen of ions like C II, C IV, N III, O III, Ne III, S II, plays an important rôle in explaining the ionization equilibrium. Good agreement between calculations and observations of most of the important lines in Table 1 have been found in this way in the case of NGC 7027 (see Pequignot et al. [1978]). This shows once again how, both the study of recombination and forbidden lines in nebulae has been stimulating for the collision physicists.

II.6. The (X^+ + e) system

Schematically, one can say that astrophysical motivation or problems have stimulated two kinds of important works in collision physics :

- the development of powerfull and accurate methods applicable to a large number of cases, in particular, to complex atoms or ions, at low electron energy, for the study of bound states as well as continuous states.

- the development of approximate methods, classical, semi-classical, or quantal, which have been very successfull for electron-ion collisions, including excitation, ionization, autoionization and many resonance effects.

Low energy physics of complex ions of low charge has been very much concerned with accurate methods. High energy physics of highly ionized atoms has stimulated the search for approximate methods, which will be discussed later.

In nebulae, we have seen that if collision processes are important in redistribution of energy between highly excited states, three processes are of fundamental importance :

- photoionization,
- recombination,
- excitation of forbidden transitions of a ground configuration

The two firsts are radiative processes, recombination being the inverse process of photoionization

$$X_i + h\nu \rightleftarrows X^+ + e \tag{22}$$

The last is an electron excitation process

$$X_i^+ + e \rightarrow X_j^+ + e \tag{23}$$

All these processes imply at low electron energy an accurate representation of the system

$$X^+ + e \tag{24}$$

(or X + e in the cases of photodetachment of negative ion X^- and of excitation of X). As known, reaction (22) obeys a definite selection rule for the dipole radiation operator and only a few values of the angular momentum of the ejected electron have to be considered. In reaction (23), all values of the angular momentum have to be considered, and in this sense one can say that a photoionization calculation is a by-product of a collision calculation.

The common problem is to find a good representation of the N + 1 electron system, to take account of the electron correlation within the target (an atomic structure problem for the N electron system) and to take account of the electron correlation due to the additional electron. The total wave function has the form

$$\Psi(N+1) = \mathcal{A}\left[\sum_i \Phi_i(N)\, F_i(r_{N+1}) + \sum_{i'} \bar{\Phi}_{i'}(N)\, F_{i'}(r_{N+1})\right] + \sum_j a_j \Phi_j(N+1) \tag{25}$$

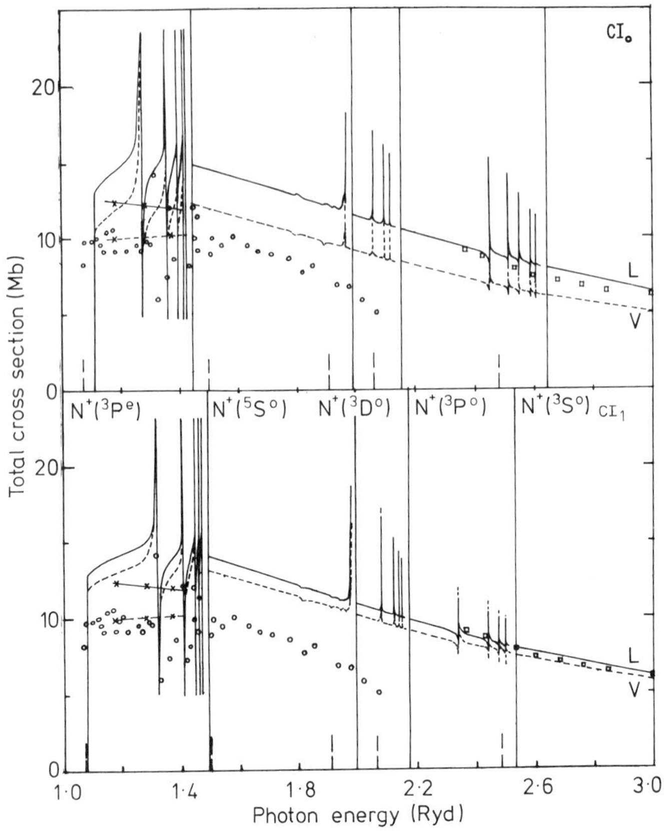

Figure 5 - Photoionization of the ground state (^4S) of nitrogen

In this antisymmetrized expansion the Φ_i(N) are accurate multiconfigurational wave functions of the target, F_i (r_{N+1}) are the radial wave functions of the colliding electron, the $\bar{\Phi}_{i'}$ (N) are judicious pseudo wave functions representing the polarization of the target by the colliding electron (important in neutral + electron systems), the Φ_j (N + 1) are known bound functions of the N + 1 system which allows for excitation of virtual orbitals to take account of capture and short range correlation effects.

When solving the Schrödinger equation, one obtains a system of integrodifferential equations for determining the F_i (r_{N+1}) whose asymptotic forms give the cross sections and the unknown a_j coefficients. In his lecture P.G. Burke has discussed the different methods and degrees of approximation for solving this problem.

Let us simply give a few examples relevant to physics in planetary nebulae. In figure 5 is given the photoionization cross section of the ground state of nitrogen. Results for two different calculations are given, both including in expansion (25) all the states of N^+ corresponding to the ground configuration $2s^2\,2p^2$ and to the excited configuration $2s\,2p^3$ (coupling with the last gives rise to the resonance structure). Both calculations include configuration interaction within the target N^+. The Φ_i (N) are multiconfigurational wave functions : C I_0 includes the"degenerate"

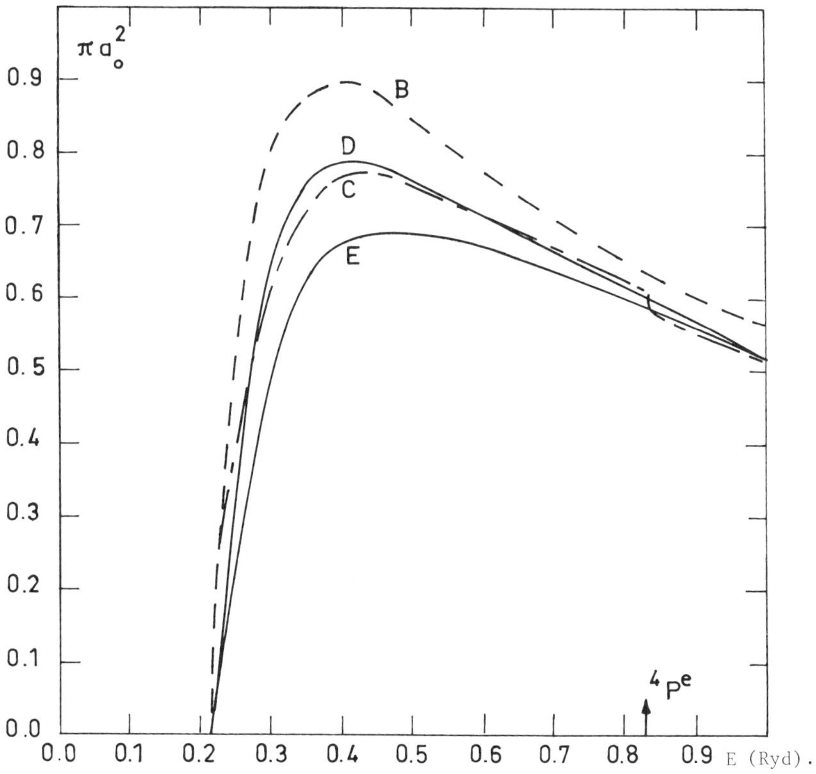

Figure 6

configurations $2s^2\,2p^2$ and $2p^4$ for the ground states, C I, includes in addition the virtual excitation of the n = 2 shell (using correlation orbitals 3s, 3p, 3d). Calculations in the length and velocity forms agree well with experiment. The spectrum is strongly perturbed by an autoionized series N^* ($2s\,2p^3\,np\;^4P_e$) which has been measured with high resolution and for which very good agreement is found with theoretical calculation (see Le Dourneuf et al. (1976).

Accurate calculations have been made for inelastic excitation within the ground configuration of C, N, O. Figure 6 shows just an example relative to the collisional excitation of the ^4S - ^2D transition in neutral nitrogen of configuration $2s^2\,2p^3$, which is important for diagnostics in nebulae (see LeDourneuf et al (1975)).

Curve B is a "close coupling" calculation which takes account of some configuration interactions in the target ($2s^2\,2p^3 + 2p^5$ configurations). Curve C is a close coupling calculation with more states included in expansion (25). In curve D, these additional states are replaced by pseudo states which improve the representation of long range correlation effects. Curve E includes both excited configuration states and some pseudo states. These curves illustrate the importance of the additional e - target correlation effects at low energies. This explains the discrepancies in the collision rates corresponding to different calculations which are given in Table 3.

Unfortunately, very few accurate calculations concern excitation of forbidden transitions in positive ions with $2s^2\,2p^q$ and $3s^2\,3p^q$ configuration like the studies on N II, O III, Ne II, Ne III by Seaton [1975] or O II and S II by Pradhan [1976] . As far as photoionization is concerned, there is an urgent need of accurate cross sections for positive ions. Astrophysicists are using simple Hartree Fock functions without taking account of any correlation effects. For photoionization from excited states, the situation is still worse and a pure hydrogenic approximation is currently made for the initial state wave function. Due to the larger spatial extension of the excited state wave function, the photoionization cross section should be less sensitive to short range correlation effects, quantum defect theory may help to represent some excited states. These photoionization cross sections are needed to improve the recombination rate to the lowest excited states, recombination being the most important population process of these levels in nebulae. Some of these calculations are also a great interest in other astrophysical applications.

On atomic processes in nebulae, special mention must be made of a review article by Seaton [1972].

III. ATOMIC PROCESSES IN THE SOLAR CORONA

III.1. The line spectrum

When Edlen, in 1942, identified the strong emission lines of the solar corona with the forbidden lines of highly ionized atoms like Fe XIV, it was demonstrated that the temperature of the coronal gas was $T = 10^6$ °K or higher, and that the densities were of the order of 10^9 cm^{-3}. At this temperature :

- O VIII, Ne IX, Mg XII, Si XIV, S XVI are hydrogen like ions.
- Fe XIV, Ni XVI are aluminium-like ions with ground configuration $3s^2\ 3p$.

In the hot part of a solar flare temperature can reach 30.10^6 K and one can observe hydrogen-like Fe XXVI, helium-like Fe XXV, lithium-like Fe XXIV. Figure (11) which gives the ionization equilibrium of Fe ions in the solar corona, shows the possibility of observing ion lines from Fe IX below $T = 10^6$ up to Fe XXVI for $T = 3 10^7$ K.

At Ne = 10^9 (even for $10^{11} < N_e < 10^{13}$ in some part of the flares) the coronal plasmas is far from beeing in LTE. Densities are higher than in the nebulae. The radiation coming from the sun is very weak the sun is a cool star, whose atmosphere just below the corona has a temperature of 5.500 K. Photoionization or photon excitation of coronal ions by the radiation flux are therefore negligible at short wavelengths. Collisional ionization is now the main process for ionizing the atoms. Collisional excitation contributes to populate most of the levels.

For such ionized atoms, very few emission lines appear in the visible. They are the fine structure transitions within the ground configuration, like the famous green line of Fe XIV $3s^2\ 3p$ $[^2P_{3/2} - {}^2P_{1/2}]$ at 5303 Å. Theory of these forbidden lines does not differ very much from the theory of forbidden lines in nebulae. Good atomic data are needed : transition probabilities as well as excitation cross sections. Configuration interactions are important. L S coupling is inadequate. Resonance effects are very strong. For the $^2P_{3/2}$ upper level, the populating processes are : radiative excitation (the solar flux at 5303 is strong), direct collisional excitation and collisional excitation to higher levels followed by cascades.

Most of these forbidden transitions in ions of configuration $2p^q$ and $3p^q$ appear now in the UV spectrum. In Table 4 is given a list of emission lines observed by Gabriel et al. [1971] during a solar eclipse and identified by Jordan [1971].

TABLE 4. Observed UV coronal emission lines.

T_e ($\times 10^6$°K)	Ion	Sequence	Transition	Obs.A
0·69	Mg VII	C $2p^2$	$^3P_1 - {}^1S_0$	1190·2
0·69	Si VII	O $2p^4$	$^3P_2 - {}^1D_2$	2147·4
0·93	Si VIII	N $2p^3$	$^4S_{3/2} - {}^2D_{3/2}$	1446·0
1·2	Si IX	C $2p^2$	$^3P_2 - {}^1D_2$	1715·3
1·2	Si IX	C $2p^2$	$^3P_1 - {}^1D_2$	1985·0
1·2	Si IX	C $2p^2$	$^3P_2 - {}^1D_2$	2149·5
1·5	Fe XI	S $3p^4$	$^3P_1 - {}^1S_0$	1467·0
1·7	Fe XII	P $3p^3$	$^4S_{3/2} - {}^2P_{3/2}$	1242·2
1·7	Fe XII	P $3p^3$	$^4S_{3/2} - {}^2P_{1/2}$	1349·6
1·7	Fe XII	P $3p^3$	$^4S_{3/2} - {}^2D_{5/2}$	2169·7
1·9	Fe XIII	Si $3p^2$	$^3P_1 - {}^1S_0$	1213·0
2·0	O VII	He $1s^2$	$^3S_1 - {}^3P_2$	1624·0
2·0	S XI	C $2p^2$	$^3P_1 - {}^1D_2$	1614·6
2·0	S XI	C $2p^2$	$^3P_2 - {}^1D_2$	1826·0
2·0	Ni XIII	S $3p^4$	$^3P_2 - {}^1D_2$	2126·0
2·0	Ni XIV	P $3p^3$	$^4S_{3/2} - {}^2D_{5/2}$	1866·9
2·2	Ni XIV	P $3p^3$	$^4S_{3/2} - {}^2D_{3/2}$	2185·1
2·5	Ni XV	Si $3p^2$	$^3P_1 - {}^1D_2$	2085·7

In the following, we shall be mainly concerned with the XUV region of the spectra, where the solar corona emits most of its energy, having a temperature in excess of 10^6 K. Figure 7 gives part of the coronal spectrum below 25 Å. This spectral range includes the resonance transitions of almost all the highly ionized coronal ions. Good spectral resolution is now possible and permits the identification of the lines.

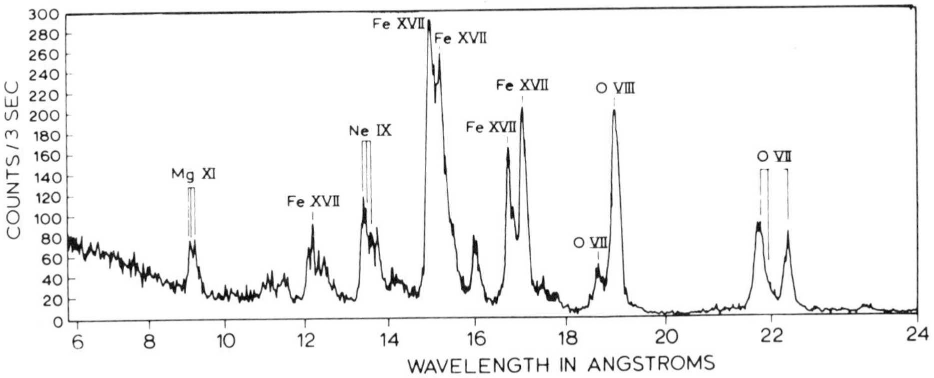

Figure 7. X-ray spectra of hot regions of the Solar Corona

In addition to permitted dipole transitions, one observes magnetic dipole, and electric quadrupole transitions (in particular in H, He, Ne-like ions) as well as "satellite" lines which appear in the vicinity of the resonance transition of the corresponding ion.

Coronal studies using X emission lines have given a lot of new information on the physical state of the plasma and on the solar activity, in particular on the flaring active regions.

III.2. Ionization equilibrium in the solar corona.

At T = 10^6 and Ne = 10^9 the ionization equilibrium, the ratio of the total population of ion X^{+q+1} to the total population X^{+q} is given by

$$\frac{N(X^{+q+1})}{N(X^{+q})} = \frac{S(X^{+q})}{\alpha_t(X^{+q})} \tag{26}$$

where S, the rate of ionization by electron impact is balanced by α_t the rate of recombination.

This assumes steady state which is valid in most of the cases. But eruptive phenomena like solar flare vary very quickly with time : these plasmas have to be considered as transient relative to ionization. The state of ionization varies with time $\frac{dN^{+q}}{dt}$.

Equation (22) is a formal one. It implies in fact that only the ground state is ionized and that the total recombination coefficient α_t is just the sum of recombination into all the individual levels of the ion (X^+q).

In fact, it is not possible to decouple the ionization equilibrium equation from the system of equations giving the population of all the levels. We have learnt from the discussion below on recombination theory, that many processes have to be taken into account, particularly the redistribution of energy between high levels due to electronic collisions. Consequences on the ionization rate S and on the recombination α_t have been studied by Burgess and Summers (1969) (see also Seaton and Storey (1976)).

In the following, for simplicity, we assume that S (X^{+q}) is the rate of collisional ionization from the ground state; S is the sum of two contributions

$$S(X^{+q}) = S_{direct}(X^{+q}) + S_{auto}(X^{+q}) \tag{27}$$

S direct begin the rate of ionization by electron impact

$$X_a^{+q} + e \rightarrow X_b^{+q+1} + e + e \qquad (28)$$

S auto being the rate of ionization via excitation and autoionization

$$X_a^{+q} + e \rightarrow X_\beta^{+q} + e \rightarrow X_b^{+q+1} + e + e \qquad (29)$$

a two step process, in which β is an autoionization state lying above the first limit of ionization.

Take the example of neutral lithium (q = 0) in state a = $1s^2\ 2s$. By electron excitation of the inner shell state β $1s\ 2s^2$ is excited, which lies above the ionization limit $1s^2$ of Li^+ and can interact with the adjacent continuum and autoionize giving Li^+ with ground $1s^2$ plus a free electron. This is represented in figure 8.

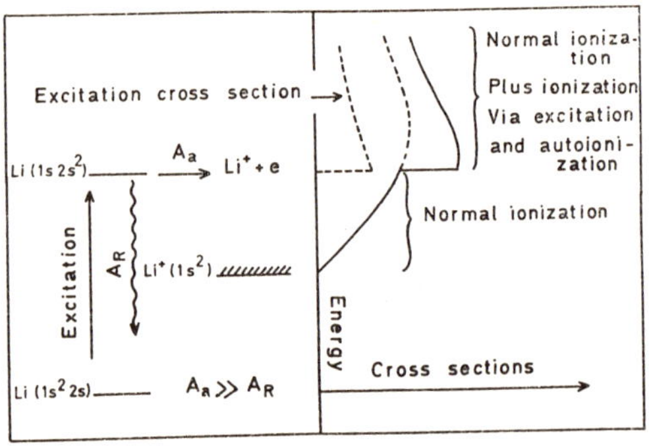

Fig. 8. Ionization via excitation and autoionization

Because of the last process of autoionization, with probability A_a, is in competition with spontaneous decay A_r into a given bound state below the ionization threshold, the cross section for ionization via excitation and autoionization has the cross section

$$Q_{auto} = \sum_\beta Q(a \rightarrow \beta) \frac{A_a^\beta}{A_a^\beta + A_r^\beta} \qquad (30)$$

summing over the target bound state β above the ionization limit. We neglect here the influence of the other continua.

For neutral atoms, we have in general $A_a \sim 10^{13}$ sec^{-1} compared to $A_r \sim 10^8$ sec^{-1} and the total ionization cross section when $A_a \gg A_r$ reduces to

$$Q_{total} = Q_{direct} + \sum_{\beta} Q\,(a-\beta) \qquad (31)$$

which is valid for neutral and ion in low state of ionization.

But the dipole transition probability A_r varies like Z^4, and on the other hand, A_a is nearly constant with Z. For highly ionized positive ions (for $Z > 10$) of interest, the correct branching ratios in formula (30) have to be known, and this implies the calculation of autoionization probabilities A_a.

In equation (26), $\alpha_t\,(X^{+q})$ the total rate of recombination, is also a sum of two contributions

$$\alpha_t = \alpha_r + \alpha_d \qquad (32)$$

$\alpha_r = \sum_j \alpha_r^j$ is the total radiative recombination rate, already introduced in the discussion of the ionization equilibrium in nebulae.

α_d is the total dielectronic recombination rate. We assume here that α_d is the sum of the α_d^j to all bound states below the ionization limit.

In the dielectronic recombination process, a free electron is first captured into a doubly excited state (the inverse process of autoionization), and recombination is due to subsequent emission of radiation (the stabilization process).

To take the example of He-like ions

$$X^{+q+1}(1s^2) + e \rightarrow X^{+q}(1s\,2p\,n\ell) \quad \text{capture}$$

$$X^{+q}(1s\,2p\,n\ell) \rightarrow X^{+q}(1s^2\,n\ell) + h\nu \quad \text{emission of radiation (in competition with)} \qquad (33)$$

$$\searrow X^{+q+1}(1s^2) + e \quad \text{autoionization}$$

At $T = 10^6$, many electrons are very energetic, the kinetic energy is of the order of 100 eV, and may have energy of the order of the excitation energy of the ion (excitation energy of 1s 2p) where many autoionization states (of type 1s 2p $n\ell$) accumulate. Dielectronic recombination can be therefore an important process for $T \simeq 10^5 - 10^7$ and is indeed important in coronal plasmas.

In thermodynamical equilibrium, electron capture is balanced by autoionization. The capture rate C_d is therefore related to autoionization A_a by a Boltzmann-Saha relation $C_d = A_a\, S\,(T)$, with

$$S(T) = \frac{h^3}{(2\pi m kT)^{3/2}} \frac{\omega(X_\beta^{+q})}{2\omega(X_\ell^{+q''})} \exp -\frac{\Delta E_{\beta\ell}}{kT} \tag{34}$$

which is valid whatever are the equilibrium conditions.

But radiative decay compete with autoionization, and the rate of dielectronic recombination is proportional to the branching ratio of A_r (the stabilization process) to the sum $A_r + A_a$ (the total probability of leaving the autoionization state)

$$\alpha_d = \sum_{j=n\ell} A_a \, S(T) \frac{A_r}{A_r + A_a} \tag{35}$$

summing over all the $n\ell$ autoionization states (for simplicity we have only taken account of one excited state 1s 2p of the ion).

For n sufficiently large, the radiative probability A_r which is independent of n is much bigger than A_a which varies like n^{-3}, and therefore emission of radiation is indeed the mechanism of depopulating the doubly excited state.

When $A_r \gg A_a$, α_d reduces to a sum of capture cross sections

$$\alpha_d = \sum_{n\ell} S(T) \, A_a \tag{36}$$

which can be very large due to the large number of states of doubly excited configurations (2p $n\ell$) which contributes to the sum.

For n large, the frequency ν of the radiative transition (1s 2p $n\ell$ — $1s^2 n\ell$) cannot be distinguished from the resonance transitions of the corresponding ion (1s 2p — $1s^2$) and the overlapping of these "satellite lines" reinforces this resonance line intensity. This contribution must be taken into account if this resonance line intensity is used as a spectroscopic diagnostic.

On the contrary for low values of n, say 2, 3, 4, these satellite lines are well separated from the resonance lines. If they are strong enough, as it is the case in heavy ions, their relative intensities can be compared to the intensity of the corresponding resonance line.

As far as the ionization equilibrium is concerned, one needs good cross sections for direct ionization, essentially good excitation cross section for calculating the ionization through autoionization with formula (31) for low ionized positive ions, as well as autoionization probabilities A_a for low doubly excited states in the case of highly ionized atoms ($Z > 10$).

In the calculation of the dielectronic recombination rate, autoionization probabilities have also to be known, but for high n doubly excited states : in this case, autoionization probabilities can be obtained by extrapolation of excitation cross section of the corresponding ion ($1s^2 \rightarrow 1s\,2p$ in our example) below the excitation threshold ($1s\,2p$).

Figure 9 shows the calculated rate coefficient of both direct ionization and autoionization of iron ions (C. Jordan, [1969]). One sees that autoionization effects can vary suddenly along the sequence of different stages of ionization of a given element as predicted by Goldberg et al. [1965] in the case of Fe XVI. It turns out that no simple extrapolation of the autoionization rates is possible.

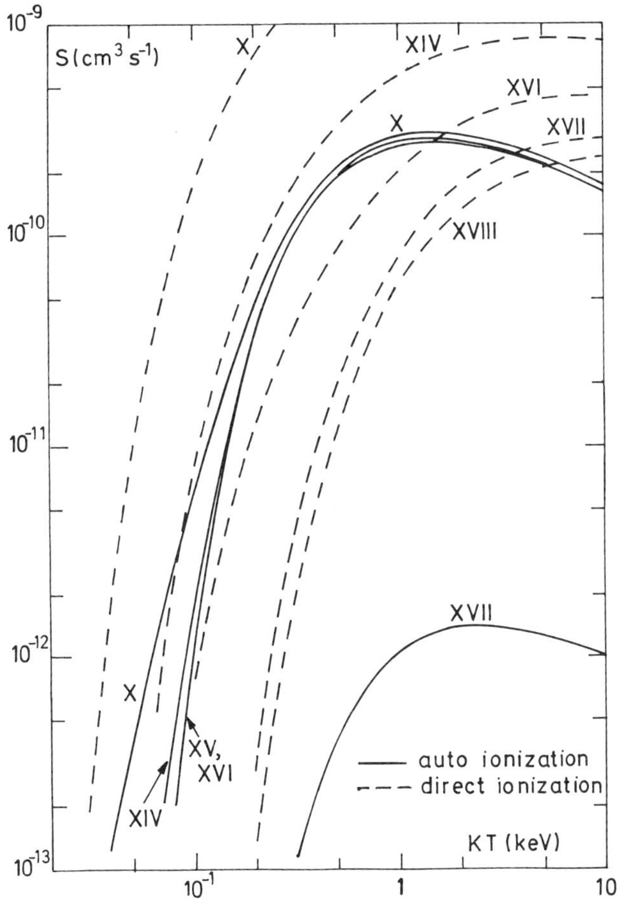

Fig. 9. Ionization rate for iron ions

In figure 10 is shown the relative importance of radiative recombination and of dielectronic recombination of He^+, from the basic work of Burgess (1964).

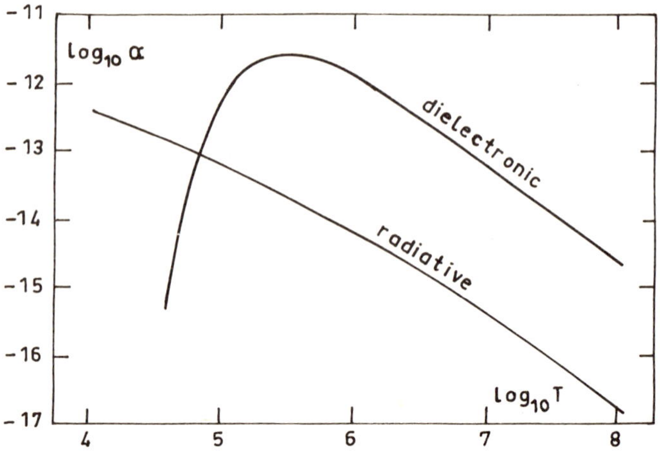

Fig. 10. $(He^+ + e)$ recombination coefficients α (cm^3 sec^{-1})

Figure 11 shows the ionization equilibrium curves for iron, in a temperature range of interest for the solar corona. One sees the large number of observable stages of ionization from $T = 10^6$ to $T = 3.10^7$, and the simultaneous existence of successive stages of ionization in the same region of the solar corona. Comparison between the two calculations of C. Jordan (1969, 1970) and of Jacobs et al. (1977), who have improved the calculation by including autoionization into excited states, shows the great influence of the accuracy of the results on α_d in ionization equilibrium calculations.

III.3. Level population mechanisms.

Contrary to the situation in planetary nebulae in which recombination population mechanisms of amny levels in the solar corona, with $T_e \sim 10^6$ and $N_e = 10^9$, many electrons are very energetic: excited states are populated by electron impact and depopulated by radiative decay.

In steady state, the system of equations giving the populations for low excited levels has the form

$$N_j \left[\sum_{i<j} (A_{ji} + N_e \alpha_{ji}) + \sum_{j<k} N_e \alpha_{jk} \right] = \sum_{i<j} N_i N_e \alpha_{ij} + \sum_{j<k} N_k (\alpha_{kj} N_e + A_{kj}) \tag{37}$$

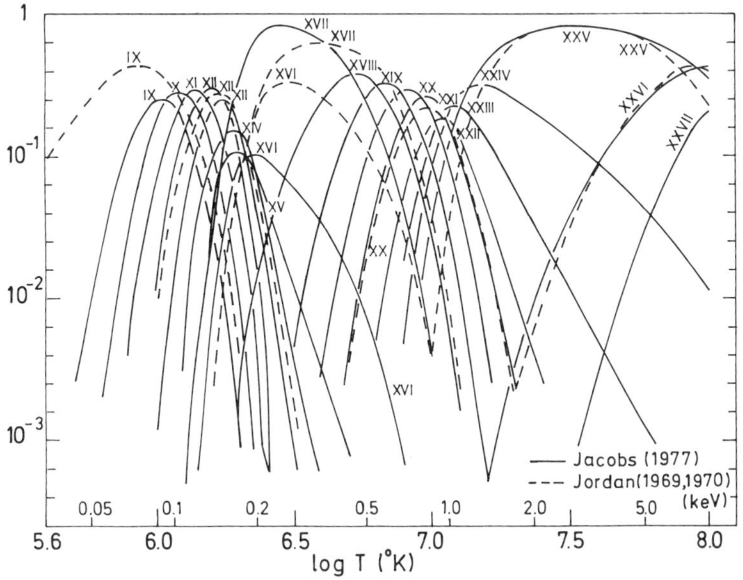

Fig. 11. The ionization equilibrium for Fe ions

Collisional excitation to levels higher than j followed by cascades play an important rôle. Proton collisions must be sometimes included. Radiative absorption of the outgoing flux from the solar photosphere can be important for some transitions. Radiative and dielectronic recombination may have some influence.

A considerable number of works have been done in using XUV lines, to obtain information on T and N_e in the solar corona. Some are concerned with the calculation of a complete X spectrum, where the corona emits most of its energy, assuming a model and a chemical composition. Most of them are using specific lines or judicious ratios of them.

If it is easy to find good temperature diagnostics, and in the discussion below we shall see that the study of satellite lines gives an additional powerful diagnostic, it is not easy to find a good spectroscopic diagnostic of the density. Coming back to the paragraph I.3., it is interesting to see why it is so.

Most of the lines are collisionally excited and radiatively deexcited. Equation (9) is valid in most of the cases : the ratio of two different lines does not depend of the density and is a function of the temperature only.

As proposed by Gabriel and Jordan (1972), let us take the example of helium-like ions represented in figure 1.

To find a density diagnostic, it is necessary to find metastable levels with very long lifetime, long enough in order that these levels are depopulated by collision even at low density.

With reference to figure 1, let f be the ground state, m be the 1s 2s ^3S the long lifetime level (A_{mf} = 3 10^{-6} Z^{10} sec^{-1}) and i be the "short" lifetime level (A_{if} = 5 Z^9 sec^{-1}).

The line intensity for transition m → f is

$$I_{mf} = N_m A_{mf} h\nu_{mf} \tag{38}$$

and the population of level m is given by

$$N_m \left[\sum_k \alpha_{mk} N_e + A_{mf} \right] = N_f \alpha_{fm} N_e \tag{39}$$

because one cannot neglect collisional depopulation (quenching) of level m compared to the small transition probability A_{mf} and provided one can neglect the radiative cascade from other levels.

On the other hand, the line intensity for transition i → f is

$$I_{if} = N_i A_{if} h\nu_{if} \quad \text{with} \quad N_i A_{if} = N_f \alpha_{fi} N_e \tag{40}$$

The ratio

$$\frac{I_{if}}{I_{mf}} = \frac{h\nu_{if}}{h\nu_{mf}} \frac{\alpha_{fi}}{\alpha_{fm}} \left[1 + N_e \frac{\sum_k \alpha_{mk}}{A_{mf}} \right] \tag{41}$$

which increases with N_e, until the forbidden line disappears.

The trouble is that the term sensitive to N_e appears only at high density, higher than 10^9 the mean density of the corona : at 7,3 10^9 for O VII, 1,5 10^{11} for Ne IX, 1, 3 10^{13} for Si XIII !

This illustrate the importance of the ionic charge q = Z − 1 and of the variation of the transitions probabilities with Z.

Let us take only one level k (which represent the quenching effect in (41)) and assume kT $\gg E_{mk}$ (a reasonable assumption at T = 10^6). α_{mk} can be estimated using an approximate expression for the electron collision rate ;

$$\alpha_{mk} = 8,63\ 10^{-6}\ \frac{\Omega}{\omega_k}\ T^{-1/2} \tag{42}$$

For an order of magnitude estimate, the collision strength Ω is taken to be 1 and the statistical weight $\omega_k = 1$.

Equation (41) is a relevant density diagnostic when

$$\mathcal{E} = \frac{\alpha_{mk} N_e}{A_{mf}} \gg 1 \tag{43}$$

For each (N_e, T) there is a critical value A_c (or a critical value of the metastability) beyond which it is difficult to find a density diagnostic.

Let us compare the situation in different astrophysical or laboratory plasmas and give A_c for (a) planetary nebulae, (b) low solar transition region between the chromosphere and the corona, (c) solar corona, (d) solar flare, (e) laser produced laboratory plasmas.

	(a)	(b)	(c)	(d)	(e)
N_e	10^4	10^{10}	10^9	10^{12}	10^{20}
T_e	10^4	10^4	10^6	10^7	10^7
A_c	10^{-3}	900	90	3000	3 10^{11}

The search for good plasmas diagnostics using high resolution spectroscopy is of crucial importance for laboratory high temperature plasmas, like the laser produced plasmas or the Tokamak plasmas for fusion (see for example, Feldman and Doschek (1977) .

Looking at the table giving A_c, one sees that in case c of the solar corona all forbidden lines of the relativistic magnetic dipole type, like 1s 2s ^3S — 1s ^{2} ^1S, for high Z (low Z ions are not present at $T = 10^6$) have transition probabilities which are too large ($A_{m\ell} \sim 900$ for O VII) and cannot be used as density diagnostics.

On the contrary in planetary nebulae, where low Z ions are present, like the beryllium-like ion C III, one finds forbidden transitions with sufficiently small transition probabilities.

In the solar corona, the density decreases with height from 10^9 near the solar limb up to values of 10^6 at 2 solar radii, 10^5 at 3 solar radii. It is possible to have information on the density using the degree of polarization of forbidden emission lines in the visible (within the ground configurations $3\,p^q$ of ions like Fe XIV or Fe XIII) observed at different distances of the solar limb (Sahal, 1974).

Polarization of these lines is due to excitation by the non isotropic radiation from the sun at a given point of the corona. But collisions are a very powerful mechanisms of depolarization giving an equilibrium distribution of the Zeeman sublevels JM_J. Direct excitation by electrons, proton collisions between Zeeman sublevels, and mainly cascade effects from higher levels contribute to depolarization. In the outer part of the corona densities are so low that radiation excitation is in competition with collisional effects. Polarization of the emission line can be observed and measured.

III.4. The satellite lines.

In paragraph III.2., we have seen that dielectronic recombination gives rise to satellite lines which appear in the vicinity of the resonance transition of the corresponding ion.

In the case of helium-like ions, they are the $1s\,2p\,n\ell - 1s^2\,n\ell$ lines in the vicinity of the resonance transition $1s\,2p - 1s^2$ of He^+ when n is small. It is important now to calculate their intensities, compared to the resonance line.

For low values of n (contrary to the case of large n discussed in III.2.), $A_a \gg A_r$ and the dielectronic recombination rate given by equation (35) is now given by

$$\alpha_d = \sum_{n\ell} S(T)\, A_2 \tag{44}$$

independent of the autoionization probability and easily calculable.

The satellite line intensity is given by

$$I_s = N(X_\beta^{+q})\, A_2\, h\nu_s \tag{45}$$

But because $A_a \gg A_r$ the population of the autoionization level β is also given by the Boltzmann-Saha relation (4). With the notation of III.2.

$$I_s = N_e N(X_a^{+q+1}) \; S(T) \; A_r \; h\nu_s \tag{46}$$

On the other hand, the resonance transition ion line intensity is

$$I_r = N_e N(X_a^{+q+1}) \; \langle v Q_r \rangle \; h\nu_r \tag{47}$$

In the case of helium ions : a = 1s ^{2}, b = 1s 2p, β = 1s 2p nℓ

The resonance line is mainly excited by electron impact. Q_r is the excitation cross section.

Therefore

$$\frac{I_s}{I_r} = \frac{S(T) \, A_r}{\langle v Q_r \rangle} \tag{48}$$

This ratio is independent of N_e and varies like T_e^{-1}. Very important is its Z^4 dependence, the satellite line being weak for light ions (Z < 10) and comparable to the resonance line in heavy multicharged ions (Z $\simeq$ 20 - 30). Resonance line of hydrogen or helium like heavy ions appear at very short X wavelengths and are observed in the hot active region of the solar corona as well as in the Tokamak fusion plasmas, at very high temperature.

For non hydrogenic ions, some satellites may be due to a competitive process, the impact excitation of an innershell electron of the recombined ion, i.e.,

$$1s^2 n\ell + e \longrightarrow 1s\,2p\,n\ell + e \tag{49}$$

The ratio is now given by

$$\frac{I'_s}{I_r} = \frac{N(X^{+q})}{N(X^{+q+1})} \; \frac{\alpha_i}{\alpha_r} \; \frac{A_r}{A_r + A_a} \tag{50}$$

where α_i is the innershell excitation rate for reaction (49). $\alpha_i = \langle v Q \text{ (innershell)} \rangle$. $\alpha_r = \langle v Q_r \rangle$ is the rate already introduced in (47). The ratio of these rates is practically the ratio of the oscillator strengths.

This process is important in hot transient plasmas like the solar flare and the Tokamak where the temperature varies so rapidly that the ionization steady state equilibrium is not valid. The number of ions N (X^{+q}) and N (X^{+q+1}) varies with time in equation (50) and consequently the intensity of the satellite line is predominantly due to this process.

These variations of intensity are effectively observed and are a powerful diagnostic of physical properties of active regions of the solar corona. An X ray polychromotor has been launched on board of the "Solar Maximum Mission" satellite, and precise measurements of these phenomena are in progress together with their theoretical interpretation. The theory of satellite line has been reviewed recently by Dubau & Volonté (1980).

III.5. Electron-ion collisions.

Physics in the solar corona is essentially concerned with electron-ion collisions and has stimulated the development of approximate methods of calculation of excitation, autoionization, ionization cross sections.

Close collisions are less important in ion-electron interactions. Because of the ion Coulomb field, the distance of closest approach R_c of the perturbing electron when its velocity tends to zero remains finite for all values of its angular momentum

$$R_c = \ell^2 [2me^2 q]^{-1}$$

using the correspondance $L = mvR = \hbar\sqrt{\ell(\ell+1)}$, q being the ionic charge

The collision cross section is usually given in terms of the collision strength

$$Q(i \to j) = \frac{1}{\omega_i} \frac{1}{k_i^2} \Omega(i,j) \quad \text{in } \pi a_0^2$$

$\Omega(i \to j) = \Omega(j \to i)$ is symmetric because of the symmetry of the S matrix and this insures the detailed balancing relation (7). k_i^2 is the initial electron kinetic energy in Rydbergs.

For positive ions, many values of ℓ contribute to Ω even at low energies, and consequently Ω (i, j) is finite at threshold. For $\ell \neq 0$ or 1, short range interactions are not important. Therefore the accuracy of approximate methods will be much better for positive ions than for neutrals. Ω (i, j) being finite at threshold and slowly dependent of the electron energies, extrapolation to negative energies will permit approximate calculations of the contribution of resonance to excitation cross sections or of autoioniation cross sections.

It is not possible to describe here all the approximate methods which have been used. Reference must be made to review articles on electron-ion collisions by Seaton (1962), (1976) and Bely and Van Regemorter (1970) which includes methods for electron excitation and ionization.

Four types of methods have been extensively used for excitation : the Coulomb Bethe Method or so-called $\bar{g}$ approximation, the semi-classical method, the Coulomb Born Method and the Distorted Wave approximation (D.W.)

The accuracy of the first has been recently discussed by Younger and Wiese (1979). The semi-classical approximation is as good as the Coulomb Born approximation. Extensive comparison between results obtained with the last and more accurate close coupling calculations has been made by Kato (1979). The D.W. results have been compared to close coupling calculations in a recent report of Los Alamos laboratory by Merts et al. (1980).

It is interesting to note that for allowed transitions, particularly for $\Delta n = 0$ transitions and for highly charged ions, very approximate methods can give very reliable results.

If for allowed transitions, the contribution of resonance, usually neglected with approximate method or with close coupling calculations without any "closed" channel, is of the order of 10%, for forbidden transitions, this contribution is of the order of 50%.

But approximate methods give the way to calculate the resonance contribution separately : the excitation collision strength (calculayed by any approximate method) can be extrapolated to negative energies. These Ω (extrapolated) permit the calculation

- of the contribution of non overlapping resonance to excitation cross sections,
- of the capture or of the autoionization rate of a given level interacting with a continuum,
- of the dielectronic recombination rate which is given in terms of autoionization probabilities of high n levels above the ionization limit.

The accuracy of this technique of using approximate methods in resonance phenomena has been tested in many calculations concerning excitation of fine structure transitions (in Fe XIV), excitation of some forbidden line (in O V), dielectronic recombination (of He^+), etc ...

The development of the approximate methods has been very usefull both in astrophysics and in plasma physics. Most of the time the modeling of a collision problem is very simple compared to the modeling of an astrophysical object.

REFERENCES

Baker, J.G. and Menzel, D.H., 1938, Astrophys. J., 88:52.
Bely, O. and Van Regemorter, H., 1970, Ann. Rev. Astron. Astrophys., 8:329.
Brocklehurst, M., 1970, M.N.R.A.S., 148:417.
Brocklehurst, M., 1971, M.N.R.A.S., 153:471.
Brocklehurst, M. and Seaton, M.J., 1972, M.N.R.A.S., 157:179.
Burgess, A., 1964, Astrophys. J., 139:776.
Burgess, A. and Summers, H.P., 1969, Astrophys. J., 157:1007.
Dubau, J & Volonte, S., 1980 "Dielectronic recombination and its applications in astronomy" in Rep. Prog., 43 : 199.
Feldman, U. and Doschek, G.A., 1977, J. Opt. Soc. Am., 67:726.
Gabriel, A., Garton, W.R.S.,Goldberg, L., Jones, T.J.L., Jordan, C., Morgan, F.J., Nicholls, R.W., Parkinson, W.H., Paxton, H.J.B., Reeves, E.M., Shenton, D.B., Speer, R.J. and Wilson, R., 1971, Astrophys. J., 169:595.
Gabriel, A.H. and Jordan, C., 1972, "Interpretation of spectral intensities from laboratory and astrophysical plasmas", in Case studies in At. Coll., 2:211.
Goldberg, L., Dupree, A.K. and Allen, J.W., 1965, Ann. d'Astrophys., 28:589.
Henry, R.J. and Williams, R.E., 1968, Publ. Astron. Soc. Pacific, 80:669.
(see also Henry, R.J., Burke, P.G., Sinfailam, A.L., 1979, Phys. Rev., 178:218.
Jacobs, V., Davis, J., Kepple, P., Blaha, M., 1977, Astrophys. J., 211:605.
Jordan, C., 1969, M.N.R.A.S., 142:501.
Jordan, C., 1971, Solar Physics, 21:381.
Kato, T., 1978, "Atomic processes in hot plasmas", in Reports of the Institute of Plasma Physics, Nagoya University.
Le Dourneuf, M., Vo Ky Lan, Hibbert, J., 1976, J. Phys. B., L359.
Le Dourneuf, M., Vo Ky Lan, Berrington, K.A., Burke, P.G., 1975, Abstr. IX ICPEAC, Seattle, 634.
(see also Le Dourneuf, M., 1976, Thesis (unpublished).
Mertz, A.L., Mann, J.B., Robb, W.D., Maggee, N.H., 1980, Informal Report, Los Alamos Laboratory.
Pequignot, D. and Aldrovandi, S.M.V., 1976, Astron. Astrophys., 50:141.
Pequignot, D. and Aldrovandi, S.W.V., 1977, Astron. Astrophys., 58:411.
Pequignot, D., Aldrovandi, S.M.V. and Stasinska, G., 1978, Astron. Astrophys., 63:313.
Penguelly, R.M., 1964, M.N.R.A.S., 127:145.
Percival, I.C. and Richards, D., 1975, Adv. Atom. Molec. Phys., 11:1
Pradhan, A.K., 1976, M.N.R.A.S., 177:31.
Sahal, S., 1974, Astron. Astrophys., 36:355.
Seaton, M.J., 1956, in "The airglow and the aurorae", Pergamon Press.

Seaton, M.J., 1962, "Excitation and ionization by electron impact" in Atomic and Molecular Processes, Academic Press.
Seaton, M.J., 1964, M.N.R.A.S., 127:177.
Seaton, M.J., 1972, "The spectra of gaseous nebulae", in Atoms and Molecules in Astrophysics, Ed. Carson and Roberts. Academic Press.
Seaton, M.J., 1975, M.N.R.A.S., 170:475.
Seaton, M.J., 1976, "Electron Impact excitation of positive ions", in Advances in At. and Mol. Phys., 11:83.
Seaton, M.J. and Storey, P.J., 1976, "Dielectronic recombination", in Atomic Processes and Applications, North Holland.
Younger, S.M. and Wiese, L.W., 1979, J.Q.R.S.T., 22:161.

PART II

COLLISIONS OF ATOMS AND MOLECULES

LOW ENERGY ATOM-ATOM COLLISIONS

M. S. Child

Theoretical Chemistry Department
1 South Parks Rd., Oxford
OX1 3TG U.K.

ABSTRACT

The semiclassical theory of atom-atom potential scattering and of low energy inelastic atom-atom scattering is reviewed. Particular attention is given to the origin and interpretation of rainbow structure, diffraction oscillations and exchange oscillations in the potential scattering differential cross-section, and to the glory structure and symmetry oscillations in the integral cross-section. Available methods for direct inversion of the cross-section data to recover the potential are reviewed in some detail.

The theory of non-adiabatic transitions is introduced by a short discussion of interaction mechanisms and of diabatic and adiabatic representations. Analytical S matrix elements are presented for two state curve-crossing (Landau-Zener-Stückelberg), Demkov and Nikitin models. The relation between Stückelberg oscillations in the S matrix and in the differential cross-section is discussed in terms of interference between trajectories belonging to two different classical deflection functions. The energy dependences of the inelastic integral cross-section for curve-crossing and Demkov type transitions are also discussed. Finally the theory is reviewed in relation to a recent close-coupled study of fine structure transitions in $F(^2P) + Xe(^1S)$ scattering.

1. POTENTIAL SCATTERING

1.1 Introduction
1.2 Scattering amplitude and differential cross-section
1.3 Integral cross-section
1.4 Extraction of $\eta(\ell)$ and $\chi(\ell)$
1.5 Extraction of V(R)

2. ELECTRONIC EXCITATION

2.1 Introduction
2.2 Semiclassical S matrices
2.3 Differential cross-section
2.4 Integral cross-section
2.5 $X(^2P) + A(^1S)$ scattering: an illustrative example

1. POTENTIAL SCATTERING

1.1 Introduction

The discussion will concentrate on certain special features of experimental atom-atom potential scattering cross sections and show how these have been used to obtain a direct inversion from the scattering data to the interaction potential V(R). This subject has been extensively reviewed in recent years 1 - 7, and only essential equations will be given.

The most important characteristic of atom-atom scattering, even at thermal energies, is that the number of significant partial waves is large, ranging from 30-50 in the case of He/He to 2000-2500 for collisions between two Xe atoms. The link between the potential and the cross-section is therefore not the individual phase shift, but the phase shift function $\eta(\ell)$ in quantum mechanics. The corresponding connection in classical mechanics is carried by the classical deflection function $\chi(\ell)$, which is related to $\eta(\ell)$ by[7]

$$\chi(\ell) = 2(d\eta/d\ell) \tag{1}$$

This is readily verified by comparison between the JWKB phase shift, which is known to be highly accurate[8]

$$\eta^{(\ell)}_{JWKB} = \lim_{R\to\infty}\left[\int_{a_\ell}^{R} k_\ell(R)dR - kR + (\ell + \tfrac{1}{2})\,\pi/2\right] \tag{2}$$

where a_ℓ is the classical turning point and

$$k_\ell(R) = \{2\mu[E - V(R) - (\ell + \tfrac{1}{2})^2\hbar^2/2\mu R^2]\}^{\frac{1}{2}}/\hbar\,, \tag{3}$$

and the classical deflection at angular momentum

$$b = (\ell + \tfrac{1}{2})/k \tag{4}$$

$$k = (2\mu E)^{\frac{1}{2}}\,/\hbar \tag{5}$$

namely

$$\chi(\ell) = \pi - 2b\int_{a_\ell}^{R} R^{-2}[1 - V(R)/E - b^2/R^2]^{-\frac{1}{2}}dR. \tag{6}$$

The characteristic forms of $\chi(\ell)$ and $\eta(\ell)$ and the connection between them are illustrated in Fig. 1. This diagram which is typical for any system is presented for a Lennard-Jones potential

$$V(R) = 4\varepsilon[(\sigma/R)^{12} - (\sigma/R)^6] \tag{7}$$

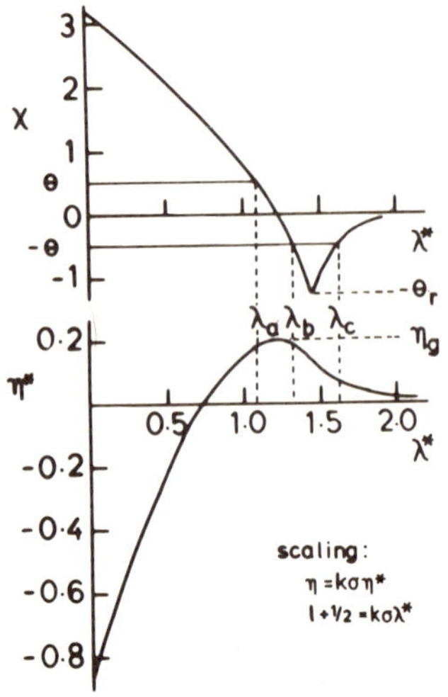

Fig. 1. Deflection function $\chi(\lambda)$ and scaled phase shift $\eta^*(\lambda)$, for a Lennard-Jones potential at reduced energy $(E/\varepsilon) = 2$

at reduced collision energy $(E/\varepsilon) = 2$ and in a scaled representation such that

$$\eta = k\sigma\eta^* \quad (8)$$

$$\ell + \tfrac{1}{2} = k\sigma\lambda^* \quad (9)$$

The first important features are the minimum in $\chi(\lambda^*)$ at the rainbow angle, $-\theta_r$, and the maximum in $\eta^*(\lambda^*)$ at the scaled glory phase shift η_g^* ; both θ_r and η_g^* tend to zero as $(E/\varepsilon) \to \infty$ [8,9]. Secondly the classical scattering at $\theta < \theta_r$ contains contributions from three angular momenta λ_a , λ_b and λ_c , and the quantum picture may be interpreted in terms of interference between these three trajectories with phases governed by the λ_ν $(\nu = a,b,c)$, the scattering angle and the phases $\eta(\lambda_\nu)$; notice that the phase difference $\eta(\lambda_a) - \eta(\lambda_b)$ is always very small. Finally the scaling given by equation (9) implies that the range spanned by the function $\eta(\ell)$ covers many multiples of π , because $k\sigma \simeq 100 - 200$ for typical atom-atom combinations. For very light systems, with $k\sigma < 30$, on the other hand the maximum $\eta(\lambda_g) < 2\pi$ and this leads to a qualitative difference in the scattering patterns; the combinations He - He and He - Ne fall into this light category for example.

The following two sections explore the consequences of Fig. 2 for the forms of the differential and integral cross-sections respectively. Sections 1.4 and 1.5 then describe various inversion procedures based on particular features of the experimental observables.

1.2 Scattering amplitude and differential cross-section.

The differential cross-section for the scattering of non-identical particles is given by standard partial wave analysis[7,10] as

$$d\sigma/d\Omega = |f(\theta)|^2 , \tag{10}$$

with $f(\theta)$ given in terms of the phase shifts $\eta(\ell)$ by

$$f(\theta) = (2ik)^{-1} \sum_{\ell=0}^{\infty} (2\ell+1)[\exp(2i\eta(\ell)) - 1]\, P_\ell(\cos\theta) . \tag{11}$$

This sum contains two types of oscillatory term, one being the exponential, and the other the Legendre polynomial because [11]

$$P_\ell(\cos\theta) \simeq [4/\pi(2\ell+1)\sin\theta]^{\frac{1}{2}} \cos[(\ell+\tfrac{1}{2})\theta - \pi/4] , \tag{12}$$

for $(\ell+\frac{1}{2})\sin\theta >> 1$. Widespread cancellation of successive groups of ℓ contributions is therefore expected, except for ℓ values close to the points at which the net phase is stationary with respect to ℓ . It is readily verified that this stationary phase condition implies, in the light of equation (1) that

$$\theta = \pm\, 2(d\eta/d\ell) = \chi(\ell) . \tag{13}$$

In other words the dominant contribution to the scattering at angle θ comes from the angular momenta giving classical deflection $|\chi(\ell)|$ of the same magnitude. Fig. 1 shows three such ℓ values for $\theta < \theta_r$ and one for $\theta > \theta_r$. This is the key to the semiclassical theory.[12-15]

Details of the reduction of equation (11) by transformation to an integral and subsequent uniform approximation [12], are beyond the scope of this discussion. Special techniques are required to handle the coalescence between contributions from λ_b and λ_c in Fig. 1 as $\theta \to \theta_r$ for example. The form of the resulting scattering amplitude may be written

$$f(\theta) = -i\Big[I_a^{\frac{1}{2}}(\theta) e^{i\gamma_a(\theta)} + \exp\{\tfrac{i}{2}[\gamma_b(\theta)+\gamma_c(\theta)]\} \{P_+\, \xi^{\frac{1}{4}} Ai(-\xi) - iP_-\, \xi^{-\frac{1}{4}} Ai'(-\xi)\}\Big] \tag{14}$$

where

$$P_\pm = \pi^{\frac{1}{2}}[I_b^{\frac{1}{2}}(\theta) \pm I_c^{\frac{1}{2}}(\theta)] \tag{15}$$

$$I_\nu(\theta) = (\ell_\nu+\tfrac{1}{2})/k^2 \sin\theta\, (d\chi/d\ell) \tag{16}$$

$$\gamma_\nu(\theta) = 2\eta(\ell_\nu) \pm (\ell_\nu+\tfrac{1}{2})\theta \tag{17}$$

$$\xi(\theta) = \left\{\tfrac{3}{4}[\gamma_b(\theta) - \gamma_c(\theta)]\right\}^{2/3} \tag{18}$$

and Ai$(-\xi)$ denotes the Airy function (see Fig. 2). The upper sign in equation (17) is taken for branches b and c , and the lower sign for branch a .

The quantities $I_\nu(\theta)$ and $\gamma_\nu(\theta)$ in these equations all have purely classical significance, the $I_\nu(\theta)$ being contributions to the classical differential cross-section, and the $\gamma_\nu(\theta)$ being derived from the classical actions along the relevant trajectories [16]. Fig. 2 shows that the Airy function oscillates for $\xi(\theta) > 0$, which corresponds to $\theta < \theta_r$. This oscillation is attributed to interference between trajectories designated by λ_b and λ_c in Figs. 1 and 3, and equation (14) predicts a further interference with branch a of the deflection function. For $\theta > \theta_r$ there are no real b and c branch trajectories such that $\chi(\ell) = -\theta$, although acceptable complex trajectories can be found; the b,c contribution therefore shows an exponential decrease with increasing $\theta - \theta_r$. The interference between this composite bc branch and branch a therefore lies within a sharply decreasing envelope, as seen in Fig. 3(a).

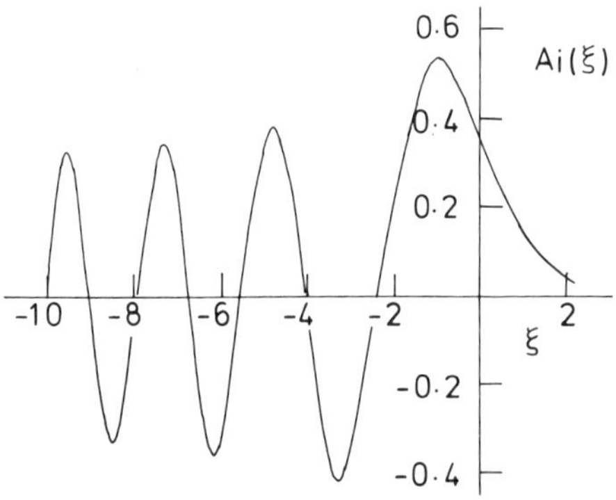

Fig. 2. The Airy function Ai(ξ)

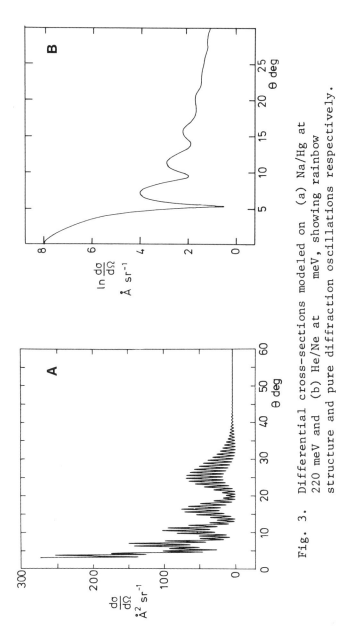

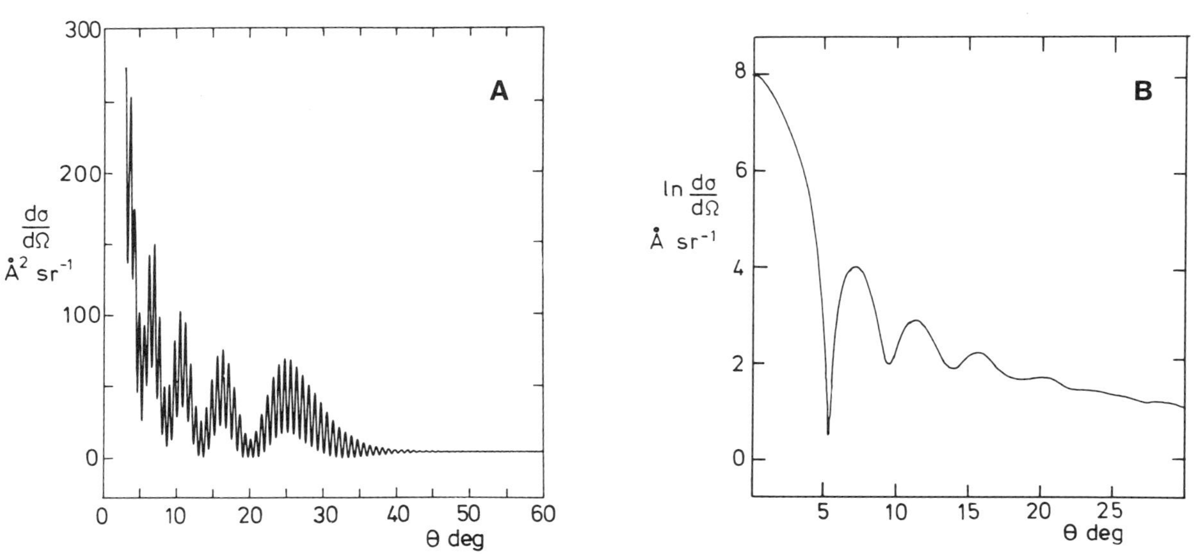

Fig. 3. Differential cross-sections modeled on (a) Na/Hg at 220 meV and (b) He/Ne at meV, showing rainbow structure and pure diffraction oscillations respectively.

The main information about the underlying deflection function $\chi(\ell)$, and hence, as will be shown in sections 1.4-1.5 about the scattering potential, is contained in the positions of the interference maxima, which are also more readily determined experimentally than are their amplitudes. The phases governing the spacing of these oscillations may be given simple graphical interpretations with the help of Fig. 4. For example the area α_r gives the phase

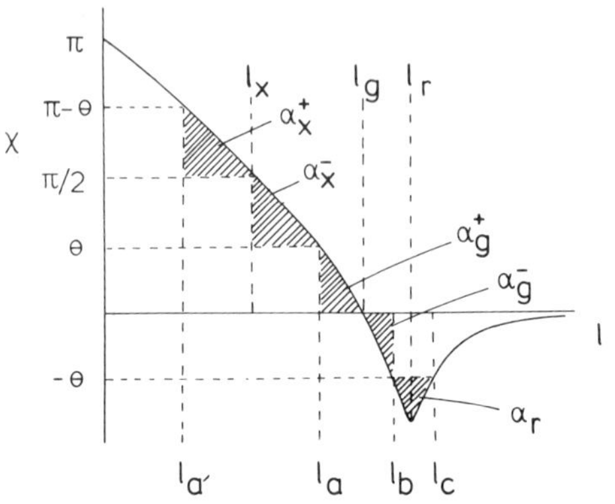

Fig. 4. Graphical illustration of the phase contributions responsible for rainbow oscillations (α_r), diffraction oscillations ($\alpha_g^{\pm}$) and exchange oscillations ($\alpha_x^{\pm}$) in the differential cross-section.

difference $[\gamma_b(\theta) - \gamma_c(\theta)]$[3] which governs the argument $\xi(\theta)$ of the Airy function and hence the spacing of the relatively slow <u>rainbow</u> oscillations in Fig. 3(a). The maximum area α_r^{max}, which therefore fixes the total number of such oscillations may also be expressed as[3]

$$\alpha_r^{max} = \gamma_b(0) - \gamma_c(0) = 2\eta_g \tag{19}$$

where η_g is the glory (maximum) phase shift in Fig. 1. A low value of η_g (say $\eta_g < 1$) therefore prevents the development of even a single rainbow maximum, and this explains the absence of any rainbow structure for light systems, as in Fig. 3(b). Turning to the faster, <u>diffraction</u>, oscillations in Fig. 3, the relevant phase difference may be expressed as[3]

$$\tfrac{1}{2}[\gamma_b(\theta) + \gamma_c(\theta)] - \gamma_a(\theta) = (2\ell_g + 1)\theta + (\alpha_+ - \alpha_-) - \tfrac{1}{2}\alpha_r \tag{20}$$

of which the dominant term is the first, because $\alpha_+ \simeq \alpha_-$ and the θ variation of α_r governs the slow oscillations. Hence the spacing of the diffraction oscillations gives a direct measure of the glory angular momentum ℓ_g, at which $\chi(\ell) = 0$.

One other type of interference gives rise to symmetry oscillations in the scattering of identical particles for which the differential cross-section is given by [14]

$$(d\sigma/d\Omega) = |\, f(\theta) \pm f(\pi - \theta)|^2 \tag{21}$$

with upper and lower signs taken for bosons and fermions respectively. The second term in equation (21) yields contribution to $(d\sigma/d\Omega)$ from a further trajectory labelled ℓ'_a, and at angles $\theta > \theta_r$, this interfere with the contribution from ℓ_a according to the phase difference

$$\gamma_a(\theta) - \gamma_{a'}(\theta) = (2\ell_x + 1)\ (\theta - \pi/2) + \alpha_x^+ - \alpha_x^- \tag{22}$$

where ℓ_x is defined such that $\chi(\ell_x) = \pi/2$ and $\alpha_x^{\pm}$ are the approximately equal areas shown in Fig. 4. The angular spacing of these oscillations, $2\pi/(2\ell_x + 1)$, is roughly twice that of the diffraction oscillations because Fig. 4 shows that $\ell_g \simeq 2\ell_x$. This exchange interference appears clearly resolved from the rainbow and diffraction patterns at angles $\theta > \theta_r$, provides $\theta_r < \pi/2$, because the scattering is symmetric about $\theta = \pi/2$. Examples are given in reference [3].

The qualitative connection between this interference structure and the scattering potential is that the classical rainbow angle fixed by the outermost point of inflexion in the rainbow structure, determines the ratio of well depth to collision energy, with a reducibility within 10% over the range of realistic potential forms [15,16]. Secondly the spacings of all interference features scale as $(k\sigma)^{-1}$, where σ is the range parameter of the potential. The extraction of finer details about the potential is postponed to sections 1.4-1.5.

1.3 Integral cross-section

The integral cross-section for the scattering of non-identical particles is given by [7]

$$\sigma(E) = \int_0^{2\pi}\!\!\int_0^{\pi} |f(\theta)|^2 \sin\theta d\theta d\phi = 4\pi k^{-2} \sum_{\ell=0}^{\infty} (2\ell+1)\sin^2\eta(\ell) \tag{23}$$

The crucial property of $\eta(\ell)$ for evaluation of this sum is the behaviour for large ℓ, namely for an inverse power potential $V(R) \simeq C_s/R^s$ as $R \to \infty$ [7].

$$\eta(\ell) \simeq a\, \ell^{-(s-1)} \tag{24}$$

with

$$a = f(s)\,(C_s/2E)k^s \tag{25}$$

where f(s) is a number of order unity, f(6) = 3π/16 for example[15]. This provides the correct cut-off at which $\eta(\ell) \ll 1$, and introduction of this form for all ℓ is tantamount to applying a random phase approximation for all $\eta(\ell) \gg 1$, with the result that[7,15]

$$\sigma(E) \simeq \sigma_{LL}(E) = p(s)\,[C_s/\hbar v]^{2/(s-1)} \tag{26}$$

where p(s) is a pure number; p(6) = 8.083 for example.

This procedure ignores the presence of a maximum in $\eta(\ell)$ however which gives rise to two types of correction. First the stationarity of $\eta(\ell)$ implies a further stationary phase contribution to $\sigma(E)$[3,7,15]

$$\sigma_g(E) = -\ell_g k^{-2}(2\pi/|d\chi/d\ell|)^{\frac{1}{2}} \cos\,(2\eta_g - \pi/4) \tag{27}$$

where ℓ_g is the glory angular momentum and $(d\chi/d\ell)$ is evaluated at $\ell = \ell_g$. This interferes with the background $\sigma_{LL}(E)$ to give a series of glory oscillations governed by the energy dependence of the glory phase shift η_g. The number of such oscillations is given by the zero-energy glory phase shift $\eta_g(0)$, which is related in turn by Levinson's theorem to the number of zero angular momentum bound states supported by the potential[7,17]. Secondly equation (26) assumes that $\eta_g \gg 1$ otherwise the cut off $\eta(\ell) \gg 1$ cannot occur over the range where equation (24) is applicable. A breakdown of this type must occur for all systems at sufficiently high energies because $\eta_g(E) \to 0$ as $E \to \infty$. The shape of $\sigma(E)$ therefore suffers a sharp change to a form dominated by the repulsive rather than the attractive branch of the potential. This general behaviour is illustrated in Fig. 5.

One other special co-operative effect in the evaluation of the cross-section arises for the scattering of identical particles, for which equation (23) is replaced by[13]

$$\sigma(E) = 4\pi\, k^{-2} \sum_{\ell=0}^{\infty} (2\ell + 1)(1 \pm (-1)^{\ell})\,\sin^2 \eta(\ell) \tag{28}$$

with upper and lower signs applicable to bosons and fermions respectively. The change from equation (23) is that only either even or odd terms contribute to the sum. Now for small ℓ values, the semiclassical equivalence between $\chi(\ell)$ and $2(d\eta/d\ell)$ and the identity $\chi(0) = \pi$ for any realistic potential allow us to write

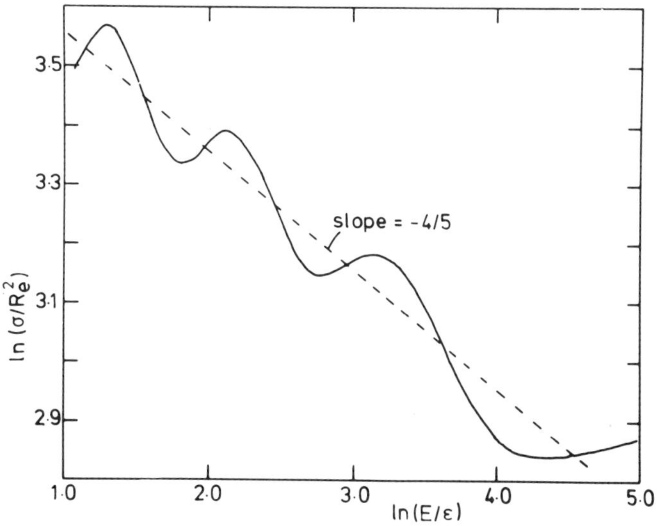

Fig. 5. Energy dependence of the integral cross-section modeled on Ar/Ar scattering.

$$\eta(\ell) \approx \eta(0) + (d\eta/d\ell)\,\ell + \tfrac{1}{2}\,(d^2\eta/d\ell^2)\,\ell^2 + \ldots$$

$$\approx \eta(0) + \ell\pi/2 + \kappa\ell^2 \qquad (29)$$

To the extent that the last term is ignored, this means that successive terms in equation (28) all contribute the same value of $\sin^2\eta(\ell)$, instead of fluctuating about the mean $<\sin^2\eta> = 1/2$ as assumed by the random phase approximation. Helbing[18] shows that this leads to an additional backward glory or symmetry oscillation contribution to the cross-section

$$\sigma_{bg}^{(\pm)}(E) = \mp\,[\kappa^{-2} + (\pi/k)^{\frac{1}{2}}\kappa^{-1} + \pi/2\kappa]^{\frac{1}{2}}\,\cos[2\eta(0)-\pi/2+\phi]\ . \qquad (30)$$

$$\tan\phi = [1 + 2(\pi\kappa)^{-\frac{1}{2}}]^{-1} \qquad (31)$$

with upper and lower signs taken for bose and fermi statistics respectively. This means that the resulting symmetry oscillations for different isotopic species will be out of phase, as seen in Fig. 6, and that the analysis of their phase variation with energy yields the energy variation of the zero angular momentum phase shift $\eta(0)$.

1.4 Extraction of $\eta(\ell)$ and $\chi(\ell)$.

Inversion of the scattering data to recover the potential V(R) usually proceeds by prior extraction of information on the phase

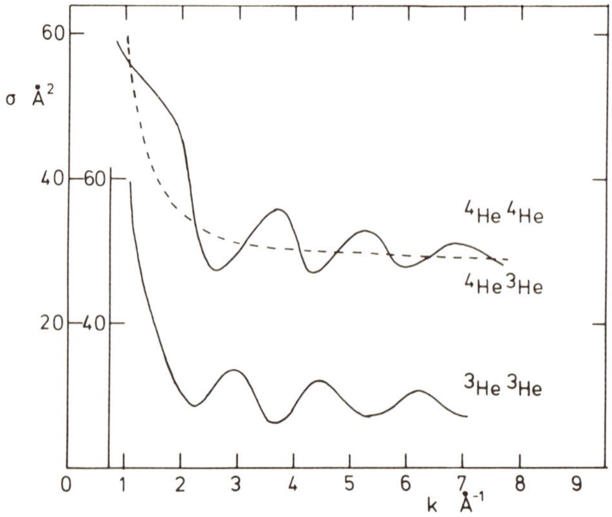

Fig. 6. Symmetry oscillations in the $^4He/^4He$ and $^3He/^3He$ integral cross-sections, compared with the $^3He/^4He$ cross-section.

shift $\eta(\ell)$ or the classical deflection function $\chi(\ell)$. The following discussion covers only those methods that have been applied in practice.

Buck-Pauly [19,20] inversion for the $\chi(\ell)$. most successful general analysis of rainbow structure in the differential cross-section is due to Buck[19] and Buck and Pauly[20]. The following six parameter piece-wise approximation is adopted

$$\begin{aligned} \chi(\ell) &= -a_1(\ell - \ell_g) && \ell_g < \ell < \ell_1 \\ &= -\theta_r + q(\ell - \ell_r)^2 && \ell_1 < \ell < \ell_2 \\ &= -c_1\ell^{-c_2} && \ell > \ell_2 \end{aligned} \tag{32}$$

with ℓ_1 and ℓ_2 chosen such that $\ell_g < \ell_1 < \ell_r < \ell_2$. Continuity equations for $\chi(\ell)$ and $\chi'(\ell)$ at ℓ_1 and ℓ_2 then reduce the number of independent parameters to four. One other constraint implied by equation (1) is that

$$\int_{\ell g}^{\infty} \chi(\ell)d\ell = -2\eta_g \tag{33}$$

with η_g obtained either from the velocity dependence of the glory oscillations in the total cross-section as given by equation (27), or less reliably by extrapolating the Airy function argument, $\xi(\theta)$ in equation (14) to $\theta = 0$. Of the remaining three parameters, ℓg is naturally fixed by the spacing of the fast diffraction oscillations (see equation (20)), and the final two parameters may be adjusted to obtain agreement between the positions of the experimental and calculated rainbow maxima using equation (14). Further information on $\chi(\ell)$ for $\ell < \ell g$ is obtained from the high angle scattering $\theta > \theta_r$, which is dominated by the first term in equation (14). Absolute values of $(d\sigma/d\Omega)$ may not be available, but the ratio in the high angle region to the magnitude of the first rainbow maximum has been found to be experimentally reliable[3]. Applications of this method are described by Buck[3] .

<u>Gerber-Shapiro[21,22] inversion for $f(\theta)$ and $\eta(\ell)$.</u> As discussed above, the scattering of particles with low reduced mass may give rise to only diffraction oscillations, with no rainbow structure (see Fig. 3(b). It is then necessary to combine features from both the amplitude and the phase information from $(d\sigma/d\Omega)$ in order to obtain $\eta(\ell)$ or $\chi(\ell)$. Gerber and Shapiro[21] employ the following unitarity condition[23] for this purpose:-

$$\text{Im } f(\theta) = (k/4\pi) \int_0^{2\pi} \int_0^{\pi} f^*(\theta')f(\theta'') \sin\theta' d\phi' d\theta' \qquad (34)$$

where

$$\cos\theta'' = \cos\theta\cos\theta' + \sin\theta\sin\theta'\cos\phi' \quad , \qquad (35)$$

or with

$$f(\theta) = R(\cos\theta)\exp[i\alpha(\cos\theta)] \qquad (36)$$

$$R(\cos\theta)\sin\alpha(\cos\theta) = (k/4\pi)\int_0^{2\pi}\int_0^{\pi} R(\cos\theta')R(\cos\theta'')$$

$$\cos[\alpha(\cos\theta')-\alpha(\cos\theta'')]\sin\theta' d\theta' d\phi' . \qquad (37)$$

Since $R(\cos\theta) = (d\sigma/d\Omega)^{\frac{1}{2}}$ this is an integral equation for the phase term $\alpha(\cos\theta)$. Despite considerable formal problems of uniqueness and convergence in any iterative procedure[23,24,25], Gerber and Shapiro[21] have shown that it is usually possible to choose a starting function $\alpha_0(\cos\theta)$ for the iteration that eliminates these difficulties. For example in the diffraction situation the first term in equation (14) dominates the scattering, and this contribution becomes overwhelming as $\theta \to \pi$. Hence equation (16) may be inverted to obtain a good approximation for $\chi(\ell)$ over the range $0 < \ell < \ell_r$, which when integrated via equation (1) gives $\eta(\ell)$ for the same range and hence via equation (17) $\alpha_0(\cos\theta) \approx \gamma_a(\theta)$. Further iteration based on the positions and amplitudes of the fluctuations in Fig. 3(b) improves $\alpha_0(\cos\theta)$ to convergence[21].

With $f(\theta)$ determined in this way, the phase shifts $\eta(\ell)$ are projected out by the following identity derived from equation (11)

$$\exp[2i\,\eta(\ell)] = 1 + ik\int_0^{\pi} f(\theta)\, P_\ell(\cos\theta)\sin\theta d\theta \qquad (38)$$

1.5 Extraction of the scattering potential V(R).

Abelian integral transformation. The JWKB phase shift and the classical deflection function given by equations (3) and (6) both involve integration of terms containing $[E-V(R)-b^2/R^2]^{\frac{1}{2}}$, where $b = (\ell + \frac{1}{2})/k$; and such integrals are amenable to the powerful abelian integral transformation, the spectroscopic RKR method[26-28] is based on the same idea. In the case of the deflection function for example, equation (6) may be expressed as

$$\chi(b) = \pi - \int_a^\infty \frac{2bdR}{R[s^2(R)-b^2]^{\frac{1}{2}}} \qquad (39)$$

where $b = (\ell + \frac{1}{2})k^{-1}$ and

$$s^2(R) = R^2[1 - V(R)/E] \quad . \qquad (40)$$

This form makes it possible to recover an inverse R(s) of the function s(R) by means of the transformation[7,29]

$$\int_{s_0}^\infty \frac{\chi(b)db}{[b^2-s_0^2]^{\frac{1}{2}}} = \pi\int_{s_0}^\infty \frac{db}{[b^2-s_0{}^2]^{\frac{1}{2}}} - \int_{s_0}^\infty\int_b^\infty \frac{(d\ell nR/ds)db^2 ds}{[s^2-b^2]^{\frac{1}{2}}[b^2-s_0^2]^{\frac{1}{2}}}$$

$$= \pi\, \ell n[R(s_0)/s_0^{\frac{1}{2}}] \qquad (41)$$

Hence on dropping the subscript from s_0

$$R(s) = s^{\frac{1}{2}} \exp\left[\frac{1}{\pi}\int_s^\infty \frac{\chi(b)db}{[b^2-s^2]^{\frac{1}{2}}}\right] \qquad (42)$$

The same result applies of course with $2(d\eta/d\ell)$ in place of $\chi(b)$[30]. The input required by this Firsov[29] method is the deflection function $\chi(b)$ for the full range of impact parameters $0 < b < \infty$. The resulting function R(s) is monotonic, provided the energy E lies above the classical orbiting limit[7], which is also a condition which allows a valid reversal of integration order in deriving the final line of equation (41). The value of s for any R may therefore be read from the graph to obtain V(R) for V(R) < E in the form

$$V(R) = E[1 - s^2(R)/R^2] \tag{43}$$

Results for several systems obtained by this method are reviewed by Buck[3].

Similar analysis[31,32,33] may be applied to knowledge of the energy dependence of the $\ell = 0$ phase shift $\eta(0,E)$ derived from symmetry oscillations in the integral cross-section shown in Fig.6. In this case it is also necessary to know the energies E(n) of the $\ell = 0$ bound states of V(R). The turning point $r_1(U)$ on the repulsive branch of V(R) is then given by the equation[31,33]

$$r_1(U) = -2(\hbar^2/2\mu)^{\frac{1}{2}} \left[\int_{-\frac{1}{2}}^{n(0)} [U-E(n)]^{-\frac{1}{2}} dn + \pi^{-1} \int_{\eta(0)}^{\eta(U)} [U-E(y)]^{-\frac{1}{2}} dy \right] \tag{44}$$

Distorted wave peeling method. A quantum mechanical alternative to these semiclassical inversions, which is also applicable to inelastic scattering situations[34,35] was first suggested by Shapiro and Gerber[22]. The essence of the method is that if a sufficiently good approximation $V_0(R)$ to V(R) is available

$$V(R) = V_0(R) + V_1(R) \quad , \tag{45}$$

the contribution $\eta_1(\ell)$ due to $V_1(R)$ will be adequately represented by the distorted wave expression[7]

$$\eta_1(\ell) = \eta(\ell) - \eta_0(\ell) = -k^{-1} \int_0^{\infty} V_1(R) [\psi_{0\ell}(R)]^2 dR \tag{46}$$

which provides an integral equation for $V_1(R)$ because $\psi_{0\ell}(R)$ is determined by $V_0(R)$. Secondly the dominant contribution to this integral comes from the region beyond the classical turning point $a(\ell)$, and $a(\ell) \to \infty$ as $R \to \infty$. Hence for sufficiently large ℓ, $V_1(R)$ may be taken as the full potential V(R). The first step is therefore to use plane wave components $\psi_{0\ell}(R)$ and $\eta_0(\ell) = 0$ in equation (46) for large ℓ and hence to determine the long range tail of V(R). This tail is then extrapolated to shorter R values and used to determine a first set of distorted waves and zero-order phase shifts $\eta_0(\ell)$ for the next range of smaller ℓ values using the discrepancy $\eta(\ell) - \eta_0(\ell)$ to determine the extent of the range.

The next layer of the potential $V_1(R)$ is then peeled off by means of equation (46). This process is repeated in stages until all ℓ values have been covered.

The practical technique adopted [22] for solution of equation (46) is to expand V(R) in a series

$$V(R) = \sum_k a_k f_k(R) \tag{47}$$

with $f_k(R)$ taken as exponential or inverse power terms. One new such term is taken at each stage of the peeling procedure, with the coefficient a_k and any parameters contained in $f_k(R)$ optimised to reproduce $\eta_1(\ell)$ over the appropriate interval.

The general approach has been combined with the unitarity based inversion for $\eta(\ell)$ discussed in section 1.4.2., and successfully applied to experimental data for He/Ne [36].

2. ELECTRONIC EXCITATION

2.1 Introduction

Valence shell excitation by atom-atom collision in conveniently interpreted as a breakdown of the Born-Oppenheimer approximation, with the adiabatic electronic wavefunctions $\phi_\alpha(q|R)$ coupled together by the nuclear kinetic energy as a result of their parametric dependence on nuclear position R. The dominant coupling term is normally the mixed radial derivative†

$$T_{\alpha\beta} = \frac{\hbar^2}{2\mu} <\phi_\alpha|\frac{\partial}{\partial R}|\phi_\beta> \frac{\partial}{\partial R} \tag{48}$$

although angular (or Coriolis) coupling may be important at energies outside the molecular range (E > 10eV) between states that become degenerate in the united atom limit [37].

The most important mechanisms causing the $\phi_\alpha(\underset{\sim}{q}|R)$ to vary with R may be identified as (i) configuration interaction, (ii) spin-orbit coupling and (iii) spin-orbit quenching by the onset of cylindrical rather than spherical symmetry. It is outside the scope of this chapter to discuss the mechanisms in any detail, but however the adiabatic electronic structure calculation is performed it may be helpful to know that explicit nuclear differentiation of the wavefunction as indicated by equation (48) is not in fact required, because [37]

† The inclusion of a translational factor[70] or switching function[71] correction may be necessary in order to remove spurious radial coupling as $R \to \infty$ and as $R \to 0$.

$$\langle\phi_\alpha|\frac{\partial}{\partial R}|\phi_\beta\rangle = -\langle\phi_\alpha|\partial H_{el}/\partial R|\phi_\beta\rangle/[V_\alpha(R) - V_\beta(R)], \tag{49}$$

where $H_{el}(\underset{\sim}{q},R)$ is the electronic hamiltonian and $V_\alpha(R)$ are the adiabatic electronic energy curves. Equation (49) also indicated the necessary conditions for strong coupling: either $\partial H_{el}/\partial R$ must be large, or the adiabatic level separation $V_\alpha(R) - V_\beta(R)$ must become small. The former is the characteristic of Demkov coupling [38-41] and the latter of curve-crossing [41-48] which are the two most important routes to electronic non-adiabaticity.

One important difference between them may be illustrated by a two state model based on diabatic electronic wavefunctions $\xi_1(\underset{\sim}{q}),\xi_2(\underset{\sim}{q})$ The R dependence of the adiabatic functions $\phi_\alpha(\underset{\sim}{q}|\underset{\sim}{R})$ then lies in a mixing angle $\theta(R)$.

$$\begin{pmatrix}\phi_\alpha(\underset{\sim}{q}|R)\\ \phi_\beta(\underset{\sim}{q}|R)\end{pmatrix} = \begin{pmatrix}\cos\theta(R) & \sin\theta(R)\\ -\sin\theta(R) & \cos\theta(R)\end{pmatrix}\begin{pmatrix}\xi_1(\underset{\sim}{q})\\ \xi_2(\underset{\sim}{q})\end{pmatrix} \tag{50}$$

where

$$\cot 2\theta(R) = \frac{H_{11}(R) - H_{22}(R)}{2H_{12}(R)} \tag{51}$$

In this model the coupling term reduces to

$$\langle\phi_\alpha|\frac{\partial}{\partial R}|\phi_\beta\rangle = -\frac{d\theta}{dR} \tag{52a}$$

$$= \frac{(dx/dR)}{2(1 + x(R)} \tag{52b}$$

with $x(R)$ used to denote $\cot 2\theta(R)$. The prototype curve-crossing case is one in which $H_{11}(R) - H_{22}(R)$ vanishes at some point while $H_{12}(R)$ remains finite. Hence $x(R)$ varies between $-\infty$ and ∞, corresponding to a change in $\theta(R)$ from $\pi/2$ to zero, and the maximum value of $(d\theta/dR)$ occurs at the crossing point $x(R) = 0$ or $\theta(R) = \pi/4$. The Demkov [38] situation on the other hand is one in which $H_{11}(R) - H_{22}(R)$ remains roughly constant, while $H_{12}(R)$ increases from zero at $R \to \infty$ to a large value $H_{12}(R) \gg |H_{11}(R) - H_{22}(R)|$ at shorter range. In this case $\theta(R) \to 0$ as $R \to \infty$ and $\theta(R) \to \pi/4$ as $R \to 0$. In an exponential approximation for $x(R)$ the maximum rate of change $d\theta/dR$ occurs at $x(R) = 1$ or $\theta(R) = \pi/8$. The difference between the two limiting angles $\theta(R) = \pi/2$ and $\theta(R) = \pi/4$ as $R \to \infty$ implies quite different compositions for the limiting adiabatic states, and this results in different boundary conditions on the coupled equations of motion [39,44]. The intermediate cases with $\pi/4 < \theta < \pi/2$ as $R \to 0$ are also possible [45] as discussed below.

A final point of practical importance, particularly in relation to the following analytical disucssion is that equations (49),(50), and (52a) may be used to construct a local two state diabatic model for the process in hand even when a proper description of the electronic structure requires a very large basis. This simplification arises because there may be only one significant peak in any relevant term $\langle\phi_\alpha|\partial/\partial R|\phi_\beta\rangle$, although ϕ_α and ϕ_β are each composed of many elements of the basis. In this situation equation (52a) may be integrated to obtain the mixing function $\theta(R)$ which provides one equation for the three functions $H_{11}(R)$, $H_{22}(R)$ and $H_{12}(R)$. The adiabatic terms

$$V_\alpha(R) = \tfrac{1}{2}[H_{11}(R) + H_{22}(R)] + \tfrac{1}{2}\{[H_{11}(R) - H_{22}(R)]^2 + 4H_{12}^2(R)\}^{\frac{1}{2}}$$
$$V_\beta(R) = \tfrac{1}{2}[H_{11}(R) + H_{22}(R)] - \tfrac{1}{2}\{[H_{11}(R) - H_{22}(R)]^2 + 4H_{12}(R)\}^{\frac{1}{2}} \quad (53)$$

complete the necessary three equations. This device due to Smith[49] has been elaborated by Baer[50].

2.2 Semiclassical scattering matrices

General semiclassical expressions for the S matrix are available in the literature for both curve-crossing[38,42-48] and Demkov[38-40] type transitions, and for an intermediate exponential model due to Nikitin[41]. Details of the derivations are outside the scope of the present discussion, but the results are of general interest, both for interpretation of experimental data, and for illustrating the differences between models. The S matrix elements are most conveniently expressed in the adiabatic representation as

$$S_{\mu\nu} = \tilde{S}_{\mu\nu} \exp\,[i(\eta_\mu + \eta_\nu)] \quad (54)$$

where η_μ and η_ν are JWKB phase shifts in the adiabatic channels, and $\tilde{S}_{\mu\nu}$ take different forms according to the model in question.

The most general form in the curve-crossing case, due originally to Stuckelberg[44] is

$$\tilde{S}_{\mu\mu} = \tilde{S}^*_{\nu\nu} = \exp(-2\delta - 2i\tau) + [1-\exp(-2\delta)]\exp(2i\phi)$$
$$\tilde{S}_{\mu\nu} = \tilde{S}_{\nu\mu} = -2i\exp(-\delta)\,[1-\exp(-\delta)]^{\frac{1}{2}}\sin(\tau + \phi) \quad (55)$$

where the real quantities τ and δ are defined by the following integral over the adiabatic wavenumbers $k_\eta(R)$ and $k_\nu(R)$, taken between the classical turning points a_η, a_ν and the complex distance R_c at which $k_\eta(R) = k_\nu(R)$;

$$\tau + i\delta = \int_{a_\mu}^{R_c} k_\mu(R)\,dR - \int_{a_\nu}^{R_c} k_\nu(R)\,dR\ . \quad (56)$$

The quantity ϕ is given in terms of $\gamma = \delta/\pi$ by [51-53]

$$\phi = \pi/4 - \gamma + \gamma \ln \gamma - \arg \Gamma(1+i\gamma) \tag{57}$$

A limitation on these equations is that the crossing point $R_x \simeq (\mathrm{Re}(R_c)$ must lie in the classically accessible region and not too close to the classical turning points.

Equation (56) has an appealing physical interpretation[54] as the complex phase difference between two trajectories arriving at the momentum conserving point R_c at which any transition must occur. The real part τ of this phase difference is responsible for interference effects, while the imaginary part governs the probability of making the transition. Integration of the necessary complex trajectory[55] is not however trivial, but fortunately equation (56) is amenable to the following reduction

$$\tau_{LZ} = \int_{a_\mu}^{R_x} k_\mu(R)\,dR - \int_{a_\nu}^{R_x} k_\nu(R)\,dR \tag{58}$$

$$\delta_{LZ} = \pi H_{12}^2 / \hbar v(F_1 - F_2) \tag{59}$$

in the diabatic formulation of the Landau-Zener model[42-44]

$$H_{ii}(R) \simeq E_x - F_i(R - R_x)$$

$$H_{12}(R) = H_{12} = \text{const.}$$

$$v = [2(E - E_x)/\mu]^{\frac{1}{2}} \simeq \text{const.} \tag{60}$$

A recent significant extension of the theory by Baranyi[48] allows an extension of equations (55) and (56) to energies even below the crossing point E_x. This is based on recognition that the physical significances of τ and δ change their roles when $k_\mu(k)$ and $k_\nu(R)$ become imaginary for $E < E_x$. The generalisation also involves small modifications to τ and ϕ in the analogue of equation (55).

Finally it may be noted that in weak diabatic coupling the following distorted wave result is available for the off-diagonal S matrix element[56,57]

$$S_{12} = -i\,(8\delta/\pi)^{\frac{1}{2}}\,\zeta^{\frac{1}{4}}\,\mathrm{Ai}(-\zeta)\,\exp[i(\eta_+ + \eta_-)] \quad , \tag{61}$$

where

$$\zeta = (+3\tau/2)^{\frac{2}{3}} \tag{62}$$

Turning to the Demkov case[38-40] the change in the range covered by the mixing angle $0 < \theta < \pi/4$, rather than $0 < \theta < \pi/2$ in the curve-crossing case, leads to a different structural form from equation (55)[39]

$$\tilde{S}_{\mu\mu} = \tilde{S}^*_{\nu\nu} = [\exp(-2\delta - 2i\tau) + 1] / [\exp(-2\delta) + 1]$$

$$\tilde{S}_{\mu\nu} = \tilde{S}_{\nu\mu} = 2i \exp(-\delta) \sin\tau / [\exp(-2\delta) + 1] \ . \qquad (63)$$

The transition and interference parameters δ and τ are however again given by equation (56). The canonical diabatic model in this case, due to Demkov[38], is

$$H_{11}(R) = \bar{E} - \Delta/2$$

$$H_{22}(R) = \bar{E} + \Delta/2$$

$$H_{12}(R) = A \exp(-\alpha R)$$

$$v = [2(E - \bar{E})/\mu]^{\frac{1}{2}} = \text{const.} \qquad (64)$$

In this approximation[38]

$$\delta_D = \pi\Delta/2\alpha\hbar v \qquad (65)$$

Finally Nikitin[41] has obtained a generalisation of this Demkov model based on the approximations

$$H_{12}(R) = A \sin\gamma \exp(-\alpha R) \ , \ H_{22}(R) - H_{11}(R) = \Delta - 2A \cos\gamma \, e^{-\alpha R} \qquad (66)$$

involving a completely general limiting mixing angle $\theta_0 = \gamma/2$. This leads in the present notation to

$$\tilde{S}_{\mu\nu} = 2 \sin\tau \, \{\sinh[\delta_D(1-\cos\theta_0)]\sinh[\delta_D(1+\cos\theta_0)]\}^{\frac{1}{2}} /$$

$$\exp(-\delta_D\cos\theta_0)\sinh 2\delta_D \ . \qquad (67)$$

This provides the machinery for the treatment of fine structure transitions. For example the $J = {}^3{}_2$ and $J = \frac{1}{2}$ components of a ^2P atomic term are split by a spin-orbit term Δ at $R \to \infty$, and the onset of cylindrical symmetry not only mixes the two $m_J = \frac{1}{2}$ spin-orbit states, but also modifies the level separation $H_{22}(R) - H_{11}(R)$. Detailed analysis[58,59] shows that $\tan\gamma = \sqrt{2}$ in this case (see equation (88)).

2.3 Differential cross-section

The main features of the differential cross-sections

$$\frac{d\sigma_{\mu\nu}}{d\Omega} = |f_{\mu\nu}(\theta)|^2 \tag{68}$$

may be understood by applying the previous semiclassical techniques to reduction of the scattering amplitudes

$$f_{\mu\nu}(\theta) = (2ik)^{-1} \sum_{\ell=0}^{\infty} (2\ell+1)(S_{\mu\nu}(\ell)-1)\, P_\ell(\cos\theta) \tag{69}$$

The main new result is that the Stuckelberg phase terms τ in equations (55), (63) and (67) are ℓ dependent and hence may give rise to additional interference effects to those due to the adiabatic phase shifts $\eta_\pm(\ell)$.

The simplest situation arises for inelastic scattering when the derivative of $(\eta_+(\ell) + \eta_-(\ell))$ decreases monotonically with increasing ℓ . It is then natural to introduce two inelastic deflection functions $\chi_a(\ell)$ and $\chi_b(\ell)$ by the definitions[60-62]

$$\chi_{a,b}(\ell) = \frac{d}{d\ell}\left[\eta_+(\ell) + \eta_-(\ell) \mp \tau(\ell)\right] \tag{70}$$

These may be associated with scattering from the lower to the upper adiabatic channel, either with a non-adiabatic transition on the inward or on the outward part of the trajectory; since different forces are applied the classical deflections are quite different. This interpretation is quite straightforward if the transition point is classically accessible, and these are normally the dominant trajectories. The same idea can also be extended[48] to cover transitions induced by tunnelling trajectories.

Child and Gerber[34] have adopted the following ansatz for $S_{12}(\ell)$ in such situations,

$$S_{12}(\ell) = A(\ell)\, \mathrm{Ai}(-\zeta(\ell)) \exp[i\eta_+(\ell) + i\eta_-(\ell)] , \tag{71}$$

where consistent with equations (55), (63) and (67)

$$\zeta(\ell) = [3\tau(\ell)/2]^{\frac{2}{3}} \tag{72}$$

This allows for a sharp cut-off around $\tau(\ell) = 0$, where the classical turning points coalesce with the crossing point, with an arbitary, but supposedly slowly varying amplitude $A(\ell)$. The following uniform approximation for $(d\sigma_{12}/d\Omega)$ is then obtained[34]

$$(d\sigma_{12}/d\Omega) = [4k_1^2\sin\theta]^{-1}\{(\pi_a^{\frac{1}{2}} + \pi_b^{\frac{1}{2}})^2\xi^{\frac{1}{2}}Ai^2(-\zeta) + (\pi_a^{\frac{1}{2}} - \pi_b^{\frac{1}{2}})^2$$

$$\xi^{-\frac{1}{2}}Ai'^2(-\zeta)\} \tag{73}$$

where

$$\xi(\theta) = \{\tfrac{3}{4}[\gamma_a(\theta) - \gamma_b(\theta)]\}^{2/3}$$

$$\gamma_\nu(\theta) = \eta_+(\lambda_\nu) + \eta_-(\lambda_\nu) \pm \tau(\lambda_\nu) - \lambda_\nu\theta$$

$$\pi_\nu = \lambda_\nu A(\lambda_\nu)\zeta^{\frac{1}{2}}(\lambda_\nu)/\chi_\nu'(\lambda_\nu) \tag{74}$$

with $\lambda_\nu(\theta)$ given by

$$\chi_\nu(\lambda_\nu) = \theta \tag{75}$$

Fig. 7 shows graphically how the twin deflection functions $\chi_a(\ell)$ and $\chi_b(\ell)$ provide the connection between the functions $S_{12}(\ell)$ and $(d\sigma_{12}/d\Omega)$, a connection that is given greater point by noting that according to equation (70),

$$\tau(\ell) = \int_\ell^{\ell max} [\chi_a(\ell) - \chi_b(\ell)]\, d\ell \tag{76}$$

while according to equations (74) and (75)

$$\gamma_a(\theta) - \gamma_b(\theta) = \int_{\theta_{min}}^{\theta} [\ell_a(\theta) - \ell_b(\theta)]\, d\theta \tag{77}$$

where $\ell_\nu(\theta)$ is derived from $\chi_\nu(\ell)$ by inversion of equation (75). Equation (76) means that the phase of the Stückelberg oscillations in $S_{12}(\ell)$ is obtained by integration of the vertical spearation between the two curves, while the analogous oscillations in $(d\sigma_{12}/d\Omega)$ are governed by the integrated horizontal separation between the same curves. Since the total included area is fixed the total number of oscillations is the same in both cases. Child and Gerber [34] have

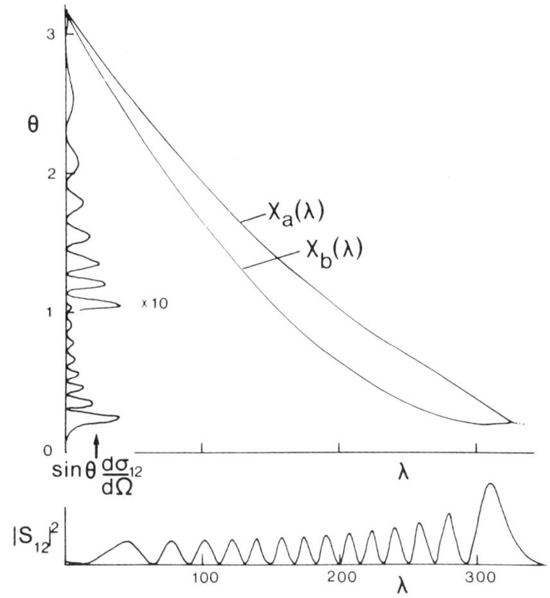

Fig. 7. Stückelberg oscillations in S_{12} (ℓ) and ($d\sigma_{12}/d\Omega$) due to interference between trajectories following classical deflection functions $\chi_a(\ell)$ and $\chi_b(\ell)$. [Taken from Child[37]].

exploited this reflection principle in showing that knowledge of the mean deflection function

$$\bar{\chi}(\ell) = [\chi_a(\ell) + \chi_b(\ell)]/2 \tag{78}$$

is sufficient to recover $S_{12}(\ell)$ from $(d\sigma_{12}/d\Omega)$, and hence $V_{12}(R)$ from $S_{12}(\ell)$.

The above simple situation in which $\chi_a(\ell)$ and $\chi_b(\ell)$ are both monotonic in ℓ serves to indicate the origin of the Stückelberg oscillations that are a special feature of the inelastic scattering, but it is not the general case. More commonly, one or even both of the $\chi_\nu(\ell)$ will pass through maxima or minima and hence give rise to normal rainbow and diffraction oscillations as well as to the Stückelberg phenomenon. The resulting complicated interference pattern[61,63,64] may be handled by obvious generalisation of the present discussion[60-62]. One general feature is that provided no rainbow angle exceeds π, the scattering will always show purely Stückelberg interference in the backward direction.

2.4 Integral cross-section

The integral inelastic scattering cross-section may be approximated as

$$\sigma_{12}(E) = \frac{2\pi}{k_1^2} \sum_{\ell=0}^{\infty} (2\ell+1)|S_{12}(\ell)|^2 \simeq \frac{2\pi}{k_1^2} \int_0^{\infty} |S_{12}(\ell)|^2 d\ell^2 \tag{79}$$

which is amenable to simple reduction by applying a random phase approximation $\sin^2\tau(\ell) \simeq \frac{1}{2}$ to the terms arising from equations (55) and (63), and also by introducing the parameterisations suggested by equations (59) and (65)

$$2\delta(\ell) = (v^*/v)\ [1 - (\ell/\ell_{max})^2]^{-\frac{1}{2}} \tag{80}$$

with the characteristic Landau-Zener and Demkov scaling velocities given by

$$v^*_{LZ} = 2\pi H_{12}^2(R_x)/\hbar(F_1 - F_2)[1 - (E_0/E)]^{\frac{1}{2}} \tag{81}$$

$$v^*_D = \pi\Delta/\alpha\hbar\ [1 - (E_0/E)]^{\frac{1}{2}} \tag{82}$$

and with E_0 taken as E_x or as $\bar{E}$ in equations (60) or (64) as the case may be. Finally the integration variable ℓ^2 in equation (79) is replaced by

$$x = [1 - (\ell/\ell_{max})^2]^{-\frac{1}{2}} \tag{83}$$

with the result that[37]

$$\sigma_{12}(E) = \pi R_0^2 \; [1 - (E_0/E)] \; G(v/v^*) \tag{84}$$

where v is the initial collision velocity,

$$R_0 = k_1 \ell_{max} \tag{85}$$

and

$$G_{LZ}(\alpha) = 4 \int_1^\infty e^{-x/\alpha} \; (1 - e^{-x/\alpha}) x^{-3} \, dx \tag{86}$$

$$G_D(\alpha) = \int_1^\infty \mathrm{sech} \; (x/2\alpha) x^{-3} dx \tag{87}$$

The characteristic behaviour of these two functions is illustrated in Fig. 8. The presence of a maximum in $G_{LZ}(v/v^*)$ may be understood on the basis that at low velocities the trajectory follows a purely adiabatic path, and at very high velocities a purely diabatic path neither giving rise to any nett change in the system. At intermediate velocities on the other hand both inward and outward motions are

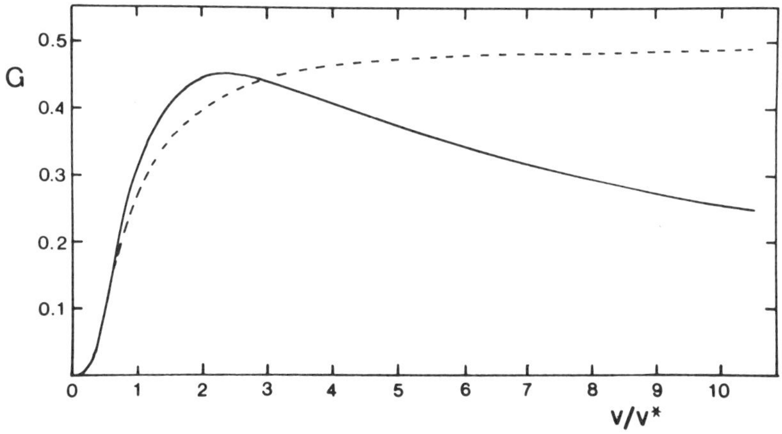

Fig. 8. Reduced inelastic integral cross-sections $G(v/v^*)$ for curve-crossing (solid line) and Demkov (broken line) transitions [taken from Child[37]].

partly adiabatic and partly diabatic leading to some inelastic scattering. The Demkov model also follows adiabatic behaviour at low energies, again leading to zero inelastic scattering. At high energies on the other hand the initial population of the pure asymptotic state is equally shared at short range between the two possible adiabatic states, and these two populations are then equally shared between the two asympototic states during the outward motion. Hence the overall inelastic transition probability is 0.5 . Similar probability arguments based on a limiting short range mixing angle $\theta = \gamma/2$ in the notation of equations (66) and (67), would lead to a high energy inelastic branching fraction of $0.5 \sin^2\gamma$.

2.5 $X(^2P) + A(^1S)$ scattering : an illustrative example

The results of a recent close coupled study of halogen, rare gas scattering by Becker et al[65] may serve to complete the picture.

Following Mies[58-59] and Reid[66-67] the potential coupling matrix between the X atom spin-orbit states $|\frac{3}{2}, \frac{3}{2}\rangle, |\frac{3}{2}, \frac{1}{2}\rangle$ and $|\frac{1}{2}, \frac{1}{2}\rangle$ was taken as

$$\underset{\sim}{V} = \begin{bmatrix} V_\pi - \Delta , & 0 , & 0 \\ 0 , & \frac{2}{3} V_\Sigma + \frac{1}{3} V_\pi - \Delta , & \frac{\sqrt{2}}{3} (V_\Sigma - V_\pi) \\ 0 , & \frac{\sqrt{2}}{3} (V_\Sigma - V_\pi) , & \frac{1}{3} V_\Sigma + \frac{2}{3} V_\pi \end{bmatrix} \tag{88}$$

with Δ fixed at the atomic spin-orbit splitting and V_Σ and V_π calculated to vary for F + Xe as shown in Fig.9. The resulting adiabatic curves

$$V\left[{}^2\Pi\left(\frac{3}{2}\right)\right] = V_\pi - \Delta$$

$$V\left[{}^2\Pi\left(\frac{1}{2}\right)\right] = 0.5 \ [V_\pi + V_\Sigma - \Delta - D]$$

$$V\left[{}^2\Sigma\left(\frac{1}{2}\right)\right] = 0.5 \ [V_\pi + V_\Sigma - \Delta - D] \tag{89}$$

$$D = \left[(V_\pi - V_\Sigma)^2 + \left(\frac{2}{3}\right) \Delta \ (V_\pi - V_\Sigma) + \Delta^2\right]^{\frac{1}{2}} \tag{90}$$

go over at short range, where $V_\pi - V_\Sigma \gg \Delta$ to

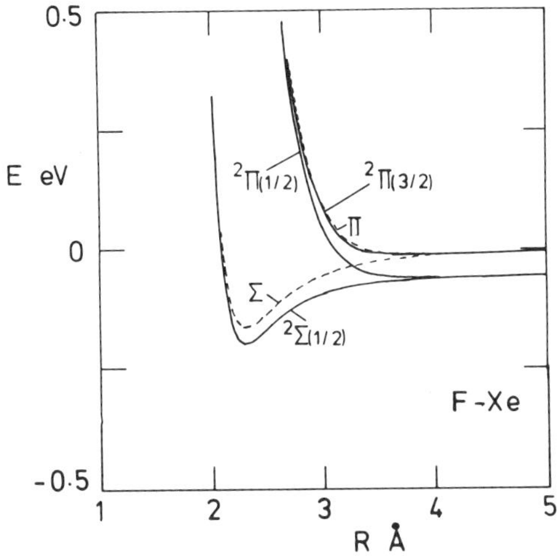

Fig. 9. Potential curves for fine structure transitions in $F(^2P) + Xe(^1S)$. Solid curves give the adiabatic potentials. The dashed curves are the electrostatic potentials V_Σ and V_π. [Adapted from Becker et al [65]].

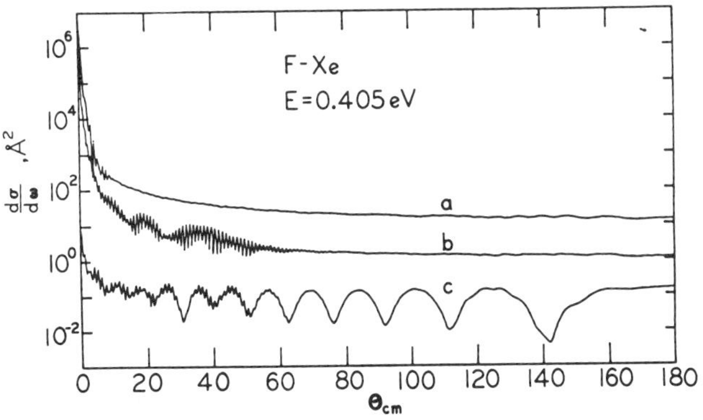

Fig.10. Differential cross-sections for (a) $^2P(\frac{1}{2}) \to {}^2P(\frac{1}{2})$ (b) $^2P(\frac{3}{2}) \to {}^2P(\frac{3}{2})$ and (c) $^2P(\frac{1}{2}) \to {}^2P(\frac{3}{2})$ transitions in the scattering of $F(^2P)$ and $Xe(^1S)$ at 0.405 eV. Curve (a) has been displaced upwards by one order of magnitude for clarity. [Taken from Becker et al[65]].

$$V\left[{}^2\Pi\left(\frac{3}{2}\right)\right] = V_\pi - \Delta$$

$$V\left[{}^2\Pi\left(\frac{1}{2}\right)\right] \simeq V_\pi - \Delta/3$$

$$V\left[{}^2\Sigma\left(\frac{1}{2}\right)\right] \simeq V_\Sigma - 2\Delta/3 \tag{91}$$

The maximum rate of change of the adiabatic electronic wavefunction occurs at approximately $R_0 = 3{,}6Å$, giving $\pi R_0^2 \simeq 30$ Å^2 for use in the analogue of equation (84).

The calculated differential cross-sections for F + Xe at E = 0.405eV are shown in Fig.10. Curves a and b refer to elastic scattering for the atomic ${}^2P(\frac{1}{2})$ and ${}^2P(\frac{3}{2})$ states respectively. Fine structure transitions during the collision appear to have little influence on the elastic scattering because the structureless form of curve a is attributable to the repulsive nature of the ${}^2\Pi(\frac{1}{2})$ potential, while the rainbow structure of curve b presumably arises purely from the ${}^2\Sigma(\frac{1}{2})$ component of the ${}^2P(\frac{3}{2}) \rightarrow {}^2P(\frac{3}{2})$ scattering. This evidence of weak coupling is supported by the relatively low magnitude of the inelastic ${}^2P(\frac{1}{2}) \rightarrow {}^2P(\frac{3}{2})$ cross-section. The main structure is due to Stückelberg interference between ${}^2\Pi(\frac{1}{2})$ and ${}^2\Sigma(\frac{1}{2})$ trajectories, with some high frequency structure at small angles due to the presence of a minimum in the ${}^2\Sigma(\frac{1}{2})$ inelastic deflection function.

This calculation[65] which was performed in the space fixed representation included both radial and Coriolis ($\Delta\Omega = \pm 1$) coupling. Aquilanti et al[72,73] have investigated a variety of decoupling approximations for similar systems. Similar calculations on the Na/Hg system by Düren et al[68,69] may also be mentioned.

APPENDIX: Semiclassical interference - a phase space picture

The unifying features of modern semiclassical developments are (i) the role of the classical action in determining the interference (and tunneling) characteristics of the system and (ii) the use of uniform approximations in removing spurious singularities at the caustics or classical thresholds for the observation. These features may be given added emphasis by visualising the nature of the relevant classical phase space picture

The simplest illustration is provided by a one dimensional continuum wavefunction as shown in Fig.A1(a), with the corresponding

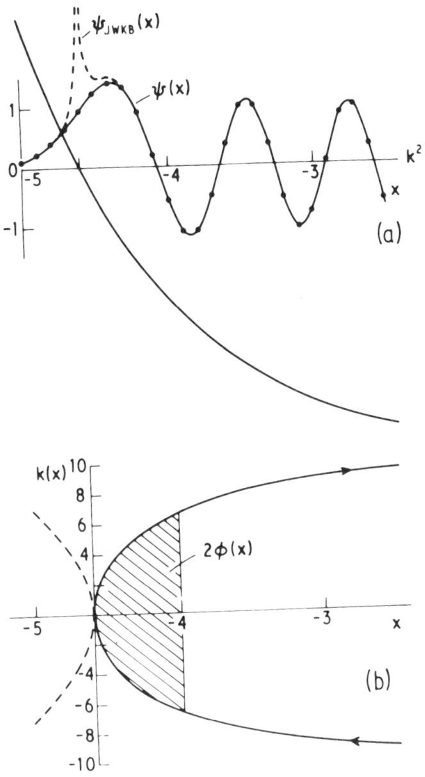

Fig.A.1 (a) The wavefunction and (b) the phase space orbit for an exponential potential, e^{-x}, at reduced energy $k^2 = 100$. The continuous curve in (a) is the exact wavefunction; the points (•) are given by the uniform approximation; the dashed curve is the JWKB approximation.

classical phase space motion in Fig.A1(b). Here the shaded area

$$\phi(x) = \int_a^x k(x)\,dx \;, \tag{A.1}$$

determines the phase of the JWKB wavefunction in the oscillatory (classically allowed) region $x > a$;

$$\psi_{JWKB}(x) = C\,[k(x)]^{-\frac{1}{2}} \sin(\phi(x) + \pi/4) \;. \tag{A.2}$$

The same definition for $\phi(x)$ may be extended into the forbidden region $x < a$, with $k(x)$ purely imaginary as illustrated by the dashed curve in Fig.A1(b). In this region

$$\psi_{JWKB}(x) = \tfrac{1}{2}C\ |k(x)|^{-\frac{1}{2}} \exp(-|\phi(x)|) \ . \tag{A.3}$$

Inclusion of the spurious singularity at $x = a$, due to disappearence of $k(x)$ is now however no longer a serious handicap in semi-classical theory, because as first shown by Langer[74] and later emphasised by Miller and Good[75], the same classical information on $k(x)$ and $\phi(x)$ may be combined in a uniform approximation, which remains well behaved at the turning point

$$\psi_{uni}(x) = \pi^{\frac{1}{2}}\ C[\xi(x)]^{\frac{1}{4}}[k(x)]^{-\frac{1}{2}}\ Ai(-\xi(x)) \ , \tag{A.4}$$

where

$$\xi(x) = [3\phi(x)/2]^{2/3} \tag{A.5}$$

with the branches $\arg k(x) = \pi/2$, $\arg \phi(x) = 3\pi/2$ being taken for $x < a$. By this definition $\xi(x)$ is real, for real x, with $\xi(x) < 0$ for $x < a$ and $\xi(x) > 0$ for $x > a$, and the divergence of $[k(x)]^{-\frac{1}{2}}$in equation (A.4) is exactly cancelled by the disappearence of $\xi(x)$ at $x = a$.

Precisely analogous arguments are used in the derivation of equation (14) for description of the rainbow pattern in elastic scattering. The (χ,ℓ) diagram illustrated in Fig. 4 is the relevant phase space, and the area α_r determines the interference pattern. Finally the uniform approximation again relies only on the same classical quantities $I_\nu(\theta)$ and $\alpha_r(\theta)$ as those required for a primitive (stationary phase) description[7].

The Stückelberg oscillations encountered in the discussion of non-adiabatic transitions also have a simple phase space interpretation. Thus for the effective potential curves illustrated in Fig. A.2(a), the phase term τ defined by equations (56) and (58) is the area enclosed by the two possible classical orbits $k_\pm(R)$, as shown in Fig. A.2(b). Any increase in the centrifugal contributions to $V_\pm(R)$ clearly brings the turning points $a_\pm$ closer together and to the crossing point, thereby causing a reduction in $\tau(\ell)$. Thus the resulting transition probability $|S_{12}(\ell)|^2$ oscillates with ℓ up to a critical value ℓ_{max} at which $a_\pm$ coalesce at the crossing point. This behaviour is shown in Fig. A.2(c). Fig. 7 shows how the (θ,λ) phase space may be used to map this behaviour from $S_{12}(\ell)$ to the differential cross-section $(d\sigma_{12}/d\Omega)$.

The same Stückelberg phenomenon may also be recognised in other situations. For example Figs. A.2(d)-(f) refer to predissociation[76], where the observable is a fluctuating level width (or lifetime). Again the phase of the oscillations depends on the phase space overlap in Fig. A.2(e). Finally Figs. A.2(g)-(h) refer to vertical

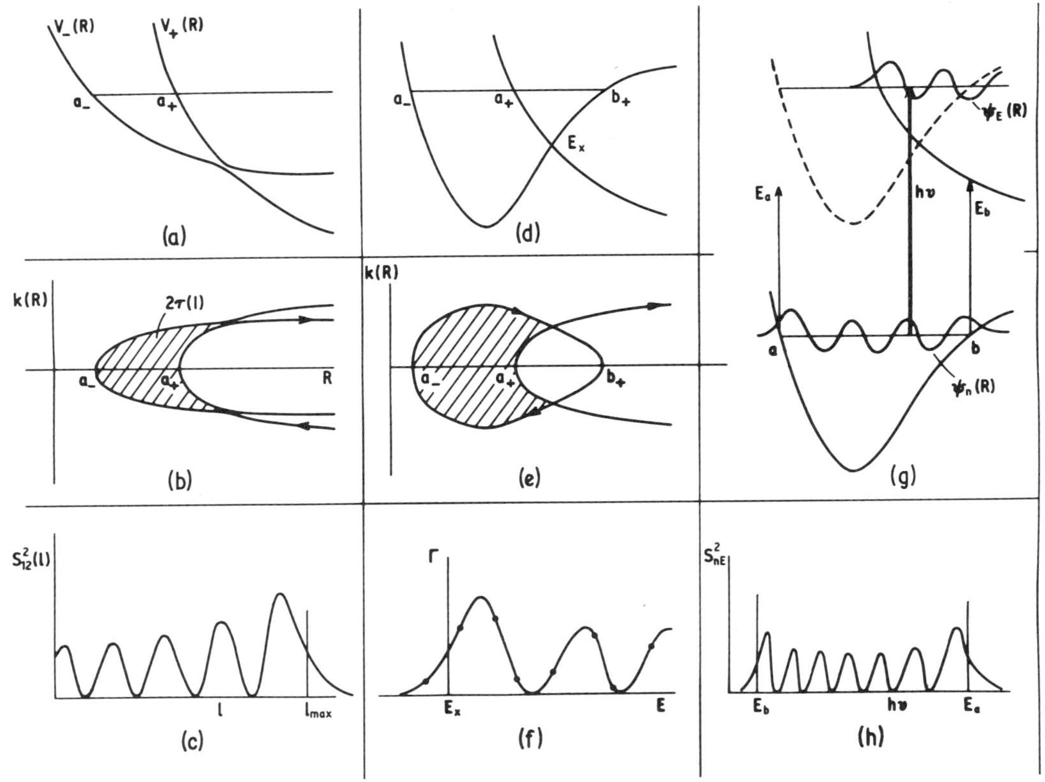

Fig.A.2 Origin and phase space interpretation of Stückelberg oscillations in various contexts (a) - (c) non-adiabatic transitions; (d) - (f) molecular predissociation; (g) - (h) Franck-Condon transitions.

Franck-Condon transitions[77]. The relevant phase space picture is again Fig.A.(g), but with the difference that in the predissociation case both bound and continuum orbits change with a change in energy; in the Franck-Condon case, one is usually interested in transitions from a given state $\psi_n(R)$, with a fixed orbit k(R), to different parts of the continuum. Nevertheless the resulting Condon reflection pattern in Fig. A.2(h) has precisely the same physical origin as the Stückelberg oscillations in Figs. A.2(c) and A.2(f).

Acknowledgements

I should like to thank Mrs. L. Cross and Mrs. J. Drew for their careful and skillful typing of the manuscript. I am also grateful to Mr. J. Hutson for assistance in preparing some of the figures.

REFERENCES

1. J.P. Toennies, Molecular beam scattering experiments on elastic, inelastic and reactive collisions in Physical Chemistry, an Advanced Treatise, Vol. VI A, Academic Press, New York, Chap. 5 (1974).
2. H. Pauly, Collision processes, theory of elastic scattering in Physical Chemistry, an Advanced Treatise, Vol. VIB, Academic Press, New York, Cahp. 8 (1974).
3. U. Buck, Elastic scattering, Adv. Chem. Phys., 30, 313-388(1975).
4. U. Buck, The inversion of molecular scattering data, Rev. Mod. Phys. 46, 369-389(1974).
5. H. Pauly, Elastic cross sections: spherical potentials, in Atom-Molecule Collisions ed. R.B. Bernstein, Plenum Press (1979).
6. M. V. Berry and K. E. Mount, Semiclassical approximations in wave mechanics, Rep. Prog. Phys., 35, 315-397(1972).
7. M.S. Child, Molecular collision thery, Academic Press, (1974).
8. R. B. Bernstein, Quantum mechanical (phase shift) analysis of differential elastic scattering of molecular beams, J. Chem. Phys., 33, 795-804(1960).
9. J. O. Hirschfelder, C. F. Curtiss and R. B. Bird, Molecular theory of gases and liquids, Wiley (1954).
10. H. Faxen and J. Holtsmark, Betrag zur Theorie des Durchgang Langsam Electronen durch Gase, Zeit. Phys., 45, 307-324(1927).
11. M. Abramowitz and I. A. Stegun, Handbook of Mathematical functions, Dover, (1967).
12. M. V. Berry, Uniform approximation for potential scattering involving a rainbow, Proc. Phys. Soc. 89, 479-490(1966).
13. N. F. Mott and H. S. W. Massey, Theory of atomic collisions, 3rd ed. Oxford, (1965).
14. F. T. Smith, Elastic and inelastic atom-atom scattering, in

Lectures in Theoretical Physics: Atomic collision processes XIC, Gordon and Breach, 95-118(1969).
15. R. B. Bernstein, Quantum effects in elastic molecular scattering, Adv. Chem. Phys., 10, 75-134(1966).
16. R. B. Bernstein and J. T. Muckerman, in Intermolecular forces ed. J. O. Hirschfelder, pp. 389-486(1967).
17. N. Levinson, On the uniqueness of the potential in a Schrödinger equation for a given asymptotic phase, Kgl. Danske. Vid. Selskab. Mat-Fys. Medd. 25, 1-29(1949).
18. R. K. B. Helbing, Additional quantum effects in atom-atom scattering: higher order glory scattering, J. Chem. Phys., 50, 493-500(1969).
19. U. Buck, Determination of intermolecular potentials by the inversion of molecular scattering data I. The inversion procedure, J. Chem. Phys., 54, 1923-1928(1971).
20. U. Buck and H. Pauly, Determination of intermolecular potentials by the inversion of molecular scattering data II. High resolution measurements of differential cross-sections and inversion of the data for Na-Hg , J. Chem. Phys., 54, 1929-1936(1971).
21. R. B. Gerber and M. Shapiro, A numerical method for determination of atom-atom scattering amplitudes from the measured differential cross-sections, Chem. Phys., 13, 227-233(1976).
22. M. Shapiro and R. B. Gerber, Extraction of interaction potentials from the elastic scattering amplitudes : an accurate quantum mechanical procedure, Chem. Phys., 13, 235-242(1976).
23. R. G. Newton, Scattering theory of waves and particles, McGraw-Hill, (1966).
24. R. G. Newton, Determination of the amplitude from the differential cross-section by unitarity, J. Math. Phys., 9, 2050-2055 (1968).
25. R. B. Gerber and M. Karplus, Determination of the phase of the scattering amplitude from the differential cross-section, Phys. Rev. D1, 998-1012(1970).
26. J. T. Rydberg, Graphische Darstellung einiger banenspektropische Ergebnisse, Z. Phys. 73, 376-385(1931).
27. O. Klein, Zur Berechnung von Potentialkurven für zeiatomige Moleküle mit Hilfe von Spektraltermen, Z. Phys. 76, 226-235 (1932).
28. A. L. G. Rees, Calculation of potential energy curves from band-spectroscopic data, Proc. Phys. Soc. (Lond) 59, 998-1003(1947).
29. O. B. Firsov, Determination of forces between atoms with use of the differential cross-section of elastic scattering, Zh. Eskp. Teor. Fiz. 24, 279-283(1953).
30. G. Vollmer, Inverse problem in atom-atom scattering in WKB approach, Z. Phys., 226, 423-434(1969).
31. W. H. Miller, WKB solution of inverse problems for potential scattering, J. Chem. Phys.. 51, 3631-3638(1969).

32. W. H. Miller, Additional WKB inversion relations for bound state and scattering problems, J. Chem. Phys., 54, 4174-4177 (1971).
33. R. Feltgen, H. Pauly, F. Torello and H. Vehmeyer, Determination of the He^4-He^4 repulsive potential up to 0.14ev by inversion of high resolution total cross-section measurements, Phys. Rev. Lett. 30, 820-823(1973).
34. M. S. Child and R. B. Gerber, Inversion of inelastic atom-atom scattering data : recovery of the interaction function, Mol. Phys. 38, 421-432(1979).
35. R. B. Gerber, V. Buch and U. Buck, Direct inversion method for obtaining anistropic potentials from rotationally inelastic and elastic cross sections, J. Chem. Phys., 72, 3596-3603.
36. R. B. Gerber, M. Shapiro, U. Buck and J. Schleusener, Quantum-mechanical inversion of the differential cross-section : determination of the He-Ne potential, Phys. Rev. Lett., 41, 236-239(1978).
37. M. S. Child, Electronic excitation : non adiabatic transitions in Atom-Molecule Collisions ed. R.B. Bernstein, Plenum(1979).
38. Yu. N. Demkov, Charge transfer at small resonance defects, Sov. Phys. J.E.T.P. 18, 138-142(1964).
39. D. S. F. Crothers, A critique of Zwaan-Stückelberg phase integral techniques, Adv. Phys., 20, 405-451(1971).
40. D. S. F. Crothers, Perturbed symmetric resonance : an exact formula, J. Phys., B6, 1418-1425(1973).
41. E. E. Nikitin, Theory of non-adiabatic transitions : recent developments with exponential models, Adv. Q. Chem. 5, 135-184(1970).
42. L. D. Landau, Zur Theorie der Energie ubertragung II, Phys. Z. Sowetunion 2, 46-51(1932).
43. C. Zener, Non-adiabatic crossing of energy levels, Proc. R. Soc. London Ser. A 137, 696-702(1932).
44. E. C. G. Stückelberg, Theorie der unelastischen Stösse zwischen Atömen, Helv. Phys. Acta 5, 369-422(1932).
45. E. E. Nikitin, Theory of non-adiabatic transitions : recent developments of the Landau-Zener (linear) model in Chemische Elementarprozesse, ed. H. Hartmann, Springer pp. 43-77(1968).
46. P. K. Janev, Non adiabatic transitions between ionic and covalent states, Adv. Atom. Mol. Phys., 12, 1-38, (1976).
47. J. B. Delos and W. R. Thorson, Studies of the potential curve crossing problem II General theory and a model for close crossings, Phys. Rev. A6, 728-745(1972), erratum 9, 1026 (1974).
48. A. Baranyi, Scattering matrix for curve-crossing collisions, J. Phys. B.11, L399-L402(1978) ; Deflection functions for curve-crossing collisions, J. Phys. B. 12,2841-2855(1979).
49. F. T. Smith, Diabatic and adiabatic representations for atomic collision problems, Phys. Rev., 179, 111-123(1969).

50. M. Baer, Adiabatic and diabatic representations for atom-molecule collisions : treatment of the three-dimensional case, Chem. Phys., 15, 49-57(1976)
51. L. P. Kotova, Angular distribution in inelastic atomic collisions, Sov. Phys. J.E.T.P. 28, 719-722(1969).
52. M. S. Child, Thermal energy scattering of alkali atoms from halogen atoms and molecules. The effect of curve-crossing, Mol. Phys. 16, 313-327(1969).
53. A. D. Bandrauk and M. S. Child, Analytical predissociation widths from scattering theory, Mol. Phys., 19, 95-111(1970).
54. W. H. Miller and T. F. George, Semiclassical theory of electronic transtions in low energy atomic and molecular collisions involving several degrees of freedom, J. Chem. Phys., 56, 5637-5652(1972).
55. Y. W. Lin, T. F. George and K. Morokuma, Semiclassical treatment of electronic transitions in molecular collisions : $H^+ + D_2 \rightarrow HD^+ + D$, J. Chem. Phys., 60, 4311-4322(1972).
56. W. H. Miller, Uniform semiclassical approximations for elastic scattering and eigenvalue problems, J. Chem. Phys., 48, 464-467(1968).
57. M. S. Child, A uniform approximation for one dimensional matrix elements, Mol. Phys., 29, 1421-1429(1975).
58. F. H. Mies, Molecular Theory of atomic collisions : fine-structure transitions, Phys. Rev. A7, 942-956(1973).
59. F. H. Mies, Molecular theory of atomic collisions : calculated cross-sections for $H^+ + F(^2P)$, Phys. Rev. A7, 957-967(1973).
60. R. E. Olson, Two-state Stückelberg-Landau-Zener theory applied to oscillatory inelastic total cross-sections, Phys. Rev. A2, 121-126(1970).
61. G. A. Delvigne and J. Los, The differential cross-section for chemionization in alkali atom - halogen molecule collisions: classical interpretation, Physica (Utrecht) 59, 61-76(1972).
62. J. B. Delos, Studies of the potential curve-crossing problem III collisional spectroscopy of close crossings, Phys. Rev. A9, 1626-1634(1974).
63. M. B. Faist and R. D. Levine, Collisional ionization and elastic scattering in alkali - halogen atom collisions, J. Chem. Phys., 64, 2953-2970(1976).
64. M. B. Faist and R. B. Bernstein, Computational study of elastic and electronically inelastic scattering of Br by ground state I atoms : role of potential curve crossing, J. Chem. Phys., 64, 2971-2984(1976).
65. C. H. Becker, P. Casavecchia, Y. T. Lee, R. E. Olson and W. A. Lester Jr., Coupled channel study of halogen (^2P) and rare gas (^1S) scattering, J. Chem. Phys., 70, 5477-5488(1979).
66. R. H. G. Reid and A. Dalgarno, Fine structure transitions and shape resonances, Phys. Rev. Lett., 22, 1029-1034(1969).
67. R. H. G. Reid, Transitions among the $3p^2P$ states of sodium induced by collisions with helium, J. Phys. B6, 2018-2039 (1973).

68. R. Düren, The interpretation of experimental scattering cross-sections with pseudopotential calculations for the alkali - mercury interaction, J. Phys. B10, 3467-3481(1977).
69. R. Düren and H. O. Hoppe, Measurement and evaluation of differential scattering cross-sections for Na $^2P(3/2)$ with Hg, J. Phys. B11, 2143-2167(1978).
70. D. R. Bates and R. McCarroll, Electron capture in slow collisions, Proc. Roy. Soc. A 245, 175-183(1958).
71. S. B. Schneiderman and A. Russek, Velocity dependent orbitals in proton-on-Hydrogen-Atom Collisions, Phys. Rev. 181, 311-320(1969).
72. V. Aquilanti and G. Grossi, Angular momentum coupling schemes in the quantum mechanical treatment of P-state atom collisons, J. Chem. Phys., 73, 1165-1172(1980).
73. V. Aquilanti, P. Casavecchia, G. Grossi and A. Laganà, Decoupling approximations in the quantum mechanical treatment of P-state atom collisions, J. Chem. Phys., 73, 1173-1180 (1980).
74. R. Langer, On the asymptotic solutions of ordinary differential equations, with an application to Bessel functions of large order, Trans. Am. Math. Soc., 33, 23-64(1931).
75. S. C. Miller and R. H. Good, A WKB-type approximation to the Schrödinger equation, Phys. Rev., 91, 174-179(1953).
76. M. S. Child in Molecular Spectroscopy II ed. R. F. Barrow, D. A. Long and D. J. Millen, (Chemical Society Specialist Periodical Report) 1974.
77. M. S. Child (ed.) Semiclassical methods in molecular scattering and spectroscopy, D. Reidel, Dortrecht (1980).

INTERNAL ENERGY TRANSFERS IN MOLECULAR COLLISIONS

Franco A. Gianturco

Istituto di Chimica Fisica
Città Universitaria
00189 Rome, Italy

MPI für Strömungsforschung
Böttingerstraße 6-8
D-3400 Göttingen, BRD

ABSTRACT

The theoretical and computational methods recently applied to the quantum mechanical calculation of various types of inelastic cross sections, those that preside over the transfer of kinetic energy into rotovibrational molecular modes, are reviewed and discussed.

Ab initio methods to obtain the necessary potential energy surfaces (PES) are presented and their effectiveness shown in a few specific cases.

The coupled representations in the space-fixed (SF) and Body-fixed (BF) frames of reference are used to expand the total wave-functions; the corresponding close coupled (CC) equations are also presented.

Decoupling methods that are aimed at reducing the fast-growing dimensions of the above rigorous equations are also summarized and discussed, together with the most recent numerical techniques employed to solve these equations and to yield the corresponding S-matrix elements. A large bibliography has been selected from the very rich, recent literature on this expanding subject.

1. INTRODUCTION

2. THE INTERACTION BETWEEN PARTNERS

2.1 Separation of nuclear and electronic motions.
2.2 The computing of intermolecular potentials.
2.3 Some atom-molecule examples.

3. THE ROTOVIBRATIONAL PROBLEM

3.1 Quantal treatment of rotational excitations.
3.2 The coupled representation in a LAB frame.
3.3 The Body-fixed form of the scattering equations.
3.4 Vibrationally inelastic collisions.

4. DECOUPLING APPROXIMATIONS

4.1 The CS approximation.
4.2 The IOS approximation.
4.3 The fixed-nuclei approximation.
4.4 The l-dominant decoupling.
4.5 The effective potential methods.

5. NUMERICAL METHODS OF SOLUTION

1. INTRODUCTION

It is now a well-established fact that intermolecular collisions are primarily responsible for most observable rate phenomena, both physical and chemical, in fluid systems. Indeed, the object of chemical physics at a microscopic level is mainly the study of the numerous processes that before, during and after a chemical reaction (or even in subreactive situations) control either the transformation of one type of molecular energy into another type or its transfer into the thermal bath.

When the events under study take place in the absence of external fields and when the molecular systems involved are in their gaseous states, one is obviously faced with a simpler situation, although it is still general enough to give rise to a great variety of phenomena. Moreover, even as the theoretical enquiry is narrowed down to examining only non-reactive processes or encounters with species mainly exhibiting 'physical' interactions the wealth and number of final outcomes is sufficient to define a rather broad field of investigation.

It has only been in recent years, however, that the experimental side of such a field has reached maturity through the development of modern electronic and vacuum instrumentation. A wealth of new and detailed experimental data on atomic and molecular collision cross sections has thus been forthcoming, covering the whole range from elastic to inelastic to reactive scattering[1]. The now old molecular beam technique, first applied to molecular scattering by Stern in the twenties, has been impressively transformed in the last twenty years into a flexible, general method for the study of molecular collisions[2]. The measurement of the angular and energy distributions of the scattering of atoms, molecules and ions are now performed as a function of collision energy in many laboratories around the world and, in some cases, even state-selected species can be prepared before the encounter or analyzed after the event[3].

The corresponding conceptual approach to the theory of molecular collisions is also well established and known for a long time. Starting with the Hamiltonian of the total 'supersystem', the electronic potential energy surfaces that control the motion of the nuclei can be computed over the whole configuration space relevant to the process under study. The various quantum, semiclassical or classical state-to-state scattering cross sections can be in turn obtained from the S-matrix generated by the numerical solution of the wave equation or of the many trajectories that provide the nuclear motion.

The number of possible processes, however, still remains very large and therefore a drastic reduction in the range covered by the present chapter was obviously necessary. Thus, only one of the simplest types of energy transfer in molecular collisions has been considered, namely that involving the conversion of translational energy for one of the partners into the internal rotational and/or vibrational energies of the other. The latter, in turn, has been considered a diatomic molecule in its closed-shell ground electronic state (Σ-state), while the former was taken to be a structureless atom or a small molecule, also in its closed-shell ground electronic state.

This is probably still one of the most common events among those which happen during a chemical reaction and it is also the only one for which a thorough, ab initio approach to its various computational aspects has been carried out for a few representative systems in the literature of recent years[4].

The fields of intermolecular collisions is one characterized by a strong admixture of theory with experiments. Thus, the sections which make up the present Chapter follow the pathway that experimental needs most naturally suggest to the theorist in order for him to provide an increased understanding of measured events. Section 2, in fact, will examine the forces at play between colliding partners: the influence of each structural property in driving along different 'trajectories' the interacting molecules and the ways and means that computational theorists have to obtain a grasp of such forces in a reasonable way or with acceptable accuracy.

The following section will then focus on the dynamical event itself, i.e. will examine the quantum mechanical formulation of the problem and the corresponding complexity of the usual functional expansion techniques that are applied to most studied systems. As a consequence of such complexity, dimensionality reduction schemes, those that can be based either on simplifications suggested by the physics of the problem or on numerical reductions that might hopefully turn out to be right, are analyzed in section 4.

Finally the numerical methods that have been applied over the years to treat the corresponding coupled differential equations that appear in the quantum mechanical collision problem are reviewed and examined in the last section 5.

2. THE INTERACTION BETWEEN PARTNERS

2.1 Separation of nuclear and electronic motions

First of all, one readily recognizes that the effects of the interaction between the electrons belonging to the scattering systems and the effects corresponding to the interactions between nuclei are switched on within different time scales due to the different velocities involved when molecular collisions at near-thermal energies are examined. Thus, if one considers that a typical velocity for a valence electron is of the order of 10^8 cm $\cdot$ sec^{-1}, the corresponding velocity for two colliding atoms will reach similar values only for collision energies larger than 100 keV.

If one therefore considers the 'supermolecule' made out of all the collision partners as a system having n electrons and N nuclei, in the laboratory, space-fixed (SF) reference frame the corresponding Hamiltonian can be written as (in atomic units $\hbar$ = m = e = 1):

$$\mathcal{H}_{TOT} = \hat{T}_n + \hat{T}_{el} + \hat{V} + \hat{V}_{SO}$$

$$= \mathcal{H}_{ES} + \hat{V}_{SO} \tag{2.1}$$

where

$$T_n = \sum_j \frac{1}{2M_j} \nabla^2_{\underline{R}_j} \; ; \qquad \hat{T}_{el} = -\frac{1}{2} \sum_i \nabla^2_{\underline{r}_i}$$

$$V = \sum_{j>j'} \frac{Z_j \cdot Z_{j'}}{|\underline{R}_j - \underline{R}_{j'}|} - \sum_{i,j} \frac{Z_j}{|\underline{r}_i - \underline{R}_j|} + \sum_{i>i'} \frac{1}{|\underline{r}_i - \underline{r}_{i'}|} \tag{2.2}$$

V_{SO} = spin-orbit and relativistic interactions.

Here the M_j's are the masses of all the interacting nuclei and the R_j's define their corresponding positions with respect to the SF origin. The vectors r_i are the coordinates of the electrons with respect to that same origin. More precisely, one should have used for each atom (or molecule) the coordinate of the electrons + nucleus (or electrons plus nuclei) system. The low mass ratio, however, allows without loss of accuracy the substitution of these coordinates with the ones defined in eq. (2.2).

Because of the large velocity differences between electrons and nuclei at near thermal energies, it is clearly useful to try to separate out the centre of mass motion of the whole system from the remaining coordinates. The dynamics of the encounters are then described by an arbitrary set of 3 (N + n − 1) independent internal variables; several choices for such coordinates can be made to suit particular problems but from the point of view of keeping the 'supermolecule' picture in discussing the interaction between partners it is very convenient to choose for all bound electrons a set of coordinates which are centred at the centre-of-mass of all the nuclei. This implies the disregard of the small displacing effects due to the electronic masses. For the simple case of two interacting atoms one can then write, as an example, that

$$\underline{R}_{CM} = \frac{1}{M}\left(M_A \underline{R}_A + M_B \underline{R}_B\right)$$

$$\underline{R} = \underline{R}_B - \underline{R}_A \qquad (2.3)$$

$$\underline{r}'_i = \underline{r}_i - \underline{R}_{CM}$$

where M is now the 'total mass' corresponding to some centre-of-mass coordinate $\underline{R}_{CM}$. All the electron-electron, electron-atom and atom-atom interactions are now included in the potential term V ($\underline{r}$, $\underline{R}$) with $r = \sum_i^m \underline{r}'_i$. The translational invariance of the total Hamiltonian implies the conservation of total linear momentum when the transformation (2.3) is applied to the coordinates in eq. (2.2). Thus, one can easily see that:

$$-i\left(\frac{\partial}{\partial \underline{R}_A} + \frac{\partial}{\partial \underline{R}_B} + \sum_i \frac{\partial}{\partial \underline{r}_i}\right) = $$

$$= -i \frac{\partial}{\partial \underline{R}_{CM}} \qquad (2.4)$$

The corresponding total kinetic energy operator is obtained now in the following form for the above atom-atom example:

$$\hat{T} = \hat{T}_n + \hat{T}_{el} =$$

$$= -\frac{1}{2M}\frac{\partial^2}{\partial \underline{R}^2_{CM}} - \frac{1}{2}\sum_i \frac{\partial^2}{\partial \underline{r}_i^2} -$$

$$-\frac{1}{2\mu}\frac{\partial^2}{\partial \underline{R}^2} - \frac{1}{2M}\left(\sum_i \frac{\partial}{\partial \underline{r}'_i}\right)^2 \qquad (2.5)$$

Now μ corresponds to the reduced mass of the two nuclei and the last term on the r.h.s of the equation is often called the 'mass polarization' term that provides cross terms between electronic coordinate operators. Its contributions are of the order of M^{-1} when compared with the $\frac{\partial}{\partial r'_i}$ terms and hence can be disregarded.

Since the potential energy term is independent of R_{CM}, the total Hamiltonian of eq. (1.1) can now be rewritten as:

$$\mathcal{H}_{TOT} = \hat{T}_{CM} + \hat{T}_n + \hat{T}_{el} + \hat{V}(\underline{r}, \underline{R}) \qquad (2.6)$$

where V_{SO} contributions have also been disregarded for the moment. The above transformation therefore allows a rigorous factorization of the part of the total wavefunction that only depends on R_{CM}:

$$\Psi_{TOT}(\underline{r}, \underline{R}, \underline{R}_{CM}) = e^{i\underline{P}\cdot\underline{R}_{CM}}\, \psi(\underline{r}, \underline{R}) \tag{2.7}$$

where $\underline{P}$ is the total linear momentum for the whole system. The general Hamiltonian for a complex system of N nuclei and n electrons can now be written in terms of the $3\cdot(N + n - 1)$ internal coordinates:

$$\begin{aligned} \mathcal{H} &= -\frac{1}{2}\sum_j \frac{1}{\mu_j} \nabla^2_{\underline{R}'_j} - \frac{1}{2}\sum_i \nabla^2_{\underline{r}'_i} + \\ &+ V(\underline{R}, \underline{r}) = \hat{T}_n + \mathcal{H}_{el}(\underline{r}, \underline{R}) \end{aligned} \tag{2.8}$$

where the primed coordinates remind us that they are now internal coordinates while $\underline{R}$ and $\underline{r}$ stand for the collective coordinates on which the potential energy term depends.

For slow molecular velocities one can look for an approximate solution to the Hamiltonian (2.8) by disregarding the effect of the $\hat{T}_n$ operator and instead writing a Schrödinger equation of simplified form:

$$\mathcal{H}_{el}\, \varphi_n(\underline{r}; \underline{R}) = W_n(\underline{R})\, \varphi_n(\underline{r}; \underline{R}) \tag{2.9}$$

The electronic Hamiltonian $\mathcal{H}_{el}$ contains now the nuclear coordinates as parameters and therefore its corresponding eigenfunctions φ_n and eigenvalues W_n will also depend parametrically on each chosen nuclear configuration $\underline{R}$. Moreover, it is generally assumed that the set of the φ_n functions, eigenfunctions of an Hermitian operator, form a complete set of electronic functions which are chosen to be orthonormal:

$$\int d^3r\, \varphi^*_K(\underline{r}, \underline{R})\, \varphi_\ell(\underline{r}, \underline{R}) = \delta_{K\ell} \qquad \text{fixed } \underline{R}, \tag{2.10 a}$$

$$1 = \sum_\ell |\varphi_\ell\rangle\langle\varphi_\ell| \qquad \text{fixed } \underline{R}, \tag{2.10 b}$$

It is worth noting that the above set of electronic functions, while depending on nuclear positions, does not constitute a complete set of nuclear functions and is not normalizable within the $\{R\}$ space.

The total wavefunction for nuclei and electrons can thus be expanded over a set of trial functions:

$$\Psi(\underline{r}, \underline{R}) = \sum_{\ell} \chi_{\ell}(\underline{R})\, \phi_{\ell}(\underline{r}, \underline{R}) \tag{2.11}$$

Each nuclear wavefunction χ ($\underline{R}$) describes the motion of the nuclei on the potential energy surface (PES) associated with the electronic state 1. The physical meaning of these surfaces changes when different basis sets ϕ_ℓ are chosen. Specifically, different choices will lead to the commonly used adiabatic or diabatic representations. The function $\chi_\ell(\underline{R})$, as $\underline{R}$ goes to ∞, will give the amplitude for the probability that the total system ends up in the electronic state ϕ_ℓ and thus directly produces the S-matrix for the transition from the initial state specified by the problem. Hence the quantum mechanical treatment of the scattering problem reduces to the solution of the Schrödinger equation for $\chi_\ell(\underline{R})$.

Writing now $\hat{T}_n$ as $\sum_K \hat{P}_K^2/2M_K$, the effect on a general electronic function $\phi_\ell(\underline{r};\, \underline{R})$ of each nuclear momentum operator is:

$$\hat{P}_K\, \phi_\ell = -i\, \frac{\partial \phi_\ell}{\partial \underline{R}_K} \tag{2.12}$$

and therefore one can write:

$$\hat{T}_n\, \phi_\ell = \sum_K \frac{1}{2M_K} \left\{ \phi_\ell\, \hat{P}_K^2 + 2\left(\hat{P}_K \phi_\ell\right)\cdot \hat{P}_K + \hat{P}_K^2\, \phi_\ell \right\} \tag{2.13}$$

Since these general electronic functions which are not yet chosen as being eigenfunctions of the $\mathcal{H}_{el}$ of eq. (2.9), also form a complete set, one can write the total Hamiltonian of eq. (2.8) as follows:

$$\mathcal{H} = \sum_{\ell, m} |\phi_\ell\rangle \langle\phi_\ell|\, \mathcal{H}\, |\phi_m\rangle \langle\phi_m| \tag{2.14}$$

substitution of (2.13) into (2.8) gives the general result:

$$\mathcal{H} = \sum_\ell |\phi_\ell\rangle \left\{ \sum_K \frac{\hat{P}_K^2}{2M_K} + \langle\phi_\ell|\mathcal{H}_{el}|\phi_\ell\rangle + \sum_K \frac{1}{2M_K} \times \right.$$
$$\left. \times \langle\phi_\ell|\hat{P}_K^2|\phi_\ell\rangle \right\} \langle\phi_\ell| + \sum_\ell \sum_{m\neq\ell} |\phi_m\rangle \left\{ \sum_K \frac{1}{2M_K} \times \right.$$
$$\times \left[2\langle\phi_m|\hat{P}_K|\phi_\ell\rangle \cdot \hat{P}_K + \langle\phi_m|\hat{P}_K^2|\phi_\ell\rangle \right] +$$
$$\left. + \langle\phi_m|\mathcal{H}_{el}|\phi_\ell\rangle \right\} \langle\phi_\ell| \tag{2.15}$$

The first summation on the r.h.s of (2.15) contains all the diagonal terms coming from the partial resolution of the identity (2.14). They correspond to the effective potential energy surfaces that govern the nuclear motion via the electronic Hamiltonian plus a correction term that involves second derivatives of the electronic functions with respect to nuclear coordinates. This last correction is usually small and can be ignored.

The adiabatic representation corresponds to choosing the electronic functions to be eigenfunctions of the Hamiltonian of eq. (2.9):

$$\langle \phi_m | \mathcal{H}_{el} | \phi_\ell \rangle = W_\ell(\underline{R})\, \delta_{\ell m} \tag{2.16}$$

The Hamiltonian of eq. (2.15) can thus be written as a zeroth order term (the sum over diagonal contributions) plus a correction term of nonadiabatic coupling, originating from the second sum over non diagonal matrix elements, which act as a perturbation:

$$\mathcal{H} = \mathcal{H}_{BO} + \hat{\lambda}_{BO} \tag{2.17}$$

The subscript refers now to the well-known Born-Oppenheimer approximation[5], where the above result was derived by essentially applying a perturbative treatment to the molecular eigenvalue problem.

The use of the complete eigenfunctions φ_n of eq. (2.9) allows one to write, for the adiabatic representation, that:

$$\mathcal{H}_{BO} = \sum_n |\varphi_n\rangle \left[\sum_K \frac{\hat{P}_K^2}{2M_K} + W_n(\underline{R}) \right] \langle \varphi_n | =$$

$$= \sum_n |\varphi_n\rangle\, \hat{h}_n\, \langle \varphi_n | \tag{2.18}$$

The operator $\hat{h}_n$ acts only on the nuclear coordinates and its eigenfunctions are the nuclear functions needed in the expansion (2.11) when the general electronic functions ϕ_ℓ are given there by the adiabatic eigenfunctions of eq. (2.9):

$$\hat{h}_n\, \chi_v^n(\underline{R}) = E_{nv}^{(0)}\, \chi_v^n(\underline{R}) \tag{2.19 a}$$

and

$$\Psi(\underline{r},\underline{R}) \Rightarrow \sum_{nv} \Psi_{nv}^{(0)}(\underline{r},\underline{R}) = \sum_{n,v} \chi_v^n(\underline{R})\, \varphi_n(\underline{r},\underline{R}) \tag{2.19 b}$$

When the electronic state considered is non-degenerate, the perturbation $\hat{\lambda}_{Bo}$ does not give a first order correction to the total energy, since only off-diagonal terms appear in the second sum of eq. (2.15). On the other hand, the second order correction is given by:

$$E_{nv}^{(2)} = \sum_{k \neq n} \sum_{s} \frac{|\langle \chi_s^k | \hat{h}_{nk} | \chi_v^n \rangle|^2}{E_{ks}^{(0)} - E_{nv}^{(0)}} \tag{2.20}$$

while the first order correction to the adiabatic expansion functions appearing in eq. (2.19 b) is given by:

$$\Psi_{nv}^{(1)} = = - \sum_{k \neq n} \sum_{s} \frac{\langle \chi_s^k | \hat{h}_{nk} | \chi_v^n \rangle \chi_s^k (\underline{R}) \varphi_k (\underline{r}, R)}{E_{ks}^{(0)} - E_{nv}^{(0)}} \tag{2.21}$$

One clearly sees then that the corrections to the adiabatic approximate wave functions of the expansion (2.19 b) can be disregarded when the off-diagonal couplings appearing in the numerator of the r.h.s. of eq. (2.21) are much smaller than the energy separation between individual electronic energies at the nuclear geometries of interest.

Non adiabatic couplings therefore tend to peak near avoided crossings, where the denominators in eq. (2.21) attain minimum values. Very often these regions arise because the adiabatic electronic states are mainly mixtures of two simple molecular orbitals (or valence bond structures) whose corresponding potential energy surfaces cross.

From a collisional viewpoint one can then say that if the relative motion of the nuclei is sufficiently slow, they will tend to follow a single adiabatic PES even when they are nearing an avoided crossing. The corresponding electronic motion, in fact, has always time to adjust to the changes brought in by the slowly altered nuclear configurations and a realistic representation of the total wavefunction of (2.11) can therefore be obtained via adiabatic electronic functions with small non-adiabatic corrections given by the second sum of eq. (2.15).

On the other hand, if the nuclei move very rapidly in the vicinity of certain nuclear configurations for which the denominators of eq. (2.21) become small, the probability of non-adiabatic transitions, i.e. of the partners to move from one adiabatic surface to another, will approach unity. The relevant bound electrons, in fact, will not have sufficient time to adjust their motion adiabatically to the rapidly changing forces due to the moving nuclei.

In such a situation, therefore, it might become more efficient to employ in the expansion (2.11) another set of electronic functions which correctly depict the physical origin of the transitions between PES.

One can then define a set of diabatic functions $\eta_n(\underline{r};\, \underline{R})$ as those for which, in eq. (2.15), the following relations are valid:

$$\langle \phi_m | P_K | \phi_n \rangle \Longrightarrow \langle \eta_m | P_K | \eta_n \rangle = $$

$$= \langle \eta_m | \frac{\partial \eta_n}{\partial \underline{R}_K} \rangle = 0 \qquad \forall\; m, n \tag{2.22}$$

This being valid for one or more of the nuclear coordinates. The above conditions automatically imply that: $\langle \eta_m | \frac{\partial^2 \eta_n}{\partial R_K^2} \rangle = 0$. The corresponding diagonal terms in eq. (2.15) do not correspond any more to eigenvalues of the electronic Hamiltonian of (2.9) and therefore do not follow the symmetry conditions implied by the non-crossing rule Moreover, the transitions between surfaces are induced primarily by the last terms in the second sum on the r.h.s. of (2.15): The off-diagonal elements of the electronic Hamiltonian $\mathcal{H}_{el}$. In cases in which these terms are small, the corresponding diabatic representation may have to be preferred over the adiabatic one, even at low collision energies.

The collision problems discussed in this chapter, however, will essentially consider the energy flow between translational motion and roto-vibrational states of one (or both) of the scattering partners as due to the interaction among them when moving over one potential energy surface that is in turn generated via an adiabatic set of electronic functions.

The theoretical methods for acquiring a reliable knowledge of such functions, and hence of the relevant PES, will be briefly discussed in the next paragraph.

2.2 The Computing of Intermolecular Potentials

In the last paragraph we have separated electronic from nuclear motions by appropriately choosing two different representations for the electronic basis functions. To deal with the nuclear dynamics and with the forces that drive it, it is convenient to further separate the coordinates describing relative motion of the partners from those describing internal motion within each partner (vibration and rotation). Quantum mechanically this is done by expanding the functions $\chi_\ell(\underline{R})$ appearing in eq. (2.11) as:

$$\chi_\ell(\underline{\rho},\underline{R}) = \sum_\alpha u_{\ell\alpha}(\underline{R})\,\xi_{\ell\alpha}(\underline{\rho};\underline{R}) \tag{2.23}$$

where $\underline{\rho}$ stands as a collective index for the internal motion coordinates and $\underline{R}$ for the relative motion coordinates. The $\xi_{\ell\alpha}$ are the nuclear wavefunctions for the αth rotovibrational state of the electronic state. As written above, they can be chosen to depend parametrically on the internal coordinates $\underline{R}$. The unknown nuclear scattering wavefunctions $u_{\ell\alpha}(\underline{R})$ are therefore the object of our study, since they directly give asymptotically the S-matrix elements for transitions between internal states belonging to different electronic PES.

The corresponding total Hamiltonian defined in eq. (2.8) can also be rewritten as:

$$\mathcal{H} = -\frac{1}{2\mu}\nabla_{\underline{R}}^2 + \sum_i \mathcal{H}_{int}^i(\underline{\rho}_i) + V(\underline{R},\underline{\rho}) \tag{2.24}$$

under the condition that the potential vanishes as $\underline{R}\to\infty$ and with the internal Hamiltonians $\mathcal{H}_{int}^i$ being also parametric functions of $\underline{R}$, the main collision coordinate in simple systems.

We are interested here in the calculation of the intermolecular potential given as:

$$V(\underline{R},\underline{\rho}) = W_n(\underline{R},\underline{\rho}) - W_n(\infty,\underline{\rho}) \tag{2.25}$$

with the W_n's being the adiabatic electronic eigenvalues of eq. (2.9) for a given geometry $\underline{\rho}$ of the colliding partners computed for both the asymptotic and the actual values of the relative internuclear distance $\underline{R}$ that controls the overall collision process.

While the problem of solving eq. (2.25) is well defined in principle, in practice not even for the simplest reaction systems is the PES V(R,$\underline{\rho}$) available accurately enough that it can be regarded as known.

All the calculations that have been made so far by treating all the terms of the Hamiltonian (2.1), and that are usually defined as being of the ab initio type, have involved an expansion of the total, unknown eigenfunctions of the $\mathcal{H}_{el}$ operator of eq. (2.8):

$$\varphi_i(\underline{r},\underline{R},\underline{\rho}) = \sum_j C_{ij}(\underline{R},\underline{\rho})\,\vartheta_j(\underline{r},\underline{R},\underline{\rho},\alpha) \tag{2.26}$$

where the ϑ_j are the n-particle composite functions of an elementary basis set $\{\beta_k\}$, depending on the electronic and nuclear coordinates and on a set of non-linear parameters α. The ϑ's are antisymmetric functions of electronic space-spin coordinates and are usually (but not necessarily) eigenfunctions of the operators which commute with the electrostatic Hamiltonian $\mathcal{H}_{ES}$ of eq. (2.1). The β_K can in turn be one-particle (orbital) functions, two-particle (geminal or pair) functions or even multi-particle functions. The expansion (2.26) is usually truncated and only a finite number of terms can be included in it. When, as it is commonly the case, the ϑ_j's do not contain explicitly interelectronic distance coordinates in a correct way, the convergence of the expansion is slow. If, however, such coordinates are directly used, the corresponding integrals in the equations which follow are much more difficult and have 3n-dimensionality, where n is the number of electrons. The way of determining the functions ϑ_j and the linear coefficients c_{ij} is therefore very important and has been extensively discussed in the quantum chemistry literature[6,7,8,9].

In most studies of systems with more than two electrons the trial functions that appear in (2.26) are expressed in a determinantal form over the elementary basis set $\{\beta_K\}$:

$$\vartheta_j = \| \beta_1(1)\, \beta_2(2) \ldots \beta_n(n) \| \tag{2.27}$$

where the β's are one-electron spin orbitals. The most common examples of them are provided by the Slater-type orbitals, or STO's by the Gaussian-type functions called GTO's and by a mixture of the two given by the so-called 'contracted GTO functions', in which a linear combination of GTO's with coefficients and exponents fixed by optimizing STO atomic calculations are used to provide a way of overcoming the poor behaviour of the GTO's near the nuclei[10]. These functions are usually of the atomic type, i.e. they are in some way each centered on one of the nuclei composing the 'supermolecule'.

Another possibility is provided by elementary basis of molecular functions. Suffice it here to mention the united-atom or one-centre expansion where the origins of the functions are fixed on one major centre of each molecular partner although their non-linear parameters are chosen to represent the distribution over the entire relevant partner. Convergence of expansions in single-centre basis sets are so slow, however, as to severely limit their usefulness for making accurate calculations[11,12].

The most common method for obtaining reasonably well-behaved expansion functions for each determinant of eq. (2.27) is provided by the Hartree-Fock (HF) self-consistent field (SCF) procedure[13]. For closed shell molecular partners one can in fact write:

$$\vartheta_j = \| u_1\alpha(1)\, u_1\beta(2) \ldots\ldots u_m\beta(2m) \| \tag{2.28}$$

where the u's are now spatial orbitals and the $\alpha(i)$, $\beta(i)$ are the spin functions. In the above, restricted, Hartree-Fock approximation (RHF), 2 m = n electrons are accommodated in m orbitals and a single Slater determinant is used to represent each expansion function ϑ_j. Moreover, at this level of approximation only one ϑ_j is used in the expansion (2.26). Each of the $u_\ell(k)$ is an eigenfunction of the Fock operator F(k):

$$\hat{F}(k)\, u_\ell(k) = \varepsilon_\ell\, u_\ell(k) \tag{2.29}$$

where the ε_ℓ's are called the orbital energies and:

$$\hat{F}(k) = \hat{H}(k) + \sum_i \left\{ 2\hat{J}_i(k) - \hat{K}_i(k) \right\} \tag{2.30}$$

with $\hat{H}(k)$ being the kinetic energy and nuclear attraction operator acting on each one-particle orbital. The $\hat{J}_i$'s are the Coulomb operators and the $\hat{K}_i$'s are the exchange operators defined as:

$$\hat{J}_i(k) = \int r_{k\ell}^{-1}\, u_i^*(\ell)\, u_i(\ell)\, d\tau_\ell \tag{2.31 a}$$

$$\hat{K}_i(k)\, u_j(k) = \int r_{k\ell}^{-1}\, u_i^*(\ell)\, u_j(\ell)\, d\tau_\ell \tag{2.31 b}$$

Eq. (2.29) is termed a pseudo-eigenvalue equation since the Fock operator depends on the lowest n eigenfunctions $u_\ell(k)$. Such an equation is therefore solved by using an iterative method that first guesses some trial orbitals $u_\ell(k)$ and then uses these to evaluate the Fock operator itself. According to the quality of the initial guess, this procedure usually converges faily rapidly for closed shell systems not far from their equilibrium geometries[14]. As the size and flexibility of the basis set is increased, the SCF orbitals and energy approach the true HF ones. Thus, this basic method, in several modifications and variants, was coupled to numerous early molecular integral programs that provided the bulk of molecular electronic energy calculations till the early 70's[15,16].

Although the RHF method provides an attractive orbital picture of the molecular wavefunctions, it has, however, a number of shortcomings that have prevented its wide-scale application to the production of PES to be used in scattering problems:

(i) Closed-shell MO wavefunctions in general do not correlate well with the separated atom states that couple to form a given molecular state. Hence, the long-range part of the interaction in reactive systems is not correctly predicted by the single-determinant approach. Nor is it sufficient to yield information on dispersion forces where correlation energy effects (strongly dependent on $\underset{\sim}{R}$) require many terms in the expansion (2.26)).

(ii) The HF method may yield a good approximation for the ground electronic state of a certain symmetry but it becomes much less reliable when electronically excited states of the same symmetry are needed and the question of their being orthogonal to lower-lying states arises[17].

(iii) The pair probability of electrons with opposite spins is given in this theory by a simple product wavefunction, while the same probability for electrons with identical spins contains a 100% negative correlation of motion via the exchange terms, as required by the Pauli principle[18]. The correlation error is therefore defined as the difference in energy between the electronic energy in the HF limit and its exact value. It really includes both the correlation error proper and the relativistic correction[18]. Since the correlation energies are normally at least as large as energies of chemical interest, the PES computed through the HF scheme can therefore be useful only if the correlation correction does not vary a great deal over the spatial regions of interest of the surface.

The method of configuration interaction (CI) provides one of the earliest techniques used to overcome the correlation problem and it is based on an extensive mixing of the states arising from different spin-orbital configurations. In principle, one can expand the exact solution (2.26) in terms of the complete (infinite) set of determinantal functions which are in turn constructed from a complete set of one-electron spin orbitals[19].

The total wavefunction must be an eigenfunction of the operators which commute with the chosen Hamiltonian of the systems and whose eigenvalues approximately characterize the molecular states of interest. The form of a symmetry-adapted φ therefore depends on the method selected for evaluating matrix elements and on the method chosen for generating eigenfunctions of the total spin operator $\hat{S}^2$. One can thus write:

$$\varphi_i = \sum_k a_{ik} D_{ik} \tag{2.32}$$

where the D_{ik} are Slater determinants of space-spin orbitals. The mixing coefficients and energies of the resulting CI states are then found by solving the secular equations:

$$\left| \mathbb{H} - \mathbb{E}^{(ci)} \mathbb{S} \right| \mathbb{C} = 0 \tag{2.33}$$

where $\mathbf{C}$ is the matrix of coefficients and the elements of the Hamiltonian matrix are given by:

$$H_{kl} = \langle D_{il} | \mathcal{H}_{ES} | D_{ik} \rangle \tag{2.34 a}$$

with

$$S_{lk} = \langle D_{il} | D_{ik} \rangle \tag{2.34 b}$$

An SCF calculation for each geometry of an n-electron closed-shell system and via an N-orbital expansion will produce n/2 doubly occupied MO's and $(N-\frac{n}{2})$ vacant MO's. A standard CI procedure will therefore form the needed D's by systematically promoting electrons from occupied MO's of the reference determinant, D_{io} , to the virtual MO's. The number of configurations thus becomes of the order of n^N and gets easily out of hand for most of the systems of interest.

According to perturbation theory, however, if the zeroth order wavefunction is taken to be the SCF wavefunction, then all triple and higher excitations make no contribution to it to the first order and therefore most of the very powerful theoretical methods dealing with the correlation problem are essentially performing CI calculations that include all Single and Double (S+D) excitations[20-22] coming out of an RHF reference configuration.

Another method which follows the same approach but is an improvement over the well-known slow convergence of the expansion (2.32) is the multiconfiguration SCF method (MC-SCF) which can be described as a procedure which forces <u>both</u> the CI coefficients of (2.32) and the atomic orbital coefficients of each $u_l(k)$ in eq. (2.29) to be varied to minimize the individual CI energies and the total electronic energy. A doubly iterative procedure is therefore needed and it yields, upon convergence, the optimum orbitals and the optimum configuration mixing coefficients for the basis sets used. In general, one will need fewer configurations as compared to a conventional CI (S+D), since the vertical orbitals are always forced to be within the same physical space as the occupied orbitals[23,24,25].

2.3 Some Atom-Molecule Examples

Several systems have been studied theoretically in recent years, using both *ab initio* and approximate procedures to generate the ground-state potential energy surfaces needed to carry out scattering calculations. Even a passing review of them is outside the scope of these notes, while some extensive discussions of methods and results have recently appeared in the literature[26,27]. What we will be discussing here are some of the examples for which several theoretical results are available and where further scattering calculations have attempted a quantitative assessment of their reliability.

1. $Li^+ - H_2$: Among non reactive systems, one for which the potential energy surface has been carefully studied is the $Li^+ + H_2$ system. It has also been the subject of several experimental studies of collisional vibro-rotational excitations[28]. The first ab initio study was done within the single-configuration HF approximation and using a flexible basis set of better than double zeta plus polarization quality[29,30].

One of the interesting features of the surface is the presence of an attractive well which, as shown in Fig. 1, is most pronounced for the Li^+ approach in the C_{2v} geometry. This aspect is also brought out when a comparison is made with the $Li^+ - N_2$ system, where SCF calculations[31] exhibit the opposite anisotropy. In this latter instance, in fact, the attractive well is the strongest for collinear $Li^+ - N_2$ geometries and the weakest for the C_{2v} approach. The charge-quadrupole anisotropic interaction turns out to be the main cause for this effect, since the molecular quadrupole moment of N_2 is negative, while that of H_2 is positive and smaller in magnitude.

As the H_2 internuclear separation is increased, the well becomes correspondingly deeper, going from a depth value of $\sim$ 0.24 eV when H_2 is fixed near its equilibrium geometry to a value of $\sim$ 0.35 eV for the H_2 bond distance of 2.0 a.u. These features are rather important when discussing vibrationally inelastic collisions and the bearing of concurrent rotational excitations on the corresponding total, integral, and partial differential cross sections for the above processes. Further studies on the effect of correlation energy correction (IEPA) to the PES calculations at the previous SCF level were carried out by Kutzelnigg and coll.[32]. In this strongly interacting system, it was found by them that the SCF and IEPA surfaces are qualitatively very similar, thus providing corresponding dynamical features that turned out to be largely independent of which of the two surfaces was used[33] in studying roto-vibrational inelasticity for this system.

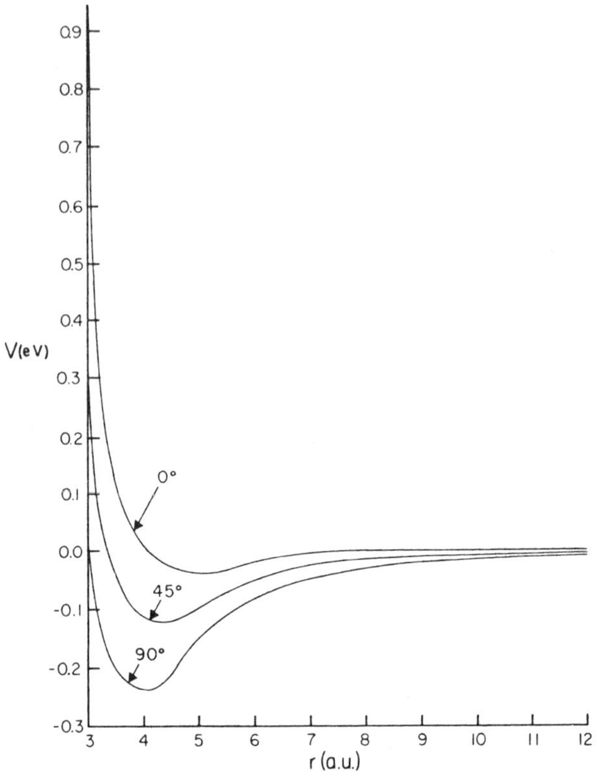

Fig. 1 $Li^+ - H_2$ potential energy surface[29] for an H_2 internuclear distance of 1.4 a.u. 0° and 90° correspond to the collinear and C_{2v} approaches.

2. $H+H_2$: This has been the simplest and oldest of the studied systems for which non-reactive and reactive collisions have been examined via some sort of computed potential surface. A recent review[34] analyzes the many interesting features of this atom-molecule system, while very recent calculations[35] appear to be of nearly quantitative accuracy. Previous SCF-CI calculations (SSMK)[36] and semiempirical calculations (PK)[37] suggested the linear transition state predicted by the classical London-Eyring-Polanyi model but substantially differed in the value of the barrier height predicted by their methods. Figure 2 presents the energy profiles generated by the above calculations and shows how the more accurate, recent results[35] are intermediate in value between the former .

3. $H^+ + H_2$: This system constitutes another outstanding example of how successive improvements of the quality of the computed PES are capable of clarifying the found discrepancies between dynamical experiments and theoretical calculations. Earlier CI calculations over various geometries were carried out[38] to study low energy $H^+ - H_2$ reactions and the minimum of the equilibrium equilateral geometry was found by them to be - 1.3373 a.u. Further calculations[39] with larger basis set and a complete CI approach found a minimum of - 1.342284 a.u. More recent, extensive CI calculations on the same system[40] provided the entire PES for rotational and vibrational inelastic collisions and found a total minimum which was ~ 1.8 kcal/mol below the earlier Czismadia results[38], but ~ 1.6 kcal/mol higher than the complete CI of ref. 39. On the other hand, they showed how important it is the role that correlation corrections play at short intermolecular distances, even with systems that correctly separate into the true fragments within the HF model as does the present one.

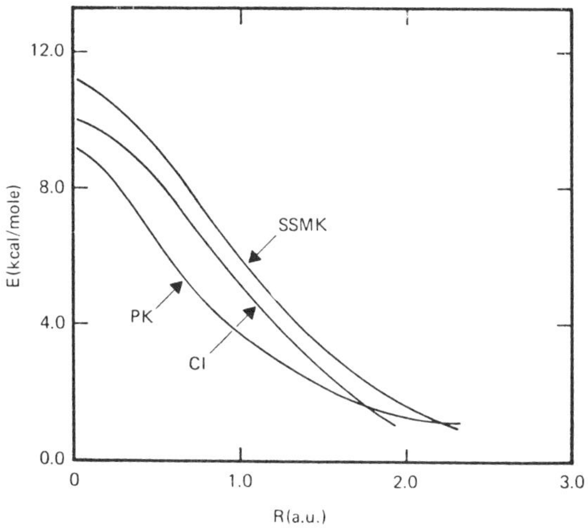

Fig. 2: Energy profiles along the collinear reaction paths for the $H + H_2$ system. The references to the various curves are: SSMK[36], PK[37] and CI[35].

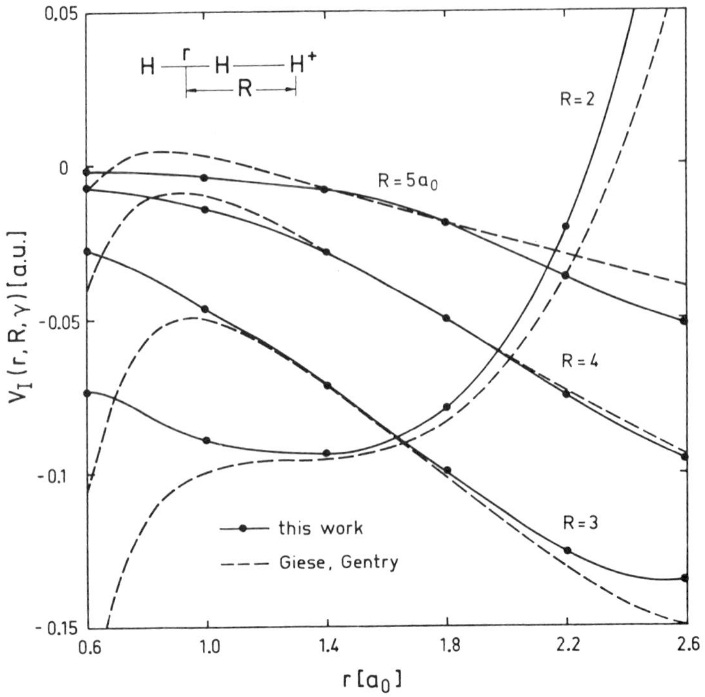

Fig. 3: Collinear geometry of the H^+ - H_2 system as a function of H_2 internuclear distance and of the intermolecular coordinate R. The continuous line is from ref. 40, while the dashed curves are from ref. 41.

This is demonstrated by the curves of Fig. 3, where the collinear approach of the proton is studied as a function of the H_2 internuclear distance and for intermolecular geometries relevant for vibrationally inelastic collisions. The dashed line is produced by fitting the calculations of ref. 38, while the continuous line refers to the recent calculations[40] with improved CI and more flexible basis for the whole potential energy surface that was mapped out over a much larger range of geometries than previously attempted.

3. THE ROTOVIBRATIONAL PROBLEM

3.1 Quantal Treatment of Rotational Excitations

Once the problem of reliably knowing the intermolecular potential of eq. (2.24) has been solved in some way for the system of interest, the quantum mechanical treatment of the dynamics can now be approached by solving, in the coordinate representation, the time-independent Schrödinger equation containing the full Hamiltonian of (2.24):

$$\mathcal{H}\chi_A(\underline{R},\underline{\rho}) = E_{TOT}\,\chi_A(\underline{R},\underline{\rho}). \tag{3.1}$$

Here A labels the totally-symmetric, ground state electronic PES of the partners involved, which are here assumed for simplicity to be closed-shell interacting species. If we consider two rigid molecules colliding with each other, then the corresponding Hamiltonian would be:

$$\mathcal{H} = \mathcal{H}_0 + V =$$

$$= -\frac{1}{2\mu}\nabla_{\underline{R}}^2 + \mathcal{H}_1 + \mathcal{H}_2 + V(\underline{R},\hat{\rho}_1,\hat{\rho}_2) \tag{3.2}$$

where the unit vectors, $\hat{\rho}_1$ and $\hat{\rho}_2$, define the orientations of the molecular subsystems with respect to some reference frame that shall be defined later. The $\mathcal{H}_1$ and $\mathcal{H}_2$ Hamiltonians describe the internal structures of separated partners, i.e. their rotational eigenvalues and eigenfunctions when $V \rightarrow 0$.

Several possible coupling schemes have been suggested when choosing the most suitable functional expansion for the unknown wavefunction $\chi_A(\underline{R},\underline{\rho})$, although the most popular one essentially relies on the limiting form that assumes the potential V to be a generally small perturbation on the internal structures of the partners and one acting only in limited regions of the $\{R\}$ space. This assumption therefore suggests the following (asymptotic) expansion in the eigenfunctions of the independent internal Hamiltonian, i.e. one writes first:

$$\mathcal{H}_i(\underline{\rho}_i)\psi_j^i(\underline{\rho}_i) = \varepsilon_j^i\,\psi_j^i(\underline{\rho}_i) \tag{3.3}$$

$$i = 1, 2, \ldots, k$$

and then

$$\chi_A(\underline{R}, \underline{\rho}) = \sum_j u_j(\underline{R})\, \psi_j^1(\underline{\rho}_1) \ldots \psi_j^k(\underline{\rho}_k) = \sum_j \alpha_j(\underline{R}, \underline{\rho}_1, \ldots, \underline{\rho}_k) \tag{3.4}$$

If, to simplify notation, we take the case of k = 1, then after substitution of (3.4) into (3.1), multiplication by $\psi_i^1(\underline{\rho}_1)^*$ and integration over $\underline{\rho}_1$, one obtains the following set of coupled equations:

$$\left[\nabla_R^2 + k_i^2\right] u_i(\underline{R}) = \sum_j U_{ij}(\underline{R})\, u_j(\underline{R}) \tag{3.5}$$

where:

$$k_i^2 = 2\mu\left(E_{TOT} - \varepsilon_i^1\right) \tag{3.6 a}$$

$$U_{ij}(\underline{R}) = 2\mu \int d\underline{\rho}_1\, \psi_i^{1*}(\underline{\rho}_1)\, V(\underline{R}, \underline{\rho}_1)\, \psi_j^1(\underline{\rho}_1) \tag{3.6 b}$$

The diagonal terms $U_{ii}(\underline{R})$ directly contribute to the elastic scattering when one of the partners is prepared in the ith internal state of (3.3). Moreover, they also affect the inelastic cross sections by distorting the various $u_j(R)$ from their outgoing plane wave forms. The off-diagonal terms $U_{ij}(R)$ which couple together the different channels, are responsible in this scheme for any inelastic process that occurs and for polarization contributions to the elastic scattering channels.

For systems where a strong interaction is present and where the electronic potential greatly distorts the structure of the colliding species over rather large regions of the $\{R\}$ space, eq. (3.4) should be replaced by a more effective expansion over a set of adiabatic internal states $\varphi_K^i(\underline{R}, \underline{\rho}_i)$ that are eigenfunctions of the full Hamiltonian:

$$\left[\mathcal{H}_i(\underline{\rho}_i) + V(\underline{R}, \underline{\rho}_i)\right] \varphi_K^i(\underline{R}, \underline{\rho}_i) = \eta_K^i(R)\, \varphi_K^i(\underline{R}, \underline{\rho}_i) \tag{3.7}$$

The expansion (3.4) can then be rewritten as:

$$\chi_A(\underline{R}, \underline{\rho}) = \sum_K \tilde{u}_K(\underline{R})\, \varphi_K^1(\underline{R}, \underline{\rho}_1) \ldots \varphi_K^n(\underline{R}\, \underline{\rho}_n) \tag{3.8}$$

Its substitution into eq. (3.1) yields now the following set of equations for the $\tilde{u}_K(\underline{R})$ unknown functions (for n = 1 in eq. (3.8)):

$$\left[\nabla^2_{\underline{R}} + k^2_K(\underline{R})\right]\tilde{u}_K(\underline{R}) = \sum_{\ell}\left[A_{K\ell}(\underline{R})\cdot\nabla_{\underline{R}} + B_{K\ell}(\underline{R})\right]\tilde{u}_{\ell}(\underline{R}) \tag{3.9}$$

where

$$K^2_K(\underline{R}) = 2\mu\left[E_{TOT} - \eta^1_K(\underline{R})\right] \tag{3.10}$$

and:

$$A_{K\ell}(\underline{R}) = -2\int \varphi^{1*}_K(\underline{R},\underline{\rho}_1)\nabla_{\underline{R}}\varphi^1_{\ell}(\underline{R},\underline{\rho}_1)\,d\underline{\rho}_1 \tag{3.11 a}$$

$$B_{K\ell}(\underline{R}) = -\int \varphi^{1*}_K(\underline{R},\underline{\rho}_1)\nabla^2_{\underline{R}}\varphi^1_{\ell}(\underline{R},\underline{\rho}_1)\,d\underline{\rho}_1 \tag{3.11 b}$$

In full analogy to the previous discussion of section 2.1, the non-adiabatic effects embodied in the integrals of (3.11 a) and (3.11 b) may arise from both the radial and angular operators, although the radial contributions are normally the most important for molecular systems[42]. Moreover, the higher order corrections in a perturbative sense look very similar to the terms appearing as corrections to the B.O. Hamiltonian of eq. (2.17). This means that, under semiclassical conditions, one can say that the adiabatic expansion becomes valid when the non-adiabatic corrections are small, i.e. when the ratio of the local energy difference between adiabatic molecular levels with the radial coupling times the local velocity $\underline{v}$ becomes large:

$$\frac{\eta_K(\underline{R}) - \eta_\ell(\underline{R})}{\langle\varphi_K|\frac{d}{dR}|\varphi_\ell\rangle\cdot\underline{v}} \gg 1 \tag{3.12}$$

Such a description is therefore favoured by large adiabatic energy differences and low collision velocities. Comparisons between these approaches have recently been carried out for rotational excitations[43], for vibrational excitation in approximate treatments[44]. The adiabatic equations (3.37) were applied a while ago to collinear collisions via model potential[45] and very recently recast in a new and very promising computational form[46].

3.2 The Coupled Representation in a LAB Frame

The procedures most generally used in the recent literature, however, rely on the asymptotic expansion of eq. (3.4) and concentrate on the study of what would be the most efficient basis set to represent the unknown continuum functions required in that expansion.

To treat the latter problem, it is convenient to define two different coordinate systems and, in doing so, we shall here confine the detailed analysis to molecular targets with a vanishing component of their electronic angular momentum along the main axis and to structureless atoms in their closed-shell, ground electronic state. This is the simplest case that one can examine and, as we shall see later, the relevant theory can be easily extended to more complex cases.

As we have seen in the previous section, the existence of a non-spherical interaction poses the problem of angular momentum coupling. The total angular momentum, and its component about a chosen axis of quantization, are of course strictly conserved but angular momentum (torque) can be exchanged between internal and translational motions. The two most popular choices for the basis sets that can be used in eq. (3.4) originate directly from the two different, albeit equivalent, descriptions proposed by the above physical situation.

The first employs a fixed quantization axis along the direction of incident motion and hence chooses a Space Fixed (SF or Lab) coordinate system. It is conceptually simple and has been extensively discussed in the recent literature. The second allows the quantization axis to rotate with the interparticle vector R and chooses a Body Fixed (BF or Body) system of reference, the relative vector being on the OZ axis. The set of equations obtained by using the latter, BF reference frame exhibits of course the same dimensionality as the set of SF equations but its structure is different. The way in which approximate procedures can reduce dimensionality is therefore shown differently according to which representation one chooses. This last point will become clearer in the following section.

Even when only rotational degrees of freedom are considered, the internal functions $\Psi_j^i(\hat{\rho}_i)$ appearing in eq. (3.4) are in general linear combinations of rotation matrices[47] $\mathcal{D}^j_{mk}(\Omega_i)$. If the molecule is a linear rigid rotor, the κ quantum number will be zero and the rotation matrix degenerates into a spherical harmonic $Y^j_m(\Omega_i)$. For spherical top and symmetric top molecules, the simple rotation matrices are eigenfunctions of the rigid-rotor Hamiltonian. For asymmetric tops, the eigenfunctions are in general linear combinations of the rotation matrices.

If the molecular target is a polyatomic molecule that therefore transforms with a specific point group representation, then linear combinations of the rotation matrices that transform like the irreducible representations of the point group symmetry are better suited as basis functions. For example, for the case of the methane molecule,

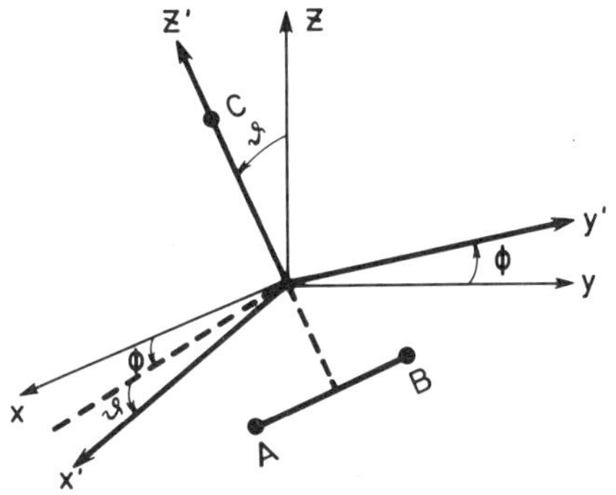

Fig. 4: SF and BF reference frames for the atom-diatomic molecule case. The primed coordinates define the Body frame of reference, while the unprimed coordinates are for the Lab frame.

the tetrahedral symmetry of the spherical top requires that the symmetry-adapted basis functions transform with the fully symmetric or A representation of that point group[48]. The best basis functions are therefore easily found by projection of an arbitrary rotation matrix $\mathcal{D}^j_{mk}(\Omega_i)$ into the A representation and further normalization.

Since the total Hamiltonian must be invariant under the molecular rotation group operations (disregarding the nuclear spin operators as yielding only small corrections), then the interaction potential cannot couple states corresponding to different representations of the relevant group. This is one of the examples of block-diagonalization that occurs in the coupled equations arising from expansion (3.4), since each block can be now independently solved.

In the simplest example mentioned above, of an atom colliding with a linear rigid rotor, the two reference frames are depicted in Fig. 4, where a rotation of Euler angles ($\vartheta, \phi, 0$) brings the (OXYZ) frame into the (OX'Y'Z') frame. In this case, the coupled representation usually chosen in an SF description of the expansion functions[49,50] starts from a set, each member of which represents a state with a definite value of the total angular momentum J and of its Z component. The analysis therefore becomes similar to that use to describe nuclear reactions[51].

One defines at the start the operator $\hat{J}$:

$$\hat{J} = \hat{L} + \hat{j} \tag{3.13}$$

where $\hat{L}$ is the angular momentum operator for the kinetic energy term in the Hamiltonian of (3.2) in polar coordinates and $\hat{j}$ the rotational angular momentum operator for the linear rigid rotor. Since, for an isolated system:

$$[\mathcal{H}, \hat{J}] = 0 , \tag{3.14}$$

then a set of compatible observables is given by $\mathcal{H}, \hat{J}$ and $\hat{J}_z$, the eigenvectors of which satisfy the equation:

$$\mathcal{H} | EJM \rangle = E | EJM \rangle \tag{3.15}$$

or, in the coordinate representation:

$$\langle \underline{R} , \hat{r} | EJM \rangle = \varphi^{E}_{JM} (\underline{R} , \hat{r}) \tag{3.16}$$

where $\hat{r}$ is now the internal coordinate of the diatomic target in the SF frame of Fig. 4.

Since the potential is non-spherical, neither ℓ nor j, eigenvalues of the operators on the r.h.s. of eq. (3.13), are good quantum numbers. One can, however, choose a set of functions where j and l are good quantum numbers in the asymptotic situation, when $V \rightarrow 0$. The projection of such a set onto the SF basis in the coordinate representation can be written as:

$$\langle \hat{r}, \hat{R} | j \ell J M \rangle = \mathcal{Y}^{JM}_{j\ell} (\hat{r}, \hat{R}) \tag{3.17}$$

where the total energy E is now taken as a constant for the isolated system. The required expansion functions of eq. (3.4) can then be written as:

$$\alpha_j(\underline{R}, \underline{\rho}_1, \cdots \underline{\rho}_K) \Rightarrow \Psi_{j\ell}^{EJM}(\underline{R}, \hat{r}) =$$

$$= \sum_{j',\ell'} u_{j'\ell'}^{Jj\ell}(R) \frac{1}{R} \mathcal{Y}_{j'\ell'}^{JM}(\hat{r}, \hat{R}) \qquad (3.18)$$

where a further summation over j' and l' has been introduced, thus preparing the representation for the effect of the anisotropic potential, which allows transitions from the initial asymptotic state $|j\ l\rangle$ to the final asymptotic state $|j'\ l'\rangle$. The above functions form a complete set in the Hilbert space and one can define a set of operators in that space which all commute with one another. Since the total Hamiltonian is invariant under rotation, the eigenfunctions of the total angular momentum are not coupled with each other via the potential. Another form of block-diagonalization has therefore been performed by making the choice of expansion (3.17). We still have, however, block dimensions controlled by the triangular relation between coupled angular momenta in the coefficients $\mathcal{Y}$'s: $|j-l| \leq J \leq j+\ell$
More specifically, one has that:

$$\mathcal{Y}_{j\ell}^{JM}(\hat{R}, \hat{r}) =$$

$$= \sum_{m_j m_\ell} C(j\ell J; m_j m_\ell M)\, Y_{m_j}^{j}(\hat{r})\, Y_{m_\ell}^{\ell}(\hat{R}) \qquad (3.19)$$

where the C's are Clebsch-Gordan coefficients[52].

For each $|j\ l\rangle$ initial conditions one can choose an expansion set set in eq. (3.18) that can be used in (3.1). By then pre-multiplying by the complex conjugate of $\mathcal{Y}_{j'\ell'}^{JM}$ and integrating over the angular variables, one obtains the familiar close-coupled (CC) equations:

$$\left[\frac{d^2}{dR^2} + K_{jj'}^2 - \frac{\ell'(\ell'+1)}{R^2}\right] u_{j'\ell'}^{Jj\ell}(R) =$$

$$= \sum_{j''\ell''} \langle j'\ell'|V|j''\ell''\rangle\, u_{j''\ell''}^{Jj\ell}(R) \qquad (3.20)$$

or, in matrix form:

$$\left[\mathbb{1}\frac{d^2}{dR^2} + \mathbb{K}^2 - \frac{\mathbb{L}^2}{R^2} - \mathbb{V}^J(R)\right] \mathbb{U}^J(R) = 0 \qquad (3.21)$$

where

$$(\mathbb{K})_{j'\ell',j''\ell''}^{j\ell} = \left[2\mu(E_j - \varepsilon_{j'})\right]^{1/2} \delta_{j'j''}\, \delta_{\ell'\ell''} \qquad (3.22\ a)$$

$$(\mathbb{L})_{j'\ell' j''\ell''}^{\ell} = \left[\ell'(\ell'+1)\right]^{1/2} \delta_{j'j''}\, \delta_{\ell'\ell''} \qquad (3.22\ b)$$

$$E_j = E - \varepsilon_j \tag{3.22 c}$$

$$\left(\mathbb{V}^J(R)\right)_{j'\ell',j''\ell''} = 2\mu \int d\hat{R}\, d\hat{r}\, \mathcal{Y}^{JM^*}_{j'\ell'}\, V(\underline{R},\hat{r})\, \mathcal{Y}^{JM}_{j''\ell''} \tag{3.22 d}$$

In this representation one therefore sees inelastic transitions taking place as caused by the spatial range of the electrostatic potential, which in turn controls the extent of distorsion due to the various terms on the r.h.s. of (3.20) as the solution propagates to its asymptotic ($R \rightarrow \infty$) form, and from the non-spherical nature of the same potential that in turn couples different components of the angular eigenfunctions $\mathcal{Y}$'s.

For the simple atom-diatom case, one can expand the PES in the following set of orthonormal functions:

$$V(\underline{R},\hat{r}) = \sum_\lambda V_\lambda(R)\, P_\lambda(\hat{R}\cdot\hat{r}) \tag{3.23}$$

hence the coupling matrix elements of (3.22 d) become:

$$\left(\mathbb{V}^J(R)\right)_{j'\ell',j''\ell''} = 2\mu \sum_\lambda f_\lambda(j'\ell',j''\ell'';J)\, V_\lambda(R) \tag{3.24}$$

where the f_λ -coefficients are known as Percival-Seaton coefficients and are given in terms of 3-j and 6-j coupling terms[52]:

$$f_\lambda(j\ell,j'\ell';J) = \int d\hat{R}\, d\hat{r}\, \mathcal{Y}^{JM^*}_{j\ell}\, P_\lambda\, \mathcal{Y}^{JM}_{j'\ell'} \tag{3.25}$$

From the structure of the angular functions in eq. (3.19), the coefficients of (3.24), for each λ value, will be non-vanishing when the usual triangular relation is satisfied by $\{j, j', \lambda\}$, hence rotational transitions will depend on the strength of the anisotropic components of the expansion (3.23). Moreover, the coupling is now independent of M, hence requires solutions for only one of the $(2J + 1)$ components.

It is clear from the above analysis that the properties of each component in eq. (3.23) are ultimately controlling the dynamics within the representation described above. This therefore means that the increase in complexity of the colliding partners directly appears in the expansion of their corresponding interaction.

Most of the calculations and discussions that have appeared in the literature, however, have essentially dealt with the simple case described by (3.23), with only a few studies referring to diatom-diatom collisions[53-55]. An example of atom-symmetric top was given by the model calculations on He-NH_3[56], while the atom-spherical top case has been examined for He-CH_4 collisions[57].

The most general form of the expansion for the PES between two non-linear, molecular partners can thus be obtained in terms of rotation matrices[58]:

$$V(\underline{R},\Omega_1,\Omega_2) = \sum_{j_1 j_2 \ell m_1 m_2} V_{j_1 j_2 \ell}^{m_1 m_2}(R) \times$$
$$\times \sum_{n_1 n_2 n} C(j_1 j_2 \ell; n_1 n_2 n) \times \mathcal{D}_{n_1 m_1}^{j_1}(\Omega_1) \times \tag{3.26}$$
$$\times \mathcal{D}_{n_2 m_2}^{j_2}(\Omega_2)\, Y_\ell^n(\hat{R})$$

where the molecular orientations and the collision coordinate are all referred to the SF reference frame.

If one of the molecules belongs to a symmetric point group ($\neq C_s$) clearly the rotation matrices will be replaced by linear combinations that block-diagonalize the coupling according to the various irreducible representations that play a role in the collisional event.

When one of the colliding partners is a linear molecule, then one of the rotation matrices for, say, the Ω_1 coordinate is replaced by a spherical harmonic $Y_{m_1 j_1}^{j_1}(\hat{r}_1)$ in eq. (3.26). In the even simpler situation of it becoming a structureless atom (assuming no unpaired electrons in its electronic state or neglecting spin-orbit terms in the full Hamiltonian) the j_1 associated with it will become zero and the coupling coefficient would disappear:

$$V(\underline{R},\Omega_2) = \sum_{\ell m n} V_{\ell m}(R)\, \mathcal{D}_{n m}^{\ell *}(\Omega_2)\, Y_n^\ell(\hat{R}) \tag{3.27}$$

The spherical harmonics phase convention plus the requirement of having a real, local potential imply that:

$$V_{\ell,-m}(R) = (-)^m\, V_{\ell,m}^*(R) \tag{3.28}$$

A further reduction takes place when the other scattering partner is a diatomic molecule, thus recovering the special expansion form of (3.23)

Since the intermolecular potential calculations are nearly always performed in a body-fixed reference frame with the Z' axis being along the collision coordinate R', the SF expansion (3.26) can be rewritten in the BF frame as:

$$V(R,\Omega_1',\Omega_2') = \sum_{j_1 j_2 \ell m_1 m_2} V_{j_1 j_2 \ell}^{m_1 m_2}(R) \sum_{n_1 n_2} C(j_1 j_2 \ell; n_1 n_2 0) \times \mathcal{D}_{n_1 m_1}^{j_1}(\Omega_1') \mathcal{D}_{n_2 m_2}^{j_2}(\Omega_2') \cdot \left[\frac{2\ell+1}{4\pi}\right]^{1/2} \tag{3.29}$$

From the properties of the C's one sees that $n_1 = - n_2$. Moreover, only relative rotations of the two molecules about Z' change the anisotropy of the potential. Thus, if $\Omega_1' \equiv (\alpha_1', \beta_1', \gamma_1')$ and $\Omega_2' \equiv (\alpha_2', \beta_2', \gamma_2')$ are the BF Euler angles, only rotations of $(\alpha_1' - \alpha_2')$ contribute to the potential.

In a similar way, the atom-molecule case of eq. (3.27) becomes:

$$V(R,\Omega_2') = \sum_{\ell, m} V_{\ell m}(R)\, Y_m^{\ell}(\beta_2', \gamma_2') \tag{3.30}$$

whereby one sees that the rotation around the BF collision coordinate has no effect on the potential. The last case, of an atom-linear molecule system, reduces the spherical harmonics of (3.30) to Legendre polynomials, hence the potential form in the BF frame becomes the now familiar expansion of (3.23), which remains the same in either frames of reference.

If we now go back to the coupled equations related to the atom-diatomic system shown in eq. (3.20), the required boundary conditions on each of the unknown radial functions are given by:

$$u_{j'\ell'}^{Jj\ell}(R) \underset{R\to 0}{\sim} 0$$

$$u_{j'\ell'}^{Jj\ell}(R) \underset{R\to\infty}{\sim} K_{jj'}^{-1/2} \left\{ \delta_{jj'} \cdot \delta_{\ell\ell'} \exp\left[-i\left(K_{jj} R - \tfrac{1}{2}\ell\pi\right)\right] - S_{j'\ell', j\ell}^{J}(E) \exp\left[i\left(K_{jj'} R - \tfrac{1}{2}\ell'\pi\right)\right] \right\} \tag{3.31}$$

The total wavefunction (3.4) can be obviously written in terms of the expansion (3.18) for the present example of one internal coordinate only, the $\hat{r}$ orientation:

$$\chi_A(\underline{R}, \hat{r}) = \sum_{J,M} \sum_{j\ell} B_{j\ell}^{JM} \Psi_{j\ell}^{EJM}(\underline{R}, \hat{r}) \tag{3.32}$$

which will then have the following asymptotic form:

$$\chi_A \underset{R\to\infty}{\sim} R^{-1} \sum_{JM} \sum_{jj'} \sum_{\ell\ell'} K_{jj'}^{-1/2} B_{j\ell}^{JM} \{\delta_{jj'}\delta_{\ell\ell'} \times$$
$$\times \exp\left[-i\left(K_{jj}R - \tfrac{1}{2}\ell\pi\right)\right] - S_{jj'\ell\ell'}^{J}(E) \times$$
$$\times \exp\left[i\left(K_{jj'}R - \tfrac{1}{2}\ell'\pi\right)\right]\} \mathcal{Y}_{j\ell}^{JM}(\hat{R}, \hat{\rho}) \tag{3.33}$$

The incident plane wave in the jth channel is: $\exp[i\underline{K}_{\alpha\alpha'} \cdot \underline{R}] \cdot Y_{\mu_j}^{j}(\hat{\rho})$ with propagation vector $\underline{K}_{\alpha\alpha'}$. When the latter is chosen as to define the Z axis in the SF frame of Fig. 4, then:

$$\exp[iK_{\alpha\alpha} Z]\, Y_{\mu_j}^{j}(\hat{\rho}) =$$
$$= \sqrt{4\pi} \sum_{\ell} i^{\ell} \sqrt{2\ell+1}\, J_{\ell}(K_{\alpha\alpha}R) Y_0^{\ell}(\hat{R}) \times Y_{\mu_j}^{j}(\hat{\rho}) \tag{3.34}$$

Using now eq. (3.19), the above expression exhibits the following asymptotic behaviour:

$$\exp[iK_{\alpha\alpha} Z]\, Y_{\mu_j}^{j}(\hat{\rho}) \underset{R\to\infty}{\sim} -\frac{\sqrt{\pi}}{K_{\alpha\alpha}R} \sum_{JM\ell} i^{\ell+1}(2\ell+1)^{1/2} \times$$
$$\times C(j\ell J; \mu_j 0 M) \times \mathcal{Y}_{j\ell}^{JM}(\hat{R}, \hat{\rho}) \times$$
$$\times \left\{\exp\left[i\left(K_{\alpha\alpha}R - \tfrac{1}{2}\ell\pi\right)\right] - \exp\left[-i\left(K_{\alpha\alpha}R - \tfrac{1}{2}\ell\pi\right)\right]\right\} \tag{3.35}$$

After the scattering event, the total wavefunction should behave as $\chi_A \underset{R\to\infty}{\sim} \chi_A^{inc} + \chi_A^{scatt}$ and one can therefore determine the scattered part by equating the coefficients of $e^{-iK_{\alpha\alpha}R}$ from eq. (3.33) and eq. (3.34). This yields the following result:

$$\sum_{JMjj'\ell\ell'} K_{jj'}^{-1/2} B_{j\ell}^{JM} \cdot \delta_{jj'} \cdot \delta_{\ell\ell'} \mathcal{Y}_{j'\ell'}^{JM} \exp\left(\tfrac{1}{2} i\ell\pi\right) = \tag{3.36}$$

$$= \frac{\sqrt{\pi}}{k_{jj}} \sum_{JM\ell} i^{\ell+1} (2\ell+1)^{1/2} C(j\ell J; m_j 0 M) \mathcal{Y}_{j\ell}^{JM} \exp\left(\tfrac{1}{2} i\ell\pi\right)$$

since the $\mathcal{Y}$'s are an orthonormal set of functions, one can obtain the coefficients of the expansion (3.32) by simple closure relations on integrating (3.36):

$$B_{j\ell}^{JM} = \begin{cases} i^{\ell+1} C(j\ell J; m_j 0 M) k_{jj}^{-1} [(2\ell+1)\pi]^{1/2} & \text{in the } |j\ell\rangle \text{ channel} \\ 0 & \text{in the other channels} \end{cases} \tag{3.37}$$

The scattered part of the total wavefunction can thus be obtained, in the asymptotic region, by making use of eq. (3.35) and (3.37) in eq. (3.33):

$$\chi_A^{SCATT} \underset{R\to\infty}{\sim} \sum_{j'} \left(\frac{k_{jj}}{k_{jj'}}\right)^{1/2} R^{-1} \exp(ik_{jj'}R) \times \\ \times \sum_{m_{j'}} q(j'm_{j'}; jm_j|\hat{R}) Y_{m_{j'}}^{j'}(\hat{\rho}) \tag{3.38}$$

which contains the essential new quantity, i.e. the scattering amplitude for the $|jm_j\rangle \to |j'm_{j'}\rangle$ process:

$$q(j'm_{j'}; jm_j|\hat{R}) = \sum_{JM\ell\ell'jj'} i^{\ell-\ell'} [\pi(2\ell+1)]^{1/2} \times \\ \times C(j\ell J, m_j 0 M) \times C(j'\ell' J\, m_{j'} m_{\ell'} M) \times T^J_{j'\ell' j\ell} Y^{\ell'}_{m_{\ell'}}(\hat{R}) \tag{3.39}$$

where the T's are elements of the reactance matrix that is related to the S-matrix by the well-known relation:

$$\mathbb{T} = \mathbb{1} - \mathbb{S} \tag{3.40}$$

The state-to-state differential cross section (DCS), for each chosen collision energy E, is thus given by:

$$\frac{d\sigma}{d\hat{R}}(jm_j \to j'm_{j'}|E) = k_{jj}^{-2} |q(j'm_{j'}, jm_j|\hat{R})|^2 \tag{3.41}$$

When one averages over the m_j components of the initial state of the spherical rotor and sums over the final $m_{j'}$ components, one obtains a degeneracy-averaged DCS for the above process:

$$\frac{d\sigma}{d\hat{R}}(j \to j'|E) = \frac{(-)^{j'-j}}{4(2j+1)k_{jj}^2} \sum_L A_L \cdot P_L(\cos\vartheta) \tag{3.42}$$

where:

$$A_L = \sum_{J_1 J_2 \ell_1 \ell_2 \ell'_1 \ell'_2} Z(\ell_1 J_1 \ell_2 J_2; j L) \cdot Z(\ell'_1 J_1 \ell'_2 J_2 \, j' L) \times$$
$$\times T^{*J_1}_{j'\ell'_1 j \ell_1} \, T^{J_2}_{j'\ell'_2 j \ell_2} \tag{3.43}$$

and:

$$Z(a,b,c,d;e,f) = (-)^{(f-a+c)/2} \times$$
$$\times \left[(2a+1)(2b+1)(2c+1)(2d+1)\right]^{1/2} \times C(a,c,f;000) \times$$
$$\times W(a,b,c,d;e,f) \tag{3.44}$$

In eq. (3.44) the W's are Racah coefficients[51]. At this stage it thus becomes easy to obtain the partial, integral cross section of the rotationally inelastic process:

$$\sigma(j \to j'|E) = \int d\hat{R} \, \frac{d\sigma}{d\hat{R}}(j \to j'|E) = \frac{(-)^{j-j'} A_0}{4(2j+1)k_{jj}^2} =$$
$$= \frac{\pi}{(2j+1)k_{jj}^2} \sum_{J\ell\ell'} (2J+1) \left| T^J_{j'\ell' j\ell} \right|^2 \tag{3.45}$$

The following microscopic reversibility relation originates from the unitarity of the S matrix in (3.40):

$$\sigma(j \to j'|E) = k_{jj}^{-2} \, k_{j'j'}^{2} \, (2j+1)^{-1} (2j'+1) \, \sigma(j' \to j \,|\, E) \tag{3.46}$$

The determination of the T^J, or of the S^J, matrix elements is therefore the basic computational problem, since all collisional observables are ultimately related to it. The full rotational character of the physical problem is contained in the coupling matrix elements of eq. (3.22 d), which therefore change their structure according to the nature of the rotational structures of the involved partners. The latter will affect in turn the blockdiagonalizations performed by the vector coupling schemes used to define the angular basis set in the wavefunction expansion of eq. (3.18).

It is also important to realize that sums over J and averages over scattering angles, energy, rotational states, etc. are very often necessary to compare computed quantities with experimental observables. Moreover, when only bulk relaxation data are available, a Boltzmann velocity average of the cross sections and proper consideration of the kinetic master equation is often required.

3.3 The Body-Fixed Form of the Scattering Equations.

In the SF frame the initial relative velocity vector was used to define the Z-axis orientation, thus introducing a simplification of the problem in that one could have an initial representation without any angular momentum of relative motion about the Z-axis, i.e. with $m_l = 0$. The final relative velocity vector, however, lies along the R direction and therefore its component along the Z-axis has to be referred to the original, fixed SF reference system, thus introducing in the coupled representation of the total wavefunction a set of eigenfunctions of $\hat{J}^2$ and $\hat{J}_z$ as in eq. (3.16).

In the helicity representation, however, the problem is treated in a symmetric manner, in that the scattered wave is described in a BF, rotating coordinate system whose Z'-axis lies along the $\hat{R}$ direction and follows the target during interaction. In this reference frame the orbital angular momentum, $\hat{L} = \underline{R} \times \underline{P}$ has a vanishing projection along $\hat{Z}'$.

In the SF representation the asymptotic quantum numbers j, l appeared in the coupled expansion (3.18) by further requiring the basis set to be formed of eigenfunctions of $\hat{j}^2$ and $\hat{l}^2$ chosen at $R \sim \infty$, as indicated by eq. (3.19). Since the helicity functions defined in the BF frame relate to the above functions via the transformation[47]:

$$\varphi^E_{JM}(\underline{R}, \hat{r}) = \sum_{\Omega} \mathcal{D}^J_{\Omega M}(\mathcal{R})\, \tilde{\varphi}^E_{J\Omega}(\underline{R}', \hat{r}') \qquad (3.47)$$

one sees that the rotation matrices are operated upon by the operator $\hat{l}^2$, therefore the new φ' s are no longer angular momentum eigenfunctions. One therefore speaks of scattering from a state of helicity Ω about the initial direction to a state of helicity Ω' about the final direction.

The main mathematical consequence of this transformation is to remove terms with $m_l \neq 0$ in the coupling matrix elements, since the

BF axis are chosen to be coincident with the vectors of $\underline{R}$ in the SF frame and therefore there can be no orbital angular momentum about $\underline{R}$.

The corresponding expansion (3.32) can now be rewritten in terms of new eigenfunctions, this time of $\hat{J}^2$, $\hat{J}_z$, $\hat{j}^2$ and $\hat{j}_z$:

$$\chi_A(\underline{R}', \hat{r}') = \sum_{JM} \sum_{j\Omega} C^{JM}_{j\Omega} \tilde{\Psi}^{EJM}_{j\Omega} (\underline{R}' \hat{r}') \tag{3.48}$$

where, in analogy to eq. (3.18):

$$\tilde{\Psi}^{EJM}_{j\Omega}(\underline{R}', \hat{r}') = \sum_{j'\Omega'} g^{Jj\Omega}_{j'\Omega'}(R) \frac{1}{R} \mathcal{Y}^{JM}_{j'\Omega'} (\hat{R}', \hat{r}') \tag{3.49}$$

The new BF angular eigenfunctions are given by:

$$\mathcal{Y}^{JM}_{j\Omega}(\hat{r}', \hat{R}') = \left[\frac{(2J+1)}{4\pi}\right]^{1/2} \mathcal{D}^{J}_{\Omega M}(\phi, \vartheta, 0) Y^{j}_{\Omega}(\hat{r}') \tag{3.50}$$

where the rotation matrices bring the internuclear coordinate unit vector onto the BF reference frame. No coupling coefficients between the $\underline{j}$ and $\underline{l}$ vectors are needed here, unlike eq. (3.19), since $\hat{R}'$ now coincides with the Z' axis. The angular momentum operator appearing in the Hamiltonian, however, will now be more complicated than in the case of its SF frame representation, since we are no longer in an inertial system and one would, classically speaking, observe the projectile coming along the Z' axis and going out along the same Z'-axis after collision.

One can also use wavefunctions that have definite parity with respect to inversion in the origin, and thus take a linear combination of the $+\Omega$ and $-\Omega$ components in eq. (3.50). These new functions will no longer be eigenfunctions of $\hat{j}_{z'}$, but of $|\hat{j}_{z'}|$, with eigenvalue $\bar{\Omega}$ [61].

The previous CC equations, given in the SF frame by (3.20), acquire now a different structure when expansion (3.49) is used and the matrix elements of the Hamiltonian (3.2) are constructed between the BF basis functions (3.50)[4,51]:

$$\left\{\frac{d^2}{dR^2} + k^2_{jj'}\right\} g^{Jj\Omega}_{j'\Omega'}(R) = \sum_{j''\Omega''} \langle \mathcal{Y}^{JM}_{j'\Omega'} | W | \mathcal{Y}^{JM}_{j''\Omega''} \rangle g^{Jj\Omega}_{j''\Omega''}(R) \tag{3.51}$$

where the operators in curly brackets on the l.h.s. of (3.51) have the same meaning as those of eq. (3.20). The coupling matrix elements on the r.h.s., however, exhibit here a different structure, i.e. one can write them as:

$$\langle \mathcal{Y}^{JM}_{j\Omega'} | W | \mathcal{Y}^{JM}_{j''\Omega''} \rangle = \left(\frac{1}{R}\right)^2 \langle \mathcal{Y}^{JM}_{j'\Omega'} | \hat{\ell}^2 | \mathcal{Y}^{JM}_{j''\Omega''} \rangle \times \qquad (3.52)$$

$$\times \delta_{j'j''} + 2\mu \langle \mathcal{Y}^{JM}_{j'\Omega'} | V | \mathcal{Y}^{JM}_{j''\Omega''} \rangle \delta_{\Omega'\Omega''}$$

One clearly sees, in this representation, the separation between couplings among different helicity states due to rotation of the relative vector and those couplings that take place among different rotational levels via the electrostatic potential.

The coupling via the angular momentum operator acts among channels where the Ω label changes by 0, +1 and -1 while j' does not change:

$$\langle \mathcal{Y}^{JM}_{j'\Omega'} | \hat{\ell}^2 | \mathcal{Y}^{JM}_{j'\Omega'} \rangle = J(J+1) + j'(j'+1) - 2\Omega'^2 \qquad (3.53\ a)$$

$$\langle \mathcal{Y}^{JM}_{j'\Omega'} | \hat{\ell}^2 | \mathcal{Y}^{JM}_{j'\Omega'\pm1} \rangle = \left\{ \left[J(J+1) - \Omega'(\Omega' \pm 1) \right] \times \right.$$

$$\left. \times \left[j'(j'+1) - \Omega'(\Omega' \pm 1) \right] \right\}^{1/2} \qquad (3.53\ b)$$

On the other hand, since the PES in the atom-molecule case only depends on the internal coordinate $\vartheta = \hat{r}', \hat{R}'$, the latter coincides with the unit vector of $\hat{r}'$ in polar coordinates. The corresponding multipolar expansion of V ($\underline{R}$, $\hat{r}$) therefore does not couple different Ω states thus eliminating the 6-j coefficients appearing in the f_λ's of eq. (3.25):

$$\langle \mathcal{Y}^{JM}_{j'\Omega'} | V | \mathcal{Y}^{JM}_{j''\Omega'} \rangle = \sum_\lambda (-)^{\Omega'} \left[(2j'+1)(2j''+1) \right]^{1/2} \qquad (3.53\ c)$$

$$\times \begin{pmatrix} j' & j'' & \lambda \\ 0 & 0 & 0 \end{pmatrix} \times \begin{pmatrix} j' & j'' & \lambda \\ \Omega' & -\Omega' & 0 \end{pmatrix} V_\lambda(R)$$

The solution of the coupled equations (3.51) proceeds now exactly as in the SF representation, and yields expressions for the scattering amplitudes of each state-to-state transition that are entirely analogous to eq. (3.39)[26].

The computational problem of searching for the solutions of the above CC equations is of the same order of complexity as the one yielded by the SF frame representation. Each rotor state j', in fact, produces (2 j' + 1) channels $|j'\Omega'\rangle$ (with $-j' \leq \Omega' \leq j'$) as in the SF scheme. The greatest difficulty arises here from the off-diagonal terms of eqs. (3.52), since they decrease very slowly as R^{-2}, hence are still coupling the radial solutions in those regions of space where the V_λ coefficients of the PES have already died off.

One can thus state that for small values of J and strong couplings over long-range potentials the BF treatment is more manageable since it does not require the summation over 1 appearing on the r.h.s. of eq. (3.20), while the SF frame approach becomes preferable at large J values and weak coupling potentials that mainly act at short range.

Moreover, the BF structure of the coupled equations allows one to spot in a more transparent way the various approximations discussed later, especially those which strive for a substantial reduction of the needed number of equations via some physical agreement on the specific dynamics of the system involved.

3.4 Vibrationally Inelastic Collisions

The close-coupling treatment of excitation of vibrations by collision is formally the same as that described before for rotational excitation. The most common practice is to include in the needed expansion of the total wavefunction a new set of functions that represent the vibrational wavefunctions of the non-interacting systems.

This has been applied thus far only to the simplest case of diatomic molecular targets, since at energies high enough to excite vibrational transitions to any appreciable extent, there are in general large numbers of open rotational channels. In the case of the H_2 molecule, for instance, the rotational energy levels are widely spaced thus only relatively few of them exist between the various vibrational levels. They nonetheless lead to an exceedingly large number of channels coupled together at even the lowest collision energies for which vibrational inelasticity is present.

For molecules other than H_2 or its isotopes, the rotational spacings become extremely small, thus rapidly leading to unmanageably large coupled systems. If one considers, for instance, the vibrational excitation of a N_2 molecule from its ground vibrational state to its first excited vibrational level, one can take advantage of its being a homonuclear target, treat it initially in the j = 0 state and thereby need only consider half of the rotational states of this example.

Since there are more than 30 j states below the first excited vibrational state of N_2, and we need only consider j = 0, 2, 4... 30 values, the total number of coupled rotational channels that are required on the r.h.s. of eq. (3.20) is given by $16^2 = 256$. This would therefore lead to 256 coupled equations without yet considering any of the rotational channels, open and closed, associated with the final vibrational state. Clearly then only via the use of some approximation can vibrationally inelastic collisions be treated quantum mechanically. Thus far, the formalism of the CC equations has been rigorously applied only to He atoms[62] and Li^+ ions[63] colliding with H_2 molecules and to the well-known $H+H_2$ collisional problem[64]. In all the above cases a large number of channels was needed to reach some acceptable level of convergence, while the reliability of the PES employed raised further questions on the accuracy of the dynamics.

For the simplest case of a diatomic molecule without electronic angular momentum along its axis, one writes down the rotovibrational functions as:

$$\varphi_j^v(\underline{r}) = r^{-1}\chi_j^v(r)\, Y_{m_j}^j(\hat{r}) \tag{3.54}$$

where the χ's are eigensolutions of the familiar radial equation:

$$\left[-\frac{1}{2m}\frac{d^2}{dr^2} + \frac{j(j+1)}{2mr^2} + V_0(r)\right]\chi_j^v(r) = \varepsilon_{v,j}\,\chi_j^v(r) \tag{3.55}$$

Once the intramolecular potential $V_o(r)$ is known, a common procedure is to then expand the χ's into an harmonic oscillator H.O. basis:

$$\chi_j^v(r) = \sum_m c_{vjm} N_m H_m(\alpha x) \exp\left(-\frac{1}{2}\alpha^2 x^2\right) \tag{3.56}$$

One can then determine the unknown coefficients variationally by solving the corresponding set of homogeneous equations:

$$\sum_{m=0}^{n_{max}} \left| H_{mn} - \varepsilon_{vj}\,\delta_{mn}\right| = 0 \qquad m = 0, 1, \ldots, n_{max} \tag{3.57}$$

where the matrix elements of the potential in (3.55) can be often obtained in closed form over the H.O. basis of (3.56). A variety of numerical techniques can then be employed to obtain target eigenfunctions for isolated diatomics, the most studied case being obviously the H_2 molecule[15,66].

In the rigorous expansion of (3.18) one therefore replaces the basis functions as:

$$\varphi_{j\ell}^{EJM}(\underline{R},\hat{r}) \Longrightarrow \varphi_{jv\ell}^{EJM}(\underline{R},\underline{r}) = \tag{3.58}$$
$$= \varphi_{\gamma\ell}^{EJM}(\underline{R},\underline{r})$$

where γ = (v, j) is the subscript for the channel index denoting the initial state of the diatomic target.

The coupling matrix elements of eq. (3.22) are correspondingly modified in the following way:

$$\langle j'\ell'|V|j''\ell''\rangle \Longrightarrow \langle \gamma'\ell'|V|\gamma''\ell''\rangle =$$

$$= 2\mu \iiint d\hat{R}\, d\hat{r}\, dr\, \mathcal{Y}_{\ell'\gamma'}^{JM*} \chi_{j'}^{v'*}(r)\, Y_{m_{j'}}^{j'*}(\hat{r}) \times \tag{3.59}$$
$$\times V(\underline{R},\underline{r})\, \mathcal{Y}_{\ell''\gamma''}^{JM} \chi_{j''}^{v''}(r)\, Y_{m_{j''}}^{j''}(\hat{r})$$

Such a triple integration needs obviously to be performed for each R value required to propagate the solutions for the unknown radial functions from the origin into the regions of vanishing electrostatic potential. The above increase in numerical complexity is indeed another indicator of how rapidly the dimensions of the rotovibrationally inelastic processes get out of reach from the computational point of view.

As a consequence of this, one of the most popular approximations in studying vibrational excitations has been the collinear approximation. In this simpler case one can treat the linear molecule as confined to a straight line, while the projectile is also considered to be an atom colliding with the target on the same straight line.

The above model has many of the features of a real collision process and useful information can be obtained from such studies, which are also rather easy to extend to the case of two linear molecules colliding with each other. It is the simplest of all inelastic quantum mechanical problems to solve and therefore it is ideally suited for testing either new numerical methods or new theoretical ideas that can thus be compared with the performances of previously existing approaches.

A somewhat more realistic treatment of vibrational inelasticity is provided by the use of a simplified form of PES, whereby its anisotropic part is entirely disregarded:

$$V(\underline{R},\underline{\rho}) \Longrightarrow V(R,\rho) = V(R,r) \tag{3.60}$$

hence no coupling exists between the diatomic rotational states (or $\hat{r}$ orientation) and the relative angular momentum of the interacting systems (related to the $\hat{R}$ orientation). Coupling is, however, still present between linear momentum and the vibrational degree of freedom. The consequent expansion of the total wavefunction can then be done much more simply than the case of the coupled representation of (3.18):

$$\Psi^{E}_{jv}(\underline{R},\underline{r}) = = \sum_{v'\ell} C^{jv}_{v'\ell} \cdot R^{-1} u^{jv}_{\ell v'0}(R)\, \chi^{j=0}_{v'}(r)\, Y^{\ell}_{0}(\hat{R})\, Y^{0}_{0}(r) \tag{3.61}$$

The now familiar substitution into eq. (3.1) leads to the following radial equations for the unknown u's:

$$\left[\frac{d^2}{dR^2} + k^2_{jvv'} - \frac{\ell(\ell+1)}{R^2}\right] u^{jv}_{\ell v'}(R) = = \sum_{v''} U_{v'v''}(R)\, u^{jv}_{\ell v''}(R) \tag{3.62 a}$$

where:

$$U_{v'v''}(R) = \int dr\, \chi^{0*}_{v'}(r)\, V(R,r)\, \chi^{0}_{v''}(r) \tag{3.62 b}$$

and:

$$k^2_{jvv'} = 2\mu\left(E_{jv} - \varepsilon_{v'0}\right) \tag{3.62 c}$$

Physically, the above equations say that one starts with the basis functions, 'prepared' in their ground rotational states and includes no other rotational state in the problem other than the initial state of the target that appears only in defining the wave vector of each set of coupled equations. Rotational excitation is therefore completely ignored. One obtains then only one channel for each vibrational state $|v''\ 0\rangle$ carried on the r.h.s. of (3.62 a), whereby the number of coupled equations is equal to the number of

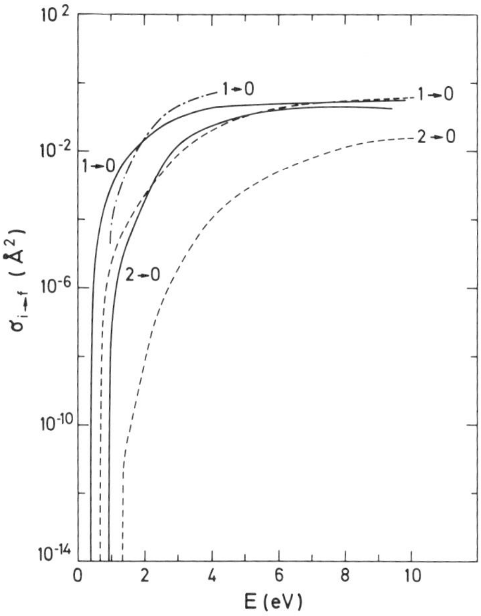

Fig. 5: Vibrationally inelastic partial total cross sections for He + HF colliding via a spherical potential. Six vibrational Morse oscillators are used in the expansion rigorously solved by the data of the continuous lines[67]. Curves (---) and (-·-·-) correspond to approximate, DWA treatments.

such levels. The total angular momentum is just the relative angular momentum of the atom-molecule system, its eigenfunctions in (3.61) being simply given by Legendre polynomials.

The corresponding scattering amplitude is simply the multichannel version of the potential scattering formula:

$$f_{v\to v'}(\vartheta, E) = \frac{1}{2\sqrt{k_v k_{v'}}} \sum_\ell (2\ell+1)\left[S^\ell_{vv'}(E) - \delta_{vv'}\right] P_\ell(\cos\vartheta) \tag{3.63}$$

with corresponding partial differential and integral cross sections:

$$\sigma_{v\to v'}(\vartheta|E) = \frac{1}{4k_v^2}\left|\sum_\ell (2\ell+1)\left[\delta_{vv'} - S^\ell_{vv'}(E)\right] P_\ell(\cos\vartheta)\right|^2 \tag{3.64}$$

$$\sigma_{v\to v'}(E) = \frac{\pi}{k_v^2}\sum_\ell (2\ell+1)\left|\delta_{vv'} - S^\ell_{vv'}(E)\right|^2 \tag{3.65}$$

The total cross section is also given by a simple generalization of the optical theorem:

$$\begin{aligned} \sigma_{TOT}(E) &= \sigma_{v\to v}(E) + \sum_{v'} \sigma_{v\to v'}(E) \\ &= \frac{2\pi}{k_v^2}\sum_\ell (2\ell+1)\left[1 - \mathrm{Re}\, S^\ell_{vv}(E)\right] \\ &= \frac{4\pi}{k_v^2}\, \mathrm{Im}\, f_{vv}(\vartheta = 0) \end{aligned} \tag{3.66}$$

Some recent examples of calculations employing the above, simplified form of the interactions are shown in Figures 5 and 6 for the He + HF system[67] and the H^+ - CO system[68]. They both show that the range of collision energies which can be spanned via such an approach, often called the Breathing Sphere Approximation (BSA), would be definitely out of reach for rigorous roto-vibrational coupling schemes.

Though the BSA is capable of giving physically useful results, it is clear that rotational degrees of freedom are important in

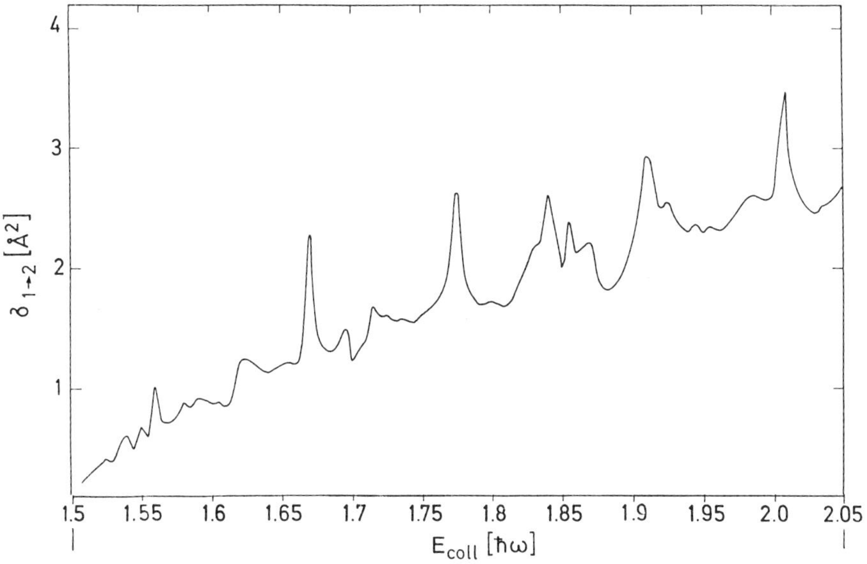

Fig. 6: Vibrational excitation of the 1 → 2 transition in H^+ + CO collisions via a spherical potential. An expansion over harmonic oscillators is used to treat the CO target[68].

vibrational energy transfer and therefore more realistic treatments should be employed when searching for simplified forms of eq. (3.20) that allow a reduction of computational labour while still preserving some coupling between rotational and vibrational degrees of freedom.

Some of the treatments most frequently applied in the recent literature will be discussed in the next Section.

4. DECOUPLING APPROXIMATIONS

4.1 The CS Approximation

In the previous section we saw how the dimensionality problem in atom-molecule inelastic collisions came from the rapid proliferation of quantum channels associated with the (2 j + 1) degeneracy of the energy levels of the rotating target. We discussed before the atom-homonuclear case. For atom-heteronuclear systems the problem splits up into the even-parity case, where the number of coupled equations is given by:

$$N_{even} = \sum_{j=0}^{j_{max}} (j+1) = (j_{max}+1)(j_{max}+2)/2 \qquad (4.1\ a)$$

and the odd-parity set of equations, the total number of which is given by:

$$N_{odd} = \sum_{j=0}^{j_{max}} j = (j_{max}+1)\, j_{max}/2 \qquad (4.1\ b)$$

For even the most efficient algorithm, the labor of solving coupled equations goes as the cube of the numbers of equations being solved, so it is obvious that the computational problem becomes rapidly intractable even for modest values of j_{max}.

As a consequence of this, the CS approximation has been extensively studied in recent years. It was called the Coupled States approximation in its earlier development[69], while its simultaneous and independent introduction by Pack[70] was called the Centrifugal Sudden approximation. Further refinements and interpretations of the CS physicsl meaning were indeed discussed by many authors[71-80] in the recent literature and several computational applications have also been carried out for atom-diatomic systems[81-84]; a recent review lists in detail most of the recent results and refinements[85].

For the purely rigid rotor target interacting with a structureless atom, it is convenient to remind ourselves of the form taken by the total Hamiltonian (3.2) in the BF frame:

$$\mathcal{H}_{BF} = -\frac{1}{2\mu}\frac{1}{R}\frac{\partial^2 R}{\partial R^2} + \frac{\hat{\ell}^2}{2\mu R^2} - \frac{1}{2m}\nabla_{r'}^2 + V(\underline{r}', R) \qquad (4.1\ c)$$

When the above operator is applied to the trial function expanded over the coupled molecular states discussed in Section 3.3, one sees that it would lead to purely radial equations on R if it were not for the $\hat{l}^2$ operator. Hence the lack of separation of the angular part of R is purely kinematic in origin and, in a sense, does not depend on the nature of the potential although the latter provides the necessary coupling of radial functions.

In order to obtain the CS approximation, one now introduces an effective orbital angular momentum eigenvalue and approximates the centrifugal potential of (4.1) by:

$$\frac{\hat{\ell}^2}{2\mu R^2} \sim L(L+1)/2\mu R^2 \tag{4.2}$$

This is also, of course, a sort of sudden approximation which essentially says that the relative kinetic energy is sufficiently large that the precise value of the centrifugal potential is not important. Whether this approximation is true or not, depends on how the different turning points for the different effective potentials are related. If the electrostatic interaction is purely repulsive, then the rate of change of the turning points with the various l is not large and therefore the relation (4.2) is expected to be valid.

For interactions exhibiting attractive wells one might encounter ranges of l values which have three turning points, all rapidly changing with l, and therefore the fixed, effective eigenvalue of (4.2) is not a reliable approximation to that situation.

If we now return to the BF equations (3.50), we see that prescription (4.2) provides a substantial simplification of those coupling matrix elements that are responsible for transitions between different helicity states:

$$\langle \mathcal{Y}^{JM}_{j\Omega} | \hat{\ell}^2 | \mathcal{Y}^{JM}_{j\Omega\pm1} \rangle \sim 0 \tag{4.3}$$

i.e. one neglects the intermultiplet couplings provided by the different eigenvalues of $\hat{l}^2$ appearing in the $\mathcal{H}_{BF}$. Moreover, it further reduces the form of the coupling between states with the same helicity index to the following simpler expression for eq. (3.53a)

$$\langle \mathcal{Y}^{JM}_{j\Omega} | \hat{\ell}^2 | \mathcal{Y}^{JM}_{j\Omega} \rangle \simeq J(J+1) \tag{4.4 a}$$

or to:

$$\simeq [J(J+1) + j(j+1) - 2\Omega^2] \tag{4.4 b}$$

according to whether one chooses in eq. (4.2) L = J or the exact value of the diagonal matrix element. The latter choice slightly complicates the long-range form of the asymptotic boundary conditions, which require now spherical Bessel functions of order λ , where $\lambda(\lambda+1) = J(J+1) + j(j+1) - 2\Omega^2$.

The corresponding coupled equations in the BF frame therefore become:

$$\left[\frac{d^2}{dR^2} - \frac{L(L+1)}{R^2} + K^2_{jj'}\right] g^{Jj\Omega}_{j'\Omega}(R) =$$

$$= 2\mu \sum_{j''} \langle \mathcal{Y}^{JM}_{j'\Omega} | V | \mathcal{Y}^{JM}_{j''\Omega} \rangle \, g^{Jj\Omega}_{j''\Omega}(R) \tag{4.5}$$

where:

$$L(L+1) = J(J+1) + j'(j'+1) - 2\Omega^2 \tag{4.6}$$

The coupling matrix elements on the r.h.s. of (4.5) exhibit now a simpler structure since they only contain the coupling between different j states via the electrostatic potential, without the Coriolis terms that cause coupling between different helicity states and that appear instead in the correct BF eqs. (3.51).

They can be written as already shown in eq. (3.53 c) and indicate that, within the CS approximation the full coupling is independent of l and is zero between states with different helicity, or different j_z-component on the chosen axis of quantization. This is therefore the reason why this method is also called j_z-conserving approximation or Coupled States approximation, in the sense that only coupling between different rotational states (and not between substates) is allowed to appear in the dynamical treatment.

Since the value of Ω is now conserved during the collision, a separate set of (4.5) exists for each Ω value consistent with $|\Omega| \leq$ min (j, j'), although the size of the set is reduced by the fact that the potential matrix containing the coupling from the electrostatic potential only depends on $|\Omega|$.

Various $\Delta\Omega$ transitions can therefore be achieved merely by projecting the approximate wavefunction onto the correct asymptotic conditions. This aspect which we mentioned here only in passing, is a very important one since the slow decrease of the Coriolis coupling potential of the BF coupled equations requires the setting of <u>near</u> boundary conditions (at those R values where the electrostatic potential has vanished) and of <u>far</u> boundary conditions (where also the centrifugal coupling has vanished). This point has been clearly discussed in the recent literature[86,87] and summarized also for the case of reactive scattering[88].

The approximate CS equations given by (4.5) represent now the entire collision process as a sum over Ω of individual transitions $|j\Omega\rangle \rightarrow |j'\Omega\rangle$ and the decoupled equations give equal weighting to each Ω value when computing the cross sections, instead of having them weighted by the intermultiplet coupling operators discussed in the previous section.

On the other hand, it is well-known that the SF angular eigenfunctions used in the expansion (3.18) are related by a unitary transformation[61] to the corresponding BF eigenfunctions of eq. (3.50):

$$\mathcal{Y}_{j\ell}^{JM}(\hat{r},\hat{R}) = \sum_{\Omega} P_{j\ell\Omega}^{JM}\, Y_{j\Omega}^{JM}(\hat{r},\hat{R}) \tag{4.7}$$

where:

$$P_{j\ell\Omega}^{JM} = (-)^{J+\Omega}\,(2\ell+1)^{1/2}\begin{pmatrix} j & J & \ell \\ \Omega & -\Omega & 0 \end{pmatrix} \tag{4.8}$$

It therefore follows that the corresponding SF angular coefficients can be simplified via the decoupling prescriptions (4.4) and written as:

$$\langle \mathcal{Y}_{j\ell}^{JM} | \hat{\ell}^2 | \mathcal{Y}_{j'\ell'}^{JM} \rangle \sim \delta_{jj'}\,\delta_{\ell\ell'}\,L(L+1) \tag{4.9}$$

where the arbitrary parameter can be given, among others, the value of J as used in eq. (4.4).

By making use of this result in rewriting eq. (3.20) one obtains the CS coupled equations in the SF frame of reference for atom-rotovibrator systems[73]:

$$\left[\frac{d^2}{dR^2} - \frac{L(L+1)}{R^2} + K_{jj'}^{vv'}\right] u_{jj'vv'}^{Lm_j}(R) =$$

$$= 2\mu \sum_{j''v''} V_{j'v'j''v''}^{m_j}(R)\, u_{jvj''v''}^{Lm_j}(R) \tag{4.10}$$

where:

$$V_{j'v'j''v''}^{m_j}(R) = \sum_{\lambda} (-)^{m_j}\left[(2j'+1)(2j''+1)\right]^{1/2} \times$$

$$\times \begin{pmatrix} j' & \lambda & j'' \\ 0 & 0 & 0 \end{pmatrix} \begin{pmatrix} j' & \lambda & j'' \\ -m_j & 0 & m_j \end{pmatrix} \times$$

$$\times \int dr\, \chi^*_{v''j''}(r)\, V_\lambda(r,R)\, \chi_{v'j'}(r) \tag{4.11}$$

Thus, one has reproduced exactly the same result of the BF equations in (4.5), this time m_j being the decoupled index of the SF magnetic substates. There is still a system of equations for each L to be solved, and there are the same numbers of L quantum numbers to be considered as J states in the CC equations (3.20). Although each coupled system is smaller, more systems need to be solved.

One has, in fact, to solve eq. (4.11) for each m_j value, which obviously runs from $-j_{max}$ to j_{max}. This point explains once more that the number of coupled equations coincides with the number of states. Moreover, only positive m_j values need to be considered, thus halving the number of blocks that are to be solved, and only those m_j values that are necessary to compute the desired cross sections. When the full CC equations are solved, on the other hand, one gets information on all the possible transitions whether desired or not.

It is interesting to note at this point that different expressions for the partial, integral and differential, state-to-state cross sections are obtained in the CS approximation depending on the choice made for the arbitrary, constant angular momentum parameter L. Moreover, different effects originate from the choice made depending on whether this is done starting from the BF or SF expressions of the scattering amplitude[26].

The structure of the T-matrix, in fact, is made to be block-diagonal in Ω when the BF form is used[61]:

$$T^J_{j\Omega j'\Omega'} = \delta_{\Omega\Omega'} \cdot T^{J\Omega}_{jj'} \tag{4.12}$$

Taking now the incident plane wave in the SF $+\hat{z}$ direction, the projections of J and j on the Z-axis are initially equal, i.e. $M = m_j$ and also, in the BF frame, $M = -\Omega$ as can be seen from Fig. 4. The CS approximation defines $M = \Omega'$ and therefore the SF T-matrix also becomes simplified in the CS scheme[90]:

$$T^J_{CS,SF}(j\ell j'\ell'|E) = (-)^{j+j'}\, i^{-(\ell+\ell'-2L)} \times$$

$$\times \sum_\Omega \left[\frac{(2\ell'+1)(2\ell+1)}{2J+1}\right]^{1/2} C(j\ell J;\Omega 0 \Omega) \times \tag{4.13}$$

$$\times C(j'\ell'J;\Omega 0 \Omega)\, T_{jj'}^{J\Omega}(E)$$

where the index $\Omega = \min(j, j', J)$.

If one chooses L = 1', i.e. the parameter to be given by the final angular momentum vector, the phase of the above eq. becomes independent of 1 and therefore the sums on J and 1 appearing in the SF scattering amplitude of (3.39) are done analytically. The final result is:

$$q_{L=\ell'}^{CS,SF}(j m_j\, j' m_{j'} | \hat{R}) = \frac{i(-)^{j+j'}}{2\sqrt{k_j k_{j'}}}\, \delta_{m_j m_{j'}} \times$$

$$\times \sum_{\ell'} (2\ell'+1)\, T_{jj'}^{\ell' \Omega = m_j}\, P_{\ell'}(\cos\vartheta_R) \tag{4.14}$$

where we see that Δm_j magnetic transitions do not appear in such a formulation of the CS scattering amplitude. Since:

$$T_{jj'}^{\ell' m_j} = \delta_{jj'} - S_{jj'}^{\ell' m_j} \tag{4.15}$$

the general, correct BF S-matrix can be related to the above, approximate CS expression:

$$S_{j\Omega, j'\Omega'}^{J} \sim \delta_{\Omega\Omega'} (i)^{j+j'} (-)^{J-\ell'} S_{jj'}^{\ell'\Omega} \tag{4.16}$$

It therefore follows that the use of the above form in the BF equation for the scattering amplitude allows one to obtain the corresponding BF scattering amplitudes in the CS approximation via the helicity form of the CS T-matrix[76,90]:

$$T_{CS,BF}^{J}(L=\ell') =$$

$$= \sum_{\ell'} C(j'\ell'J, \Omega' 0 -\Omega')\, C(j'\ell'J, \Omega 0 -\Omega)\, T_{jj'}^{\ell'\Omega} \tag{4.17}$$

which is now non-diagonal in the helicity index via the weighting Clebsch-Gordan coefficients and therefore allows the evaluation of $\Delta\Omega$ magnetic transitions via scattering amplitudes $q_{L=\ell'}^{CS,BF}(j\Omega\, j'\Omega'|\hat{R})$ similar to those of eq. (4.14) but without the selection rule present in the latter.

One can finally obtain degeneracy-averaged partial cross sections, integral and differential, by the use of either of these scattering amplitudes and performing the usual sum over $m_{j'}$ (Ω') and averaging over m_j (or Ω):

$$\sigma^{CS}_{L=\ell'}(j\to j'|E) = \frac{\pi}{k_j}\sum_J \sigma^J_{j\to j'}(CS, L=\ell') \tag{4.18}$$

where:

$$\sigma^J_{j\to j'}(CS; L=\ell') = \frac{1}{2j+1}\sum_{\ell' m_j}(2\ell'+1)\cdot C^2(\ell' j' J; 0 m_{j'} -m_j)\cdot\left|T^{\ell' m_j}_{j,j'}\right|^2 \tag{4.19}$$

and:

$$\frac{d\sigma^{CS}}{d\Omega}(j\to j') = \frac{1}{(2j+1)4k_j^2}\sum_{\ell,\ell',m_j}(2\ell+1)(2\ell'+1)\cdot P_\ell(\cos\vartheta_R)\cdot P_{\ell'}(\cos\vartheta_R)\cdot T^{\ell m_j *}_{j,j'}\cdot T^{\ell' m_j}_{j,j'} \tag{4.20}$$

where we see again that the dynamics does not change the projection value of j in the computed T-matrix elements.

The other choice of the arbitrary parameter, i.e. L = 1, the initial angular momentum value, provides slightly different expressions for the scattering amplitudes in both frames of reference, as amply discussed in the recent literature[85,90]. Moreover, it has been shown[79] that the two different forms of such amplitudes are related by the transformation:

$$q^{CS}_{L=\ell}(j m_j\to j' m_{j'}|\hat{R}) = \sum_{\lambda,\lambda'}\mathcal{D}^{j'*}_{\lambda' m_{j'}}(\vartheta,\varphi,0)\, q^{CS}_{L=\ell'}(j m_j\to j' m_{j'}|\hat{R})\cdot\mathcal{D}^{j}_{\lambda m_j}(\vartheta,\varphi,0) \tag{4.21}$$

If one remembers that the $q^{CS,SF}_{L=\ell'}$ is diagonal in m_j, the above result can be rewritten as:

$$q^{CS,SF}_{L=\ell}(j m_j\to j' m_{j'}|\hat{R}) = \frac{i}{2(K_j K_{j'})^{1/2}}\times\sum_\ell(2\ell+1)\left[\sum_\lambda \mathcal{D}^{j'*}_{m_{j'}\lambda}(\hat{R})\times T^{\ell\lambda}_{jj'}\,\mathcal{D}^{j}_{m_j\lambda}(\hat{R})\right]P_\ell(\cos\vartheta_R) \tag{4.22}$$

which shows that the SF amplitudes obtained with this prescription within the CS approximation are not any more diagonal in m_j as in eq. (4.14) and therefore can directly provide state-to-state, magnetic transitions for the system involved. The same scattering amplitudes defined in the BF frame of reference also show that the T-matrix elements are not diagonal in Ω, hence $\Delta\Omega$ transitions are given by Clebsch-Gordan coefficients.

It therefore follows that the choice of the initial angular momentum for the arbitrary parameter L produces scattering amplitudes that allow for magnetic transition calculations in both the BF and SF reference frames.

The corresponding opacities can be computed via the q's of (4.22) and one obtains an expression which slightly differs from the previous eq. (4.19):

$$\sigma^{J}_{j\to j'}(CS, L=\ell) = \frac{1}{2j+1}\sum_{\ell m_j}(2\ell+1)\times C^2(\ell j J;\ 0\ m_j\ -m_j)\left|T^{\ell m_j}_{jj'}\right|^2 \quad (4.23)$$

that is now a weighted sum over T-matrix elements indexed via the initial channel angular momentum. For large Δj transitions the coefficients in (4.23) and (4.19) would then be different, hence producing different values for the degeneracy cross sections (4.18). The choice of L = 1, on the other hand, produces the same expression for the partial, integral cross sections as the former choice.

Finally, if one identifies the arbitrary L parameter with the <u>total</u> angular momentum J, the corresponding opacities are given by yet another expression[90]:

$$\sigma^{J}_{j\to j'}(CS, L=J) = \frac{1}{2j+1}\sum_{\ell'}\sum_{m_j}(2\ell'+1)\times C^2(\ell' j' J;\ 0\ m_j\ -m_j)\left|T^{\ell' m_j}_{jj'}\right|^2 \quad (4.24)$$

This last result differs from (4.23) and (4.19) since the summation over m_j runs over the index $\pm m = \min(j, j', J)$, while the previous summations were only restricted by $m = \min(j, j')$. Hence, for large (j, j') values the final opacities will be different here than in the two previous cases. In any event, this last choice of L = J also allows magnetic transitions $\Delta m_j \neq 0$ to be computed from the scattering amplitudes given by the CS approximation.

In conclusion, one can briefly say that the validity of the CS approximation strongly depends on which region of the PES one is considering. As stated before, the primary regions where CS breakdown occurs are in the vicinity of the turning points where the potential is a strong function of l. Moreover, for rotationally inelastic transitions, the approximate conservation of Ω requires that the rotor orientation must change. In the long range region this suggests that CS will only be accurate for rotationally elastic processes, while being bad for those inelastic transitions which are controlled by the large-R region of interaction.

4.2 The IOS Approximation

The infinite-order sudden approximation has a long history in the molecular collision literature[91-95]. It can be viewed as an energy sudden approximation to the results of the CS approximation discussed before.

In the latter case, in fact, the $\hat{l}^2$ operator is replaced by an effective eigenvalue form, thereby allowing one to use the closure property of the complete set of the partial wave expansion to simplify the sums over l, m_l In the IOSA one further applies the same sort of approximation to the rotational energy. In the case of a linear molecule, therefore, $\hat{j}^2/2I$ is replaced by another effective eigenvalue form j (j+1). Thus one has that:

$$K^2_{jj'} \sim K^2_{j\bar{j}} \qquad \text{for all } j' \tag{4.25}$$

The above condition, combined with condition (4.2) yields a constant term within brackets on the l.h.s. of (3.20). One can then apply a continuous, angle-dependent transformation[73] to the rigorous CC equations and obtain the following simpler form of IOSA equations for a linear rotovibrator target:

$$\left[\frac{d^2}{dR^2} - \frac{L(L+1)}{R^2} + K^2_{j\bar{J}}\right] f^{LJv}_{v'}(R,\hat{r},\hat{R}) =$$

$$= 2\mu \sum_{v''} \langle v'|V|v''\rangle^L_{\bar{J}} f^{LJv}_{v''}(R,\hat{r},\hat{R}) \tag{4.26}$$

where the new radial wavefunctions now depend parametrically on the SF orientation of R and r. Since the coupling potential only depends on γ = R · r, i.e. on their relative orientation, the IOS wavefunction which is now diagonal in the new rotational index j, does the same. Eqs. (4.26) are, however, still rigorously coupled in the vibrational expansion given on the r.h.s. by the index v".

The above solutions must satisfy the usual boundary conditions:

$$f_{v'}^{L\bar{J}v}(R\,\gamma) \underset{R \to 0}{\sim} 0 \tag{4.27 a}$$

$$f_{v'}^{L\bar{J}v}(R\,\gamma) \underset{R \to \infty}{\sim} K_{v'\bar{J}}^{1/2} \Big\{ \delta_{vv'} \exp\left[-i\left(K_{v\bar{J}}R - \frac{L\pi}{2}\right)\right] - S_{vv'}^{L\bar{J}}(\gamma) \exp\left[i\left(K_{v'\bar{J}}R - L\pi/2\right)\right]\Big\} \tag{4.27 b}$$

where one clearly sees, from eq. (4.27 b) the formal analogy with the asymptotic wavefunctions that solve the vibrationally inelastic problem for a spherical potential, as discussed in Section 3.4, for each given value of the orientation j.

The corresponding S-matrix elements for each rotovibrational channel can therefore be obtained by coupling the angular functions of (3.19) via the new, approximate S-matrix:

$$S^{J}_{j\ell vj'\ell'v'}(E) \sim i^{\ell+\ell'-2L} \langle \mathcal{Y}^{JM}_{j'\ell'} | S^{\bar{J}L}_{vv'}(\gamma) | \mathcal{Y}^{JM}_{j\ell} \rangle \tag{4.28}$$

where the phase factor is needed to give the approximate solutions of (4.27 b) the same phases as those required by the radial asymptotes of the correct CC solutions[90]. The above integration, on the other hand, is usually accomplished by Gaussian quadrature over γ, by computing solutions of (4.26) for values of γ for which $\cos\gamma$ are Gaussian quadrature evaluation points. At higher collision energies, where vibrational transitions are important, and several coupled equations need to be solved over the contributing $|v\rangle$ states, it is usually found that the phases of the S-matrix elements are a rapidly varying function of molecular orientation.

The required scattering amplitudes can now be worked out by making use of (4.28) in the exact CC form of eq. (3.39):

$$q^{IOSA}(jm_j v \to j'm_{j'}v' | \hat{R}) = \tag{4.29}$$

$$= \sum_{\ell\ell' m_{\ell'}} (-)^{\ell-L} \left[\pi(2\ell+1)\right]^{1/2} \times$$

$$\times\, T^{IOSA}_{vv'}(l'm_{l'}\, j'm_{j'} | l0\, jm_j)\; Y^{l'}_{m_{l'}}(\hat{R})$$

where:

$$T^{IOSA}_{vv'}(l'm_{l'}\, j'm_{j'} | l0\, jm_j) = \qquad (4.30)$$

$$= \int d\hat{R}\, d\hat{r}\, Y^{l'*}_{m_{l'}}(\hat{R})\, Y^{j'}_{m_{j'}}(\hat{r})\, T^{JL}_{vv'}(\gamma)\, Y^{j}_{m_j}(\hat{r})\, Y^{l}_{0}(R)$$

and the γ-dependent T-matrix elements are related to the S-matrix elements of (4.28) by the usual relationship:

$$T^{LJ}_{vv'}(\gamma) = \delta_{vv'} - S^{LJ}_{vv'}(\gamma) \qquad (4.31)$$

The integration of (4.30) acquires now different and simpler forms depending on the choice that one makes for the centrifugal effective quantum number L, in complete analogy with what was discussed before for the CS approximation.

For the choice of L = 1, the initial angular momentum, the sum over 1' and m_1', can be carried out outside the integral which now becomes a two-dimensional integral over $\hat{R} = (\Theta_R, \Phi_R)$. Since the potential coupling does not depend on the chosen frame, a transformation back into the BF frame of reference allows one to perform the Θ_R-integration easily, thus yielding a one-dimensional integral over $\gamma = \Theta_r$' to obtain the T-matrix element needed in (4.29):

$$T^{IOSA}_{vv'}(BF, L=l) = (2l+1)\,2\pi \int Y^{j'}_{\Omega}(\gamma, 0)\,\cdot \qquad (4.32)$$

$$\times\, T^{lJ}_{vv'}(\gamma)\, P_l(\cos\vartheta_R)\, Y^{J}_{\Omega}(\gamma, 0)\, \sin\gamma\, d\gamma$$

These matrix elements, in a way entirely similar to the CS results of the previous section, are diagonal in Ω and cannot be used to compute state-to-state magnetic transitions. When the above rotation back into the BF frame is not performed, however, the IOSA (SF) scattering amplitudes allow one to obtain $\Delta m_j \neq 0$ inelastic cross sections [90].

When the L = 1' choice is applied, on the other hand, the corresponding T-matrix elements can be obtained from (4.30) without having to rotate back into the BF frame of reference[76], i.e. one can write directly a simple expression for the IOSA matrix elements:

$$T_{vv'}^{IOSA}(SF, L=\ell') = 2\pi(2\ell'+1) \times$$

$$\times \int Y_{m_j}^{j}(\gamma,0)\, T_{vv'}^{\ell' J}(\gamma)\, P_{\ell'}(\cos\vartheta_R)\, Y_{m_j}^{j'}(\gamma,0)\, d(\cos\gamma) \tag{4.33}$$

where the integration is again over the internal angular variable and the Δm_j transitions are not allowed any more. Correspondingly, the $\Delta\Omega$ transitions can be obtained in the helicity representation as shown before in the CS approximation.

One can obviously compute the corresponding degeneracy-averaged of the amplitudes from eqs. (4.32) and (4.33). Because of the properties of the rotation matrices, the corresponding cross sections are given by the same equations, just as in the CS case:

$$\frac{d\sigma^{IOSA}}{d\vartheta_R}(jv \to j'v'|E) = \frac{1}{4(2j+1)k_{jv}^2} \sum_{m_j} \Big| \sum_L 2\pi(2L+1) \times$$

$$\times \int_0^{2\pi} Y_{m_j}^{j'}(\gamma,0)\, T_{vv'}^{LJ}(\gamma)\, P_L(\cos\vartheta_R)\, Y_{m_j}^{j}(\gamma,0) \sin\gamma\, d\gamma \Big|^2 \tag{4.34}$$

therefore the degeneracy-averaged DCS are independent of the L-choice. The corresponding integral quantity is indeed a rather simple expression:

$$\sigma_{jv\to j'v'}^{IOSA}(E) = \frac{\pi}{k_{jv}^2} \sum_L (2L+1)\, P_L(jv \to j'v') \tag{4.35}$$

with:

$$P_L(jv \to j'v') = \frac{1}{2j+1} \cdot$$

$$\times \sum_{m_j} \Big| 2\pi \int_0^{2\pi} Y_{m_j}^{j'}(\gamma,0)\, T_{vv'}^{JL}(\gamma)\, Y_{m_j}^{j}(\gamma,0) \sin\gamma\, d\gamma \Big|^2 \tag{4.36}$$

The above IOSA opacities can be rewritten in terms of total angular momentum J, similarly to what was done in the CS approximation[76], in order to compare them more easily with exact CC results. Thus, one needs to use either eq. (4.32) or eq. (4.33) and, although the final expression turns out to be the same, each outcoming J-opacity is weighted by the sum over l or over l'.

The third obvious choice for L, i.e. L = J, provides different contributions to the P_L appearing in (4.35), although magnetic transitions can be evaluated in the SF frame. There is of course no a priori method available to determine the best choice for L. An empirically obvious criterion is that the computed cross sections agree within a reasonable degree of accuracy with the correct CC results. Thus, the proper choice is going to be determined, in the end, by the particular nature of the interaction potential, the relative strengths of the magnetic transitions and the relative collision energy used in the experiments as compared to the interaction time between the partners.

Several comparisons have appeared in the literature and have been recently reviewed[26,85].

In the analysis of experimental data the target might either be 'prepared' in a specific rotational state (and vibrational state) or one might not be able to resolve individual, rotationally different final states. In both cases, the IOS approximation provides simple expressions for the corresponding cross sections.

It is expedient to write the T-matrix elements as:

$$T^{JL}_{vv'}(\gamma) = \sum_{\lambda} \tau^{JL\lambda}_{vv'} P_{\lambda}(\cos\gamma) \tag{4.37}$$

where:

$$\tau^{JL\lambda}_{vv'} = \left(\lambda + \frac{1}{2}\right) \int_{-1}^{1} T^{JL}_{vv'}(\gamma) P_{\lambda}(\cos\gamma)\, d(\cos\gamma) \tag{4.38}$$

then (4.36) becomes:

$$P_L(jv \to j'v') = \sum_{\lambda} (2\lambda+1)^{-1} C^2(j\lambda j'\,000) \left|\tau^{JL\lambda}_{vv'}\right|^2 \tag{4.39}$$

Substituting (4.39) into (4.35) and using the special transition cross section from the $|0v\rangle$ state, one obtains a very useful factorization of the whole matrix of cross sections by only computing one of its columns[96]:

$$\sigma^{IOSA}_{jv \to j'v'}(E) = \frac{k^2_{0v}}{k^2_{jv}} \sum_{j''} C^2(jj''j'\,000)\, \sigma^{IOSA}_{0v \to j''v}(E) \tag{4.40}$$

Moreover, if one does not resolve the final state j' and therefore eq. (4.40) is obtained after a further sum over j', the rotationally summed integral and differential cross sections are given by remarkably simple expressions:

$$\sum_{j'} \frac{d\sigma^{IOSA}}{d\vartheta_R}(jv \rightarrow j'v') = \sum_{j''} \frac{k_{ov}^2}{k_{jv}^2} \frac{d\sigma^{IOSA}}{d\vartheta_R}(ov \rightarrow j''v') \tag{4.41 a}$$

$$\sum_{j'} \sigma^{IOSA}_{jv\rightarrow j'v'}(E) = \sum_{j''} \frac{k_{ov}^2}{k_{jv}^2} \sigma^{IOSA}_{ov\rightarrow j''v'}(E) \tag{4.41 b}$$

These results imply that, apart from the detailed balance energy factor, the total (rotationally-summed) integral and differential cross sections are independent of the initial state of the rotor.

Another simple form for the above integral total (rotationally total) cross sections can be obtained if we go back to eq. (4.34) and choose the j parameter to be the initial rotational state, thus ensuring that at least the large elastic component in each set of coupled equations exhibits the correct channel wavenumber. After simple passages, one can then write for the DCS:

$$\frac{d\sigma^{IOSA}}{d\vartheta_R}(jv \rightarrow v') = \sum_{j'} \frac{d\sigma^{IOSA}}{d\vartheta_R}(jv \rightarrow j'v') =$$

$$= \frac{1}{4k_{jv}^2}\left|\sum_L (2L+1)\int_{-1}^{1} T^{jL}_{vv'}(\gamma)\, P_L(\cos\vartheta_R)\, d(\cos\gamma)\right|^2 \tag{4.42}$$

The above result is independent of the L-choice, since both 1 and 1' selections lead to the eq. (4.42).

Integrating over Θ_R yields now the γ-dependent integral cross section:

$$\sigma_j^{IOSA}(v \rightarrow v';\gamma) = \frac{\pi}{k_{jv}^2} \sum_L (2L+1)\left|T^{jL}_{vv'}(\gamma)\right|^2 \tag{4.43}$$

an expression which is entirely similar to the partial wave summation that appears in the cross sections for scattering from spherical potentials that are γ-dependent. The corresponding quadrature over γ now gives the total, integral cross section for the $(v \rightarrow v')$ process:

$$\sigma^{IOSA,j}_{v\rightarrow v'}(E) = \frac{1}{2}\int_{-1}^{1} \sigma_j^{IOSA}(v \rightarrow v',\gamma)\, d(\cos\gamma) \tag{4.44}$$

When only rotational excitations are considered, the eq. (4.26) simply looks like the differential equation of potential scattering theory. This means that the corresponding matrix elements can be written as:

$$S_J^L(\gamma) = \exp\left[2i\,\eta_{JL}(\gamma)\right] \tag{4.45}$$

where the 'phase shifts' η are a function of γ at each fixed selection of (j, L). The corresponding total integral cross section is then simply analogous to the scattering result from structureless partners:

$$\sigma^{IOSA}(k_J) =$$

$$= \frac{2\pi}{k_J^2} \sum_L (2L+1) \int_{-1}^{1} \sin^2 \eta_{JL}(\gamma)\, d(\cos\gamma) \tag{4.46}$$

One can further surmise some simple γ-dependence of η_{jL} and therefore perform the (4.46) quadrature over a very limited number of γ values. Moreover, though the phases do not vary greatly from one L value to the next, they can vary considerably over several L values. If another function could be found which followed the phases closely, then one could solve the eqs. (4.26) only at selected L values and interpolate for the rest. For large L values, where inelastic transitions are becoming less important, the WKB method gives an accurate absolute phase for the elastic S-matrix elements, thus even allowing analytic solutions for the eqs. (4.26).

The simplicity of the above results naturally lends itself to the application of the present approximation to the evaluation of thermally averaged integral cross sections and rates[97]. Thus, one can directly evaluate degeneracy-averaged relaxation cross sections, diffusion coefficients and viscosity coefficients in terms of the relevant collision integrals[98].

The general use of the WKB approximation in computing the phases of (4.46) is expected to be valid when the distance ΔR over which the potential varies is large as compared with the associated wavelength of the projectile, $(2ME_j)^{-1/2}$. This means that the ratio τ_{WKB} must satisfy the following inequality:

$$\tau_{WKB} = \frac{1}{\Delta R \sqrt{2ME_j}} \ll 1 \tag{4.47}$$

This clearly happens as the projectile mass M increases, or for higher collision energies, or for that region of the interaction where the potential is a very slow function of R, thus making ΔR larger. Moreover, when marked tunnelling effects are expected, the choice of a WKB turning point becomes more questionable, as is the case for strongly interacting collisional partners. In the latter instance the centrifugal sudden approximation, implied by both the CS and the IOS prescriptions, is likely to fail thus requiring correct treatment of intermultiplet coupling.

4.3 The Fixed-Nuclei Approximation

As we just discussed in the previous section, an approximation scheme that can be viewed as complementary to the CS and IOS decoupling methods reported above has been recently discussed in the literature as applied to atom-molecule[73] collisions[99-101], although it has already had a long history in the literature of electron-molecule collisions[91]. This approximation can be essentially viewed as a closure approximation to the rotor states, in the sense that only the energy-sudden condition of eq. (4.25) is applied to the correct CC equations, while the centrifugal potential is treated in its full form.

Thus, such a reductive scheme can be used when the rotor level spacings are small as compared to the relative kinetic energy and the energetically closed rotational states are not strongly coupled to the states involved in the studied transition.

In order to report briefly the derivation of this decoupling scheme (also called the l_z-conserving approximation) we recall that one can, in principle, perform the SF expansion of eq. (3.18) using a decoupled basis set of known angular functions. Hence, each component of the above equation is given by:

$$\Psi_{jm_j}^{\ell m_\ell}(\underline{R},\hat{r}) =$$

$$= \sum_{\ell' m_{\ell'} j' m_{j'}} R^{-1} u_{\ell' m_{\ell'} j' m_{j'}}^{\ell m_\ell j m_j}(R)\, Y_{m_{\ell'}}^{\ell'}(\hat{R})\, Y_{m_{j'}}^{j'}(\hat{r}) \qquad (4.48)$$

Then the correct CC equations can be obtained as in (3.20), with the difference that the coupling matrix elements on its r.h.s. are given by:

$$\langle \ell' m_{\ell'} j' m_{j'} | V | \ell'' m_{\ell''} j'' m_{j''} \rangle =$$

$$= \int Y_{\ell' m_{\ell'}}^{*}(\hat{R})\, Y_{j' m_{j'}}^{*}(\hat{r})\, V(R,\hat{R},\hat{r})\, Y_{\ell'' m_{\ell''}}(\hat{R}) \qquad (4.49)$$

$$\times Y_{j'' m_{j''}}(\hat{r})\, d\hat{R}\, d\hat{r}$$

If one now replaces the j-state wavevector, for all j values, with a constant value K^2, one can then apply a decoupling transformation to the radial unknown functions of eq. (4.48)[73]:

$$u_{\ell' m_{\ell'}}^{\ell m_\ell}(r, \hat{r}, \hat{r}') =$$

$$= \sum_{j m_j\, j' m_{j'}} Y_{j' m_{j'}}^{*}(\hat{r}')\, g_{\ell' m_{\ell'}\, j' m_{j'}}^{\ell m_\ell\, j m_j}(r)\, Y_{j m_j}(\hat{r}) \qquad (4.50)$$

This substitution into the correct CC equations, in their decoupled form, leads to a new set of simpler equations when the energy-sudden condition is maintained for the wave vectors within brackets on their l.h.s.:

$$\left[\frac{d^2}{dR^2} - \frac{\ell'(\ell'+1)}{R^2} + K^2\right] u_{\ell' m_{\ell'}}^{\ell m_\ell}(R\, \hat{r}) =$$

$$= 2\mu \sum_{\ell'' m_{\ell''}} \langle \ell' m_{\ell'} | V | \ell'' m_{\ell''} \rangle\, u_{\ell'' m_{\ell''}}^{\ell m_\ell}(R, \hat{r}) \qquad (4.51)$$

where the radial functions are decoupled in the $(\hat{r}, \hat{r}')$ of (4.50) and the coupling matrix elements explicitly depend on the SF orientation of the molecule, $\hat{r}$:

$$\langle \ell' m_{\ell'} | V | \ell'' m_{\ell''} \rangle =$$

$$= \int d\hat{R}\, Y_{\ell' m_{\ell'}}^{*}(\hat{R})\, V(R, \hat{r}, \hat{R})\, Y_{\ell'' m_{\ell''}}(\hat{R}) \qquad (4.52)$$

A continuous transformation relates now these new radial functions among themselves for each value of the $\hat{r}$ orientation[100]. Since the potential only depends on the relative orientation $\hat{r} \cdot \hat{R}$, one can clearly define a set of equations where the R-dependence of the unknown functions is separated from their $\hat{r}$ dependence by bringing the molecular orientation to a fixed position with respect to the SF position of initial $\hat{R}$, $\bar{m}$:

$$\left[\frac{d^2}{dR^2} - \frac{\ell'(\ell'+1)}{R^2} + K^2\right] u_{\ell\ell'}^{\bar{m}_\ell}(R) =$$

$$= 2\mu \sum_{\ell''} V_{\ell'\ell''}^{\bar{m}_\ell}(R)\, u_{\ell\ell''}^{\bar{m}_\ell}(R) \qquad (4.53)$$

with:

$$V_{\ell'\ell''}^{\bar{m}_\ell} = 2\pi \int Y^*_{\ell''\bar{m}_\ell}(\gamma,0)\, V(R,\gamma)\, Y_{\ell'\bar{m}_\ell}(\gamma,0)\, d(\cos\gamma) \tag{4.54}$$

where the structure of the equations and of the coupling matrix elements is very similar to what was obtained in section 4.1 for the centrifugal sudden approximation.

The corresponding definition of asymptotic S-matrix elements from the solutions of (4.54) allows one to obtain the state-to-state inelastic scattering amplitudes from the S-matrix elements at each orientation:

$$S_{\ell m_\ell \ell' m_{\ell'}}(\hat{r}) = \sum_{\bar{m}_\ell} \mathcal{D}^{*\ell'}_{\bar{m}_\ell m_{\ell'}}(\hat{r})\, S^{\bar{m}_\ell}_{\ell\ell'}\, \mathcal{D}^{\ell}_{\bar{m}_\ell m_\ell}(\hat{r}) \tag{4.55}$$

since the latter produces the desired former quantities by inverse transformation and integration:

$$f(jm_j \to j'm_{j'} \mid \hat{R}) = \frac{1}{2K} \sum_{\ell\ell' m_{\ell'}} \frac{i^{\ell-\ell'+1}}{4\pi\sqrt{2\ell+1}} \times$$

$$\times \int d\hat{r}\, Y^*_{j'm_{j'}}(\hat{r}) \left[\delta_{\ell\ell'}\delta_{m_{\ell'}0} - S_{\ell 0 \ell' m_{\ell'}}(\hat{r})\right] \times \tag{4.56}$$

$$\times\, Y_{\ell' m_{\ell'}}(\hat{R})\, Y_{jm_j}(\hat{r}) =$$

$$= \int d\hat{r}\, Y^*_{j'm_{j'}}(\hat{r})\, f(\hat{r},\hat{R})\, Y_{jm_j}(\hat{r})$$

which is the expression provided by the so-called adiabatic approximation in electron-molecule collisions[91].

By using eq. (4.55), one can also obtain the state-to-state scattering amplitudes as sums over Clebsch-Gordan coefficients that multiply the $S^{\bar{m}_l}_{l,l'}$ matrix elements produced by the solutions of 4.53).

The degeneracy-averaged quantities can be directly obtained from (4.56) by using the completeness relation among spherical harmonics[100], and one finds that the factorization relations discussed before within the IOS scheme apply exactly within the fixed-nuclei approximation, even if the specific state-to-state scattering amplitudes are obviously given by a very different expression. In particular, the averaged cross section of eq. (4.41) appears in very similar form within the present approximation, via the orientation-dependent scattering amplitudes of eq. (4.50):

$$\frac{1}{2j+1}\sum_{jm_j j'm_{j'}}\frac{d\sigma}{d\Omega}(jm_j\rightarrow j'm_{j'}) = = \frac{1}{4\pi}\int d\hat{r}\,|f(\hat{R},\hat{r})|^2 \qquad (4.57)$$

The significant difference in using this approximation as opposed to those previously discussed resides in the enormous range of values of $\bar{m}_l$ for which the coupled equations (4.53) need to be solved, since many l, l' values are used in the sums of (4.56). A recent computational suggestion[100] reduces the number of equations by referring the $S^{\bar{m}_l}_{l,l'}$ matrix elements to a specific (0 → j') sets of matrix elements, although the number of equations to be solved still increases with m_j and their dimensionality is given by coupling (2 j_{max} + 1) l values in (4.53). On the other hand, the CS approximation needs to solve only (j_{max} + 1) coupled equations. The present approximation, however, can be seen as complementary to the latter, since it is clearly expected to be more useful for heavier systems and for long range anisotropies. In this last instance, in fact, it is acceptable to disregard level spacings among rotational states while the intermultiplet coupling needs to be correctly treated. The price one pays, however, is that even approximate fixed-nuclei calculations are roughly four times slower than CS calculations[100].

4.4 The l-Dominant Decoupling

In case of short-range coupling the corresponding total and partial cross sections, integral and differential, are dominated by those encounters that take place at small impact parameters. The corresponding centrifugal barriers are therefore small and relatively unimportant with respect to the coupling due to the electrostatic potential terms. Hence the previous simplifications are likely to work at their best.

The opposite physical situation, on the other hand, is the one dominated by the centrifugal terms, whereby long-range forces still act at large l values although the latter are controlling the dynamics over substantial ranges of the radial variable. Therefore, an alternative scheme has been recently suggested to deal with this type of situation[102-104] where the centrifugal potential plays a dominant role.

In the long-range region, in fact, $V(\underset{\sim}{R})$ varies rather slowly, and as a result different values of the orbital momentum quantum numbers lead to significantly different classical turning points.

It was observed in CC and CS calculations[69] that, for a given value of J, the (j+1) value is approximately conserved during a transition. This means that, for a given Δj value, the most important contributions come from $l = J$ and $l' = J - \Delta j$. The two impact parameters classically associated with (l, l'), in fact, produce the lowest centrifugal barriers hence permitting more of the coupling potential to be sampled.

Physically, this implies that the preferred orientation for rotational inelasticity in collisions will come from the constructive addition of l and j, i.e. from a nearly coplanar collision with molecular and orbital motions in the same direction.

A new channel index can thus be introduced, by defining an alignment quantum number $\tau = l + j - J$. It measures the degree to which J and j are aligned. By using the asymptotic behaviour of the f_λ coefficients of eq. (3.25), the final decoupling in the l-dominant situation (DLD) yields the following equations:

$$\left[\frac{d^2}{dR^2} + k_{jj'}^2 - \frac{(J+j'+\tau)(J-j'+\tau+1)}{R^2}\right] u_{jj'}^{J\tau}(R) =$$

$$= 2\mu \sum_{j''} V^{J}_{j'\tau j''\tau}(R)\, u^{J\tau}_{j''j'}(R) \qquad (4.58)$$

where (for rotational inelasticity only):

$$V^{J}_{j\tau j'\tau}(R) =$$
$$= \sum_{\lambda} V_\lambda(R)\, f_\tau(j, J-j+\tau; j'\, J-j'+\tau | J) \qquad (4.59)$$

One sees that the f_τ coefficients appear uncoupled in τ, hence alignment coupling is neglected in the potential. In the above DLD equations the centrifugal potential coupling is included while the electrostatic potential coupling is simplified, as opposed to what is done in the CS case where it is the l^2 operator that is written in approximate form.

The degree of coupling turns out to be the same in both cases, since one equation per each coupled j state appears in (4.58) as in (4.5). Moreover, since τ ranges up to j_{max} the number of coupled equations decreases as τ increases, a fact also known in the CS coupling scheme.

For vibrational inelasticity, however, or for rotationally inelastic cross sections where the amount of energy transfer is nearly equal to the total energy, the alignment situation surmised by DLD is no longer valid and therefore the method is found to fail. It still remains, however, a useful complement to CS calculations for those special systems where the CS approach only is likely to be unrealistic, e.g. for ion-molecule collisions or when long-range dispersion appears in the interaction[105].

4.5 The Effective Potential Methods

All the schemes for reducing dimensionality which have been discussed above, are essentially based on some suggestion of a physical reason for which the $\hat{l}^2$ operator can be treated in a simplified manner or for which the IK matrix can be reduced to a constant. The main coupling is therefore transferred to the matrix elements of the electrostatic potential, that are less dependent on orientation parameters and which are correctly treated.

Other quantum mechanical approaches at dimensionality reduction have focussed instead on the selection of a general Effective Potential where in some way the orientation averaging has been performed already, thereby essentially coupling rotational or/and vibrational states with l as a good quantum number.

In one of these methods[106] an effective potential EP preaverages over degenerate m_j states to reduce the coupling matrix elements of the correct CC equations (3.20), for the case of rotovibrational processes, to the following form:

$$\begin{aligned} \langle j'l'v'|V|jlv\rangle_J &= V^J_{ll'}(jv \rightarrow j'v';R) = \\ &= \sum_\lambda V^\lambda_{vv'}(R)\, f_\lambda(jl,j'l'J) \sim \\ &\sim \sum_\lambda V^{\lambda,\mathrm{eff}}_{vv'}(R)\, f^{\mathrm{eff}}_\lambda(jj') \end{aligned} \tag{4.60}$$

where:

$$V^{\lambda,\mathrm{eff}}_{vv'}(R) = \int dr\, \chi^{\mathrm{eff}*}_{v'}(r)\, V_\lambda(r,R)\, \chi^{\mathrm{eff}}_{v}(r) \tag{4.61}$$

and:

$$f_{\lambda}^{eff}(j,j') = \exp\left[i\pi/2\left(|j-j'| + j + j'\right)\right] \times$$
$$\times \left[(2j+1)(2j'+1)\right]^{1/4} (2\lambda+1)^{-1/2} \begin{pmatrix} j & \lambda & j \\ 0 & 0 & 0 \end{pmatrix} \tag{4.62}$$

The anisotropy of the potential only couples j with j' without acting on the two phase angles of the corresponding rotor states before and after the collision. The coefficients appearing in f_λ are thus indicating some sort of simple weighting for the coupling coefficients associated with the (2j+1) or (2j'+1) degeneracy of orientations.

The resulting coupled equations therefore acquire the following very simple form:

$$\left[\frac{d^2}{dR^2} + K_{jv}^2 - \frac{\ell(\ell+1)}{R^2}\right] u_{jv}^{\ell}(R) =$$
$$= \sum_{j'v'} \langle j'v' | V_{eff} | jv \rangle \, u_{v'j'}^{\ell}(R) \tag{4.63}$$

since l is here again a good quantum number, the equations are exactly like those of section3.4 for a spherical potential. The angular part of V($\underline{R}$, $\underline{r}$) in fact does not act during the collision but simply gives rise, before the dynamical calculations are performed, to the weighting coefficients of (4.62). The corresponding partial integral cross section is therefore given by the usual sum over partial waves:

$$\sigma_{jv \to j'v'}^{EP}(K_{jv}^2) = \frac{\pi}{K_{jv}^2} \sum_{\ell} (2\ell+1) \left| T_{vj\,v'j'}^{\ell} \right|^2 \tag{4.64}$$

Because of the forced averaging, the above cross sections satisfy the following detailed balance equation:

$$K_{vj}^2 \, \sigma_{vj \to v'j'}^{EP} = K_{v'j'}^2 \, \sigma_{v'j' \to vj}^{EP} \tag{4.65}$$

instead of the correct equality that accounts for rotational multiplicity and which contains an extra factor:

$$K_{vj}^2 \, \sigma_{vj \to v'j'}^{CC} = \frac{2j'+1}{2j+1} K_{v'j'}^2 \, \sigma_{v'j' \to vj}^{CC} \tag{4.66}$$

The latter factor goes to one for $\Delta j \ll j$ and $j \gg 1$, while it should be introduced in eq. (4.64) when low-j transitions are involved.

The number of coupled equations has thus been strongly reduced by washing out all the m_j-information contained in the correct f_λ-coefficients of (3.25) and only keeping in a numerical factor that weights, according to each (j,j') couple, each ($v \rightarrow v'$) transition. The EP coupled equations are thus given by $N_{EP} = (j_{max}+1)\cdot(j_{max}+1)$ for each partial wave needed in (4.64). This is a significant reduction of the corresponding number of equations needed, for each J, by the correct CC result of (3.20) since there one has that: $(N^J_{CC})^{1/2} = N^{\ell}_{EP}$. For pure rotationally inelastic transitions, therefore, the ratio of CC calculations expense with the corresponding expense for EP calculations goes as $\sim (j_{max}+1)^2$ which makes the latter approach very attractive for the evaluation of inelastic cross sections over large energy ranges or for many (Δj, Δv) values as those required in rate constants calculations.

The present EP modelling of the interaction, however, forces the coupling information to be distributed among fewer channels, thus enhancing the off-diagonal coupling contributions. This is especially true for strongly anisotropic systems[107] where the EP approach exhibits an unrealistic rigidity in providing the correct, detailed partitioning of flux among the physical $|jm_j\rangle$ channels, which are instead properly coupled in the full CC approach.

An alternative form of effective potential has also been provided[108-110] under assumptions similar to those necessary for the above EP modelling to be valid, i.e. for physical systems where the potential-dominated situation shows only weak dependence of the dynamics of m_j (or m_l) orientations, hence no special 'propensity' rules appear when generating detailed S-matrix elements within each m_j- (or m_l) manifold. It is in fact well-known that several experimental situations exist where interpreting bulk data mainly requires the detailed knowledge of degeneracy-averaged, rotationally-summed cross sections for transitions between different vibrational levels. Therefore, even the additional weighting factors that appear for the j-index in eq. (4.68) could be included implicitly in the potential, thereby coupling only $(v_{max}+1)$ equations for each partial wave.

Thus, if for the atom-molecule case the potential is written in the now familiar form:

$$V(\underline{r},\underline{R}) = \sum_\lambda V_\lambda(r,R)\, P_\lambda(\hat{r}\cdot\hat{R})$$

$$= V_0(r,R) + \sum_{\lambda\neq 0}{}' V_\lambda \cdot P_\lambda(\hat{r}\cdot\hat{R}) \qquad (4.67)$$

one can define a new potential V_{SA} that depends on the collision energy considered, $E_{coll} = E - E_v$, and that contains a Statistical Average of the anisotropic terms appearing in the Σ' summation of above:

$$V_{SA}(r, R, E_{coll}) = V_0(r, R) + \bar{V}_{asym}(r, R, E_{coll}) \tag{4.68}$$

where:

$$\bar{V}_{asym}(r, R, E_{coll}) = \mathcal{N}^{-1} \sum_{\lambda \neq 0}' \langle V_\lambda \cdot W_\lambda \rangle_\gamma \tag{4.69}$$

the integration on the r.h.s. of (4.69) is over the internal variable j = arcos (r·R) and the coefficients W are now the weighting factors:

$$W_\lambda(\underline{r}, \underline{R}, E_{coll}) = \exp\left[- \frac{V_\lambda}{E_{coll}} P_\lambda(\hat{r} \cdot \hat{R}) \right] \tag{4.70}$$

The normalization factor, needed for the flux conservation, is given by:

$$\mathcal{N}^{-1} = \int_{-1}^{1} \prod_{\lambda \neq 0} W_\lambda(\underline{r}, \underline{R}, E_{coll})\, d(\cos\gamma) \tag{4.72}$$

The corresponding coupled equations are then given by:

$$\left[\frac{d^2}{dR^2} - \frac{\ell(\ell+1)}{R^2} + K_{v'}^2 \right] u_{v'}^{\ell v}(R) = \\ = 2\mu \sum_{v''} V_{SA}^{\ell v' v''}(R)\, u_{v''}^{\ell v}(R) \tag{4.73}$$

This expression reduces to the case of a pure spherical potential of Section 3.4 when only the first coefficient on the r.h.s. of eq. (4.67) is used. The weighted averaging of the full potential via a Boltzmann-like factor therefore modifies the strength of the potential according to the magnitude of the $V_{\lambda \neq 0}$ contributions and to the collision energy that allows, in a classical sense, different regions of such coefficients to be sampled during each trajectory. The net effect of these new factors will be one of generating spherical potentials with energy-dependent r-gradients, hence with different contributions to the vibrational couplings needed in (4.73).

5. THE NUMERICAL METHODS OF SOLUTION

After having decided which form of the coupled equations is most suitable to treat our specific physical problem, the task is now reduced to the formulation of a numerical procedure that could efficiently yield the relevant cross sections at minimum cost. Several reviews have appeared in the recent literature[26,50,111,112] and we will therefore only briefly summarize which methods have been most popular in the computational approach at solving eq. (3.20) or any of its simpler variants reported in Section 4.

One of the advantages of the presently discussed problem comes from the fact that the CC equations are linear in the unknown continuum functions $u_i(R)$ and therefore these linear differential equations may be solved by non-iterative numerical methods. The general form of our equations is given by:

$$\left[\mathbb{1}\frac{d^2}{dR^2} + \mathbb{K}^2 - \mathbb{V}(R)\right]\mathbb{U}(R) = 0 \tag{5.1}$$

where $\mathbb{V}(R)$ is the Hermitian matrix of the coupling potential plus centrifugal barrier and $\mathbb{K}^2$ is the diagonal wavevector matrix. One can also define another Hermitian matrix as given by:

$$\mathbb{W}(R) = -\mathbb{K}^2 + \mathbb{V}(R) \tag{5.2}$$

and therefore write down the following set of linear equations:

$$\frac{d^2}{dR^2}\mathbb{U}(R) = \mathbb{W}(R)\,\mathbb{U}(R) \tag{5.3}$$

which exhibits some interesting properties that are often exploited in finding its solutions:

(i) the collision energy is a scalar, hence different values for it change the eigenvalues of $\mathbb{W}$ all by the same amount.

(ii) No first derivative, nor coupling containing first derivative, appears for the radial matrix $\mathbb{U}(R)$.

(iii) The coupling matrix is symmetric, it can be a sparse matrix, and it is a slowly-varying function of R hence amenable to analytic, piecewise, solutions as dicussed below.

If the dimensions of the matrices in eq. (5.1) is $N \times N$, the problem then consists in finding N linearly independent solutions, for each required collision energy, which satisfy the usual boundary conditions in each channel and for those regions of R in which the interaction is presumed to be such that: $\lim_{R \to \infty} RV(R) = 0$. Moreover, since the second order matrix equation of above admits $2 \cdot N$ solutions that are linearly independent, N of them are eliminated by further imposing another condition for each sought solution: $u_{ii}(0) = 0$. This is the usual requirement that solutions be regular at the origin. Physically this means that, since the interaction potentials are always repulsive when $R \to 0$, therefore the corresponding probability density of any of the particles being found located at the origin should be zero.

Since the eqs. (5.1) are homogeneous in the components $u_{ii'}(R)$ the multiplication of a solution by an arbitrary factor is also a solution. This implies that the asymptotic amplitudes produced by the real solutions (K-matrix) are arbitrary and are then commonly fixed in the present problem by setting:

$$\frac{d}{dR} \mathbb{U}(0) = \mathbb{D}_0 \tag{5.4}$$

where D_0 is an arbitrary, non-singular matrix. The solution problem is therefore reduced to be, from an apparent two-points boundary value problem, a simpler initial-value problem.

There are basically two different numerical approaches commonly in use to solve the latter problem for the $N \times N$ multichannel situation. One attempts to solve numerically eq. (5.1), or its equivalent integral form, under the boundary conditions discussed above, hereby generating what one may call an <u>approximate solution.</u>

The other method tries to approximate $\mathbb{W}(R)$ in some realistic way and then solves the corresponding coupled equations exactly. This is usually called the <u>approximate potential</u> approach.

In either of the above approaches the only quantity that one is really trying to obtain is the corresponding matrix element necessary to define a cross section (T-matrix or S-matrix), this being independent of any complete knowledge of the solutions for the eqs. (5.1). It therefore follows that, for the special purpose of the present physical problem, the above matrix elements can be obtained with different methodologies within each of the approaches defined before, since the common aim is one of yielding coefficients in the asymptotic expansion of each channel that in turn can give flux distribution after the scattering event.

One main methodology is based on propagating the solution matrix chosen by the initial conditions (5.4) from an R point taken to be well into the non-classical region of the problem (where the potential energy is greater than the total energy) and proceeding then to follow the solution step by step into the asymptotic region. This is probably the most common technique that has been used for the approximate solution approach and is exemplified by the De Vogelaere method[50], the Sams and Kouri method[113] and the method of Choi and Tang[114].

In general, this technique is plagued by instabilities that require measures to be taken from time to time during the propagation of the solution in order to ensure overall stability. In any event, its implementation depends on the problem under study since it can be also applied via the approximate potential approach[115-117] that exhibits a different set of difficulties, many of which strongly depend on the physics to be treated.

The approximate solution approach is capable of higher accuracy while the chief advantage of the approximate potential treatment is that it allows large integration step sizes. The latter, however, requires more work per step than the approximate solution approach for the first step and therefore they are of comparable order of difficulty for one solution of low accuracy. If, however, a large number of solutions to the scattering problem are needed, and at different collision energies, the approximate potential approach is ideal since much of the work done for the first solution can be saved and the problem may be solved at another energy with very little extra effort.

The numerical technique of propagating the approximate solutions into the asymptotic region, as briefly mentioned above, strongly requires that these solutions remain independent during the propagation process, having started at the origin with a complete independent set of solutions. This is difficult to achieve sometimes, since all solutions are growing exponentially as we integrate toward large R and each set will contain some contribution from the most rapidly growing solution. The latter will then rapidly dominate and all columns of the $\mathbb{U}(R)$ matrix will start looking like the fastest growing column solution. One therefore needs to alter this precision-losing process by performing every so often, and early on in the propagation, linear combinations of the found solutions which are orthogonal and hence strongly independent.

The approximate potential approach, on the other hand, breaks up the real potential into piecewise continuous polynomials, or into constant steps, according to the method used. This simple form therefore allows an exact solution of the equation in each region where the piecewise form has been found. There is still coupling between equations and therefore it is necessary to diagonalize the potential matrix at each step. In general different transformations diagonalize the potential in each region and therefore it becomes necessary to transform the results at each region boundary. The stabilizing of solutions also remains necessary as in the approximate solution approach. The approximate potential, however, requires step sizes that are independent of the collision energies, while approximate solutions require smaller steps for the shorter wavelength situations.

In general one can say that the best approach is to use the approximate potential method in those regions of the interactions where the potential is slowly varying (thus allowing for rather large step sizes) and then use the approximate solution approach in the rapidly varying regime of the potential, where the latter becomes a more suitable method due to the smaller amount of work needed for each solution.

The other major technique that has been used to solve the equations (5.1), either through finding simpler solutions or simpler potentials, has been recently called the invariant imbedding technique[111]. It consists of solving the scattering for a specific piece of the potential where often a special basis set expansion can be employed to represent the eigensolutions. A connection technique is then used for combining the R matrices for part of the potential into R matrices for larger regions, till such a matrix is developed for the whole range of the potential.

When one seeks an approximate solution for each channel, the methods that have been employed are the amplitude density[118] and the log-derivative method[119]. The use of this technique via a simpler potential form gives rise instead to the R-matrix method[120]. In either case, the invariant imbedding technique breaks the radial region of integration into different segments and poses the potential to be zero outside each of them. For each small segment one solves then the scattering problem in one step and obtains the R-matrix at, say, the first of them. The next adjacent region is then solved and another R-matrix generated at its boundaries. The two R matrices are then combined to yield a new R-matrix for the whole bigger segment, this being repeated till the asymptotic region is reached and the problem is solved.

The combining of R-matrices requires matrix inversion procedures at each step but its solutions are completely stable and free of the problems of the propagation methods discussed before. This means that, although the procedure of generating solutions (or of approximating potentials) in artificial regions of space to be later extended to the whole space is time-consuming and costly, it is possibly the only way for, say, some curve crossing problems where one curve is well inside the non-classical region of the other, so that stable solutions cannot be obtained for it by propagation methods.

Recently, two very different techniques have appeared in the literature and promise to provide fresh impetus to the numerical solutions of large numbers of coupled equations. One is an iteration technique that looks for only some of the wanted solutions by starting from a reasonable guess[121], the other is the finite element method[122] which appears to be well suited to some problems which are too difficult for the methods discussed before. Both approaches are still at the developing stage in current literature[112] and therefore only time will clearly tell of their improved power in solving the scattering problem as opposed to the one of the other techniques that have been so widely tested in recent years.

Acknowledgements

I am grateful to Professor J. Peter Toennies for his kind hospitality in Göttingen, where this work was completed. I also thank Dr. F. Battaglia for his careful proof-reading of the final text.

REFERENCES

1. H.S.W. Massey, "Electronic and Ionic Impact Phenomena, Vol. 3, Slow Collisions of Heavy Particles", Clarendon Press, Oxford (1971).

2. N.F. Ramsey, "Molecular Beams", Clarendon Press, Oxford (1956).

3. P.R. Brooks and E.F. Hayes, Eds. "State-to-State Chemistry", ACS Symposium Se.n. 56, Am. Chem. Soc., Washington, DC (1977).

4. e.g. see: F.A. Gianturco "The Transfer of Molecular Energies by Collision", Springer Verlag, Berlin (1979).

5. M. Born and J. R. Oppenheimer, Ann. Phys., 84:457 (1927).

6. C. Eckart, Phys. Rev., 36:878 (1930).

7. H. Conroy, J. Chem. Phys., 47:5307 (1967).

8. S.F. Boys and N.C. Handy, Proc. Roy. Soc., Ser. A 310:43 (1969).

9. C.C.J. Roothaan, Rev. Mod. Phys., 23:69 (1951).

10. E. Clementi and D.R. Davis, J. Comp. Phys., 1:223 (1966).

11. E.F. Hayes and R.G. Parr, Progr. Theor. Phys., Suppl. 40:78 (1967).

12. F.A. Gianturco and D.G. Thompson, Chem. Phys., 14:111 (1976).

13. J.C. Slater "Quantum Theory of Atomic Structure", Vol. 2, McGraw-Hill, New York, (1960).

14. C.C.J. Roothaan and P. Bagus, Methods Comput. Phys., 2:47 (1963).

15. R.K. Nesbet, Advan. Quantum. Chem., 3:1 (1967).

16. R.K. Nesbet, Advan. Chem. Phys., 14:1 (1969).

17. I.H. Hillier, V.R. Saunders and M.H. Wood, Chem. Phys. Lett. 1:323 (1970).

18. R. McWeeny and B.T. Sutcliffe "Methods of Molecular Quantum Mechanics", Academic Press, New York (1969).

19. J.C. Slater, "Quantum Theory of Molecules and Solids", McGraw-Hill, New York (1963).

20. W. Meyer, J. Chem. Phys., 58:1017 (1973).

21. C.E. Dykstra, H.F. Schaefer and W. Meyer, J. Chem. Phys. 65:2740 (1976).

22. J.A. Pople, R. Seeger and R. Krishnan, Int. J. Quantum. Chem., 115:149 (1977).

23. G. Das and A.C. Wahl, Adv. Quantum Chem., 5:261 (1970).

24. P.J. Hay, T.H. Dunning and W.A. Goddard, J. Chem. Phys. 62:3912 (1975).

25. K. Kirby-Docken and B. Lin, J. Chem. Phys., 66:4309 (1977).

26. F.A. Gianturco "The Transfer of Molecular Energy by Collisions", Chapter 2, Springer Verlag, Berlin (1979).

27. H.F. Schaefer III, in "Atom-Molecule Collision Theory" R.B. Bernstein Ed., Plenum Press, New York (1979).

28. J.P. Toennies, Chem. Soc. Rev., 3:407 (1974).

29. W.A. Lester Jr., J. Chem. Phys., 53:1511 (1970).

30. W.A. Lester Jr., J. Chem. Phys., 54:3171 (1971).

31. V. Staemmler, Chem. Phys., 7:17 (1975).

32. W. Kutzelnigg, V. Staemmler and C. Hoheisel, Chem. Phys., 1:27 (1973).

33. J. Schaefer and W.A. Lester Jr., J. Chem. Phys., 62:1913 (1975).

34. R.F. Bader and R.A. Gangi, Theoret. Chem., Vol. II, The Chemical Society, London (1975).

35. P. Siegbahn and B. Liu, J. Chem. Phys., 68:2457 (1978).

36. I. Shavitt R.M. Stevens, F.L. Minn and M. Karplus, J. Chem. Phys., 48:2700 (1968).

37. R.N. Porter and M. Karplus, J. Chem. Phys., 40:1105 (1964).

38. I.G. Czismadia, R.E. Kari, J.C. Polanyi, A.C. Roach and M.A. Robb, J. Chem. Phys., 52:6205 (1970).

39. C.E. Dykstra, A.S. Gaylord, W.D. Gwinn, W.C. Swope and H.F. Schaefer III, J. Chem. Phys., 68:3951 (1978).

40. R. Schinke, M. Dupuis and W.A. Lester Jr., J. Chem. Phys., 72:3909 (1980).

41. C.F. Giese and W.R. Gentry, Phys. Rev., A 10:2156 (1974).

42. W.R. Thorson, J. Chem. Phys., 39:1431 (1963).

43. N.A. Mullaney and D.G. Truhlar, J. Chem. Phys., 39:91 (1979)

44. L. Eno and G.G. Balint-Kurti, J. Chem. Phys., 71:1447 (1979)

45. C. Nyeland and A. Hunding, Chem. Phys. Lett., 5:143 (1970)

46. M. Baer, G. Drolshagen and J.P. Toennies, J. Chem. Phys., August 1980.

47. A.R. Edmonds "Angular Momentum in Quantum Mechanics", Princeton U.P., Princeton N.J. (1974).

48. L.N. Smith, D.J. Malik and D. Secrest, "Rotational Compound State Resonances for an Argon and Methane Scattering System", J. Chem. Phys., 71:4502 (1979).

49. A.M. Arthurs and A. Dalgarno, The Theory of Scattering by a Rigid Rotator, Proc. R. Soc. London, A 256:540 (1960).

50. W.A. Lester Jr., The N Coupled-Channel Problem, in "Modern Theoretical Chemistry", W.H. Miller ed., Vol. 1, Plenum Press, New York (1976).

51. J.M. Blatt and L.C. Bierdenham, Rev. Mod. Phys., 24:258 (1952).

52. M. Rose "Elementary Theory of Angular Momentum", J. Wiley & Sons, New York (1957).

53. S. Green, J. Chem. Phys., 62:2271, 3568 (1975).

54. J. Schaefer and W. Meyer, Theoretical Studies of H_2-H_2 Collisions, J. Chem. Phys., 70:344 (1979).

55. A.E. De Pristo, M.H. Alexander, J. Chem. Phys., 66:1334 (1977).

56. S. Green, P. Thaddeus, Ap. J., 205:766 (1976).

57. T.G. Heil, D. Secrest, J. Chem. Phys., 69:219 (1978).

58. G.G. Gray, Can. J. Phys., 46:135 (1968).

59. C.F. Curtiss and F.T. Adler, J. Chem. Phys., 20:249 (1952).

60. R.T. Pack, J. Chem. Phys., 60:633 (1974).

61. J.M. Launay, J. Phys. B, 9:1823 (1976).

62. J.P. Toennies, Chem. Soc. Rev., 3:407 (1974).

63. J. Schaefer and W.A. Lester Jr., J. Chem. Phys., 62:1913 (1975).

64. G.C. Schatz and A. Kuppermann, J. Chem. Phys., 65:4668 (1976).

65. M.H. Alexander, J. Chem. Phys., 61:5167 (1974).

66. J. Schaefer and W.A. Lester Jr., Chem. Phys. Lett., 20:575 (1973).

67. F. Battaglia and F.A. Gianturco, unpublished results.

68. G. Drolshagen, F.A. Gianturco and J.P. Toennies, J. Chem. Phys., Nov. 1980.

69. P. McGuire and D.J. Kouri, J. Chem. Phys.,60:2488 (1974)

70. R.T. Pack, J. Chem. Phys., 60:633 (1974).

71. M. Tamir and M. Shapiro, Chem. Phys. Lett., 31:166 (1975)

72. R.B. Walker and J.C. Light, Chem. Phys., 7:84 (1975).

73. D. Secrest, J. Chem. Phys., 61:710 (1975).

74. Y. Shimoni and D.J. Kouri, J. Chem. Phys., 65:3372 (1976).

75. Y. Shimoni and D.J. Kouri, J. Chem. Phys., 66:675 (1977).

76. G.A. Parker and R.T. Pack, J. Chem. Phys., 66:2850 (1977).

77. R. Goldflam, S. Green and D.J. Kouri, J. Chem. Phys., 67:4149 (1977.

78. R. Goldflam and D.J. Kouri, J. Chem. Phys., 66:542 (1977).

79. V. Khare, J. Chem. Phys., 67:3897 (1977).

80. V. Khare, D.J. Kouri and R.T. Pack, J. Chem. Phys., 69:4419 (1978).

81. D.J. Kouri and P. McGuire, Chem. Phys. Lett., 29:414 (1974).

82. M. Shapiro and M. Tamir, Chem. Phys., 13:215 (1976).

83. M. Alexander and R. McGuire, Chem. Phys. Lett., 12:31 (1976).

84. S. Green, Chem. Phys. Lett., 38:293 (1976).

85. D.J. Kouri, in "Atom-Molecule Collision Theory", R.B. Bernstein Ed., Plenum Press, New York (1979).

86. B.H. Choi, R.T. Poe and K.T. Tang, J. Chem. Phys., 69:411 (1978).

87. B.H. Choi, R.T. Poe and K.T. Tang, J. Chem. Phys., 69:422 (1978).

88. G.D. Barg and G. Drolshagen, Chem. Phys., 47:209 (1980).

89. D.J. Kouri, T.G. Heil and Y. Shimoni, J. Chem. Phys., 65:1462 (1976).

90. R. Schinke and P. McGuire, Chem. Phys., 28:129 (1978).

91. D.M. Chase, Phys. Rev., 104:838 (1956)

92. K. Takayanagi, Progr. Theor. Phys. Suppl., 25:1 (1963).

93. R.B. Bernstein and K.H. Kramer, J. Chem. Phys., 40:200 (1964).

94. T.P. Tsien and R.T. Pack, Chem. Phys. Lett.,6:54 (1970)

95. R.D. Levine, Chem. Phys. Lett., 4:211 (1969).

96. R. Goldflam, S. Green and D.J. Kouri, J. Chem. Phys., 67:5661 (1977).

97. G.A. Parker and R.T. Pack, J. Chem. Phys., 68:1585 (1978).

98. L. Monchick, K.S. Yun and E.A. Mason, J. Chem. Phys., 39:654 (1963).

99. S.I. Chu and A. Dalgarno, Proc. R. Soc. London, A 342:191 (1975).

100. V. Khare, J. Chem. Phys., 68:4631 (1978).

101. G. Bergeron, C. Leforestier and J.M. Launay, Chem. Phys. Lett., 59:129 (1978).

102. A.E. DePristo and M.H. Alexander, J. Chem. Phys., 64:3009 (1976).

103. A.E. DePristo and M.H. Alexander, Chem. Phys. Lett., 44:214 (1976).

104. A.E. DePristo and M.H. Alexander, J. Phys. B., 9:2713 (1976).

105. A.E. DePristo and M.H. Alexander, J. Chem. Phys., 66:1334 (1977).

106. For a thorough review, see: H. Rabitz, in 'Modern Theoretical Chemistry, Vol.I', W.-H. Miller Ed., Plenum Press, New York (1976).

107. S. Green, J. Chem. Phys., 62:3568 (1975).

108. F.A. Gianturco and U.T. Lamanna, Chem. Phys., 3:110 (1974).

109. F.A. Gianturco and U.T. Lamanna, Chem. Phys., 17:255 (1976).

110. F.A. Gianturco and U.T. Lamanna, Chem. Phys., 25:401 (1977).

111. D. Secrest, in R.B. Bernstein Ed., Plenum Press, N.Y. (1979).

112. D. Secrest, in "Atom-Molecule Collision Theory", R.B. Bernstein Ed., Plenum Press, N.Y. (1979).

113. W.N. Sams and D.J. Kouri, J. Chem. Phys., 51:4809 (1969).

114. B.H. Choi and K.T. Tang, J. Chem. Phys., 63:1775 (1975).

115. R.G. Gordon, J. Chem. Phys., 51:14 (1969).

116. J.C. Light, in "Methods of Computational Physics", Vol. 10, B. Alder and M. Rotenberg Eds., Academic Press, N.Y. (1971).

117. A.S. Chung and D.J. Wilson, J. Chem. Phys., 51:3448 (1968).

118. B.R. Johnson and D. Secrest, J. Math. Phys., 7:2187 (1966).

119. B.R. Johnson, J. Comp. Phys., 13:445 (1973).

120. J.C. Light and R.B. Walker, J. Chem. Phys., 65:4272 (1976).

121. L.D. Thomas, J. Chem. Phys., 70:2979 (1979).

122. H. Rabitz, A. Askar and S. Cakmak, Chem. Phys., 26:61 (1978).

CHEMICAL PHOTOPHYSICS⁺

R.D. Levine

Department of Physical Chemistry
The Hebrew University
Jerusalem, Israel

ABSTRACT

Chemical photophysics is the study of the chemical consequences of selective optical excitation. This lecture is an eclectic review of the new types of phenomena whose study became possible by the introduction of lasers and of the new concepts required to interpret such experiments.

1. INTRODUCTION

2. PHOTOSELECTIVE CHEMISTRY

   2.1 The Discrete Region
   2.2 The Quasicontinuum
   2.3 Radiationless Electronic Transitions
   2.4 The Dissociative Continuum
   2.5 Multiphoton Ionization

3. THE CONSTRAINED PHASE SPACE

   3.1 Entropy and Surprisal
   3.2 Maximum Entropy
   3.3 Fragmentation Patterns in Multiphoton Ionization
   3.4 Future Developments

⁺Work supported by the Air Force Office of Scientific Research (AFOSR).

The discussion here will be on the introductory level. Three, very recent, references[19-21] provide much additional material.

Lasers have contributed significantly to our understanding of both inter- and intra-molecular dynamics not only by the preparation of the initial state but also by their use to probe its subsequent evolution and/or the distribution of final states. The availability of tunable laser sources, and of ultra-short laser pulses (down to the picoseconds region) provides powerful tools for this purpose. Examples of such applications will be noted. The primary division of the lecture is however in order of increasing level of molecular excitation.

2.1 The Discrete Region

The low lying vibrotational states of small and medium sized molecules are discrete and comparatively sparse. The molecule

1. INTRODUCTION

Collision theory often starts with a well defined initial state. This is seldom the case in chemistry. Indeed, traditionally chemical reactions were studied under conditions which insured a thermal equilibrium. The initial state is then not a pure state but a mixture where the weight of each quantum state is given by the Boltzmann factor. The only parameter subject to experimental control is the mean energy (i.e. the temperature) of the mixture. As in physics, one can however probe chemical reactions on a molecular level[1]. To do so it is necessary to select the initial state and/or resolve the distribution of the final, post collision, states. Over the last twenty years there has been a considerable activity in this direction on both the experimental and theoretical fronts. The resulting observations and inferences have been very extensively reviewed[1-18] and the key results are now to be found in the standard textbooks of physical chemistry.

2. PHOTOSELECTIVE CHEMISTRY

An important and recent development has been the use of lasers to selectively prepare the initial state. Currently available lasers provide photons over the entire range of the spectrum of chemical interest. Moreover, some lasers can provide high powers sufficient for the absorption of more than one photon per molecule, thereby preparing states that are not readily accessible by other excitation methods. This lecture examines the nature of the excited states of molecules that can be pumped by laser radiation and the types of dynamical processes that such excited states can engage in.

absorbs light only at definite frequencies and the absorption spectrum is a series of sharp spikes. Our thinking about this region is conditioned by the level structure of diatomic or triatomic molecules, shown for HCl in Fig. 1. Since the radiative widths of neighboring states do not overlap it is possible, using a laser, to excite a particular state. Except for spontaneous radiative decay (which in the infrared is quite slow) such a prepared state can decay only via a bimolecular collisional mechanism.

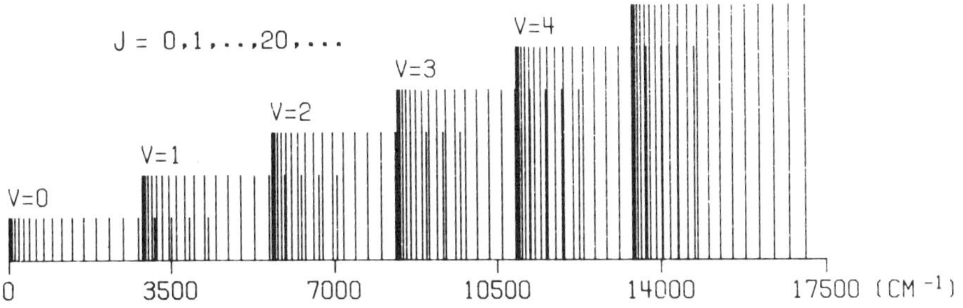

Fig. 1. Vibrotational energy levels of HCl illustrating the discrete spectrum of small molecules. The diagram shows the first twenty five rotational levels in each of the five vibrational levels. (For clarity, levels in different manifolds are designated by lines of different heights).

Consider the Cl + HI reaction

$$Cl + HI \rightarrow I + HCl(v), \qquad (1)$$

using thermal reactants which (cf. Fig. 2) produces HCl molecules in the vibrational state v=3 in preference to v=2,1 or 0. At equilibrium, the rate of formation of HCl(v) molecules by the reaction (1) equals the rate of their removal by the reversed reaction

$$I + HCl(v) \rightarrow Cl + HI. \qquad (2)$$

At equilibrium, the concentration of HCl in v=3 is very small, most molecules being in v=0. Despite the low concentration, the rate of reaction (2) is higher for v=3 than for v=2,1 or 0. This, detailed balance argument[19,23,24] shows that one can significantly enhance the rate constant* of the I + HCl reaction by pumping HCl molecules to higher vibrational states. Fig. 3 shows the rate constant for the

$$Br + HCl(v) \rightarrow Cl + HBr \qquad (3)$$

reaction. Relative rate constants for v=1-4 were determined using

*Recall that the rate is a product of the concentrations of I and of HCl(v) times the rate constant.

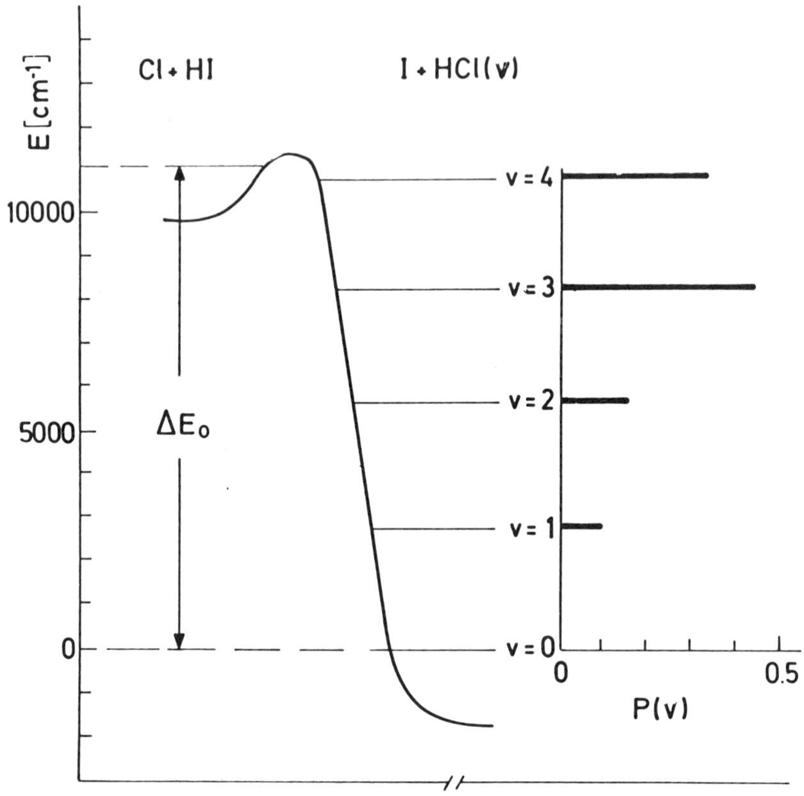

Fig. 2. Potential energy profile and vibrational energy disposal of the reaction (1). Shown on the right are the energies and the final state populations of the vibrational states of HCl. (Experimental results from ref. 22).

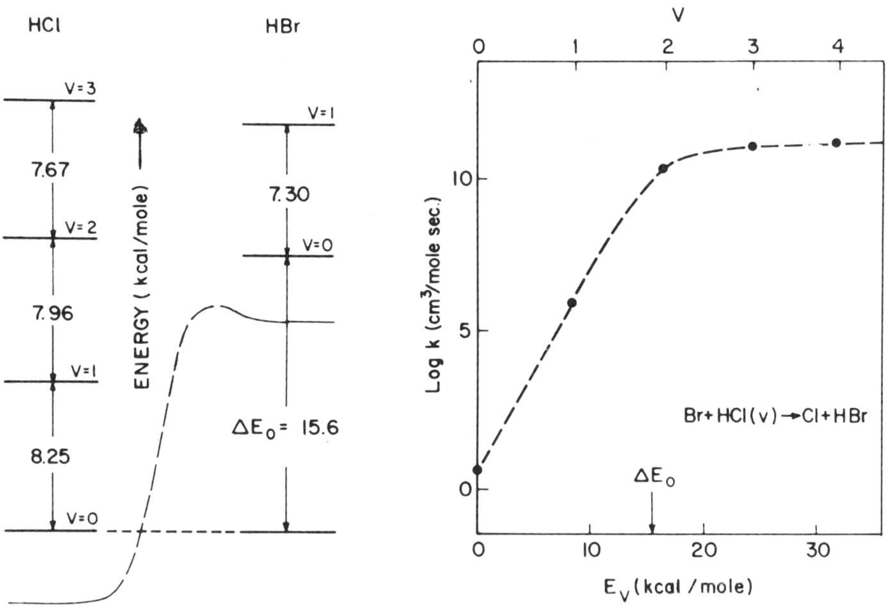

Fig. 3. Energy level diagram (left) and the rate constants (logarithmic scale) of the reaction (3) vs. the vibrational excitation of HCl (right). Note the change from an exponential behaviour in the endoergic regime to a much more moderate increase in the exoergic regime. (For experimental data see refs. 18,25-27).

excited HCl molecules formed in the Cl+HI reaction (cf. Fig. 2). The decrease in the HCl(v) concentration was monitored[25] by following the infrared luminescence. Absolute rate constants for v=1 and 2 were determined using laser excited HCl molecules by measuring the rate of appearance of Cl atoms[26,27]. Another method to obtain the endoergic rate constants is to convert the rates of the reverse exoergic reaction Cl + HBr → HCl(v) + Br via the detailed balance principle, as is discussed in refs. 23 and 24.

One notable feature of Fig. 3 deserves comment. This is the much more moderate effect of HCl vibrational excitation past the v=2 level. This reflects the rule of thumb (known[1] as 'the exponential gap rule') that direct molecular collisions distinctly disfavor changes in the translational energy. For the Br + HCl(v) reaction, as long as v < 2, the internal energy of the reactants is lower than the ground state energy of the products (cf. Fig. 3). The energy required for the reaction is provided by the translational energy. Increasing the HCl vibrational excitation decreases the energy defect (makes the reaction less endothermic) and leads to an exponential enhancement of the reaction rate. For v=2 the process is about isoenergetic. The gap has been closed. For v > 2, the internal energy of the reactants exceeds the ground state of the products, and kinetic energy can be released. Since the rate does not decline upon increase of the HCl vibrational excitation beyond v=2 it is reasonable to suggest that the excess HCl vibrational energy is preferentially released not as translation but rather as internal excitation of HBr.

The preferential conversion of reactant vibrational energy into vibrational rather than kinetic energy of the products is demonstrated in Fig. 4 for the

$$F + HCl(v) \rightarrow Cl + HF(v') \tag{4}$$

reaction. The vibrational excitation of HCl leads to an increase in

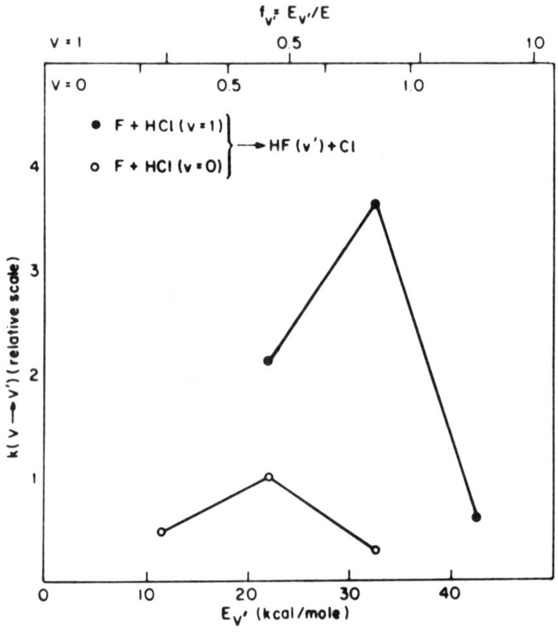

Fig. 4. Relative rate constants of the reaction (4) for reactants in two different vibrational states. The product state distributions are shown vs. the products vibrational energy, $E_{v'}$, vibrational state, v', and the fraction of the total reaction energy (E) appearing as products vibration, $f_{v'}$. (Adapted from ref. 28).

the mean vibrational energy of the product HF molecules.

Upon collision, a vibrationally excited HCl molecule can be deactivated. Such energy transfer processes can compete with the reactive collisions. In other words, the rate of depletion of HCl(v) molecules due to collisions with a Br atom is due to two distinct processes,

$$HCl(v) + Br \begin{array}{l} \xrightarrow{k_r} Cl + HBr \\ \xrightarrow[k_{nr}]{} HCl(v') + Br \end{array} \tag{5}$$

only one of which leads to the desired products. Given the measured vibrational energy disposal in the Cl + HBr reaction, detailed balance can be used to infer the dependence of k_r on the vibrational excitation of HCl. The enhancement of k_r with vibrational energy does not necessarily imply that k_{nr} is small. In fact such energy transfer processes can effectively compete with and reduce the efficiency of laser driven processes.[29,30]

To successfully compete with collisional deactivation it is thus necessary to achieve a high reaction rate which requires reagents excited past the threshold energy for reaction. Single photon excitation from the ground (v=0) vibrational state to some high lying (v>1) state is only possible for anharmonic vibrations. Near v=0 the anharmonicity is small so that the, so called, 'overtone', absorption is weak but can be observed given a sufficiently high flux of photons. One way of realizing the required conditions is to perform the pumping inside a laser cavity.[31]

Chemical reactivity is governed not only by energetic but also by steric requirements. Reaction may occur much more readily when the reagents approach along a preferred configuration[32]. Lasers can be used to prepare oriented molecules, as follows: The transition probability for light absorption depends on the orientation of the molecular transition dipole with respect to the electrical field, (i.e. as $\underset{\sim}{\mu}\cdot\underset{\sim}{E}$), being maximal when the two are parallel. Molecules which have absorbed polarized light are thus preferentially aligned.

The reaction

$$Sr + HF(v) \rightarrow SrF + H \tag{6}$$

is endothermic by about 6.5 kcal/mole for v=0 and becomes exothermic by about 5 kcal/mole for v=1. The rate constant of (6) has been determined using an HF laser to excite the reagent molecule and the laser-induced-fluorescence method (LIF, ref. 33) to monitor the SrF(v') products. It was found that while for HF(v=0) the reaction rate was negligible, a high yield was observed with HF(v=1).

Using a polarized radiation of an HF laser, it is possible to prepare the HF(v=1) molecules in a selected orientation with respect to the direction of the Sr atom propagation, Fig. 5. Thus, choosing the proper polarization, parallel ($\parallel$) or perpendicular ($\perp$) to the Sr beam, the Sr + HF(v=1) collision occurs, preferentially along a collinear or a side-way approach, Fig. 5. The fraction of SrF(v'=2) products is found to be larger in sideway collisions.

Electronically excited states of diatomic molecules often also belong to the classification of 'discrete spectrum'. The higher reactivity of electronically excited molecules has been used for, e.g. isotopic separation. Thus[35], ^{35}Cl and ^{37}Cl can be separated when $I^{37}Cl$ is laser pumped to the $A^3\Pi$ state in the presence of scavengers

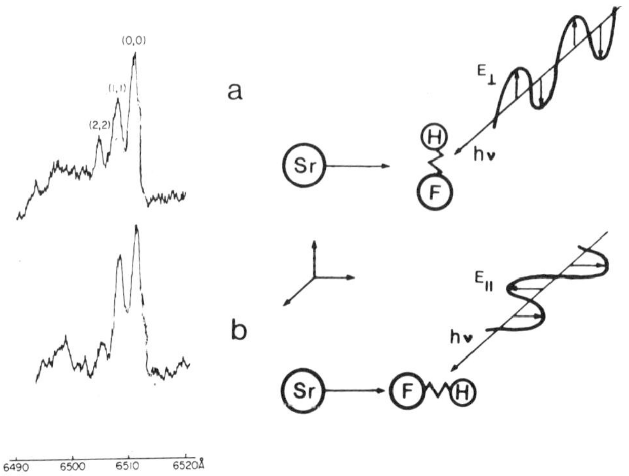

Fig. 5. The distribution of $SrF(X^2\Sigma^+)$ molecules formed by the reaction Sr + HF(v=1,J=1), where HF molecules are preferentially aligned a) perpendicular, b) parallel to the approaching Sr atom. The alignment is achieved by selecting the polarization of the exciting HF laser (right panel). The left panel shows the excitation spectrum of the $SrF(A^2\Pi_{3/2} \rightarrow X^2\Sigma^+)$ transition. A dye laser excites a particular vibronic transition, and the total resulting fluorescence is monitored as a function of the excitation wavelength. Vibrational population ratios are obtained from known Franck Condon factors showing that a sideway attack favors higher vibrational excitation in the products. (Adapted from ref. 34).

which do not react with the ICl molecule in its ground electronic state. A portion of the visible absorption spectrum of ICl is shown in Fig. 6. It is evident that despite the heavy masses, the isotope shift is large enough, making even a broadband laser suitable for preferential excitation of one isotopic species in the presence of the other.

Because of the small energy gaps, cf. Fig. 6, energy transfer between different isotopic species is usually very rapid, tending to spoil the selectivity established in the initial excitation stage. When the reaction rate with the scavenger is larger than the scrambling rate, high overall isotopic enrichment can be achieved, as shown in Fig. 7.

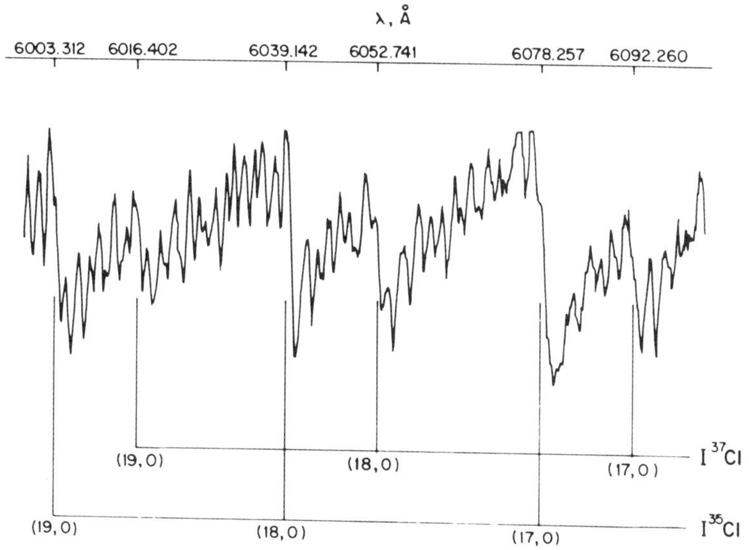

Fig. 6. A portion of the medium resolution electronic absorption spectrum, illustrating the isotopic shift between $I^{37}Cl$ and $I^{35}Cl$. Even a broadband laser ($\Delta\nu \sim 10$ cm^{-1}) can be used to selectively excite one of the isotopic species. (Adapted from ref. 35).

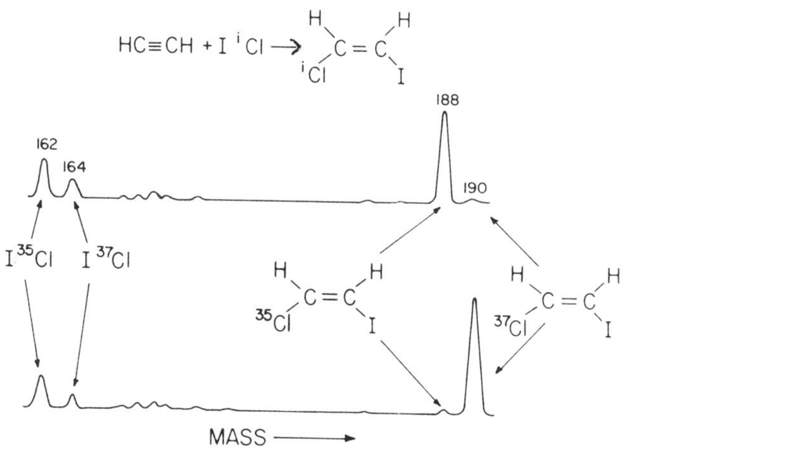

Fig. 7. Mass spectrometric analysis of the products of the reaction of photoexcited ICl with acetylene. By tuning the laser to an absorption line of either chlorine isotope, natural abundance starting material can be enriched to more than 97% purity in a single step. (Adapted from ref. 36).

We have seen that energy requirements[37] and energy disposal in chemical reactions is very strongly (i.e. exponentially) dependent on the gap (or energy mismatch) between the initial and final states. This gives rise to two questions, one practical and the other theoretical. The gap can be reduced by using lasers to excite one (or both) of the reagents before the collision. Energy transfer processes tend however to degrade some of the excitation. Can lasers be used to close the gap during the actual collision? This possibility is discussed both in this volume and elsewhere[38]. The second question is: Can one offer a more quantitative characterization of the exponential gap in particular and, in general, of the dynamical constraints that operate during the collision. This question, which has been extensively discussed elsewhere[19,39,40] is taken up in section 3.

2.2 The Quasicontinuum

For larger molecules and/or at higher excitation energies the number of molecular states per energy interval rapidly increases. Fig. 8 illustrates the congestion which is revealed upon improved resolution in a large but not giant molecule (SF_6) at a comparatively low ($\sim$1000 cm^{-1}, one quantum in the ν_3 mode) energy. Considerable simplification of the observed spectra is possible by cooling such molecules to the exceedingly low temperatures made possible using supersonic nozzles. Under these conditions only the very lowest rotational states are populated, and hence most of the transitions possible at room temperature are of negligible intensity. Even so, as the excitation energy is increased, so does the number of final states*. Eventually the manifold of final states appears to be a continuum on the intrinsic energy resolution of the experiment, (determined, for instance, by the exciting laser bandwidth). It is then referred to as the quasicontinuum. Note that the definition is an operational one - the quasicontinuum is reached when it is no longer possible to excite a single molecular eigenstate under the given experimental conditions.

Direct, single (typically visible) photon absorption into the quasicontinuum region of large molecules and emission from higher lying states into the quasicontinuum have been observed[42]. Most eigenstates in the quasicontinuum have only very small optical transition probabilities from the ground state. The state prepared by optical excitation is thus a superposition of eigenstates and hence

---

*In the classical limit, where internal energy can vary continuously, the number δN of excited vibrotational states in the energy range E to E+δE increases with E as $\delta N \propto E^{\alpha-1}\delta E$.Here α=s+r/2 where s and r denote the number of vibrational and rotational degrees of freedom of the molecule, respectively.

**The superposition of delocalized vibrational eigenstates to describe a localized vibration is analogous to the superposition of the delocalized states of a free particle to describe a localized wave packet.

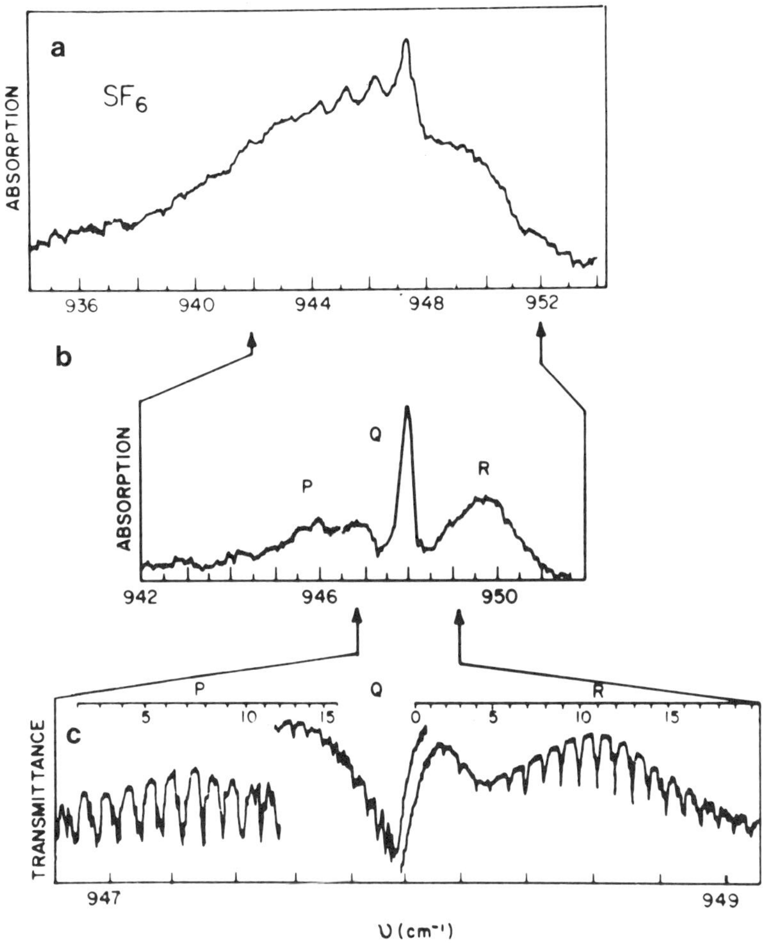

Fig. 8. A portion of the vibrational absorption spectrum of SF_6, corresponding to single photon absorption in the ν_3 mode. a) A low resolution spectrum at room temperature. b) A grating spectrum at 123K. P,Q,R branches can now be identified. (P,Q and R transitions correspond to $v,J \rightarrow v+1,J'$ with $J'=J-1,J$ and $J+1$, respectively). c) A high resolution spectrum of SF_6 at 123K obtained with a tunable diode laser having a resolution of 0.001 cm^{-1} (adapted from ref. 41, where the original experiments are cited). Each of the P,Q and R "lines" can be shown to be further split and is thus really a manifold of lines. The splittings are mainly due to rotation-vibration interaction (Coriolis coupling).

has a character similar to the overtone spectra of diatomic molecules i.e. the excitation is localized in a particular mode**[43]. A superposition of eigenstates of somewhat different energies is no longer an eigenstate of the molecular Hamiltonian and hence the states prepared by optical excitation into the quasicontinuum are not stationary but are coupled (by the molecular potential) to other states. Energy can be pumped into the quasicontinuum but only indirectly, i.e. via these 'doorway' states. This theme, that the optically prepared state in the quasicontinuum is typically non-stationary will recur throughout this section.

Fig. 9 shows the CH overtone absorption spectra of CH_2=CH-CH_2NC. The three overtone bands corresponding to the three types of CH sites (CH_2=, =CH- and -CH_2-) are identified. Very many other states are present in this energy range, but because of very unfavorable transition probabilities (due to the need to simultaneously displace several modes from equilibrium), contribute only very weakly as a background absorption.

The overtone bandwidths are much larger than the band contours in the fundamental CH stretch region, Fig. 9. This is taken to reflect the non-stationary character of the optically prepared state. If we think of this state as n phonons localized in the CH stretch mode then it is not an eigenstate of the molecular Hamiltonian. It can undergo intramolecular transitions, say to a state with n-1 quanta at the original site plus one CH stretch quantum on a different site. (The energy mismatch due to the anharmonicity of the vibration must be compensated by energy exchange with other modes). Such states in turn are coupled to other states and ultimately the energy deposited by the optical excitation is spread throughout the molecule. The large widths indicate (by the uncertainty principle) that the initial evolution of the prepared state will be rapid, of the order of few vibrational periods. It is therefore not necessarily the case that one can obtain bond selective chemistry using laser excitation into the quasicontinuum.

Allyl isocyanide can isomerize to allyl cyanide.

$$CH_2 = CHCH_2NC \rightarrow CH_2 = CHCH_2CN \tag{7}$$

The available experimental results, Fig. 9, do show however that while the rate constant for unimolecular isomerization (represented by the overall conversion efficiency) depends primarily on the total energy, it does show some dependence on the particular CH site where this energy was initially present. While the optically prepared state is not stationary, it does not, on the time scale of the unimolecular reaction (which is quite slow) quite reach a distribution of the excess energy which is independent of the site of the original excitation.

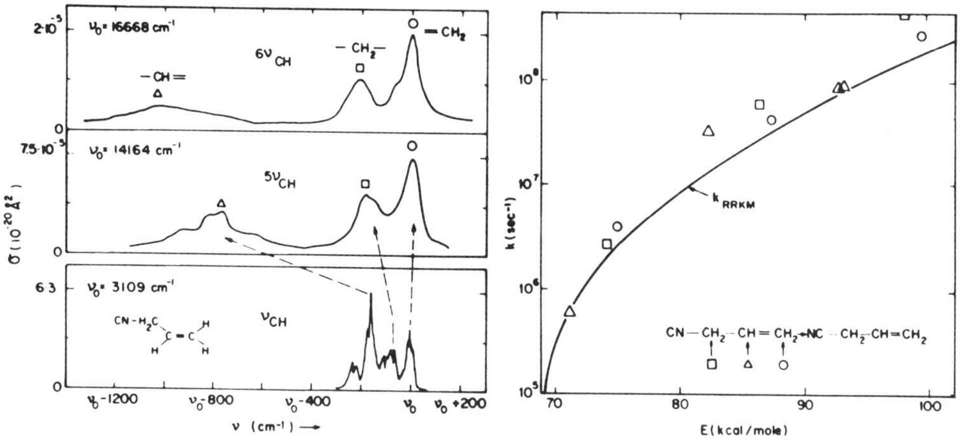

Fig. 9. Left: The fundamental, ν_{CH}, the fifth overtone, $5\nu_{CH}$, and the sixth overtone, $6\nu_{CH}$, spectra of allyl isocyanide as measured in intracavity dye laser experiments[43]. The overtone spectra are broadened and shifted with respect to the fundamental but the local mode structure is preserved even when the molecule is excited directly to the quasicontinuum. Right: The rate constant for the unimolecular photoisomerization of allyl isocyanide as a function of the excitation energy and the site excited (marked by □, Δ and o for the three types of CH bonds). The solid curve is the prediction of the RRKM theory, which assumes that energy randomization in the molecule is complete. (Adapted from ref. 31).

An observation that generated considerable activity[44-48] is that polyatomic molecules can be pumped to the quasicontinuum region and ultimately to dissociation by high power infrared lasers even under collision free conditions. Moreover, the process is species selective. It is possible to dissociate only those molecules corresponding to a particular isotopic content by using the shifts in the absorption spectrum of different isotopic molecules. Fig. 10 shows the fraction of SF_6 molecules which have undergone dissociation as a function of the laser fluence. The laser energy determines the average number of laser quanta $\langle n \rangle$ absorbed by the dissociating molecules. The process can be described as

$$SF_6 + nh\nu \rightarrow SF_5 + F \qquad (8)$$

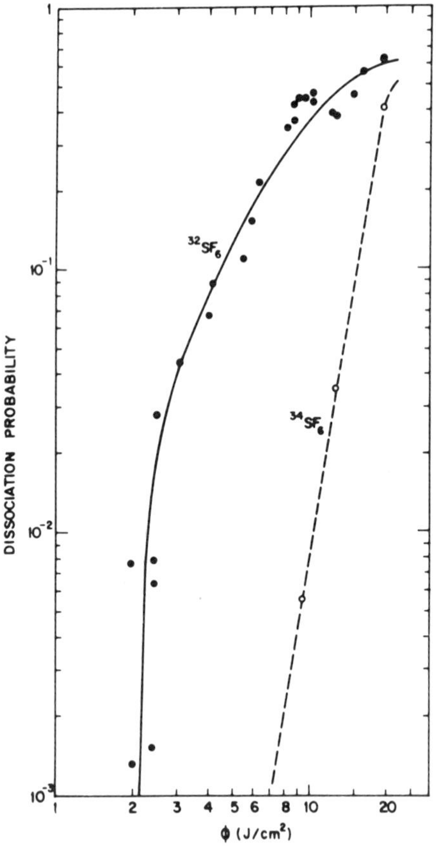

Fig. 10. The fraction of $^{32}SF_6$ and $^{34}SF_6$ molecules undergoing dissociation by multiphoton absorption as a function of the flux, ϕ, of the CO_2 laser radiation. (ϕ, often called the 'fluence' is the total pulse energy, per unit area of the laser beam). The data were obtained by irradiating low pressure (<1 torr) samples of SF_6 static gas cells with the P(20) line (10.6 μm) of a pulsed CO_2 laser. The product compositions were analyzed in a mass spectrometer. Below 0.3 torr the dissociation yields of the two isotopes are pressure independent. Above 1 torr the isotopic selectivity is lost due to scrambling collisions. (Adapted from ref. 44).

Species selectivity is achieved primarily via absorption into the low lying, discrete regime which serves as a bottleneck for entry into the quasicontinuum. This interplay between the two regimes has been studied in some detail and the following is only a brief introduction: Tens of infrared photons need be absorbed in order to dissociate a typical chemical bond. Since the laser photons have all the same frequency, the question is what mechanism allows their absorption by an isolated molecule where due to anharmonicity, levels are not equally spaced. For a large molecule at higher temperatures, the quasicontinuum provides the qualitative answer. The mean energy per (classical) vibration is $\underset{\sim}{k}T$. Hence if there are many modes and/or the temperature is high, a mean energy content of the molecule is high enough so that many states of the quasicontinuum are populated to begin with. For any initial state there will thus be a final state where their energy gap is matched by the photon frequency. Of course, this answer is somewhat facile since it disregards the question of the magnitude of the transition probability (the 'oscillator strength'), but it does provide for the more fundamental problem of a frequency matching.

For smaller molecules, the absorption of the first few photons requires a near coincidence between the laser frequency and a sequence of molecular transitions. The coincidence should be close but need not be exact due to power broadening. Fig. 11 shows the results of a pump and probe experiment on the ν_3 vibrational mode of SF_6. A Q-switched high power CO_2 laser is used to pump SF_6 molecules from the ground to the first excited state of the ν_3 normal mode. Due to the pumping, there is a depletion of ground state molecules which can absorb light at the frequency of the pump laser. A tunable, low power infrared (diode) laser is then used to measure the absorption. The absorption spectrum thus shows a transmission window, a phenomenon termed 'hole burning' or induced transparency. As the pump laser power is increased so does the width of the hole. Molecules which are not in exact resonance, can at high field strengths, absorb energy from the field. The hole width at half dip is known as the Rabi frequency and, depending on the magnitude of the transition probability, can reach 1 cm^{-1} or even higher, for currently available CO_2 lasers.

At higher pressures, collisions will serve to replenish any state depleted due to absorption. The sublevels of a given J are very close in energy and so are very effectively coupled by collisions. At somewhat higher pressures or over longer times, collisions will also couple states of different J's. The 'double resonance' experiment of Fig. 11 can be used to monitor this population transfer by introducing a delay between the pump and the probe lasers. The relaxation time is reflected by the rate of replenishing the hole burnt by the pump laser. The rotational relaxation time due to SF_6-SF_6 collisions corresponds to an 'effective' collision cross section of 170 $Å^2$.

The absorption of a second laser photon by SF_6 can follow a similar route: a $\nu_3 \rightarrow 2\nu_3$ transition where now, because of anharmonicity, a J → J (so called 'Q-branch') transition offers the closest frequency match. Fine tuning is again provided by the power broadening and the Coriolis splitting of the upper rotational level. Alternatively, at high powers, two photons can be coherently absorbed. Because the first photon is very nearly resonant with molecular transitions, the cross section for such a two photon process can be

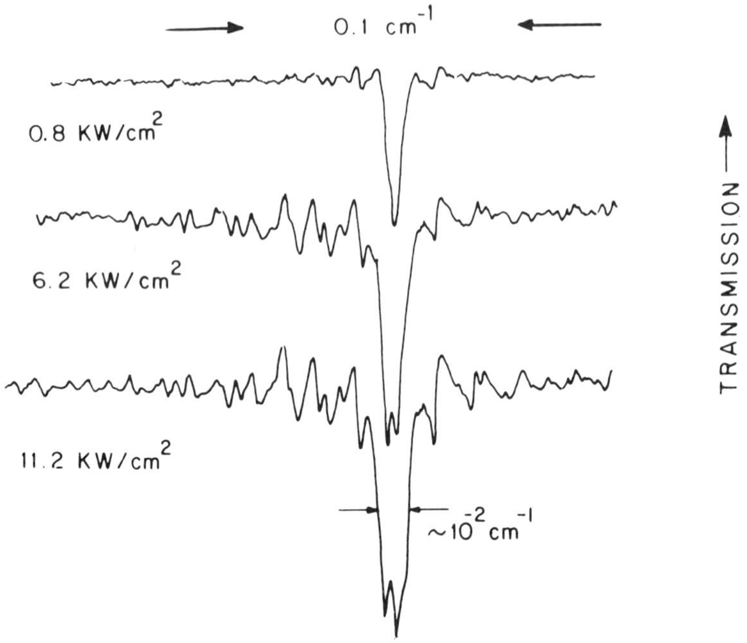

Fig. 11. Power broadening of the infrared spectrum of 0.08 Torr SF_6. The P(18) line of a high power CO_2 laser is used to pump SF_6 molecules from the ground to the first excited level of the ν_3 vibrational mode. The changes in the ground level population due to the pumping are reflected in the transmission spectrum of a continuous, low power infrared (diode) laser whose spectral width is less than 0.001 cm^{-1}. The three traces are the transmission spectra corresponding to three different power levels of the pump laser. The range of frequencies excited increases with the power as is reflected by the widths of the dips ('holes') in the transmission spectrum. The width of these power broadened dips is the order of the Rabi frequency, $\mu\varepsilon_o/h$, and are thus proportional to $\mathcal{P}^{1/2}$. μ is the transition dipole, ε_o the electric field amplitude and $\mathcal{P} \propto \varepsilon_o^2$ is the laser power. (Adapted from ref. 45).

molecule up in the ladder of states. Due to anharmonicity the excitation at any given energy is being rapidly dispersed by the intramolecular coupling among the many accessible isoenergetic states. The two processes are in competition and state-selective pumping requires that the pumping up rate be faster. One way of approaching this goal is to take larger steps - i.e. to use higher energy laser photons.

Past the lowest threshold for dissociation the up-pumping competes also against the depletion of molecules by dissociation. For a large molecule quite a few photons can be absorbed before the rate of dissociation is dominant*. The fragments will then carry quite a bit of internal excitation. Such fragments may then be in the quasicontinuum and can themselves absorb the laser radiation, leading ultimately to secondary or even tertiary dissociation processes. Extensive fragmentation of polyatomic molecules has indeed been observed using high power lasers. (See also Sec. 2.5).

The up-pumping of molecules beyond the lowest threshold is particularly evident when there are several dissociation channels. An example is

$$C_2F_5Cl \rightarrow C_2F_5 + Cl \tag{9}$$

where the threshold is at about 83 kcal/mole and

$$C_2F_5Cl \rightarrow CF_3 + CF_2Cl \tag{10}$$

with a somewhat higher threshold (the C-C bond energy is, roughly, 100 kcal/mole). At lower photon flux the first process dominates but at higher energy fluence the C-C bond rupture reaction becomes increasingly important, Fig. 12. Analysis of the products energy distribution verifies that the C_2F_5Cl molecule can indeed be pumped to energies considerably above its lowest dissociation threshold. significant. At high powers even the simultaneous absorption of three or more photons is possible.

Having absorbed three or more photons, an SF_6 molecule is essentially in the quasicontinuum region. One is then no longer worried about frequency matching but a new worry arises. In the quasicontinuum, states with say n quanta in the ν_3 mode are essentially at the same energy with states with n-1 quanta in ν_3 and the balance in other modes or with n-2 quanta in ν_3 and the balance in other modes, etc. Any one such state is coupled by the anharmonicity to the other states and is not stationary. The pumping moves the

*The reason is that dissociation requires the localization of energy along the coordinate which is broken[1,9,12,13]. At low excess energy, this is a low probability event.

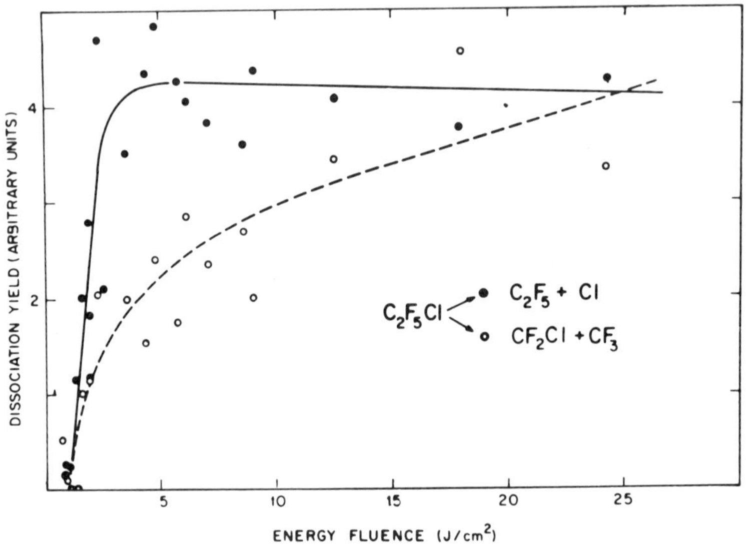

Fig. 12. Multiphoton dissociation yield of C_2F_5Cl by a CO_2 laser vs. the laser energy fluence. At low fluence Cl elimination is the dominant process. At higher fluence the elimination saturates while the rupture of the stronger C-C bond becomes increasingly important. (Adapted from ref. 46).

2.3 Radiationless Electronic Transitions

Above the ground potential energy surface lie the potential energy surfaces of electronically excited states. The single photon transition probability from a low lying vibrotational state to a dissociative state on the same surface is exceedingly small*. Hence optical, single photon, excitation typically prepares a state which is initially confined to the upper surface. The novel feature associated with such surfaces for polyatomic molecules is that any vibrotational state of an excited surface is essentially isoenergetic with one or more states belonging to a lower surface**. If the minimal electronic excitation energy exceeds the dissociation energy on the ground surface then such a coincidence is assured: any state on the upper surface is exactly isoenergetic with a dissociative state of the lower surface. Since the Born-Oppenheimer separation into electronic and nuclear motion is only an approximation, such electronically excited states are not stationary. The coupling to the lower surface will ultimately result in dissociation. This, 'predissociation' is further mentioned in Sec. 2.4.

Often however the electronic excitation energy is sufficiently low so that the lower lying vibrotational states of the excited surface coincide in energy with the bound quasicontinuum region of the ground surface. This provides for an alternative route for entering the quasicontinuum and one which permits the potentially selective deposition of fairly large amounts of energy.

For a system at equilibrium all isoenergetic states are equally populated. In a large polyatomic molecule excited to the lower lying region of an excited electronic state there are overwhelmingly many more states in the quasicontinuum of the ground surface than there are in the same energy range in the excited surface. The intramolecular, interstate, relaxation will thus appear to be practically irreversible. The population in the excited surface will be severely depleted at a rate competitive or even exceeding that of its fluorescence (hence the name 'radiationless' transitions[49,50]). Of course, unless the population in the quasicontinuum is further drained*** by additional decay modes, the transition is not strictly irreversible. However, in a large molecule, the equilibrium is very much one sided.

---

*This is another reflection of the extreme reluctance of the nuclei to undergo large changes in their momenta. We have previously seen this reluctance in the role of energy in collisions. The propensity for such changes that require the least change in momenta is known in spectroscopy as the Franck-Condon principle.

**This energy matching is due to the quasicontinuum on the lower surface. For, say, a diatomic this coincidence is not necessarily the case.

***Such will be the case if it is not a quasicontinuum but a real, dissociative continuum, or if collisional deactivation or dephasing takes place, etc.

It is also possible to reach electronically excited states by the inverse process, e.g. by multiphoton pumping to the quasicontinuum of the ground potential energy surface. Transitions into an excited electronic surface can be detected both by the appearance of electronically excited fragments in multiphoton dissociation and by emission from low lying electronically excited states of the parent molecule. As discussed before, it is preferable to use a smaller molecule, with a low lying electronic state if the process inverse to radiationless electronic transitions is to be observed in practice.

2.4 The Dissociative Continuum

High above the well of a potential energy surface lies the continuum - a region where unbound motion of the nuclei is possible in at least one direction. A collision between two reactants always takes place in the continuum of some potential energy surface. If the unbound motion is possible only in one direction - the collision cannot lead to any rearrangement but can of course be inelastic. If there is more than one exit channel then the corresponding collision can be reactive*. The two equivalent ways of thinking about the continuum - as a dissociative state of a parent molecule or as a state formed by collision of two fragments - are often summarized by referring to dissociation as a 'half collision'.

If the potential energy surface is purely repulsive, it has only states in the continuum**. Except if the potential is everywhere repulsive, states in the continuum of a polyatomic system can be isoenergetic with discrete states. If the surface has a deep basin these discrete states will form a quasicontinuum. As discussed in Sec. 2.2, a molecule pumped to such a quasicontinuum will have a finite lifetime and need not dissociate within a vibrational period. Similarly, when two polyatomic fragments collide, they can be trapped in the multitude of bound states of the quasicontinuum and spend considerable time in the interaction region before they separate.

*Of course, a given potential energy surface may allow only inelastic collisions at lower energies but will allow reactive collisions at higher energies. This corresponds to a parent molecule having two dissociation modes, e.g. Fig. 12. Reactive collisions are possible past the threshold of the higher energy mode.

**Such surfaces are rare. It is typically the case that there is at least a shallow well due to van der Waals forces. These interactions may be weak in comparison with chemical binding but are strong enough to insure the existence of a bound spectrum even for such 'inert' systems as a molecule colliding with a rare gas atom.

Photodissociation from an excited potential energy surface is possible when the optically prepared state is isoenergetic with some continuum*. The dissociative continuum may belong to the same surface or to some other electronic state (predissociation). In the latter case, observable dissociation requires that the coupling be not too weak. Besides purely electronic considerations the magnitude of the coupling depends, as always, on the changes in the momenta of the nuclei. Efficient crossing requires a minimal change. This condition can be realized when the two surfaces cross, i.e. at that configuration of the nuclei where the two (electronic) potential energy surfaces have the same value. The spectroscopy of dissociative states provides therefore valuable input towards the mapping of ground and excited potential energy surfaces.

Due to surface crossing the dissociation products can appear in more than one electronic state. In the photodissociation[51] of CS_2, Fig. 13, the initial optical excitation is to a predissociating state, crossed by two dissociative states, one leading to $CS(X^1\Sigma)$ and $S(^1D)$ and the other to $CS(X^1\Sigma)$ and $S(^3P)$.

The photodissociation of alkyl iodides to yield electronically excited I atoms is the pumping mechanism for the iodine atomic laser. Fig. 14 shows the observed translational energy distribution in the photodissociation of CH_3I and of $n\text{-}C_3H_7I$. The possible pathways are

$$RI \xrightarrow{h\nu} \begin{cases} R + I^*(^2P_{1/2}) \\ R + I(^2P_{3/2}). \end{cases} \tag{11}$$

Due to the electronic excitation carried by the I* atom, the available energy (for distribution among translation and vibrotational excitation of R) is lower in the first path. The experimental results are that for CH_3I, about 20% of the products are formed with more translational energy than is consistent with the first path. Not so for $n\text{-}C_3H_7I$. There are practically no products with translational energy in excess of that allowed by the first path. This does not rule out the second path (since the $n\text{-}C_3H_7$ radical can soak up the energy not in translation). It does however suggest that formation of I* is more dominant for $n\text{-}C_3H_7I$ than for CH_3I.

*Direct excitation into the continuum of a polyatomic molecule is rare because of the small Franck-Condon factors.

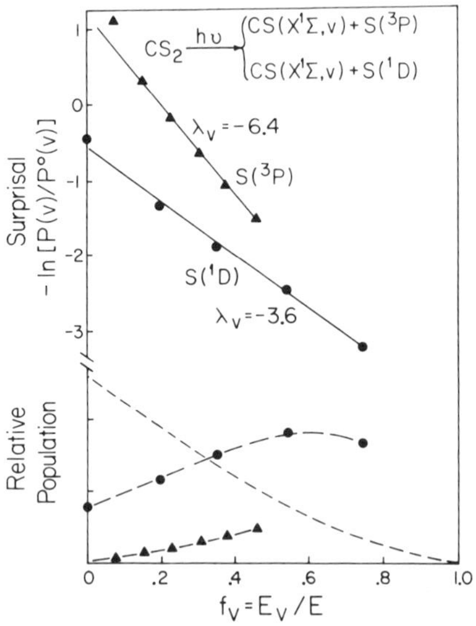

Fig. 13. The vibrational distribution (bottom panel) and its surprisal (upper panel, cf. Sec. 3) of CS in the photodissociation of CS_2. (Adapted from ref. 51). The dashed line is the vibrational distribution in the absence of any constraints, cf. Eq. (14) below.

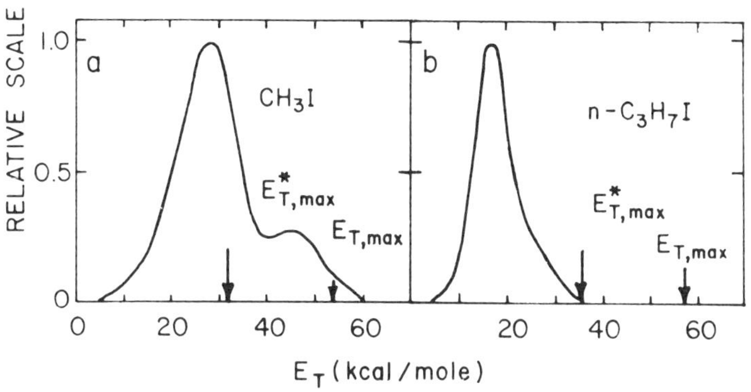

Fig. 14. The distribution of relative translational energy in the products in the photodissociation of RI. a) R = CH_3, b) R = n-C_3H_7. The maximal translational energy possible is thus $E_{T,max} = h\nu - D$ or $E^*_{T,max} = h\nu - D - E^*$ depending on whether a ground state or excited iodine atom is formed. E* is the electronic excitation. The experimental results show evidence of the formation of ground state I atoms only for CH_3I. The sharp decline of $P(E_T)$ towards $E^*_{T,max}$ in the case of n-C_3H_7I suggest that most iodine atoms are electronically excited. (Adapted from ref. 52).

2.5 Multiphoton Ionization

Highest on the chemical energy scale is ionization. Since the molecular ion is often formed with high internal excitation, ionization is frequently followed by fragmentation.

The final state in the ionization process is, as in dissociation, a continuum state. Here, a state corresponding to the unbounded motion of an electron and a molecular ion. For polyatomic molecules such states are isoenergetic with different quasicontinua of bound states. These include states where the electron is bound by an internally excited ion. (The energy required for the ejection of the electron is provided by the deexcitation of the ion). There may also be bound states of other, stable*, electronic surfaces. Optical excitation is often to one or more of these non-stationary quasicontinua, known as 'autoionizing' states. In short, direct photoionization is not necessarily the rule in polyatomics.

Using high power lasers in the visible or near uv one can provide the energy required for ionization by multiphoton absorption. Each photon provides a substantial energy and, particularly at high powers, the up pumping can compete with the lateral decay processes which deplete molecules which could otherwise continue to absorb. The fragmentation pattern of benzene, Fig. 15, and in particular the appearance of ions which require very high energies (such as C^+, whose minimal appearance requires 26 eV or about 8 photons) indicates that considerable extent of up pumping that can be achieved and/or that absorption by the primary ion followed by a secondary fragmentation is also possible. Note also that (Fig. 15) the fragmentation pattern is qualitatively different from that observed following ionization by (70 eV) electron impact. In Sec. 3.3 we shall however argue that the fragmentation patterns shown in Fig. 15 are largely statistical in character and differ primarily in the mean energy content per parent molecule[54], Fig. 18.

3. THE CONSTRAINED PHASE SPACE

What do we mean by 'selective'? Consider first a molecular collision. Because of the internal structure of the projectiles, it

*These are, of course, only nominally stable. The coupling between different electronic surfaces enables them to ultimately ionize.

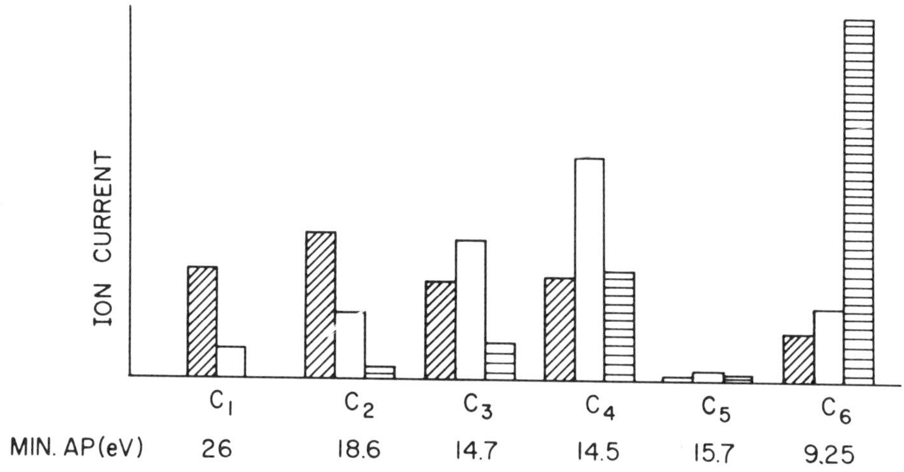

Fig. 15. Formation of fragment ions by dye laser (hν = 3.17 eV) induced multiphoton ionization of benzene. The bars show the relative abundance of fragments containing the indicated number of carbon atoms. The bars on the right are for electron impact ionization at 70 eV. Blank bars for weak focusing of the laser beam, and dashed bars for tight focusing. Fragmentation is much more extensive in the case of multiphoton ionization and the C^+ ions which are absent from the electron impact spectrum, can be made the most abundant fragment by using high enough laser power. Minimum appearance potentials are indicated for each fragment. (Adapted from ref. 53).

follows that unless the total energy is exceedingly low there will be more than one possible quantum initial (or final) state in any energy interval. These states differ in the way the total energy is partitioned between the kinetic and internal contributions*. A selective preparation is one where only one (or, in general, a subset) of the possible states is selected. A collision can be regarded as a preparation of the final state. If in this final, postcollision, state the accessible quantum states are not equally probable, we regard the collision as selective. Explicitly, in a selective collision, the isoenergetic final quantum states are not produced at the same rate. By microscopic reversibility, in a selective collision the isoenergetic initial quantum states do not react with the same rate.

By 'selective' we thus mean that at a given energy (and other good quantum numbers) the quantum states of the system are not equally probable**.

The quasicontinuum spectral region (Sec. 2.2) enables one to apply this definition equally well to energy rich isolated molecules. As discussed in Sec. 2.2 one can selectively excite the molecule into this region but intramolecular processes tend to degrade the initial selectivity. Hence for both collisional and unimolecular processes the key dynamical question is how far can the selective excitation of the initial state be used to influence the course of the event[19,55]. Lest we fall prey to the fashionable current (1980) pessimism in this area, one should reinspect Fig. 4, showing the reflection of the initial selectivity in the final state.

3.1 Entropy and Surprisal

As a quantitative measure of selectivity we use the entropy[19,23,39,40] of the distribution over quantum states,

$$S = -\sum_j P(j)\ell nP(j), \tag{12}$$

which is maximal if and only if all quantum states are equally probable. Typically however one does not resolve single quantum states. Let $P(\alpha)$ be the probability of a group of g_α states

$$P(\alpha) = \sum_{j\varepsilon\alpha} P(j), \quad g_\alpha = \sum_{j\varepsilon\alpha} 1 . \tag{13}$$

*They can also differ in their internal angular momentum orientation ('magnetic') quantum numbers, cf. Fig. 5, etc.

**This definition is independent of the basis set used.

The distribution of groups is, in the absence of any selectivity, $P^o(\alpha)$, the, so called, prior distribution,

$$P^o(\alpha) = g_\alpha/\sum_j 1 = g_\alpha/\sum_\alpha g_\alpha \ . \tag{14}$$

In terms of the observed distribution over groups, the entropy can be expressed as

$$\begin{aligned} S &= -\sum_j [P(\alpha)/g_\alpha]\ell n[P(\alpha)/g_\alpha] \\ &= -\sum_\alpha P(\alpha)\ell n[P(\alpha)/g_\alpha] \\ &= \ell n[\sum_\alpha g_\alpha] - \sum_\alpha P(\alpha)\ell n[P(\alpha)/P^o(\alpha)] \ . \end{aligned} \tag{15}$$

The second term in (15) is non negative and vanishes if and only if $P(\alpha) = P^o(\alpha)$ for all α. We thus have a diagnostic test known as[19,23,39,40] 'surprisal analysis': Examine the surprisal

$$I(\alpha) = -\ell n[P(\alpha)/P^o(\alpha)] \ . \tag{16}$$

If, (to within the experimental or computational accuracy of $P(\alpha)$), it is non-zero then the process is selective.

It is important to emphasize that it is the deviation of $P(\alpha)$ from the prior distribution $P^o(\alpha)$, which by no means need be uniform, which is the signature of selectivity. In dynamical terms, the transition rate to any group α is given by $\sum_f' |T_{if}|^2\rho_f$ where ρ_f is the translational density of final states and summation is restricted to final states in the group α. Now

$$\sum_f' |T_{if}|^2\rho_f = [\sum_f' |T_{if}|^2\rho_f/\sum_f'\rho_f]\sum_f'\rho_f \ . \tag{17}$$

Hence, with

$$P^o(\alpha) = \sum_f'\rho_f/\sum_f\rho_f \tag{18}$$

the surprisal is a logarithmic measure of the variation in the average transition rate among the different groups.

Details of the state counting required in (18) have been extensively discussed[19,40] and will not be repeated here.

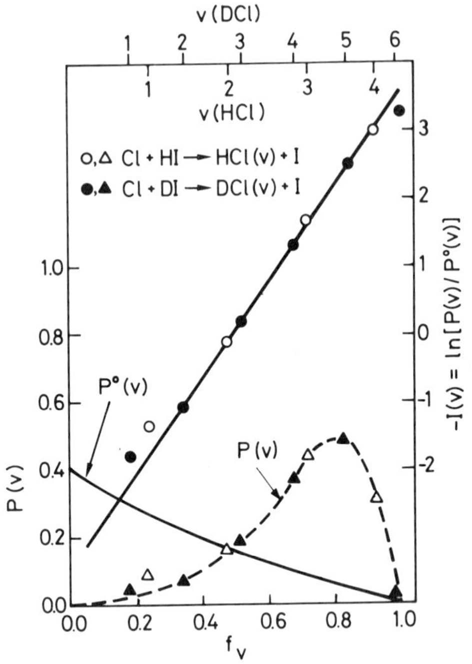

Fig. 16. Observed [Ref. 22, triangles] and prior ($P^{O}(v)$ solid line) product vibrational state distribution in the Cl+HI (open triangles) and the Cl+DI (solid triangles) reactions vs. v. (Note that on prior grounds v=0 is the most probable final vibrational state). The abscissa shows $f_V = E_V/E$, the fraction of available energy present as product vibration. The circles show the surprisals of the different product vibrational states. The dashed line is the distribution of exactly linear surprisal.

The results of surprisal analysis of the final vibrational state distribution for the reaction (1) are shown in Fig. 16. The surprisal is seen to be quite large and in the direction indicated by the exponential gap law (cf. the energy profile in Fig. 2). Results for a photodissociation process have been shown in Fig. 13 and Fig. 17 shows an analysis for a photoelimination process[55]

$$RHCl \xrightarrow{h\nu} R + HCl(v). \tag{19}$$

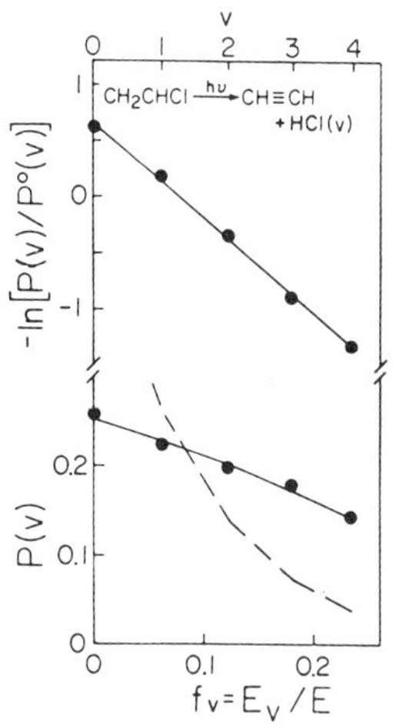

Fig. 17. The measured[55] HCl vibrational distribution (dots; bottom panel) and its surprisal (dots, upper panel) for the photoelimination reaction (19) vs. f_V. The prior distribution for E = 133 Kcal/mol is drawn as a dashed line. The solid lines are the distribution of maximal entropy (20) subject to the constraint (21), and its surprisal determined as discussed in section 3.2. The optimal value of λ_V is -8.4.

As in Fig. 16, the slope of the surprisal is quite negative indicating strong preferential population of the higher HCl vibrational states. For the reaction (19) there are however many more modes where the balance of the energy, $E-E_V$, can be placed than for the reaction (1). Hence $P^O(v)$ is a strongly decreasing function of E_V, and despite the negative slope, the distribution, P(v),

$$P(v) = P^O(v)\exp[-\lambda_V(E_V/E)-\lambda_0], \quad (20)$$

of HCl vibrational states declines with v for reaction (19).

3.2 Maximum Entropy

In section 3.1 the entropy and the surprisal were introduced as measures in a phenomenological analysis of a given distribution. Here we consider the complementary question of inferring the distribution. The formalism[57] we shall use (which can be grounded in dynamics[58] and which is variational in character[59]), is: For a given set of constraints which do not suffice to specify a unique state, choose that distribution which is of maximal entropy subject to the given constraints.

A very simple example is if the only constraint is normalization. Then, the entropy, (12), is maximal when all accessible quantum states are equally probable or, (cf. (13) and (18)), $P(\alpha) = P^{o}(\alpha)$. To account for the deviance from the prior we can, e.g. impose the condition of a given mean vibrational excitation

$$\sum_{v} E_{v} P(v) = \langle E_{v} \rangle \ . \qquad (21)$$

That distribution which is (a) normalized and (b) has a given value of $\langle E_v \rangle$ and is of maximal entropy can be found by the method[57] of Lagrange multipliers. The result is (20). The value of λ_v is determined by the implicit equation (21) and the value of λ_0 by the condition that P(v) is normalized. A suitable numerical procedure has been described[60].

The origin of the constraint condition (21) can be discussed from a dynamical point of view. It turns out that as in thermodynamics, constraints reflect constants of the motion[61,62]. A simple but not trivial example follows.

3.3 Fragmentation Patterns in Multiphoton Ionization

Extensive fragmentation of polyatomic molecules is, Fig. 15, observed[53,63] in (resonance enhanced[53]) multiphoton ionization. The only obvious constants of motion beside the energy are the number of atoms of the different elements and the charge. Let X_ℓ be the number of species of type ℓ and $a_{k\ell}$ be the number of atoms of type k in the species of type ℓ (or the charge). Then the constraints are the X_k's

$$X_k = \sum_{\ell} a_{k\ell} X_{\ell}, \qquad k = 1,2,.. \qquad (22)$$

where the range of k is the number of distinct elements in the parent molecule (two for benzene, Fig. 15), plus one (for conservation of charge). The simplest account for the energy provided by the multiphoton absorption is to specify the mean number of absorbed photons (or, equivalently, the mean energy). One can then determine the fragmentation pattern by maximising the entropy subject to these few constraints.

The results[54], shown for benzene in Fig. 18, are in qualitative accord with experiment. In particular, the extensive fragmentation at higher energies and the transition from larger (and more stable) ions at lower energies to smaller (and energy expensive) ions at higher energies are well reproduced. Indeed the shift evident in Fig. 18 can be summarized by the thermodynamic rule of thumb: At low energies, energetic factors dominate while at higher energies entropic factors are more significant.

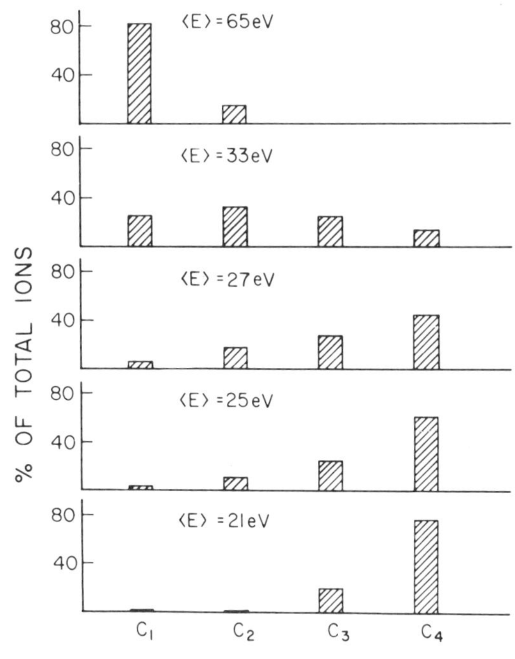

Fig. 18. Computed[54] branching fractions for the formation of the four dominant ion groups (C_1 to C_4) in benzene at different mean energies deposited per parent molecule. The bars are normalized to unity for the four groups. Note the gradual transition towards the smaller ions (which have higher thresholds) as the mean energy is increased. The panel at 25 eV is to be compared with the high focusing case of Fig. 15. (For the laser used in the experiment[53], 33 eV corresponds roughly to 10 photons).

3.4 Current Developments

A question of considerable current interest[64] is the gradual reduction in the selectivity of the initially excited state due to intramolecular energy transfer processes. Such processes are always possible when the excited state can dissociate and hence are typical of both energy rich species formed in a collision or by photoactivation. Their onset is however in the quasicontinuum, i.e. typically below the threshold for dissociation. To characterize such processes, consider a division of the phase space accessible to the molecule into a finite number of disjoint regions (which together span the entire space). Say now that at time $t=0$ we start following several classical trajectories which originate in the same region. The initial conditions for these trajectories are therefore similar but not identical. At a time Δt later we determine the region where each trajectory is located. Because of the dispersion in the initial conditions, the trajectories need not be all found in the same region. Repeating this procedure after each time step, each trajectory generates a list (i.e. a sequence) of regions wherein it was located. The key issue is how 'similar' are the sequences generated by distinct trajectories (which originated however in the same region). If two such initially neighboring trajectories tend to rapidly diverge then the erasure of the initial selectivity will be fast.

Starting with all possible trajectories out of a given region it proves useful to examine $N(m)$, the number of distinct sequences generated after m time steps. The (logarithmic) rate of increase of $N(m)$ with m

$$h \equiv \lim_{m \to \infty} m^{-1} \ln N(m)$$

provides a measure of the 'stochasticity' of the dynamics. Kolmogorov, who introduced this measure[65] was also able to show that among all finite partitions of phase space h attains a least upper bound.

Computer experiments for realistic molecular potentials provide estimates for h and show that it is finite*. The as yet unanswered, practical question, is how to design excitation processes (i.e. selections of initial regions of phase space) for which h is small on the chemical time scale (i.e. on the time required for dissociation or for an encounter with another reagent in the case of a bimolecular process). Rules of thumb based on an intuitive understanding of the dynamics can be formulated** but the gap between the formal ergodic theory[65] and the chemical practice is yet to be bridged.

*h can be zero, if $N(m)$ increases less than exponentially with m. When this is the case, the time evolution is 'deterministic'.

**The most obvious example is that modes which differ considerably in their frequency are not going to be strongly coupled.

REFERENCES

1. Levine, R.D. and Bernstein, R.B., "Molecular Reaction Dynamics", Clarendon, Oxford (1974).
2. Nikitin, E.E., "Theory of Elementary Atomic and Molecular Processes in Gases", Clarendon, Oxford (1974).
3. Eyring, H., Jost, W. and Henderson, D., eds. "Physical Chemistry-An Advanced Treatise", Vols. VIA, B. "Kinetics of Gas Reactions", Vol. VIII "Scattering Theory", Academic Press, N.Y.
4. Brooks, P.R. and Hayes, E.F., eds., "State-to-State Chemistry", A.C.S. Symposium Ser. No. 56, Am. Chem. Soc., Washington D.C. (1977).
5. Bernstein, R.B., ed. "Atom-Molecule Collision Theory: A Guide for the Experimentalist", Plenum, N.Y. (1979).
6. Smith, I.W.M., ed. "Physical Chemistry of Fast Reactions II. Reaction Dynamics", Plenum, N.Y. (1979),
7. Miller, W.H., ed., "Dynamics of Molecular Collisions", Parts A and B, Plenum, N.Y. (1976).
8. Lawley, K.P., ed., "Molecular Scattering: Physical and Chemical Applications", J. Wiley, N.Y., Adv. Chem. Phys., 30 (1975).
9. Fluendy, M.A.D. and Lawley, K.P., "Chemical Applications of Molecular Beam Scattering", Chapman and Hall, London (1973).
10. Faraday Discussions of the Chemical Society 55 (1973), 62 (1977), 67 (1979).
11. Child, M.S., "Molecular Collision Theory", Academic, N.Y. (1974).
12. Laidler, K.J., "Theories of Chemical Reaction Rates", McGraw Hill, N.Y. (1969).
13. Weston, R.E. and Schwarz, H.A., "Chemical Kinetics", Prentice-Hall, N.J. (1972).
14. Jordan, P.C., "Chemical Kinetics and Transport", Plenum, N.Y. (1979).
15. Levine, R.D. and Jortner, J., eds., "Molecular Energy Transfer", Wiley, N.Y. (1976).
16. Lambert, J.D., "Vibrational and Rotational Relaxation in Gases", Clarendon Press, Oxford (1977).
17. Gianturco, F.A., "The Transfer of Molecular Energies by Collision", Springer, Berlin (1979).
18. Zewail, A.H., ed., "Advances in Laser Chemistry", Springer-Verlag, Berlin (1978).
19. Ben-Shaul, A., Haas, Y., Kompa, K.L. and Levine, R.D., "Lasers and Chemical Change", Springer, Berlin (1980).
20. Jortner, J., Levine, R.D. and Rice, S.A., eds., "Photoselective Chemistry", Wiley, N.Y. (1980).
21. Vetter, R., ed., "Dynamique Reactionnelle des Etats Excites", Les Houches, CERN, Paris (1979).
22. Maylotte, D.H., Polanyi, J.C. and Woodall, K.B., J. Chem. Phys. 57:1547 (1972).
23. Levine, R.D., in "The New World of Quantum Chemistry", Reidel, Holland (1976).

24. Kaplan, H., Levine, R.D. and Manz, J., Chem. Phys. 12:447 (1976).
25. Douglas, D.J., Polanyi, J.C. and Sloan, J.J., Chem. Phys. 13:15 (1976).
26. Arnoldi, D. and Wolfrum, J., Ber. BunsenGes. Phys. Chem. 80:892 (1976).
27. Leone, S.R., McDonald, R.G. and Moore, C.B., J. Chem. Phys. 63:4735 (1975).
28. Ding, A.M.G., Polanyi, J.C. and Schreiber, J.L., Disc. Faraday Soc. 55:252 (1973).
29. Bergmann, K., Leone, S.R., MacDonald, R.G. and Moore, C.B., Isr. J. Chem. 14:105 (1975).
30. Zamir, E., Haas, Y. and Levine, R.D., J. Chem. Phys. 73 (1980).
31. Reddy, K.V. and Berry, M.J., Chem. Phys. Lett. 66:223 (1979).
32. Brooks, P.R., Science 193:11 (1976).
33. Kinsey, J.L., Ann. Rev. Phys. Chem. 28:349 (1977).
34. Karny, Z., Estler, R.C. and Zare, R.N., J.Chem. Phys. 69:5199 (1978), see also Karny, Z. and Zare, R.N., J. Chem. Phys. 68:3360 (1978).
35. Brenner, D.M., Datta, S. and Zare, R.N., J. Am. Chem. Soc. 99:4554 (1977).
36. Stuke, M. and Marinero, E.E. in ref. 48.
37. Bauer, S.H., Ann. Rev. Phys. Chem. 30:271 (1979). Kenba, M. and Wolfrum, J., ibid. 31 (1980).
38. George, T.F., Zimmermann, I.H., Yuan, J.M., Laing, J.R. and DeVries, P.L., Acc. Chem. Res. 10:449 (1977).
39. Levine, R.D., Ann. Rev. Phys. Chem. 29:59 (1978).
40. Levine, R.D. and Ben-Shaul, A., in "Chemical and Biochemical Applications of Lasers", Vol. II, p. 145, Academic Press, N.Y. (1976); Levine, R.D. and Kinsey, J.L. in ref. 5.
41. Houston, P.L. and Steinfeld, J.I., in "Laser and Coherence Spectroscopy", Steinfeld, J.I., ed., Plenum, N.Y. (1978).
42. Albrecht, A. in ref. 18; Smith, D.D. and Zewail, A.H., J. Chem. Phys. 71:540 (1979).
43. Henry, B.R., Accts. Chem. Res. 10:207 (1977).
44. Fuss, W. and Cotter, T.P., Appl. Phys. 12:265 (1977).
45. Jensen, C.C., Anderson, T.G., Reiser, C. and Steinfeld, J.I., J. Chem. Phys. 71:3648 (1979).
46. Sudbø, Aa. A., Kranjovich, D.J., Schultz, P.A., Chen, Y.R. and Lee, Y.T., in "Multiphoton Excitation and Dissociation of Polyatomic Molecules", Cantrell, C.D., ed., Springer, N.Y. (1979).
47. Moore, C.B., ed., "Chemical and Biochemical Applications of Lasers", Vol. III, Academic Press, N.Y. (1977).
48. Kompa, K.L. and Smith, S.D., eds., "Laser Induced Processes in Molecules", Springer, Berlin (1979).
49. Freed, K.F., Topics Appl. Phys. 15:1 (1976). Siebrand, W. in ref. 7. Mukamel, S. and Jortner, J., in ref. 15. Avouris, P., Gelbart, W.M. and El-Sayed, M.A., Chem. Rev. 77:793 (1977). McDonald, J.D., Ann. Rev. Phys. Chem. 30:29 (1979).
50. Lim, E.C., ed., "Excited States", Academic Press, N.Y. (1974-6).

51. Yang, C.S., Freedman, A., Kawasaki, M. and Bersohn, R., J. Chem. Phys. 72:4058 (1980).
52. Riley, S.J. and Wilson, K.R., Disc. Faraday Soc. 53:132 (1972).
53. Zandee, L. and Bernstein, R.B., J. Chem. Phys. 71:1359 (1979).
54. Silberstein, J. and Levine, R.D., Chem. Phys. Lett. 74:6 (1980).
55. Rice, S.A., in ref. 18; Levine, R.D., in ref. 20; George, T.F., Opt. Eng. 18:167 (1979); Thiele, E., Goodman, M.F. and Stone, J., ibid 19:10 (1980).
56. Berry, M.J., J. Chem. Phys. 61:3114 (1974) and in ref. 17.
57. Levine, R.D. and Tribus, M., eds., "The Maximum Entropy Formalism", M.I.T. Press, Mass. (1979).
58. Alhassid, Y. and Levine, R.D., Phys. Rev. A18:89 (1978).
59. Alhassid, Y. and Levine, R.D., J. Chem. Phys. 67:4321 (1977).
60. Agmon, N., Alhassid, Y. and Levine, R.D., J. Comput. Phys. 30:250 (1979).
61. Alhassid, Y. and Levine, R.D., Phys. Rev. C20:1775 (1979).
62. Levine, R.D., in ref. 20.
63. Reilly, J.P. and Kompa, K.L., J. Chem. Phys. 73 (1980). Boesl, U., Neusser, H.J. and Schlag, E.W., J. Chem. Phys. 72:4327 (1980); Lubman, D.M., Naaman, R. and Zare, R.N. J. Chem. Phys. 72:3034 (1980); Cooper, C.D., Williamson, A.D., Miller, J.C. and Compton, R.N., J. Chem. Phys. 73 (1980); Seaver, M., Hudgens, J.W. and DeCorpo, J.J., J. Chem. Phys. 72 (1980); Fisanick, G.J., Eichelberger IV, T.S., Heath, B.A. and Robin, M.B., J. Chem. Phys. 72:5571 (1980).
64. Nordholm, K.S.J. and Rice, S.A., J. Chem. Phys. 61:203 (1974). Hansel, K.D., J. Chem. Phys. 70:1830 (1979). Duff, J.W. and Brumer, P., J. Chem. Phys. 71:2693 (1979). Heller, E.J., J. Chem. Phys. 72:1337 (1980). Rice, S.A., in ref. 18. Rice, S.A., in ref. 20. Levine, R.D., in ref. 20. Brumer, P., in ref. 20. Levine, R.D., Chem. Phys. Lett. (1980).
65. Kolmogorov, A.N., Dokl. Akad. Nauk. SSSR 124:774 (1959). Arnold, V.I. and Avez, A., Ergodic Problems of Classical Mechanics, Benjamin, New York, 1974. Sinai, Ya. G., Dokl. Akad. Nauk. SSSR 124:768 (1959). Rohlin, V.A. and Sinai, Ya. G., Sov. Math. Dokl. 2:1611 (1962).

PART III

COLLISIONS UNDER SPECIAL CONDITIONS

COLLISIONS OF CHARGED PARTICLES WITH HIGHLY EXCITED ATOMS

Ian C. Percival

Department of Applied Mathematics
Queen Mary College (University of London)
Mile End Road, London E1 4NS, U.K.

ABSTRACT

Theories based on classical mechanics and correspondence principles are introduced ant their ranges of validity and the connection with quantal theories are discussed. The correspondence principle approximations to first order perturbation theory and the sudden approximation receive particularly detailed treatment.

1. INTRODUCTION

2. CORRESPONDENCE PRINCIPLES

3. TIME-DEPENDENT COLLISION THEORIES

4. REGIONS OF VALIDITY

5. SOME OTHER THEORIES, COLLISIONS AND LITERATURE

1. INTRODUCTION

By highly excited we mean an atom or ion which has an electron whose binding energy is very small compared with its binding energy in the ground state. Properties of such atoms differ considerably from those of other atoms or ions.

Highly excited atoms are found in plasmas, where they play an important role in recombination. These plasmas may be found in the laboratory, in stellar atmospheres or in interstellar space. They may not always be observed in all these regions but they are directly observed in highly ionized regions (HII regions) of space, in the neighbourhood of hot stars and also in the HI regions which are cooler and have a relatively low degree of ionization. In interstellar space, radiofrequency transitions are observed between close-lying states using radio-telescopes, for example, the transition H(109α), which is

$$H(n = 110) \rightarrow H(n = 109) + h\nu$$

where ν is in the radiofrequency range. To my knowledge the highest excitation so far observed is the transition $n = 301 \rightarrow n = 300$ by Pedlar, Davies and Hart at Jodrell Bank and Shaver at Groningen.

Transitions between highly excited states of atoms affect the recombination of laboratory plasmas, but they are not observed directly. The observation of these states in the laboratory is generally by transition from the ground state or a state of low excitation through the absorption of ultraviolet or visible radiation. The highest state observed in the laboratory is above $n = 100$

The observations we can make in the laboratory are complementary to those made by astronomers. The former involve transitions between a highly excited state and a low-lying state; the latter, transitions between highly excited states where observations are made of the "radio-recombination lines". The distinction between the above observations is an important one. A highly excited state is characterized by quantum numbers (ℓ, m) in addition to the principal quantum number n, and the important (ℓ, m) states in the two cases differ:

for radio-recombination $\ell, m \lesssim n$

in the laboratory $\ell, m \sim 0, 1, 2, 3.$

Because the properties of highly excited states differ so much from those of low-state atoms, the appropriate theory also differs. In particular, for highly excited atoms the classical model of the atom can be taken seriously. Methods can be used which are at best doubtful for low states of excitation.

Consider the Bohr model of a classical electron moving in a circular orbit around a proton to represent the ground state $(n = 1)$ of the hydrogen atom, where the model is evidently very crude.

Table 1. Elementary Properties of Highly Excited States

I Property	II n = 1	III Arbitrary n	IV n = 110
Binding energy U_n	1 Ry = 13.6 eV	Ry/n^2	1.1×10^{-3} eV
Radius a_n of Bohr orbit	$a_0 = a_1$ $= 5.3 \times 10^{-9}$ cm	$n^2 a_0$	6.4×10^{-5} cm
Geometric cross-section πa_n^2	8.8×10^{-17} cm^2	$n^4 \pi a_0^2$	1.3×10^{-8} cm
rms velocity of electron v_n	$c\alpha = v_1 = 2.2 \times 10^8$ cm/s	v_1/n	2.0×10^6 cm/s
Period T_n	1.5×10^{-16} s	$n^3 T_1$	2.0×10^{-10} s
Fundamental classical frequency ν_c	$\nu_1 =$ 6.576×10^{15} Hz	$n^{-3}\nu_1$ $[(n-\frac{1}{2})^{-3}\nu_1]$	4.941×10^9 Hz $[5.009 \times 10^9$ Hz$]$
Classical wave number λ_c^{-1}	$\lambda_1^{-1} = 2R_H$ $= 2.194 \times 10^5$ cm^{-1}	$2n^{-3}R_H$ $[2(n-\frac{1}{2})^{-3} R_H]$	0.1648 cm^{-1} $[0.1671$ $cm^{-1}]$
Quantal $n\alpha$ frequency	None	$\frac{n-\frac{1}{2}}{[n(n-1)]^2} \nu_1$	5 009 MHz

The geometric cross-section gives a good idea of the effective size of the atom, the rms velocity is exactly right, but the classical frequency has no meaning. The classical properties of a highly excited state can be obtained from those of the ground state by certain scaling laws, as seen in columns III and IV of Table 1.

Starting with the quantal expression for the energy we find the H(110) atoms are so very weakly bound (1.1×10^{-3} eV) that particles at room temperature (1/40 eV) have enough energy to destroy them. Only low density can save them.

According to Coulomb's law the orbital radius of a charged particle bound to another one typically varies inversely with the energy, so the mean Bohr radius for an H atom varies as n^2. The size of an H(110) atom is typical of a small biological organism. If it were solid it could be seen easily with an optical microscope. It is as big when compared to a typical ground-state atom as a ground-state atom is compared to its nucleus. By analogy, the structure of the ionic core of a highly excited atom can usually be neglected. The core is distinguished only by its charge and mass, except for relatively very small effects analogous to the hyperfine structure of ordinary atoms. To this extent, all highly excited atoms are the same, their properties depending primarily on the charge of the core and the quantum numbers of the highly excited electrons. Even a molecular ion core makes little difference.

From Kepler's laws of planetary motion, the period of the electron in its orbit scales as the 3/2 power of the mean radius. From this we obtain the fundamental classical frequency for a highly excited atom, viz.

$$\nu_n^c = n^{-3}\nu_1^c \tag{1}$$

where ν_1^c is the 'frequency' of the classical atomic model for the ground state. For circular orbits with maximum angular momentum for fixed energy U_n, only the fundamental frequency is emitted (this corresponds to the quantum mechanical rule that if $\ell = \ell_{max}$, a transition with $\Delta\ell = +1$ is impossible). For elliptic orbits with smaller angular momentum, there are Fourier components of the motion at the harmonic frequencies $2\nu^c$, $3\nu^c$, etc., which correspond to quantal transitions with change Δn in the principal quantum number of $\Delta n = 2, 3, \ldots$ etc. For the quantal atom, these transitions have frequencies which are not exact multiples of the fundamental frequency, but it is not too bad an approximation when $\Delta n \ll n$.

From classical frequencies, following Bohr, we can obtain an estimate of the energy difference between initial and final states of an optical transition by using the Planck relation

$$\Delta E = h\nu = \hbar\omega . \tag{2}$$

This is a direct application of Bohr's correspondence principle for frequencies.

The correspondence principle is valid in the limit of high quantum numbers, and from Table I we see that it is clearly much better for $n = 110$ than for $n = 1$, as we would expect.

In quantum or wave mechanics, the wave functions of highly excited states have very many oscillations (nodes), about 10^6 for $n = 110$. Traditional quantal methods, such as close-coupling, for treating the collisions with such atoms require the numerical representation of the wave functions in a computer, and the number of real numbers required to represent such a wave function is at least as large as the number of oscillations. Furthermore one needs to consider approximately 10^6 different wave functions (n in each of the three dimensions) in order to solve the collision problem. Obviously when we are dealing with a non-separable system this is a very great handicap. There are methods of simplifying the quantal problem by taking advantage of the large number of oscillations. These methods are either relatively rough, or they result in expressions which can often be obtained more simply by classical methods or methods based on the correspondence principle.

The correspondence principles that we shall use can all be derived from the Schrödinger or Heisenberg forms of quantum mechanics. When this is done they can be applied directly to a wide variety of physical problems. Classical and correspondence principle methods should not be regarded as an alternative method to quantum mechanics, but as a method of applying quantum mechanics which is appropriate to highly excited states.

The original theory of the classical motion of a particle under the action of an inverse square attractive force is due to Newton and was applied by him to explain the approximately elliptic Kepler orbits of the planets around the Sun. Rutherford used a newtonian planetary model for his nuclear atom but the more recent classical mechanics of Hamilton, Jacobi and others was needed for the Old Quantum Theory of Bohr and Sommerfeld and is needed now for the theory of highly excited atoms.

Unlike modern textbooks, Born (1927) treats the relevant parts of classical mechanics in his "Mechanics of the Atom".

Consider the motion of a particle of mass μ and position r around a fixed centre of force derived from the mechanical potential

$$V(\underset{\sim}{r}) = -Ze^2/r \tag{3}$$

For the hydrogen atom $Z = 1$, and for hydrogen-like ions Z has a higher integer value. We frequently use "atom" to refer to either. If the nucleus of the atom is assumed to be infinitely heavy, then $\mu = m_e$, the mass of the electron; if not, then the analysis applies to the relative motion of the electron 2 with respect to the nucleus 3, where $\underset{\sim}{r} = \underset{\sim}{r}_2 - \underset{\sim}{r}_3$ is the relative coordinate and

$$\mu = m_e m_3/(m_e+m_3), \quad m_2 = m_e \tag{4}$$

is the reduced mass. For a nonhydrogenic atom the potential (3) is approximate, but it is a good approximation for angular momenta large compared to $\hbar$, a necessary requirement of classical mechanics, provided only one electron is highly excited.

Because the potential is central the motion is confined to a plane and because it varies inversely with r and relativistic effects are neglected the orbit is elliptic. Let a be the semi-major axis of the ellipse and ε its eccentricity. Let ξ, η be cartesian coordinates in the plane, with $\hat{\underset{\sim}{\xi}}$ chosen in the direction of the perihelion of the orbit, and $\hat{\underset{\sim}{\eta}}$ chosen so that the particles moves around the origin O in a positive sense in the plane, as shown in Fig 1.

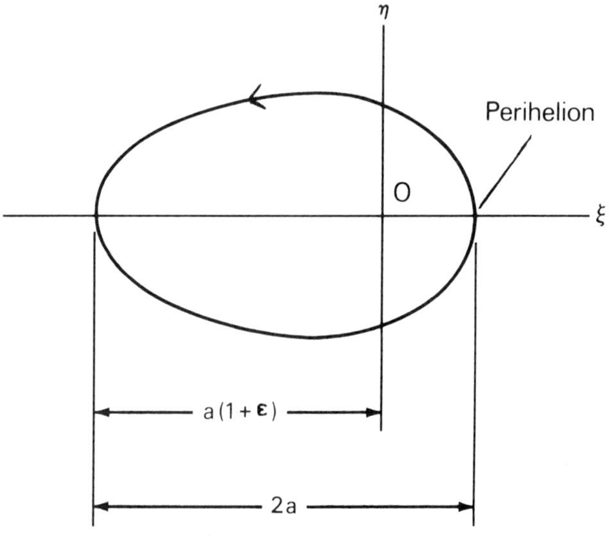

Fig. 1. The orbit in the ξ, η plane.

The motion around the orbit is defined by expressing the physical quantities as elementary functions of a parameter u named the eccentric anomaly:

$$
\begin{aligned}
\omega t + \delta &= \omega\tau = \theta_n = u - \varepsilon \sin u \\
\xi &= a(\cos u - \varepsilon), \qquad \eta = a(1 - \varepsilon^2)^{1/2} \sin u \\
\dot{\xi} &= v_\xi = -v_0 \sin u/(1 - \varepsilon \cos u) \\
\dot{\eta} &= v_\eta = v_0(1 - \varepsilon^2)^{1/2} \cos u/(1 - \varepsilon \cos u) \\
r &= a(1 - \varepsilon \cos u)
\end{aligned}
\tag{5}
$$

In these equations t is the time measured from a fixed origin independent of the motion, τ is a time or epoch measured from an instant when the particle was at perihelion, $\omega = 2\pi/T$ is the fundamental angular frequency of the motion, where T is the period and δ is the phase of the classical motion; θ_n, the mean anomaly, is an angle variable discussed in the next section. The velocity $v_0 = a\omega$ is the rms velocity around the orbit.

The validity of equations (5) may be verified by substitution into the equations of motion; derivations are given by Born (Section 22 of his book) and by Landau and Lifshitz in their "Mechanics". The constants are expressed in terms of the binding energy $U = -E$ of the particle and the magnitude ℓ_c of its classical angular momentum by the equations

$$
\begin{aligned}
a &= Ze^2/2U \\
\mu v_0^2 &= 2U \\
\omega &= (2U)^{3/2}/Ze^2\mu^{1/2} \\
1 - \varepsilon^2 &= (2U/\mu)(\ell_c/Ze^2)^2
\end{aligned}
\tag{6}
$$

Notice that a, v_0, ω and the period T depend on U but are independent of ℓ_c. The relations (6) are simple to verify for the circular orbits with $r = a$, which are the orbits of maximum ℓ_c for a given energy U; these correspond to quantal states with $\ell = n - 1$. At the other extreme, with $\varepsilon = 1$ and $\ell_c = 0$, are the "straight line" or "free fall" orbits for which the particle moves from rest at $r = 2a$ to infinite velocity at $r = 0$, and back; these correspond to quantal s-states.

Lagrange's theory of planetary perturbation required a Fourier expansion and this led him to obtain the "Bessel" coefficients or functions (before Bessel was born). Exactly the same theory is required for the application of the correspondence principle to collisions. The complex Fourier series for $\xi(t)$ and $\eta(t)$ are given by

$$\xi(t) = \sum_{s=-\infty}^{\infty} \xi_s e^{-is\omega t}$$

$$\eta(t) = \sum_{s=-\infty}^{\infty} \eta_s e^{-is\omega t} \qquad (7)$$

where, by Fourier's theorem and the integral definition of the Bessel function,

$$\xi_s = \frac{a}{s} J'_s(s\varepsilon)e^{-is\delta} \qquad |s| \geq 1$$

$$\eta_s = \frac{ia(1 - \varepsilon^2)^{1/2}}{s\varepsilon} J_s(s\varepsilon)e^{-is\delta}, \qquad |s| \geq 1 \qquad (8)$$

$$\xi_0 = -\frac{3}{2}a\varepsilon$$

$$\eta_0 = 0 \quad .$$

For most applications we can use the epoch τ instead of the time t and $\delta = 0$.

To complete the picture consider the orientation of the elliptic orbit or, equivalently, the Euler angles (α, β, γ) which define the orientation of the $O\xi\eta\zeta$ coordinate frame with respect to a fixed standard frame Oxyz. We suppose that the two frames are initially coincident so that the orbit is in the Oxy plane and that the $O\xi\eta\zeta$ frame, and the orbit with it, are rotated successively through angles γ, β, and α about the Oz, Oy, and Oz axes, respectively.

The intercept of the Oxy and $O\xi\eta$ planes is named the line of nodes. The angle β is the angle between the angular momentum vector $\underset{\sim}{L}_c$ and the z axis and $\pi/2 - \gamma$ is the angle between the perihelion and the line of nodes.

The six quantities

$$U \quad \ell_c \quad \alpha \quad \beta \quad \gamma \quad \tau \qquad (9)$$

define the position and velocity of the particle uniquely; all but the last one are conserved. The last five are used together with an impact parameter to define the initial conditions for a single collision of the Monte Carlo method.

Any bound time-independent separable system like the hydrogen atom has a set of standard coordinates named the angle-action variables. For each degree of freedom there is one angle variable θ_k and a conjugate momentum action variable I_k, such that the Hamiltonian is a function $H(I_k)$ of the action variables alone so that by Hamilton's equations the θ_k are linear in time. The angle variables are normalized so that the classical state of the system is periodic in θ_k with period 2π. The linear relation

$$\theta_k = \omega_k t + \delta_k \tag{10}$$

then defines the fundamental angular frequencies ω_k of the motion. For degenerate systems some of the ω_k may be zero and in particular for hydrogen all except one are zero. The motion of the electron may be separated classically in polar coordinates (r, Θ, ϕ), giving action variables (I_r, I_Θ, I_ϕ). It is more convenient to use the action variables:

$$I_m = I_\phi \tag{11a}$$

$$I_\ell = I_\Theta + I_\phi \tag{11b}$$

$$I_n = I_r + I_\Theta + I_\phi \tag{11c}$$

These new action variables are the classical analogs of the quantum numbers m, ℓ, n.

For the particular case of the attractive Coulomb potential the energy is given by

$$E = -U = -\frac{\mu(Ze^2)^2}{2I_n^2} , \tag{12}$$

which is independent of I_ℓ and I_m.

For the attractive Coulomb potential the action-angle variables are related to the six variables of (9) and (12) and

$$\ell_c = I_\ell \tag{13a}$$

$$\alpha = \Theta_m , \tag{13b}$$

$$\omega\tau = \Theta_n . \tag{13c}$$

Correspondence Principles

The correspondence principle of Niels Bohr relating quantal phenomena to classical mechanics led to the discovery of quantum mechanics. It thereby fulfilled its original purpose and is no longer considered in detail in courses on quantum mechanics.

For the theory of highly excited atoms this is unfortunate because the correspondence principle provides excellent approximations for systems of high quantum number. However the application of the principle for a quantal system is subtle and must be carried out carefully. It is necessary to solve the ordinary classical equations of motion for the corresponding classical system and then to use the correspondence principle to relate quantum mechanical quantities to the classical values obtained from the solution of the equations of motion. This gives exactly the same result as the application of the J.W.K.B. approximation in quantum mechanics (although the mathematics is simpler). This is not as surprising the relations given by the correspondence principle can themselves be obtained by the application of J.W.K.B. to quantum mechanics. These relations may be considered as a part of semiclassical mechanics.

We discuss the problems of the correspondence between classical and quantum mechanics by using collisions with highly excited hydrogen-like atoms as an example. For bound states the classical energy "spectrum" is continuous whereas the quantal spectrum is discrete.

Suppose a classical electron collides with a classical H-atom with a nucleus of infinite mass. U is the binding energy of the atom which in this classical model is not quantized (i.e. we have a continuum of bounded states).

Let E be the initial energy of the incident charged particle, say electron and suppose that

$$E > U \tag{14}$$

For ionization to take place, the bounded electron must gain enough energy to leave the proton, but the incident electron must not lose so much energy that it becomes bound to the proton.

We can classify classical collisions of electrons with H atoms according to the range of values of the energy transfer

$$\Delta E = E - E' \tag{15}$$

from the incident electron to the initially bound electron. Here E' is the final energy of the incident electron.

We have the possibilities:

$\Delta E < U$ Direct scattering; includes both excitation and de-excitation (16a)

$U < \Delta E < E$ Ionization (16b)

$E < \Delta E$ Exchange (16c)

For excitation there is a differential cross-section

$$d\sigma/d\Delta E \tag{17}$$

defined by the relation

$$\sigma(\Delta E_1 < \Delta E < \Delta E_2) = \int_{\Delta E_1}^{\Delta E_2} d\Delta E \; \frac{d\sigma}{d\Delta E}(\Delta E) \tag{18}$$

where the left hand side represents the cross-section for all collisions that cause a transfer of energy ΔE to the atomic electron between ΔE_1 and ΔE_2.

The classical differential cross-section $\partial\sigma/\partial\Delta E$ with respect to energy transfer must be related to the total quantal cross-section $\sigma(n \to n')$ from level n to level n' of the quantized bound system.

If $\partial\sigma/\partial\Delta E$ varies by only a small fraction over values of ΔE corresponding to several neighbouring final excited states n', then there is little difficulty in obtaining from it unambiguous approximate excitation cross-sections, by assuming a smooth distribution among the final states n', as shown in Fig 2 and as follows

$$\sigma(n \to n') \simeq \frac{\partial\sigma}{\partial n'} = \left|\frac{\partial\sigma}{\partial\Delta E}\right| \left|\frac{\partial\Delta E}{\partial n'}\right| \tag{19}$$

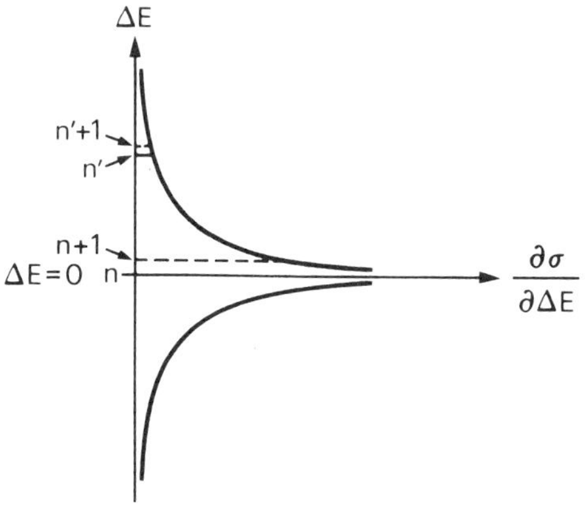

Fig. 2. Excitation cross-sections as a function of energy difference.

Here n is supposed to be fixed, ΔE is the energy transfer

$$\Delta E = U_1(n^{-2} - n'^{-2}) \tag{20}$$

required to excite the atom from level n to level n', and

$$\left|\frac{\partial \Delta E}{\partial n'}\right| = 2U_1/n'^3 \tag{21}$$

is the mean energy range between states labelled by n', or the inverse mean density of levels in the neighbourhood of n'.

This density-of-states correspondence principle states

CPO. The probability of reaching a given final quantal state or group of states is given by the corresponding classical probability weighted by the ratio of the quantal to the classical volumes of phase space, with limiting processes where necessary

(22)

The volume of phase space occupied by a quantal state takes its usual value of $(2\pi\hbar)^N$, where N is the number of degrees of freedom.

The theoretical condition for the validity of this density-of-states correspondence principle involves classical action functions. In practice it is valid when

$$\text{all} \quad n, n', |\Delta n| \gg 1 \tag{23}$$

and is sometimes valid even when Δn is comparable with unity, in particular for low energy collisions with electrons, where adiabatic theory is valid, the density of states correspondence principle applies even for $\Delta n = 1$.

The classical theory with the density-of-states correspondence principle is sometimes inadequate for small Δn. Because of the strong singularity in $|\partial\sigma/\partial\Delta E|$ at $\Delta E = 0$, the assignment of the energy range is ambiguous; and even the form of the cross-section can be wrong.

For sufficiently high incident-particle energies, the Bethe-Born approximation is valid and provides a quantal cross-section which has the form

$$\sigma^q(n \to n') \approx \frac{A\ln E + B}{E} \qquad (E \text{ large}) \tag{24}$$

For large n and Δn the logarithmic term is insignificant, but for small Δn it may be dominant for values of E of physical interest. However, the classical results have no logarithmic term.

The additional logarithmic term comes from weak collision with large impact parameters where the classical energy transferred to the bound electron is less than the energy splitting between the quantal states. The neighbouring energy states are thus classically inaccessible. Two clues enable us to overcome this problem: Williams (1931) pointed out that classical theory can give a correct value for the energy loss of charged particles, even for these weak collisions, so that the mean classical energy transfer behaves as if it were shared out among the quantum-mechanical states.

Bohr's correspondence principle for radiation deals with a similar situation, for the mean energy loss of a classical radiating electron is shared out among the quantum-mechanical states. In this case the energy transfer to each quantal state is determined by the Fourier components of the interaction with the radiation field. Thus we should expect the $(n \to n')$-transition probability $P(b)$ as a function of the impact parameter b to be related to the Fourier components of order s and $-s$, where $s = n' - n$.

This is correct, so to deal with transitions involving small changes in quantum number we need a correspondence principle like that of Bohr, relating quantum phenomena to the Fourier components of the classical motion. This comes in two parts, the first for frequencies and the second for intensities.

The correspondence principle for frequencies states

CP1. The frequency given by the quantal relation between energies and frequencies

$$\omega(n_2, n_1) = (E(n_2) - E(n_1)/\hbar = \Delta E/\hbar \tag{25}$$

tends to coincide with the multiple

$$s\omega^c, \quad s = n_2 - n_1 \tag{26}$$

of the fundamental frequency ω^c of the classical motion, in the limit where the quantum number increases and the harmonic s remains fixed

$$s\omega^c \approx \omega(n_2, n_1). \tag{27}$$

Bohr's correspondence principle for intensities of radiation was generalized by Heisenberg to arbitrary processes by relating quantal matrix elements to classical Fourier components. Any quantal formula that is expressed in terms of matrix elements of operators can then be approximated semiclassically by using the correspondence principle. Among the more important formula are the Born approximation and the sudden approximation for scattering.

Heisenberg's form of the correspondence principle is expressed in terms of the action-angle variables of classical mechanics. For a bound system of one degree of freedom they are the conjugate pair θ, I where the Hamiltonian depends on the action variable alone.

$$H = H(I) . \tag{28}$$

Hamilton's equations of motion then give

$$\dot{I} = \frac{dI}{dt} = -\frac{\partial H}{\partial \theta} = 0 \tag{29}$$

$$\dot{\theta} = \frac{d\theta}{dt} = \frac{\partial H}{\partial I} = \omega^c(I) \tag{30}$$

defining the classical angular frequency ω^c. Thus the action is a constant of the motion, and as $\ddot{\theta} = 0$, θ is linear in time

$$\theta = \omega^c t + \delta \tag{31}$$

where δ is a phase shift. The scaling is chosen so that θ changes by 2π when the system passes once around its closed orbit. Thus the action variable labels the orbit, and θ is a parameter that defines a position on the orbit. The action variable can also be written

$$I = \frac{1}{2\pi}\oint p\,dq . \tag{32}$$

where p and q are arbitrary conjugate variables and the integral is taken once around the orbit.

Now suppose the system obeys the laws of quantum mechanics, with quantum number n. According to the Bohr-Sommerfeld quantization rule, for high quantum number the action is given by

$$I = n\hbar . \tag{33}$$

Heisenberg's form of the correspondence principle relates quantal transitions from initial high state n_1 to a close final state n_2 to the properties of a classical orbit. It is convenient to choose the orbit labelled by the mean action

$$I = \frac{n_1 + n_2}{2}\hbar . \tag{34}$$

Any function $F^c(x)$ of the classical position $x(I, \theta)$ of the particle is periodic of period 2π in the angle variable θ and so may be expressed as a Fourier series in θ.

$$F^c(x) = \sum_{\text{integer } s} F^c_s(I) \exp -is\theta \tag{35}$$

Heisenberg's form of the correspondence principle then states

CP3. If $\hat{F}$ is the quantal operator corresponding to the function $F(x)$, then in the limit as the quantum numbers n_1, n_2 increase and s remains fixed, the matrix element

$$<n_2|\hat{F}|n_1> \simeq F_s^c(I) \tag{36}$$

where $s = n_2 - n_1$ and I is given by (33).

The generalization of the correspondence principle CP3 to higher dimensions is not difficult for separable systems, in particular for the hydrogen atom. In that case function of the motion of the electron with vector coordinate $\underset{\sim}{r}$ is expressed as a multiple Fourier expansion in terms of the vector angle variables $\underset{\sim}{\theta} = (\theta_n, \theta_t, \theta_m)$, depending on the integer vector

$$\underset{\sim}{s} = \underset{\sim}{n}_2 - \underset{\sim}{n}_1 = (n_2 - n_1, \ell_2 - \ell_1, m_2 - m_1) \tag{37}$$

If $\underset{\sim}{\omega}^c = (\omega_1, \omega_2, \omega_3)$ is the classical vector angular frequency, then according to CP2

$$\underset{\sim}{s}.\underset{\sim}{\omega}^c \simeq \omega^q(\underset{\sim}{n}_2, \underset{\sim}{n}_1) = (E(\underset{\sim}{n}_2) - E(\underset{\sim}{n}_1)/\hbar \tag{38}$$

and according to CP3

$$<\underset{\sim}{n}_2|\hat{F}|\underset{\sim}{n}_1> \approx F_{\underset{\sim}{s}}^c(\underset{\sim}{I}) \tag{39}$$

where

$$\underset{\sim}{I} = \frac{\underset{\sim}{n}_2 + \underset{\sim}{n}_1}{2}\hbar \quad , \quad \underset{\sim}{s} = \underset{\sim}{n}_2 - \underset{\sim}{n}_1 \tag{40}$$

For completeness we now follow Kramers' "Quantum Mechanics" and show how the correspondence principles follow from the usual JWKB approximation to quantum mechanics, and for simplicity we choose to consider a rotating system with coordinate ϕ. For example the Hamiltonian might be

$$H(p, \phi) = \frac{p^2}{2m} + V(\phi) , \tag{41}$$

as in the hindered rotation of a molecule. Classically the system rotates with non-uniform speed. Quantally the time-independent wave function can be approximated by

$$\psi(\phi) = \frac{c}{|p(\phi)|^{1/2}} \exp \frac{i}{\hbar} S(\phi) \tag{42}$$

where

$$S(\phi) = \int^{\phi} p(\phi') \, d\phi'. \tag{43}$$

for ψ to be single-valued $S(\phi)/\hbar$ must change by an integral multiple of 2π when ϕ increases by 2π, so the action integral

$$2\pi\ I = [S(\phi)]_0^{2\pi} = \int_0^{2\pi} p(\phi')\ d\phi' = 2\pi n\hbar . \tag{44}$$

From equation (30) the classical frequency is

$$\omega^c = \frac{\partial H(I)}{\partial I} \approx \frac{\Delta H}{\Delta I}$$

$$\approx \frac{E(n_2) - E(n_1)}{n_2\hbar - n_1\hbar}$$

$$s\omega^c \approx (E(n_2) - E(n_1))/\hbar \tag{45}$$

which is CP1.

For Heisenberg's principle for matrix elements we have

$$\langle n_2|\hat{F}|n_1\rangle \approx \frac{1}{2\pi}\int_0^{2\pi} d\phi\ \frac{c}{|p_2(\phi)|^{1/2}}(\exp - iS_2(\phi)/\hbar)$$

$$F(\phi)\ \frac{c}{|p_1(\phi)|^{1/2}}(\exp\ iS_1(\phi)/\hbar) \tag{46}$$

Now note $\omega\frac{d\phi}{d\theta} = \dot{\phi}$ and $\frac{\dot{\phi}}{p}$ = constant where $p \approx p_1 \approx p_2$.

$$\text{Also}\quad \frac{S_2 - S_1}{\hbar} \approx s\,\frac{S(I_2) - S(I_1)}{I_2 - I_1} \approx s\frac{\partial S}{\partial I} = + s\Theta \tag{47}$$

as Θ, I are conjugate variables, so

$$\langle n_2|\hat{F}|n_1\rangle \approx \frac{1}{2\pi}\ c'\int_0^{2\pi} d\theta\ F(\phi(\Theta))\ \exp\ is\Theta$$

$$= c'\ F_s^c, \tag{48}$$

where c' is independent of F and s.

By putting $F = 1$ we see that $c' = 1$.

A relatively straightforward application of CP3 is to the calculation of oscillator strengths. By using equations (8) and averaging over m states we obtain

$$f(n\ell \to n',\ \ell \pm 1) = \frac{n_c}{3s}[J_s'(s\varepsilon) \pm (\varepsilon^{-2}-1)^{1/2}J_s(s\varepsilon)]^2 \tag{49}$$

where $s = |n - n'|$, $\varepsilon^2 = 1 - (\ell_c^2/n_c^2)$ and where n_c and ℓ_c are appropriate combinations of n, n' and ℓ, ℓ', respectively; in the first approximation $n_c = n$ and $\ell_c = \ell$.

There is no unique choice for n_c and ℓ_c; the ambiguity comes from the fact that one classical orbit is being used to obtain matrix elements connecting two states. In Eq.(34) we choose the mean value of the initial and final state; but often a different combination is found to give better results. Naccache (1972) considered various combinations and came to the conclusion that some of the most accurate results are given by

$$n_c = \frac{n(n-s)}{n-\frac{1}{2}s} = \frac{2nn'}{(n+n')}, \qquad \ell_c = \max(\ell, \ell') \tag{50}$$

In Table 2, we compare values of the squares of the radial matrix element calculated by quantum mechanics and by the correspondence principle (Naccache, J.Phys.B 5, 1308-19, 1972)

Table 2. Comparison of Correspondence and Quantum Mechanical Numerical Values of Various Dipole Moments Squared for a Coulomb Potential

Transition	Quantum mechanical	Correspondence value $n_c = n-\frac{1}{2}s$	% difference	Correspdence value $n_c = n(n-s)/(n-\frac{1}{2}s)$	% difference
2s-3p	9.393	11.022	17.31	9.270	3.48
4s-5p	72.553	76.395	5.30	73.181	0.86
6s-7p	274.19	281.22	2.56	275.25	0.38
4p-5d	121.86	127.86	4.92	123.18	1.083
4d-5f	197.83	207.00	4.64	200.46	1.33
(13,7)-(20,8)	39.51	47.18	19.4	37.96	-3.93

$$(R_{n\ell}^{n'\ell'})^2 \simeq \left(\frac{n_c^2}{2s}\right)^2 \left[\left(1+\Delta\ell\frac{\ell_c}{n_c}\right) J_{s+1}(s\varepsilon) - \left(1-\Delta\ell\frac{\ell_c}{n_c}\right) J_{s-1}(s\varepsilon)\right]^2, \tag{51}$$

where $\Delta\ell = \ell' - \ell$, $s = n - n'$.

Next we consider oscillator strengths between levels. These can be obtained by averaging over ε to give

$$\begin{aligned} f(n \to n') &\simeq \int_0^1 d\varepsilon^2 \{f(n,\ell \to n', \ell+1) + f(n, \ell \to n', \ell-1)\} \\ &= \frac{4n_c}{3s^3} \int_0^s dy \frac{d}{dy}(yJ_s(y)J'_s(y)) \end{aligned}$$

$$f(n \to n') \approx \frac{4n_c}{3} \frac{J_s(s)\, J_s'(s)}{s^2} \tag{52}$$

This result has been obtained by many authors including Mezger who put $n_c = n$; Percival and Richards who enforced the detailed balance relation

$$n^2 f(n \to n') = -n'^2 f(n' \to n) \tag{53}$$

by choosing

$$n_c = \frac{2n'^2}{n+n'} \; . \tag{54}$$

The fractional errors for (54) are of order s^2/n^2.

Time-dependent Collision Theories

In these theories the incident particle is supposed to follow a classical path and to produce a time-dependent electric field at the atom. For total cross-sections this assumption can be made when the de Broglie wavelength of the incident particle is small by comparison with the size of the atom, so that

$$E_1 \gg \frac{m_e U_n}{m_1 n^2} \tag{55}$$

where m_1 and E_1 are the mass and the energy of the incident charged particle, U_n is the ionization energy of the atom in a state with quantum number n.

For highly excited states this condition is broken only when the incident particle energy is very low indeed.

A further constraint on this method is that the incident particle is assumed to move along a classical path unperturbed by the bound electron. Thus energy and angular momentum are not conserved and detailed balance is not satisfied; adjustments can often be made taking account of these defects. However, since energy is not conserved, for this approximation to be reasonable it is necessary that

$$E_1 \gg |\Delta E| \quad , \tag{56}$$

where ΔE is the energy transferred from the incident particle to the internal energy of the atom, i.e. the excitation energy.

In practice time-dependent theories break down because of three-body effects which are generally insignificant when

$$E_1 >> U_n, \tag{57}$$

although this condition may be relaxed for large impact parameters.

If $V(t)$ is the potential due to the incident particle and H_o the Hamiltonian of the unperturbed atom, Schrodinger's equation is

$$i\hbar\frac{\partial}{\partial t}|\Psi\rangle = (H_o+V)|\Psi\rangle \quad . \tag{58}$$

On changing to the Heisenberg representation (Messiah, 1964) the integral form of Schrödinger's equation becomes

$$|\Phi(t)\rangle = |\Phi(-\infty)\rangle - \frac{i}{\hbar}\int_{-\infty}^{t} dt' V_I(t')|\Phi(t')\rangle \ , \tag{59}$$

where

$$V_I(t) = \exp\left(\frac{iH_o t}{\hbar}\right) V \exp\left(\frac{-iH_o t}{\hbar}\right) \tag{60}$$

Equation (59) can be iterated to give the time-dependent perturbation expansion.

$$|\Phi(t)\rangle = \left[1 - \frac{i}{\hbar}\int_{-\infty}^{t} dt_1 V_I(t_1) + \left(\frac{-i}{\hbar}\right)^2 \int_{-\infty}^{t} dt_1 V_I(t_1) \int_{-\infty}^{t_1} dt_2 (V_I(t_2) \ldots\right] |\Phi(-\infty)\rangle . \tag{61}$$

If $|\Gamma\rangle$ is an eigenstate of H_o,

$$H_o|\Gamma\rangle = E_\Gamma|\Gamma\rangle \tag{62}$$

and if $|\gamma\rangle$ is the initial state the transition amplitude to a final state $|\gamma'\rangle$ is defined to be

$$S(\gamma',\gamma) = \lim_{t\to\infty} S(\gamma',\gamma;t): \quad S(\gamma',\gamma;t) = \langle\gamma'|\Phi(t)\rangle \tag{63}$$

Taking matrix elements of (59) we obtain the usual equations for the matrix elements

$$i\hbar \frac{d}{dt}S(\gamma',\gamma;t) = \sum_\alpha \langle\gamma'|V_I(t)|\alpha\rangle S(\alpha,\gamma';t)$$

$$= \sum_\alpha \langle\gamma'|V|\alpha\rangle \exp i \frac{(E_{\gamma'} - E_\alpha)t}{\hbar} S(\alpha,\gamma;t). \tag{64}$$

For collisions between particles of charge Z_1e and hydrogen-like atoms and core with charge Z_3e situated at the origin, the interaction potential is

$$V(r_2, r_1(t)) = \frac{-Z_3Z_1e^2}{r_1} + \frac{Z_1e^2}{|r_2 - r_1|}$$
$$= \frac{-Z_1(Z_3 - 1)e^2}{r_1} + Z_1e^2\sum_{\lambda=1}^{\infty} \frac{r_2^{\lambda}}{r_1^{\lambda+1}} P_{\lambda}(\cos\theta), \qquad (65)$$

where we have assumed $r_1 > r_2$ and here θ is the angle between r_2, and r_1, the position of the incident particle, which is a known function of time. Frequently the dipole approximation to (65) is made, so that

$$V(\underset{\sim}{r}_2,\underset{\sim}{r}_1(t)) = \frac{-Z_1(Z_3 - 1)e^2}{r_1(t)} + \frac{Z_1e^2\underset{\sim}{r}_2\cdot\underset{\sim}{r}_1(t)}{r_1^3} .$$

The motion of the incident particle is determined by the Hamiltonian

$$H_1 = \frac{p_1^2}{2m_1} - \frac{Z_1(Z_3 - 1)e^2}{r_1} .$$

For collisions with neutrals $Z_3 = 1$, and the unperturbed orbit of an incident particle with impact parameter b is given by

$$\underset{\sim}{r}_1(t) = b\,\hat{\underset{\sim}{x}} + V_1 t\,\hat{\underset{\sim}{y}} \qquad (66)$$

with appropriate choice of Cartesian coordinates. For ions $Z_3 > 1$ and the orbit is a hyperbola. The various time-dependent theories are often called impact parameter approximations; they are also called semiclassical, or classical path approximations. The cross-section is given by

$$\sigma(\gamma \to \gamma') = \pi \int_0^{\infty} db^2 |S(\gamma', \gamma)|^2 \qquad (67)$$

the integral over b replacing the sum over angular momenta of the time-independent theory.

If the potential is small enough for its square and higher powers to be neglected the final state is given approximately by the first term in the series

$$S(\gamma',\gamma) = \frac{-i}{\hbar}\int_{-\infty}^{\infty} dt\langle\gamma'|V|\gamma\rangle \exp i\,\frac{(E_{\gamma'} - E_{\gamma})t}{\hbar}, \quad \gamma \neq \gamma' \qquad (68)$$

This is the classical path first-order perturbation (f.o.p.) approximation.

If the collision is rapid, so that the time of the collision is short compared to the characteristic times of the atom

$$T_{col} \ll T_{atom}, \quad b/V_1 \ll \hbar/|E_{\gamma'} - E_{\gamma}|, \qquad (69)$$

then to a good approximation all the $V_I(t)$ of equation (61) commute with one another, and the equation may be written

$$|\Phi(\infty)\rangle = \left[\exp \frac{-i}{\hbar} \int_{-\infty}^{\infty} dt V(t)\right] |\Phi(-\infty)\rangle \tag{70}$$

so that the S-matrix is

$$S(\gamma',\gamma) = \langle\gamma'|\exp \frac{-i}{\hbar} \int_{-\infty}^{\infty} dt V(t)|\gamma\rangle, \qquad \gamma \neq \gamma' \text{ (sudden).} \tag{71}$$

It is interesting to note that the contribution to the cross-section coming from the region where the dipole approximation and the sudden approximation are valid is

$$\sigma(\gamma \to \gamma') = \pi \int_{B_{min}}^{B_{max}} db^2 \left|\langle\gamma'|\ \exp \frac{-2iZ_1 e^2 x_2}{bv\hbar}|\gamma\rangle\right|^2 \tag{72}$$

where B_{max} is the maximum impact parameter for which the sudden approximation is valid and B_{min} is the minimum impact parameter for which the dipole approximation is valid. Introducing the variable K,

$$K = \left(\frac{2Z_1 e^2}{v\hbar}\right) \frac{1}{b} \tag{73}$$

this becomes

$$\sigma(\gamma \to \gamma') = 8\pi\left(\frac{Z_1 e^2}{hv_1}\right) \int_{K_{min}}^{K_{max}} \frac{dK}{K^3} |\langle\gamma'|e^{-iKx}|\gamma\rangle|^2 \tag{74}$$

This has a formal similarity to the Born approximation, but does not mean that these approximations are the same. In fact they are quite different. The sudden-dipole approximation is valid when strong coupling is important while the Born approximation is not. The former is an approximation in time-dependent theory while the latter comes from time-independent theory; the impact parameter of the sudden approximation is not simply related to the momentum transfer of the Born approximation.

Heisenberg's form of the correspondence principle applied directly to these approximations gives the following S-matrices

$$S(\gamma',\gamma) = \frac{-i\omega}{2\pi\hbar} \int_0^{2\pi/\omega} d\tau e^{is\omega\tau} \int_{-\infty}^{\infty} dt V^c(x(t+\tau),t) \quad \text{(f.o.p.)} \tag{75}$$

$$S(\gamma',\gamma) = \frac{\omega}{2\pi} \int_0^{2\pi/\omega} d\tau \exp i \left\{s\omega\tau - \frac{i}{\hbar}\int_{-\infty}^{\infty} dt V^c(x(\tau),t)\right\} \tag{76}$$

(Sudden)

$$S(\gamma',\gamma) = \frac{\omega}{2\pi}\int_0^{2\pi/\omega} d\tau \exp i(s\omega\tau - Kx(\tau)) \qquad \text{(Born)} \tag{77}$$

where $x(t)$ represents the motion of the bound system.

The additional error in using the correspondence principle (C.P.) is of order $|\Delta\gamma|/\gamma = |\gamma' - \gamma|/\gamma$, so that for the important processes in which the change of quantum number is small compared to the quantum number this additional error is negligible. The CP cross-sections are much easier to evaluate than the quantal cross-sections. Often they can be evaluated analytically, so that a formula is obtained for the transition probability. Where it has been possible to check against quantal calculations the agreement has been excellent.

We now present as an example of the use of the correspondence principle the explicit derivation of the transition probability in the dipole approximation using first order perturbation theory.

The probability $P(n \rightarrow n')$ for a transition from an initial level n to a final level n' is an average over initial states and a sum over final states

$$P(n \rightarrow n') = n^{-2}\sum_{\ell m}\sum_{\ell' m'} P(n\ell m \rightarrow n'\ell' m') \tag{78}$$

Spin is neglected. The classical equivalent of this sum is an integral over the classical variables defining the corresponding elliptic orbits. This is

$$P(n \rightarrow n') = \frac{1}{8\pi^2}\int_0^{\pi} d\beta \sin\beta \int_0^{2\pi} d\alpha \int_0^{2\pi} d\gamma \int_0^{1} d\varepsilon^2 |S(n \rightarrow n')|^2 \tag{79}$$

where ε is the eccentricity of the orbit, (α,β,γ) are the Euler angles defining the orientation of the orbit. $S(n \rightarrow n')$ is the S-matrix calculated using equation (75), where the motion in the (ξ,η)-plane of the orbit is given in terms of the "atomic" time τ by equations (5). The degeneracy of the hydrogen atom allows us to use the one-dimensional form of the correspondence principle. The S-matrix element is

$$S(n \rightarrow n') = \frac{-i\omega}{2\pi h}\int_0^{2\pi/\omega} d\tau \exp is\omega\tau \int_{-\infty}^{-\infty} dt V(\underset{\sim}{r}_2(\tau),t) \times \exp(-is\omega t), \tag{80}$$

where $s = n' - n$. Using the dipole form of the interaction potential

$$V(\underset{\sim}{r}_2,t) = -Z_1 e^2 \frac{\underset{\sim}{r}_2 \cdot \underset{\sim}{r}_1(t)}{r_1(t)^3}, \qquad (81)$$

where $Z_1 e$ is the charge on the incident particle, position $\underset{\sim}{r}_1(t)$, and $\underset{\sim}{r}_2$ is the position of the bound electron, we get

$$S\ (n \to n') = \frac{i\omega Z_1 e^2}{2\pi\hbar} \int_0^{2\pi/\omega} d\tau \exp is\omega\tau$$

$$\left\{ x_2(\tau) \int_{-\infty}^{\infty} dt \frac{b \cos s\omega t}{(b^2+V_1^2 t^2)^{3/2}} - iy_2(\tau) \int_{-\infty}^{\infty} dt \frac{V_1 t \sin s\omega t}{(b^2+V_1^2 t^2)^{3/2}} \right\} \qquad (82)$$

The coordinates $x_2(\tau)$ and $y_2(\tau)$ are given in terms of $\xi(\tau)$ and $\eta(\tau)$ and the Euler angles. The integral over τ gives the Fourier coefficients of the motion of the bound electron, whereas the integral over t is the form of a modified Bessel function.

The matrix element is

$$S(n \to n') = \frac{iZ_1 e^2 2\omega a}{\hbar V_1^2} \Big[\{K_1(s\omega b/V_1) A J_s'(s\varepsilon) + K_o(s\omega b/V_1) B' \frac{(1-\varepsilon^2)^{1/2}}{\varepsilon} \times J_s(s\varepsilon)\}$$

$$+ i\ \{K_1(s\omega b/V_1) B \frac{(1-\varepsilon^2)^{1/2}}{\varepsilon} J_s(s\varepsilon) - K_o(s\omega b/V_1) A' J_s'(s\varepsilon)\} \Big] \qquad (83)$$

where the coefficients A, A' and B, B' depend on the Euler angles, as follows

$$\begin{aligned} A(\alpha,\beta,\gamma) &= \cos\alpha \cos\beta \cos\gamma - \sin\alpha \sin\gamma \\ B(\alpha,\beta,\gamma) &= -\cos\alpha \cos\beta \cos\gamma - \sin\alpha \cos\gamma \\ A'(\alpha,\beta,\gamma) &= A(\alpha - \tfrac{1}{2}\pi,\beta,\gamma), \quad B'(\alpha,\beta,\gamma) = B(\alpha-\tfrac{1}{2}\pi,\beta,\gamma) \end{aligned} \qquad (84)$$

This expression is substituted into equation (79) for the transition probability and then integrated over the Euler angles and the eccentricity to give for impact parameter b.

$$P(n \to n') = \left(\frac{2Z_1 n}{Z_3} \frac{v_n^2}{V_1^2} \right)^2 \frac{2J_s(s)J_s'(s)}{3s}$$

$$\{[K_1(s\omega b/V_1)]^2 + [K_o(s\omega b/V_1)]^2\} \quad \text{(f.o.p.)} \qquad (85)$$

where $v_n = \sqrt{2m_e U_n}$ is the rms velocity of the atomic electron,

$\omega \approx (U_{n+1}-U_n)/\hbar$ is its angular orbital frequency and V_1 is the velocity of the incident particle.

The sudden dipole approximation can also be evaluated, but a truncated Taylor series is required before it can be obtained explicitly. The justification for the truncation and the details of the evaluation are given by Richards (J.Phys.B. 6 823-836,1973).

In this case the partial cross-section

$$\sigma_{<b}(n \to n') = 2\pi \int_0^b db.b\; P(n \to n') \tag{86}$$

is given by

$$\sigma_{<b}(n \to n') = 2\pi a_n^2 \left(\frac{2nZ_1 v_n}{Z_3 V_1}\right)^2 \left\{\frac{2}{3}\,\frac{J_s(s)J_s'(s)}{s^3}\left[\ln\frac{sV_1 b}{2nv_n a_n} + \frac{5-6\ln 2}{3}\right] + \frac{I(s)}{s^2}\right\} \tag{87}$$

The Born approximation is also given by a similar closed form. The function $I(s)$ is given to within 1% by the formula

$$I(s) = \frac{1}{6|s|} - \frac{.04}{s^2}\ln|s| - \frac{.115}{s^2} \tag{88}$$

Regions of Validity

The estimation of the regions of validity of approximations is absolutely crucial to the theory. A wide variety of approximations can be evaluated, but they are of little use without a careful estimate of where they are reliable. Fortunately this is not difficult, and we find that reasonable estimates of most of the important cross-sections can be made. But different approximations are needed for different values of the parameters, leading to a "patchwork" estimate of the cross-sections that can be confusing to the beginner. Frequently the patches overlap, so that approximations based on different assumptions can be used to check one another.

A whole range of purely classical theories are required for ionization, charge transfer and when $|s|$ is comparable to n. We have not the space to consider these theories, so we limit our consideration to processes for which $|s| = |n' - n| \ll n, n'$. These dominate for most astrophysical and laboratory plasmas, though not for the lower energy ranges.

It is very helpful to consider the probability $P(n \to n')$ as a function of impact parameter b and the incident velocity V_1, or normalized energy

$$\mathcal{E}_1 = E_1 m_e / m_1 \tag{89}$$

which is the energy of an electron moving with the velocity of the incident particle. If P is determined to within (say 20%) by a certain approximation within some region of the (b, V_1) plane then we refer to this as the "region of validity" of the approximation.

Consider first order dipole perturbation theory. From equation (85) the region of validity of this approximation for any transition from the level n is bounded by the line where the probability of loss from that level approaches unity, given approximately by

$$\left(\frac{Z_1 n v_n^2}{Z_3 V_1^2}\right)^2 \left[K_1^2\left(\frac{\omega b}{V_1}\right) + K_o^2\left(\frac{\omega b}{V_1}\right)\right] = 1 \tag{90}$$

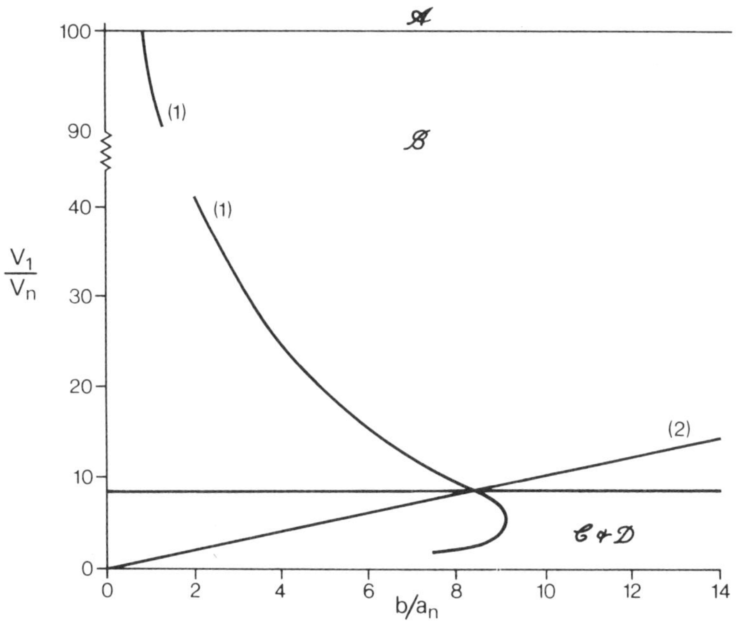

Fig.3. The b-V_1 plot for $n = 100$. Curve (1) is the boundary between the strong and weak coupling regions, Eq.(90); curve (2), the boundary between the sudden and adiabatic region. Eq.(90). The demarcation of the energy regions. $\mathcal{A}$, $\mathcal{B}$, $\mathcal{C}$, and $\mathcal{D}$ are shown by the horizontal lines.

and by the curve sketched in Fig. 3. For $\omega b/V_1$ small compared to unity, (90) simplifies to

$$V_1 = \frac{nZ_1}{Z_3}\left(\frac{a_n}{b}\right) v_n \tag{91}$$

To the right of the curve f.o.p. is approximately valid, to the left strong coupling is important and perturbation theory breaks down.

Since the dipole approximation has been used, Eq.(90) is invalid for $b < a_n$; putting $b = a_n$ into (91) we see that the lowest velocity at which perturbation theory can be used for all impact parameters greater than the atomic radius is

$$V_1 = \frac{Z_1}{Z_3} nv_n \quad \text{or} \quad \mathcal{E}_1 = Z_1^2 R, \tag{92}$$

where R is a Rydberg, half the atomic unit of energy. Notice that for incident electrons and protons this energy is a Rydberg, independent of n, s and the charge Z_3e on the nucleus.

As the incident particle energy decreases from this value there is an ever increasing range of impact parameters for which weak coupling theories are invalid; for higher energies it is reasonable to assume that weak coupling theories are valid for all impact parameters greater than the atomic radius a_n. For $n = 100$, and incident electrons, strong coupling is insignificant only when the incident particle energy is greater than half a million times the threshold for the $\Delta n = 1$ transition.

The Born approximation cannot be used for lower $\mathcal{E}_1$, and in general, for $\mathcal{E}_1 < Z_1^2 R$, a strong coupling theory is necessary. The sudden approximation assumes that the interaction potential is assumed to be peaked in time; this is the case when

$$\text{collision time} < \text{orbiting time:} \quad \frac{b}{a_n} < \frac{V_1}{v_n} \tag{93}$$

In the $b - V_1$ plane this is the straight line shown (Fig.3); to the right of this line the collisions are adiabatic and give rise mainly to small energy transfers; to the left the collisions are sudden and produce mainly large energy transfers

The intersection of the boundaries of (93) and (90) is approximately at

$$V_1 = \left|\frac{nZ_1}{Z_3}\right|^{1/2} \quad \text{or} \quad \mathcal{E}_1 = 2\left|\frac{Z_1 Z_3}{n}\right| R \tag{94}$$

For energies significantly above this there is a range of impact parameters for which both the sudden approximation and f.o.p. are valid.

For energies significantly below (94) there is a negligible contribution from the range of b in which f.o.p. is valid. Classical theory and the density-of-states correspondence principle are adequate.

It is useful to give names to the energy ranges where different combinations of approximations must be used. There are four of them, $\mathcal{A}, \mathcal{B}, \mathcal{C}, \mathcal{D}$ starting from the highest, $\mathcal{A}$, where the Born approximation is valid to the lowest, $\mathcal{D}$, where classical perturbation theory breaks down and full scale numerical integration of the classical orbits, the Monte Carlo method, is necessary. These regions are given in Table 3

Table 3. Energy ranges $\mathcal{A}$, $\mathcal{B}$, $\mathcal{C}$, $\mathcal{D}$

Region	Energy range
$\mathcal{A}$	$Z_1^2 \le \frac{\mathcal{E}_1}{R}$
$\mathcal{B}$	$\frac{2\lvert Z_1 Z_3\rvert}{n} \le \frac{\mathcal{E}_1}{R} \le Z_1^2$
$\mathcal{C}$	$4\frac{Z_3^2}{n^2} \le \frac{\mathcal{E}_1}{R} \le \frac{2\lvert Z_1 Z_3\rvert}{n}$
$\mathcal{D}$	$\frac{\mathcal{E}_1}{R} \le \frac{4Z_3^2}{n^2}$

The regions of validity of various approximations are shown in Table 4

Consideration of these ranges of validity enables us to obtain an interesting general property of the behaviour of the $n \to n+1$ cross-sections as a function of the scaled energy $\mathcal{E}_1 = \frac{1}{2} m_e V_1^2$.

Apply the classical path approximation in the region $\mathcal{A}$. It is not so accurate as the Born approximation, but adequate for our purposes. The probability P has the form

$$P(n \to n+1) = C_1/(b^2 \mathcal{E}_1), \qquad (95)$$

with constant C_1, for impact parameter b in the range

$$a_n < b < a_n V_1/v_n \qquad (\text{region } \mathcal{A}), \tag{96}$$

where a_n is the radius of the atom, and the upper limit is given by the straight line in figure 3. For larger impact parameters the collision becomes adiabatic and decreases exponentially, whereas smaller impact parameters cannot contribute more than πa_n^2 to the cross-section.

Table 4. Approximations and "regions of validity" for large n, $|s| = |\Delta n| = 1$

Approximation	Conditions of validity	Region of validity $\lvert\Delta n\rvert = 1$
Born	Weak coupling	$\mathcal{A}$
Dipole	Bulk of cross-section comes from $b > 2a_n$	$\mathcal{B}, \mathcal{C}$
Classical rectilinear path of 1	Negligible recoil	$\mathcal{A}, \mathcal{B}, \mathcal{C}$
First-order perturbation (f.o.p.)	Probability of loss from initial state for direct and reverse transition $\ll 1$	$\mathcal{A}$ $\mathcal{B}$: $bV_1/(a_n v_n) > n$ $\mathcal{C}$: (negligible)
Sudden	Time of collision less than characteristic atomic frequency, $\omega = \Delta E/\hbar$	$b\omega \ll V_1$ $(\mathcal{A}, \mathcal{B}, \mathcal{C})$
Classical with density-of-state correspondence principle	Insignificant proportion of cross-section from classically inaccessible region	$\mathcal{B}$: $n > bV_1/(a_n v_n)$ $\mathcal{C}$ $\mathcal{D}$

The contribution from the range (96) is obtained from the integral (86) between theupper and lower limit, giving the familiar Bethe form

$$\sigma(n \to n+1) \approx \frac{C_1}{\mathcal{E}_1} \ln \frac{\mathcal{E}_1}{U_n} + \frac{C_2}{\mathcal{E}_1} \qquad (\text{region } \mathcal{A}) \tag{97}$$

for the cross section. As the incident energy decreases down to $\mathcal{B}$, strong coupling becomes important for impact parameters $b > a_n$ up to the curve in figure 3. This curve then becomes a rough effective lower limit in b to significant contributions to the cross-section, so the range of significant impact parameters is now

$$\frac{a_n n v_n}{V_1} < b < \frac{a_n V_1}{v_n} \qquad \text{(region } \mathcal{B}\text{)} \qquad (98)$$

and the integration of (86) between these limits gives

$$\sigma = \frac{2C_1}{\mathcal{E}_1} \ell n \frac{\mathcal{E}_1}{U_n} + \frac{C_3}{\mathcal{E}_1} \qquad \text{(region } \mathcal{B}\text{)} \qquad (99)$$

Thus, because both limits change, the slope of the "Bethe plot" of $\mathcal{E}_1 \sigma$ against $\ell n(\mathcal{E}_1)$ has twice the slope in region $\mathcal{B}$ as in region $\mathcal{A}$. In practice, the discontinuity in slope is smoothed out in the neighbourhood of the boundary.

Some other Theories, Collisions and Literature

This is not a comprehensive list of theories, but only of those related to the material of the lectures.

The theories discussed here are described in more detail in a series of four papers[1] and a review article[2], which also discusses some of the other theories, including the strong coupling correspondence principle[3], which is the correspondence equivalent of the solution of the classical path close coupling equations, the Presnyakov-Urnov equally spaced energy level approximation[4], the Seaton-Saraph first order quantal approximation and the very large number of Born and Bethe-Born calculations. It also describes various classical theories, which are discussed in more detail as follows. The full orbit integration Monte Carlo theory is treated in[5] and an interesting variational adaptation for collision rates is given in [6]. These are necessary when the velocity of the incident particle is in region $\mathcal{D}$, comparable with that of the bound electron. Very low energies, for which the energy of the incident particle is only just enough to excite a few upward transitions have not been treated adequately but they are unlikely to be important. The combined theory[7] is an analytical classical theory that can be used in region $\mathcal{C}$.

These theories are patched together and checked against each other in [2], but improved versions have since been produced for electron [8] and proton [9] excitation. Reference [10] contains other frequently used formulae for cross-sections and rates, that are discussed in [8] and in more detail in [11], which also contains simplified formulae for cross-sections.

Cross sections or rates for change of angular momentum have been obtained for non-hydrogenic atoms and for hydrogen [12].

References

1. Percival I C and Richards D, J.Phys.B: Atom.Molec.Phys. 3: 315 (1970), 4: 918 (1971), 4: 932 (1971). Richards D, J.Phys. B:Atom.Molec.Phys. 6: 823 (1973).

2. Percival I C and Richards D, Theory of collisions between charged particles and highly excited atoms, Adv.Atom.Molec.Phys. 11:1 (1975).

3. Percival I C and Richards D, J.Phys.B: Atom.Molec.Phys. 3: 1035 (1970).

4. Presnyakov L P and Urnov A M, J.Phys.B: Atom.Molec.Phys. 3: 1267 (1970).

5. Abrines R and Percival I C, Proc.Phys.Soc.London 88: 861 (1966), 88: 973 (1966). Abrines R, Percival I C and Valentine N A, Proc.Phys.Soc. London 89: 515 (1966). Boesten L G J, Bonsen T F M and Banks D, J.Phys.B: Atom.Molec.Phys. 8: 628 (1975). Percival I C, J.Phys.B: Atom.Molec.Phys.6: 93 (1973), Comp.Phys.Communications 6: 347 (1973). Olsen R E and Salap A, Phys.Rev. A16: 513 (1977). Olsen R E, J.Phys.B: Atom.Molec. Phys. 13: 483 (1980).

6. Mansbach P and Keck J C, Phys.Rev. 181: 275 (1969). Keck J C, Adv.Atom.Molec.Phys. 8: 39 (1972).

7. Percival I C, J.Phys.B: Atom.Molec.Phys. 6: 2236 (1973).

8. Gee C S, Percival I C, Lodge J and Richards D, Mon.Not.Roy. Astron.Soc. 175: 209 (1976). Percival I C and Richards D, Mon.Not.Roy.Astron.Soc. 183: 329 (1978).

9. Lodge J G, Percival I C and Richards D, J.Phys.B: Atom.Molec. Phys. 9: 239 (1976).

10. Johnson L C, Astrophys. J.174: 227 (1972).

11. Vriens L, Phys.Rev.A to be published.

12. Percival I C and Richards, J.Phys.B: Atom.Molec.Phys. 10: 1497 (1977), 12: 205 (1979).

ELECTRON-ATOM SCATTERING IN THE FIELD OF A LASER

Marvin H. Mittleman

Physics Department
The City College of the
City University of New York

ABSTRACT

The problem of electron-atom scattering in the field of a laser has been studied in only a very limited range of laser parameters. The case of a low frequency, not too intense, laser is discussed for the case in which there is no electron-atom resonance in the absence of the laser. The case in which a resonance exists is also briefly described. The situation in which the laser is weak and tuned to a transition of the free atom is also discussed and the possibility of extending this to more intense lasers is raised. The situation in which the laser is tuned to resonate between a pair of compound states of electron plus atom is also discussed in two possibilities, one in which both states are resonances, and the other in which one of the states is a stable negative ion.

As an introduction the Born approximation for potential scattering is discussed and the kinematics of the projectile, as modified by the laser is presented.

1. INTRODUCTION
2. BORN APPROXIMATION FOR POTENTIAL SCATTERING
3. LOW FREQUENCY LASER
4. RESONANT LASER

1. INTRODUCTION

The field of electron atom-scattering in the presence of an intense electromagnetic field (a laser) is of interest for applied reasons such as problems arising in plasma heating by lasers or the workings of gas lasers. It is also of interest because of the fact that it allows for observation of electron-atom scattering parameters which would not otherwise be observable. This can be made plausible by noting that the photons of the laser can play the role of a third body during the scattering and absorb energy and momentum so that the e-atom T matrices which describe the process will be off-shell. The coupling and dynamics of photons are particularly simple so that the laser is an attractive "third body" to use in the scattering. It is the latter point of view which will be emphasized here.

The laser field is properly described as a collection of photons (a quantum electrodynamic description) but it has been shown[1] that if the occupation number of each of the laser modes is large and if the fractional change of these numbers is small during the process then a classical description of the electromagnetic field is a good one. Both of these conditions are well satisfied for any experiment which we shall consider here so the classical description will be adopted. Moreover for simplicity we shall describe the laser field as a single-mode, linearly polarized, traveling wave. The polarization can be varied with no significant change in the physics or mathematics and the traveling wave statement will not be of great significance since we shall be able to make a dipole approximation for the field in all cases. The single mode description does however require some additional discussion. The laser will in many cases consist of a collection of closely spaced modes such that $\Delta\omega/\omega \ll 1$. Here ω is the central (angular) frequency of the laser and $\Delta\omega$ is the mode spacing. In that case the field can be recast as a wave of frequency ω with an amplitude which is slowly varying in time. Then any calculation which has been made previously for a single mode laser can be adapted to the multimode case by simply taking the cross section as a function of the laser amplitude for the single mode case, and averaging it over the adiabatic variation in time of the laser amplitude[2]. It is therefore sufficient for our purposes to deal with this restricted laser.

The introduction of the laser into the scattering event brings new parameters into the problem; laser frequency, intensity, polarization mode structure, etc. Of these the frequency and intensity are the important ones, and the problem has been investigated for only a limited range of these. We shall discuss only the low frequency and resonant frequency cases for "not too intense" fields. (A precise meaning for these statements will be given below). However for pedagogic reasons we will first discuss the simple case in which the internal degrees of freedom of the atom are neglected and the e-atom interaction is approximated by a local potential. For further simplicity we shall then deal with the case in which this potential energy is much less than the kinetic energy of the electron so that a Born approximation is sufficient.

The third section is devoted to the low frequency case and the fourth to the resonant frequency case. In both of these the theorists yearning for higher laser intensities will run through the text. We should point out here that there are practical limitations to these desires beyond the trouble and expense incurred by higher power lasers. The target atom will always be in the beam for some time before the collision takes place. If the multiphoton ionization probability during that time is significant then the collision cross section will depend (critically) upon that time and this is a complication which has not been considered in any analysis and will not be discussed here. Since multiphoton ionization probabilities go up with intensity (for low intensities) this will limit the intensity of the laser beam which can be used.

II. Born approximation for potential scattering.

In a typical experiment an electron will enter the laser field from vacuum and move many laser wavelengths in the field before it scatters from the target. Therefore the dipole approximation must be made, if at all, with great care. On the other hand, in almost all cases, the range of the scattering potential will be much shorter than the laser wavelength and we should be able to exploit this fact.

The Schrodinger equation describing an electron moving in a linearly polarized laser potential $(\hbar = c = 1)$

$$\vec{A}(\vec{r},t) = \frac{\vec{E}(\vec{r})}{\omega} \cos(\vec{k}\cdot\vec{r} - \omega t) \tag{2.1}$$

and a static potential V(r) is

$$\left[i\frac{\partial}{\partial t} - \frac{1}{2m}\left(\vec{P} + e\vec{A}(\vec{r},t)\right)^2 - V(r) \right] \psi(\vec{r},t) = 0\,. \tag{2.2}$$

In order to discuss transitions described by this equation it is first necessary to obtain the "unperturbed" set of electron states in the absence of V, analogous to the plane wave states of conventional scattering theory. These satisfy

$$\left[i\frac{\partial}{\partial t}-\frac{1}{2m}\left(\vec{p}+e\vec{A}(\vec{r},t)\right)^2\right]\chi_{\vec{q}}(\vec{r},t)=0 \tag{2.3}$$

where $\vec{q}$ is the wave number vector of the electron outside the laser. In the realistic situation the laser amplitude, $\vec{E}(r)$, changes slowly over the electron wavelength and the laser wavelength is large compared to the electron wavelength. This is expressed by the inequalities

$$\frac{1}{E}\frac{dE}{dr}\ll q \quad , \quad k\ll q\ . \tag{2.4}$$

In order to use these we rewrite (2.3) as

$$\Big[i\frac{\partial}{\partial t}-\Big(\frac{p^2}{2m}+\frac{e^2E^2(\vec{r})}{4m\omega^2}+\frac{e}{m\omega}\cos(\vec{k}\cdot\vec{r}-\omega t)\,\vec{E}(\vec{r})\cdot\vec{p} + \frac{e^2E^2(\vec{r})}{4m\omega^2}\cos 2(\vec{k}\cdot\vec{r}-\omega t)\Big)\Big]\,\chi_{\vec{q}}=0\ . \tag{2.5}$$

The term immediately following the kinetic energy operator is the ponderomotive potential which is repulsive, tending to expel the electron from the laser. It can be rewritten as

$$\frac{1}{2}R_y\left(\frac{R_y}{\hbar\omega}\right)^2\frac{I(r)}{I_0} \tag{2.6}$$

where I(r), is the laser intensity and I_0 is an atomic unit of intensity

$$I_0=\frac{c}{4\pi}\left(\frac{e}{2a_0^2}\right)^2\simeq 1.8\times 10^{16}\,W/cm^2\ . \tag{2.7}$$

It is clear from (2.7) that this term is small except for high power lasers focused into a small cross section. In that case this potential plays a significant role in altering the electron kinetic energy from its value outside the laser to a lower value inside the laser. We shall assume that this term is small and slowly varying (2.4) and treat it and the remaining terms of (2.5) in an eikonal approximation. This is accomplished by assuming a form

$$\chi_{\vec{q}}=\exp i\left(\vec{q}\cdot\vec{r}-\mathcal{E}_q t-\Lambda(\vec{r},t)\right) \tag{2.8}$$

where $\epsilon_q=q^2/2m$. This leads to

$$\dot{\Lambda} + \frac{\vec{q}}{m}\cdot\vec{\nabla}\Lambda - \frac{e}{m\omega}\vec{q}\cdot\vec{E}\cos(\vec{k}\cdot\vec{r}-\omega t) - \frac{e^2E^2}{4m\omega^2}\left(1+\cos 2(\vec{k}\cdot\vec{r}-\omega t)\right)$$
$$-\frac{1}{2m}(\nabla\Lambda)^2 + \frac{1}{2mi}\nabla^2\Lambda + \frac{e}{m\omega}\cos(\vec{k}\cdot\vec{r}-\omega t)\,\vec{E}\cdot\vec{\nabla}\Lambda = 0 \qquad (2.9)$$

The last three terms of (2.9) are neglected on the assumption that Λ is slowly varying, (2.4). The remaining linear first order partial differential equation can be integrated by the method of characteristics to give.

$$\Lambda(\vec{r},t) = \int_{-\infty}^{t} d\tau \left[\frac{e^2E^2(\vec{r}-\vec{v}(t-\tau))}{4m\omega^2}\left(1+\cos 2(\vec{k}\cdot\vec{r}-\vec{k}\cdot\vec{v}(t-\tau)-\omega\tau)\right)\right.$$
$$\left. + \frac{e\vec{q}\cdot\vec{E}(\vec{r}-\vec{v}(t-\tau))}{m\omega}\cos\left(\vec{k}\cdot\vec{r}-\vec{k}\cdot\vec{v}(t-\tau)-\omega\tau\right)\right] \qquad (2.10)$$

where $\vec{v} = \vec{q}/m$. The integrals can be simplified with the use of the additional approximation that the laser amplitude varies slowly over a laser wavelength

$$\frac{1}{E}\frac{dE}{dr} \ll k\,. \qquad (2.11)$$

This is satisfied except for the case when the laser is very highly focussed. With the use of (2.11) we get

$$\Lambda(\vec{r},t) = \frac{e^2}{4m\omega^2}\int_0^{\infty} d\tau\, E^2(\vec{r}-\vec{v}\tau) + \frac{e\vec{q}\cdot\vec{E}(\vec{r})}{m\omega^2}\sin(\vec{k}\cdot\vec{r}-\omega t)$$
$$+ \frac{e^2E^2(\vec{r})}{8m\omega^3}\sin 2(\vec{k}\cdot\vec{r}-\omega t)\,. \qquad (2.12)$$

If (2.12) is substituted back into (2.8) we obtain the wave function inside the laser which is the adiabatic deformation of the plane wave $e^{i\vec{q}\cdot\vec{r}}$ outside the laser. If we place the origin of coordinates at the scattering center and expand the first term of (2.12) in powers of $\vec{r}$ and retain only up to terms linear in r then (discarding irrelevant phase factors) the effective momentum in the vicinity of the origin is

$$\vec{q}\,' = \vec{q}\left(1 - \frac{e^2E^2(0)}{8m\omega^2\epsilon_q}\right) - \frac{e^2}{4m\omega^2}\left(\nabla_\perp \int_0^{\infty} d\tau\, E^2(\vec{r}-\vec{v}\tau)\right)_{r=0} \qquad (2.13)$$

where $\vec{\nabla}_\perp = \vec{\nabla} - \hat{k}\hat{k}\cdot\vec{\nabla}$. Thus the effective momentum at the scattering center is shifted both in magnitude and direction from that outside the laser. This shifts the interpretation of the observed cross section.

We illustrate with a simple example. Suppose that the laser amplitude varies only in a direction ($\hat{x}$) perpendicular to the direction of propagation. Then (2.13)) becomes

$$\vec{q}\,' = \vec{q}\left(1 - \frac{e^2 E^2(0)}{8m\omega^2 \epsilon_q}\right) - \left(\hat{x} - \hat{q}\hat{q}\cdot\hat{x}\right)\frac{e^2 E^2(0)}{4\omega^2 q_x} \tag{2.14}$$

This relation must be applied to both initial and final momenta in order to connect the momenta at the target with the observed momenta outside the laser. If the apparent scattering angle is given by

$$\cos\Theta = \hat{q}_i \cdot \hat{q}_f \tag{2.15}$$

it can be related to the true scattering angle (at the target) defined by

$$\cos\Theta' = \hat{q}_i' \cdot \hat{q}_f' \, . \tag{2.16}$$

A little algebra with the assumption that the difference is small yields

$$\Theta - \Theta' = \frac{e^2 E^2(0)}{8m\omega^2 \epsilon_i \epsilon_f \sin\Theta}\left(\cos\Theta\,(\epsilon_i + \epsilon_f) - \sqrt{\epsilon_i \epsilon_f}\left(\frac{q_{ix}}{q_{fx}} + \frac{q_{fx}}{q_{ix}}\right)\right). \tag{2.17}$$

The first term in (2.12) is then accounted for by the shifts in energy and momentum at the target. With this proviso it can be dropped. The remaining terms in (2.12) oscillate rapidly in time so that their effect is reduced compared to the first one. Moreover, the last term in (2.12) is usually small compared to the second since their ratio is of order

$$\frac{1}{q a_0}\left(\frac{R_y}{\hbar\omega}\right)\left(\frac{I}{I_0}\right)^{1/2}$$

which is usually small compared to unity. Finally we anticipate that the wavefunction $\chi_{\vec{q}}$ will be used in an integral which contains V(r) whose range we take to be R. If, as is usually the case, we have

$$\frac{R}{E}\frac{dE}{dr} << 1 \quad , \quad kR << 1$$

then the remaining term in Λ can be evaluated at $\vec{r}=o$ to give

$$\chi_{\vec{q}} = \exp i(\vec{q}\cdot\vec{r} - \epsilon_q t - \vec{q}\cdot\vec{\alpha}_o \sin \omega t) \tag{2.18}$$

where

$$\vec{\alpha}_o = e\vec{E}(o)/m\omega^2 . \tag{2.18a}$$

This is a solution of the equation

$$\left[i\frac{\partial}{\partial t} - \left(\frac{p^2}{2m} + \frac{e p\cdot E(o)}{m\omega}\cos \omega t\right)\right]\chi_{\vec{q}} = o \tag{2.19}$$

which is the dipole approximation of (2.5) with the E^2 terms removed by a gauge transformation.

We can expand (2.18) in the form

$$\chi_{\vec{q}} = e^{i\vec{q}\cdot\vec{r}} \sum_{n=-\infty}^{\infty} J_n(\vec{q}\cdot\vec{\alpha}_o)\, e^{-i(\epsilon_q + n\omega)t} \tag{2.20}$$

where it is evident that we have a superposition of energies each of which differs from ϵ_q by an integer number of photon energies. This allows us to interpret $J_n(\vec{q}\cdot\vec{\alpha}_o)$ as the amplitude for finding n additional photons in the field due to the presence of the free electron. These are virtual processes since a free electron can not emit or absorb energy from a plane wave field. The virtual processes can however be made real by a scattering and we shall illustrate this by a Born approximation treatment of the scattering process. The exact S matrix[3] for scattering from $\vec{q}_i$ to $\vec{q}_f$ is given by

$$S_{fi} = -i\langle \chi_{\vec{q}_f}, V\,\psi^{(+)}_{\vec{q}_i}\rangle \tag{2.21}$$

where $\psi^{(+)}_{q_i}$ satisfies (2.2) with causal boundary conditions. The Born approximation is obtained by the replacement $\psi^{(+)}_{q_i} \to \chi_{q_i}$. The use of (2.20) for both initial and final states then results in

$$S^B_{fi} = -2\pi i \sum_{\ell} \delta(\epsilon_f - \epsilon_i + \ell\omega)\, T^B_{\vec{q}_f,\vec{q}_i}(\ell) \tag{2.22}$$

where

$$T^B_{\vec{q}_f\vec{q}_i}(\ell) = J_\ell(\Delta\vec{q}\cdot\vec{\alpha}_o)\,\tilde{V}(\Delta\vec{q}) \tag{2.23}$$

where $\Delta\vec{q} = \vec{q}_f - \vec{q}_i$ and

$$\tilde{V}(\Delta\vec{q}) = \int d^3r\, e^{-i\Delta\vec{q}\cdot\vec{r}}\, V(\vec{r}) \tag{2.24}$$

and we have used the relation

$$\sum_n J_n(x)\, J_{n-\ell}(y) = J_\ell(x-y)\ . \tag{2.25}$$

Clearly $T^B(\ell)$ can be interpreted as the T matrix, in Born approximation, for the scattering of an electron with the stimulated emission of ℓ photons. (Negative ℓ represents absorbtion of photons from the laser). The final energy of the electron can be observed so the processes with different ℓ add incoherently. Then the cross section for ℓ photons is

$$\begin{aligned}\frac{d\sigma^B}{d\Omega}(\vec{q}_f(\ell), \vec{q}_i; \ell) &= \frac{q_f(\ell)}{q_i}\, J_\ell^2(\Delta\vec{q}\cdot\vec{\alpha}_0)\left|\frac{m}{2\pi}\tilde{V}(\Delta\vec{q})\right|^2 \\ &= J_\ell^2(\Delta\vec{q}\cdot\vec{\alpha}_0)\left(\frac{d\sigma^B}{d\Omega}(\vec{q}_f(\ell), \vec{q}_i)\right)_{\alpha_0=0}\end{aligned} \tag{2.26}$$

where the last cross section is evaluated in the absence of the laser. We see from this, that nothing new with respect to the scattering process is obtained from the presence of the laser.

III. Low Frequency Laser

It has been known[4] for some time that the interaction with low frequency radiation modifies the scattering process in a simple way so that the amplitude for scattering in the presence of the radiation is a multiple of the amplitude in its absence. Kroll and Watson[3] showed that this is true in a modified way in the presence of a low frequency laser. We shall not follow their treatment here but instead present one which is simpler and which can readily be extended to higher orders in ω.

The words "low frequency laser" require some definition here. For the moment we simply require that the photon energy be small compared to the projectile energy, $\omega \ll \epsilon_p$, and that the parameter, $\alpha_0 = eE/m\omega^2$ be held finite in this limit.

The choice of gauge in which one works can be of significance here. For example, in the dipole approximation to

(2.2) we can go to the Kramers gauge[5] with the transformation function

$$\exp i\left(\vec{p}\cdot\vec{\alpha}(t) + \frac{e^2}{2m}\int^t dt' A^2(t')\right) \tag{3.1}$$

where

$$\vec{\alpha}(t) = \int^t dt' \frac{e}{m}\vec{A}(t') = \vec{\alpha}_0 \sin\omega t \tag{3.2}$$

The new form of (2.2) is

$$\left[i\frac{\partial}{\partial t} - \frac{p^2}{2m} - V(\vec{r} - \vec{\alpha}(t))\right]\psi = 0 \tag{3.3}$$

which can be written in momentum space and a Fourier <u>series</u> in time as

$$(\epsilon_q - \epsilon_k + n\omega)\tilde{\psi}_{\vec{q}}(\vec{k},n) - \sum_n \int \frac{d^3k'}{(2\pi)^3}\tilde{V}(\vec{k}-\vec{k}')J_{n-n'}(k-k'\cdot\alpha_0)\tilde{\psi}_{\vec{q}}(\vec{k}',n') = 0. \tag{3.4}$$

The laser frequency now appears only in the first term and so will occur only in energy denominators in which simple expansion is possible. This formulation has been exploited to reproduce the result of Kroll and Watson and to extend it one order of ω further[6]. However Kruger and Jung[7] have pointed out that it is even simpler to work in the p.A gauge where the electron-laser coupling is of order $\omega\alpha_0\cdot p$ and so can be treated by perturbation theory except in the initial and final state. The exception arises because the perturbation has a long time to act on these states and so in effect eliminates one power of ω and as we have seen results in states

$$\chi_{\vec{q}} = \exp i(\vec{q}\cdot\vec{r} - \epsilon_q t - \vec{q}\cdot\vec{\alpha}_0 \sin\omega t) \tag{3.5}$$

Our starting point for this calculation is the exact S matrix, (2.21) where in the dipole approximation the exact wave function can be taken to satisfy

$$\left[i\frac{\partial}{\partial t} - \left(\frac{p^2}{2m} + \frac{e}{m}\vec{p}\cdot\vec{A}(t) + V(\vec{r})\right)\right]\psi^{(+)}_{\vec{q}_i} = 0 \tag{3.6}$$

which can be solved as

$$\psi^{(+)}_{\vec{q}_i} = \chi_{\vec{q}_i} + \bar{G}^{(+)} V \chi_{\vec{q}_i} \tag{3.7}$$

where $\bar{G}^{(+)}$ is the Greens function for the operator in (3.6). The discussion above shows that an expansion in powers of the laser-electron interaction will be useful for intermediate states so we write

$$\bar{G}^{(+)} = G^{(+)} + \sum_{n=1}^{\infty} G^{(+)} \left(\left(\frac{e}{m} p \cdot A \right) G^{(+)} \right)^n \tag{3.8}$$

where $G^{(+)}$ is the Greens function in the absence of the laser

$$\left[i\frac{\partial}{\partial t} - \left(\frac{p^2}{2m} + V \right) \right] G^{(+)} = 1 \tag{3.9}$$

which can be written

$$G^{(+)}(\vec{r}t, \vec{r}'t') = \lim_{\eta \to 0^+} -i\,\theta(t-t') \sum_n \phi_n(\vec{r})\, \phi_n^*(\vec{r}')\, e^{-i(W_n + \eta)(t-t')} \tag{3.10}$$

where ϕ_n is one of the complete orthonormal set defined by

$$\left(W_n - \frac{p^2}{2m} - V \right) \phi_n(\vec{r}) = 0. \tag{3.11}$$

The use of these relations in the S matrix results in a series

$$S_{fi} = -2\pi i \sum_{\ell} \delta(\epsilon_f - \epsilon_i + \ell\omega) \sum_{j=0}^{\infty} T^{(j)}_{\vec{q}_f, \vec{q}_i}(\ell) \tag{3.12}$$

which is a series in powers of ω. The leading term is

$$T^{(0)}_{\vec{q}_f, \vec{q}_i}(\ell) = \sum_{s=-\infty}^{\infty} J_{\ell-s}(\vec{q}_f \cdot \vec{\alpha}_0)\, J_s(-\vec{q}_i \cdot \vec{\alpha}_0)\, (\vec{q}_f | V | \vec{q}_i) \tag{3.13}$$

where the last factor is merely another notation for (2.24). The next term can be evaluated in a straightforward manner. It is

$$T^{(1)}_{\vec{q}_f \vec{q}_i}(\ell) = \sum_s J_{\ell-s}(\vec{q}_f \cdot \vec{\alpha}_0)\, J_s(-\vec{q}_i \cdot \vec{\alpha}_0)\, (\vec{q}_f | V G^{(+)}(\epsilon_i - s\omega) V | \vec{q}_i) \tag{3.14}$$

where $G^{(+)}(E)$ is the conventional time-independent Greens function

$$G^{(+)}(E) = \sum_n \frac{\phi_n)(\phi_n}{E + i\eta - W_n} \tag{3.15}$$

Combination of these two terms yields

$$T^{(0)} + T^{(1)} = \sum_s J_{\ell-s}(\vec{q}_f \cdot \vec{\alpha}_0) J_s(-\vec{q}_i \cdot \vec{\alpha}_0) (\vec{q}_f | T(\epsilon_i - s\omega) | \vec{q}_i) \tag{3.16}$$

where T(E) is the conventional T matrix in the absence of the laser.

$$T(E) = V + V G^{(+)}(E) V. \tag{3.17}$$

If T in (3.16) is sufficiently slowly varying that it can be expanded in powers of ω then the use of (2.25) results in

$$T^{(0)} + T^{(1)} = J_\ell(\Delta\vec{q} \cdot \vec{\alpha}_0) (\vec{q}_f | T(\epsilon_i) | \vec{q}_i) + O(\omega). \tag{3.18}$$

The next order in ω arises from $T^{(2)}$ and the term of order ω which is neglected in (3.18). The algebra is straightforward, with only one new relation needed:

$$\sum_s s J_{\ell-s}(\vec{q}_f \cdot \vec{\alpha}_0) J_s(-\vec{q}_i \cdot \vec{\alpha}_0) = \frac{\ell(-\vec{q}_i \cdot \vec{\alpha}_0)}{\Delta\vec{q} \cdot \vec{\alpha}_0} J_\ell(\Delta\vec{q} \cdot \vec{\alpha}_0) \tag{3.19}$$

After a little algebra the results can be summarized as

$$\begin{aligned} \sum_{j=0}^{2} T^{(j)}_{\vec{q}_f,\vec{q}_i}(\ell) = J_\ell(\Delta\vec{q} \cdot \vec{\alpha}_0) (\vec{q}_f | T(\epsilon_i) - V G^{(+)}(\epsilon_i) \frac{\ell\omega\, \vec{q}_i \cdot \vec{\alpha}_0}{\Delta\vec{q} \cdot \vec{\alpha}_0} G^{(+)}(\epsilon_i) V \\ + V G^{(+)}(\epsilon_i) \frac{\ell\omega\, \vec{\alpha}_0 \cdot \vec{\mathbb{P}}}{\Delta\vec{q} \cdot \vec{\alpha}_0} G^{(+)}(\epsilon_i) V | \vec{q}_i) + O(\omega^2). \end{aligned} \tag{3.20}$$

As we have seen, obtaining this result (and even extending it to higher orders) requires the use of only straightforward algebra and very little ingenuity. The insight of Kroll and Watson showed that the three terms in the last factor of (3.20) could be combined to yield a single on-shell T matrix. This is most easily seen by starting with the relation

$$\begin{aligned} (\vec{q}_f + \vec{\gamma} | T(\epsilon_{\vec{q}_i + \vec{\gamma}}) | \vec{q}_i + \vec{\gamma}) = (\vec{q}_f + \vec{\gamma} | V | \vec{q}_i + \vec{\gamma}) \\ + (\vec{q}_f + \vec{\gamma} | V G^{(+)}(\epsilon_{\vec{q}_i + \vec{\gamma}}) V | \vec{q}_i + \vec{\gamma}). \end{aligned} \tag{3.21}$$

The vector $\vec{\gamma}$ is taken to be of order ω. The first term on the right hand side of (3.21) is independent of $\vec{\gamma}$ provided that V is a local potential. The second can be written as

$$\int \frac{d^3k\,d^3k'}{(2\pi)^6}\,\tilde{V}(\vec{q}_f-\vec{k})(\vec{k}+\vec{\gamma}\,|\,G^{(+)}(\epsilon_{\vec{q}_i+\vec{\gamma}})\,|\vec{k}'+\vec{\gamma})\,\tilde{V}(\vec{k}'-\vec{q}_i)$$

and expanded in powers of $\vec{\gamma}$ with the aid of the equation defining the Greens function.

$$(\epsilon_{\vec{q}_i+\vec{\gamma}}-\epsilon_{\vec{k}+\vec{\gamma}})(\vec{k}+\vec{\gamma}\,|\,G^{(+)}(\epsilon_{\vec{q}_i+\vec{\gamma}})\,|\vec{k}'+\vec{\gamma})$$
$$-\int\frac{d^3k''}{(2\pi)^3}\,\tilde{V}(\vec{k}-\vec{k}'')\,(\vec{k}''+\vec{\gamma}\,|\,G^{(+)}(\epsilon_{\vec{q}_i+\vec{\gamma}})\,|\vec{k}'+\vec{\gamma}) = \delta(\vec{k}-\vec{k}') \quad (3.22)$$

Expansion in powers of $\vec{\gamma}$ yields.

$$(\vec{k}+\vec{\gamma}\,|\,G^{(+)}(\epsilon_{\vec{q}_i+\vec{\gamma}})\,|\vec{k}'+\vec{\gamma}) = (\vec{k}\,|\,G^{(+)}(\epsilon_{\vec{q}_i})\,|\,\vec{k}')$$
$$+\int\frac{d^3k''}{(2\pi)^3}\,(\vec{k}\,|\,G_0^{(+)}(\epsilon_{q_i})\,|\vec{k}'')\,\frac{\vec{\gamma}\cdot(\vec{k}''-\vec{q}_i)}{m}\,(\vec{k}''\,|\,G^{(+)}(\epsilon_{q_i})\,|\,\vec{k}') \quad (3.23)$$

The choice

$$\vec{\gamma} = m\omega\ell\,\vec{\alpha}_0/\vec{\Delta q}\cdot\vec{\alpha}_0 \quad (3.24)$$

results in (3.21) being identical with the last factor of (3.20). Then we have

$$T^{(\ell)}_{\vec{q}_f,\vec{q}_i} = J_\ell(\vec{\Delta q}\cdot\vec{\alpha}_0)\,(\vec{Q}_f\,|\,T(\epsilon_{Q_i})\,|\,\vec{Q}_i) \quad (3.25)$$

where

$$\vec{Q}_i = \vec{q}_i+\vec{\gamma}$$
$$\vec{Q}_f = \vec{q}_f+\vec{\gamma} \quad (3.26)$$

and the important relation

$$Q_i^2/2m = Q_f^2/2m \quad (3.27)$$

which places the T matrix on the right hand side of (3.25) on-shell. The cross section for the transfer of ℓ photons can then be written as

$$\frac{d\sigma}{d\Omega}(\vec{q}_f, \vec{q}_i; \ell) = \frac{q_f(\ell)}{q_i} J_\ell^2(\Delta\vec{q}\cdot\vec{\alpha}_0)\, \frac{d\sigma}{d\Omega}(\vec{Q}_f, \vec{Q}_i)\Big|_0 \qquad (3.28)$$

where the cross section on the right hand side is that in the absence of the laser. Again we obtain the negative result that (neglecting order ω^2) the cross section in the presence of the laser is simply expressible in terms of that in the absence of the laser. We shall simply mention in passing here that it has been shown that the ω^2 terms of (3.25) contain terms which are off-shell contributions from T matrices in the absence of the laser[6]. However these appear to be too small to be observable at this time.

Experiments[8] have been done with 11 ev electrons scattered on argon in the presence of a CO_2 laser ($\omega \simeq 0.1\,ev$) in which the energy loss spectrum clearly showed the presence of seven ℓ values between ± 3. The experiment was only qualitative, the only quantitative result being the confirmation of the sum rule

$$\sum_\ell \frac{d\sigma}{d\Omega}(\vec{q}_f, \vec{q}_i; \ell) = \frac{d\sigma}{d\Omega}(\vec{q}_f, \vec{q}_i)\Big|_0 \qquad (3.29)$$

which says that the cross section for all possible values of ℓ, sums to the scattering in the absence of the laser. This result can be obtained in the $\omega = 0$ limit of (3.28) with the use of the sum rule

$$\sum_\ell J_\ell^2(x) = 1 . \qquad (3.30)$$

This limit is reasonable in the experiment and the energy and angle corrections described in the preceding section are too small to be observed at the power levels (about 10^7 W/cm^2) and accuracy of this experiment.

Another sum rule that is of practical interest in the problem of plasma heating can be obtained by asking for ℓ_{ave}, the average number of photons transferred during the scattering process. This is defined by

$$\ell_{ave} = \frac{1}{\sigma_T} \sum_{\ell=-\infty}^{\infty} \ell \int d\Omega_f \frac{d\sigma}{d\Omega}(\vec{q}_f, \vec{q}_i; \ell) \qquad (3.31)$$

where σ_T is the total elastic cross section in the absence of the laser. This can be evaluated with the use of (3.28). For a central potential the cross section on the right side of (3.28) can be written as

$$\frac{d\sigma}{d\Omega}\left(\frac{Q_i^2(\ell)}{2m}, \hat{Q}_i(\ell)\cdot\hat{Q}_f(\ell)\right)$$

and then expanded in powers of ω with the use of (3.24) and (3.26). The result after some calculation is

$$\ell_{ave} = -\frac{2m\omega\alpha_0^2}{\sigma_T}\left(\left(1+(\hat{\alpha}_0\cdot\hat{q}_i)^2\right)\frac{1}{4}\sigma_\delta + \frac{1}{2}(\hat{\alpha}_0\cdot\hat{q}_i)^2\epsilon_{q_i}\frac{\partial\sigma_\delta}{\partial\epsilon_{q_i}}\right) \tag{3.32}$$

where the momentum transfer cross section, σ_δ , is

$$\sigma_\delta = \int d\Omega_f\,(1-\hat{q}_i\cdot\hat{q}_f)\frac{d\sigma}{d\Omega}(\epsilon_{q_i}, \hat{q}_i\cdot\hat{q}_f). \tag{3.33}$$

It, and the total cross section are defined in the absence of the laser and are functions only of the projectile energy. The first term in (3.32) is negative definite and represents electron heating in the sense that it tends to make the final electron energy larger than the initial one. The second term depends on the slope of σ_δ as a function of energy. It can have either sign and therefore result in either a net absorption or emission of photons by the laser. The result (3.32) is written for a projectile beam but can be converted to a form more appropriate to a gas by averaging over the electron distribution function. For an isotropic electron distribution this becomes

$$\overline{\ell}_{ave} = -\frac{m\omega\alpha_0^2}{3\overline{\sigma}_T}\int d^3q\,\sigma_\delta(\epsilon_q)\,\epsilon_q\left(-\frac{\partial}{\partial\epsilon_q}f(\epsilon_q)\right) \tag{3.34}$$

where $f(\epsilon_q)$ is the distribution function and $\overline{\sigma}_T$ is the total cross section averaged over it. Evidently a distribution function which is monotonically decreasing in energy (such as a Maxwellian) will give a net heating of the electrons by the laser.

The result (3.32) can also be applied to the case in which the electron has a high enough kinetic energy so that the Born approximation can be used to obtain the cross sections. These can then be written

$$\begin{aligned} \sigma_T &= \int d^3q\; g(\epsilon_q(1-\mu)) \\ \sigma_\delta &= \int d^3q\,(1-\mu)\,g(\epsilon_q(1-\mu)) \end{aligned} \tag{3.35}$$

where we have used the fact that the differential cross section, in this approximation, is a function of only one variable, the square of the momentum transfer. Some simple algebra then results in

$$\ell_{ave}^{B} = -2m\omega\alpha_0^2\left(\frac{1}{4}\left(1+(\hat{\alpha}_0\cdot\hat{q}_i)^2\right)\frac{\sigma_\delta}{\sigma_T} + \frac{4\pi}{\sigma_T}\left(\frac{d\sigma}{d\Omega}\right)_{\mu=-1}\cdot\frac{1}{2}(\hat{\alpha}_0\cdot\hat{q}_i)^2 - \frac{1}{2}(\hat{\alpha}_0\cdot\hat{q}_i)^2\right) \tag{3.36}$$

where $(d\sigma/d\Omega)_{\mu=-1}$ is the differential cross section in the backward direction. In the usual case the differential cross section will be strongly peaked in the forward direction so that the backward cross section and the momentum transfer cross section will be much smaller than the total cross section. In that case the last term of (3.36) dominates with the result

$$\ell_{ave}^{B} \simeq m\omega\alpha_0^2(\hat{q}_i\cdot\hat{\alpha}_0)^2 \geqslant 0 \tag{3.37}$$

and the electrons will tend to amplify the laser as they scatter.

The result (3.32), and those following from it are linear in the laser intensity and so will not be changed by spatial or temporal inhomogeneities of the laser beam[9]. This will not be the case for higher order (ω^3) corrections to (3.32) which might conceivably be used to investigate these effects.

One can also obtain the average value of ℓ^2 in a manner analogous to the way the average value of ℓ was obtained in (3.31). It is easily shown to be of order $(\Delta q\cdot\alpha_0)^2$ which shows that the distribution in ℓ is much broader than its average value which is proportional to ω, (3.32).

The expansion in powers of ω which was used in obtaining (3.18) from (3.16) depends, for its convergence, upon some average energy transfer, $\langle \ell\omega\rangle$, being small compared to the projectile energy. If we use $\sqrt{\langle \ell^2\rangle}\,\omega$ for this average and simply use the order of magnitude, then the smallness parameter is

$$\frac{\sqrt{\langle \ell^2\rangle}\,\omega}{\epsilon_i} \simeq \frac{\Delta q\,\alpha_0\omega}{\epsilon_i} \simeq \left(\frac{I}{I_0}\,\frac{R_y}{\epsilon_i}\right)^{1/2}\frac{R_y}{\hbar\omega} \tag{3.38}$$

so that the method outlined above will not work for very high laser intensities.

Another limitation on the method arises from the fact that

the T matrix in the absence of the laser may be a rapidly varying function of energy so that the smallness of (3.38) or ($\hbar\omega/\epsilon_i$) may not be sufficient for convergence. Rapid variation of the T matrix with energy is the same as a resonance in the scattering cross section. The resonance will occur in a particular partial wave at an energy $\mathcal{E}_R$ and near that energy

the phase shift in that partial wave will be described by

$$\tan\delta = \tan\delta_B + \Gamma/2 \big/ \mathcal{E}_R - E \tag{3.39}$$

where δ_B , the background phase shift is a slowly varying function of energy. The parameter Γ is the resonance width, the width of the region around the energy $\mathcal{E}_R$ in which the phase shift jumps by π. Typical widths of electron-atom scattering resonances are several milli-electron-volts and for example, a CO_2 laser photon energy is about 0.1 ev so it is clear that an expansion is not possible. This was originally emphasized by Kruger and Jung[7].

The problem can be handled in the following way. Suppose that the resonance energy is related to the initial energy by

$$\epsilon_0 = \epsilon_{q_i} + m_0\omega \simeq \mathcal{E}_R \tag{3.40}$$

where m_0 is an integer. (The approximate equality is taken to mean an equality within a range Γ .) Therefore the absorption of m_0 photons by the incident electron brings its energy to the resonance energy. The term $s=-m_0$ in (3.16) is rapidly varying in ω and no expansion is possible but for all other $s \neq -m_0$ the expansion can still be made. The presence of the resonance generates additional terms on the right side of (3.20) which are

$$J_{l+m_0}(\vec{q}_f\cdot\vec{\alpha}_0)J_{m_0}(\vec{q}_i\cdot\vec{\alpha}_0)(\vec{q}_f|T(\epsilon_0)-T(\epsilon_{q_i})-m_0\omega\frac{\partial}{\partial\epsilon_{q_i}}T(\epsilon_{q_i})|\vec{q}_i). \tag{3.41}$$

These vanish (to order ω^2) if there is no resonance so that $T(\epsilon_0)$ is expandable. Higher order terms such as $T^{(2)}$ can be handled similarly but the problem of convergence becomes more complicated by the fact that the Greens function occurring in the T matrix, (3.17) can become large in the vicinity of the resonance. The Greens function near this energy can be written

$$G^{(+)}(E) = G_{NR}^{(+)} + \tilde{u}_R)(u_R \big/ E - \mathcal{E}_R + i\Gamma/2 \tag{3.42}$$

where $G_{NR}^{(+)}$ is a slowly varying function of E and U_R may be interpreted as the resonance state wave function. Near resonance the energy denominator is of order Γ so that both G and T(E) can be of order Γ^{-1} which should be counted as

order ω^{-1}. This makes the development of the series too complicated to be included here. It has been obtained to order ω but resonant terms from $T^{(3)}$ must be included. The terms (3.41) can be used to obtain a modification of the sum rule (3.29) with the result

$$\sum_{\ell} \frac{d\sigma}{d\Omega}(\vec{q}_f, \vec{q}_i; \ell) = \left(1 - J_{m_0}^2(\vec{q}_i \cdot \vec{\alpha}_0)\right) \frac{d\sigma}{d\Omega}(\vec{q}_f, \vec{q}_i)_{NR;0}$$

$$+ J_{m_0}^2(\vec{q}_i \cdot \vec{\alpha}_0) \frac{d\sigma}{d\Omega}(\vec{q}_f, \vec{q}_i)\Big|_0 \qquad (3.43)$$

where both cross sections on the right are obtained in the absence of the laser and the one subscripted NR means only the non-resonant part which arises from the background phase shift. The other one comes from the total phase shift and is the observable one in the absence of the laser. The fact that these occur separately is the first point at which the intrusion of the laser into the scattering process yields information which is not available in the absence of the laser.

In the experiment mentioned above 11 ev electrons were scattered from argon in the field of a CO_2 laser. This is a known resonance energy of argon so that the question of why (3.29) rather than (3.43) applies must be discussed. The answer lies in the numbers. For this experiment m_0=o and a power level of about 10^7 W/cm^2 or less was used. Therefore $q_i \cdot \alpha_0 \sim 0.3$ so that $J_{m_0}^2(q_i \cdot \alpha_0) \simeq 1$ so the difference between the two sum rules is very small. Clearly the difference should be observable at higher powers. Additional experiments[10] have been performed but these are still only qualitative in nature.

A word of caution is now in order. In the preceding discussion the atom has been replaced by a structureless entity and the electron-atom interaction has been approximated by a local-energy-independent potential. The cross section in the presence of the laser is then related to the cross section in its absence. It is well known that the electron atom interaction potential is both non-local and energy dependent and that the internal structure of the atom can be deformed by the interaction with the laser. All of these effects can change the results obtained above and so will have to be considered.

Our starting point is the generalization of the S matrix, (2.21), to the case in which the target is an atom. It is

$$S_{fi} = -i \langle \Omega_f(0), V(0)\, \psi_i^{(+)} \rangle \tag{3.44}$$

where the argument (o) means that the coordinate $\vec{r}_0$ is singled out as the projectile. That is, the (z+1) electron coordinates are numbered 0, 1,...z and the function Ω_f(o) is a product of a state with the "free" electron numbered "o" in the state (3.5) and another state with all the other electrons numbered 1 to z in the state "f" which is an atomic state in the field of the laser. That is

$$\Omega_f(0) = \chi_{q_f}(\vec{r}_0)\, \phi_f(\vec{r}_1 \cdots \vec{r}_z, t) = \chi_f(0)\, \phi_f(0) \tag{3.45}$$

and V(o) is the Coulomb interaction potential of the projectile, $\vec{r}_0$, with all the other charged particles. The state ϕ_f is very difficult to find in general but fortunately we can resort to a perturbation theory solution of the problem. The reason is the same as that given in association with (3.8). It is that the coupling to the laser is proportional to $\omega\alpha_0 \cdot p$ which is first order in ω. The Schrodinger equation for the atom in the laser is

$$\left(i\frac{\partial}{\partial t} - \sum_{j=1}^{z} \left(\frac{p_j^2}{2m} + \omega \vec{\alpha}_0 \cdot \vec{p}_j \cos\omega t \right) - U \right) \phi(0) = 0 \tag{3.46}$$

where U is the internal potential energy operator of the atom. The complete set of orthonormal unperturbed atomic states are defined by

$$\left(W_n - \sum_{j=1}^{z} \frac{p_j^2}{2m} - U \right) u_n(0) = 0 \,. \tag{3.47}$$

These can be used to solve (3.46) as a perturbation theory in ω. We must first enlarge the definition of "low frequency laser" to include

$$\omega \ll | W_n - W_{n'} | \tag{3.48}$$

where atomic states n and n' are relevant in the scattering.

Then the solution to (3.46) can be written as

$$\phi_n(0) = \left(1 - im\omega\vec{\alpha}_0\cdot\vec{R}(0)\cos\omega t\right)\phi_n^{(0)}(0) \tag{3.49}$$

where $\vec{R}(0) = \sum_{j=1}^{z}\vec{r}_j$ and

$$\phi_n^{(0)}(0) = u_n(0)\, e^{-iW_n t} \tag{3.50}$$

The remaining symbol, $\psi_i^{(+)}$, in (3.44) is the full wave function which can be written as

$$\psi_i^{(+)} = A(0)\left(1 + \bar{G}^{(+)}V(0)\right)\Omega_i(0) \tag{3.51}$$

where $\bar{G}$ is the full Greens function of the problem and A(o) is the antisymetrization operator

$$A(0) = 1 - \sum_{j=1}^{z} X_{j0} \tag{3.52}$$

where X_{j0} exchanges coordinates "o" and "j". It is included in (3.51) to account for the Pauli Principle. The time dependent Greens function in (3.51) can be expanded in powers of the laser interaction with the result

$$\bar{G}^{(+)} = G_0^{(+)} + G_0^{(+)}\omega\vec{\alpha}_0\cdot\vec{P}\cos\omega t\, G_0^{(+)} + \cdots \tag{3.53}$$

whre $G_0^{(+)}$ is the (time dependent) Greens function for all the particles in the absence of the laser and

$$\vec{P} = \sum_{j=0}^{z}\vec{p}_j = \vec{p}_0 + \vec{P}(0)\,. \tag{3.54}$$

The substitution of these results back into the S matrix and expansion in powers of ω yields

$$S_{fi} = -2\pi i\sum_{\ell}\delta\left(E_f - E_i + \ell\omega\right)T_{fi}(\ell) \tag{3.55}$$

where $E_i = \epsilon_i + W_i$ with a similar relation for E_f. After some algebra the T matrix, to order ω can be written

$$T_{fi}(\ell) = J_\ell(\Delta\vec{q}\cdot\vec{\alpha}_0)\Big\langle\vec{q}_f, u_f\Big|\, T(E_i) + \frac{\omega\ell\vec{q}_i\cdot\vec{\alpha}_0}{\Delta\vec{q}\cdot\vec{\alpha}_0}\frac{\partial}{\partial E_i}T(E_i) + T(E_i)\frac{im\omega\ell\vec{\alpha}_0\cdot\vec{r}_0}{\Delta\vec{q}\cdot\vec{\alpha}_0} - \frac{im\ell\omega\vec{\alpha}_0\cdot\vec{r}_0}{\Delta\vec{q}\cdot\vec{\alpha}_0}T(E_i)\,\Big|\vec{q}_i, u_i\Big\rangle \tag{3.56}$$

where the conventional scattering T operator is

$$T(E) = V(0)A(0)\left(1 + G_0^{(+)}(E)V(0)\right) \tag{3.57}$$

and where $G_0^{(+)}(E)$ is the time independent Greens function for the particles in the absence of the laser. The result, (3.56) is only the first two orders in a power series in ω, and to that order it is clear that it can be rewritten as

$$T_{fi}(\ell) = J_\ell(\vec{\Delta q}\cdot\vec{\alpha}_0)\langle \vec{q}_f+\vec{\gamma}, u_f | T(E_i') | \vec{q}_i+\vec{\gamma}, u_i\rangle \tag{3.58}$$

where $\vec{\gamma}$ is given by (3.24) and

$$E_i' = W_i + \epsilon_{\vec{q}_i} + \omega\ell \frac{\vec{q}_i\cdot\vec{\alpha}_0}{\Delta\vec{q}\cdot\vec{\alpha}_0} \simeq W_{q_i} + \epsilon_{\vec{q}_i+\vec{\gamma}} \tag{3.59}$$

This is the generalization of the Kroll-Watson result for potential scattering. Again only the on-shell T matrix, in the absence of the laser, is observable to this order. It is clear that off shell effects will enter in the next order but these, as well as the effects of electron-atom scattering resonances have yet to be published.

Finally we should point out the existence of a series of papers[11] by L. Rosenberg in which the laser field is treated quantum mechanically in which some of the results described here are also obtained.

IV. Resonant Laser

We consider first the situation in which the laser is tuned so that it is resonant with a ground to excited state transition in the free atom. The fact that the laser must be tuned usually precludes a high intensity situation and this will be our assumption. Our first step in calculating a cross section for scattering in the presence of this kind of laser must be to obtain the laser-distorted atomic states. We shall use the notation defined in (3.47) to describe the unperturbed atomic states which will be used to construct the atomic states in the laser. Taking the states u_0 and u_1 to be the ground and laser coupled excited states respectively we may construct two new states in the laser. The technique used is the two state rotating wave approximation[12] which will not be described here. We simply quote the results for the laser-coupled states. They are

$$\phi_\pm = \frac{1}{\sqrt{2\cosh\mu}}\left(e^{\pm\frac{\mu}{2}} u_0 e^{i\omega t/2} \pm e^{\mp\frac{\mu}{2} - i\Theta} u_1 e^{-i\omega t/2}\right) e^{-i\int_0^t (W_0 + W_1 \pm \mathcal{E})\, dt/2} \tag{4.1}$$

where the parameters are defined as follows: The dipole coupling matrix element is

$$\Lambda = \frac{e\vec{E}}{m\omega}\cdot(u_0|\vec{P}|u_1) = |\Lambda| e^{i\Theta} \tag{4.2}$$

where E is the amplitude of the laser electric field and ω is its frequency. The detuning parameter is defined by

$$\sinh\mu = \frac{\Delta\omega}{|\Lambda|} = \frac{\omega - W_{10}}{|\Lambda|} \tag{4.3}$$

is assumed to be of the order of unity and the Rabi frequency is given by

$$\mathcal{E} = \left((\Delta\omega)^2 + |\Lambda|^2\right)^{1/2} = |\Lambda| \cosh\mu\ . \tag{4.4}$$

The time integral in the last factor of (4.1) is present to allow for a slow (adiabatic) time dependence of the laser intensity in which case (4.1) is the adiabatic solution in the two-state rotating wave approximation. The order of magnitude of the error incurred by this approximation is $\Delta\omega/W_{10}$ which is usually very small. The remaining states of the atom are assumed to be unaffected by the laser since we assume no resonant connection between any other bound pair of states. We could include the lowest order modification of these states by the A.C. Stark effect but this is a very small effect and adds nothing new. Then the complete set of states of the atom is taken to be

$$\{\phi_n\} = \{\phi_\pm,\ u_n e^{-iW_n t} (n \neq 0, 1)\}\ . \tag{4.5}$$

In the experiment contemplated here the atom enters the laser beam and is then struck by the electron. It is difficult to make this switching (in the frame of the atom) from no laser to full laser field in anything but an adiabatic fashion so the states (4.1) do indeed describe the atomic states before the collision. It is also easy to see that the atom in its ground state adiabatically becomes ϕ_+ (ϕ_-) for $\Delta\omega > 0$ ($\Delta\omega < 0$) as it enters the field. Therefore the atomic state encountered by the electron will be ϕ_+ ($\Delta\omega > 0$). This state is very different from a bare atomic state so we expect that the scattering cross section will also be very different. For example, the state has an oscillating dipole and a permanent quadrupole moment so the long range forces[13] will be very different from that which the electron "sees" from an unperturbed atom. We shall denote the cross section for scattering from the initial state ϕ_+ to a final state which is one of the set (4.5) ϕ_f, by

$$\frac{d\bar{\sigma}_\ell}{d\Omega}(f,+)$$

The index ℓ has the same meaning that it had in the previous section. It denotes the number of photons transferred in some way between laser field and projectile. The transfer can occur directly as in the preceding section or it can occur with the atom acting as an intermediary. The two processes are indistinguishable.

If the above description was the entire description of the process then we could indeed expect to observe exotic effects in the scattering. It is however only part of the story. When the atom enters the laser it is pumped up and down between u_0 and u_1 (in the states ϕ_+ or ϕ_-) many times. For a typical experiment it will spend perhaps 10^{-5} sec in the laser before the collision takes place and roughly half of that time will be spent in the upper state u_1. A typical natural decay time will be of the order of 10^{-8} sec. So we can expect of the order of a thousand such decays before the collision (this is the process of resonance fluorescence). These decays will couple ϕ_+ and ϕ_- so the atomic state will no longer be purely one or the other but instead some superposition. Moreover the fluorescence process is a stochastic one so we expect that the superposition will be an incoherent one. This expectation can be rigorously derived by a method for treating the resonance fluorence photon field given by Mollow[14]. The only approximation that is necessary is that the number of fluorescent photons is large compared to unity. We shall not present the derivation here since it is lengthy and the result is almost obvious. It is that the observed cross section for a given final atomic state is

$$\frac{d\sigma_\ell(f)}{d\Omega} = P_+ \frac{d\bar{\sigma}_\ell}{d\Omega}(f,+) + P_- \frac{d\bar{\sigma}_\ell}{d\Omega}(f,-) \tag{4.6}$$

where P_+ (P_-) is the probability of finding ϕ_+ (ϕ_-) in the incoherent superposition of the two that forms the pre-collision description of the atom. These can also be obtained from Mollow's formalism but they emerge even more simply by simple detailed balancing arguments. Again we merely quote the results

$$P_{\pm} = e^{\pm 2u}/2\cosh 2u \, . \tag{4.7}$$

The utility of (4.6) is that the cross sections occurring on the right make no reference to the complicated fluorescent processes whose effects are completely described by the P factors. Finally we should also point out that the final states f=+ or - are indistinguishable since they will be inextricably mixed by the fluorescent process in the final state before the atom leaves the laser. For that reason the observable is

$$\frac{d\sigma_{\ell}(+)}{d\Omega} + \frac{d\sigma_{\ell}(-)}{d\Omega} \, . \tag{4.8}$$

Before preceding we should also point out that (4.6) has been derived[15] by another physical argument. In the usual experiment the time at which the collision takes place is not observed and so must be averaged over. This simply means that the phase in the Rabi pumping cycle of the atom at which the collision takes place is not measured. Thus ensemble average also results in an effective incoherence which produces the result (4.6).

The cross sections occurring on the right side of (4.6) may be simplified when the laser is not too intense. First we note that the effect of the coupling of the laser to the projectile is determined by parameter

$$\vec{\alpha}_0 \cdot \vec{\Delta q} \simeq \hat{E} \cdot \hat{\Delta q} \left(\frac{I}{I_0}\right)^{1/2} \left(\frac{R_y}{\hbar\omega}\right)^2 \tag{4.9}$$

as we have seen in Chapter II. This is small for a typical dye laser so we neglect this coupling. We have already neglected the coupling of the the laser to all but the atomic states and $\phi_{\pm}$ by the choice of the complete set (4.5) so the only remaining coupling to the laser occurs when the atom is in the states $\phi_{\pm}$. It is in these states initially and possibly finally and in intermediate states. A typical collision time is of the order of 10^{-16} sec which is much shorter than the Rabi period, $\mathcal{E}^{-1}$. (The Rabi period must be shorter than the natural decay time in order to saturate the $u_0 \leftrightarrow u_1$ transition but will still be much longer than the collision time). So an intermediate state which certainly exists for less than the collision time will not last long enough to be recoupled into the states $\phi_{\pm}$ by the laser. Therefore the laser may be neglected in its coupling to the atom in all but the initial and final states.

These approximations can now be built into the S matrix as follows. The exact S matrix between states

$$\Omega^{(0)}_{\vec{q},n} = \chi_{\vec{q}}(0,t)\,\phi_n(1\cdots z,t) \tag{4.10}$$

where $\chi_{\vec{q}}$ are laser-distorted projectile states, (3.5) and ϕ_n are laser distorted atomic states, (4.5) is given by (3.44) which we can write, (3.51), as

$$S(\vec{q}'n',\vec{q}n) = -i\left\langle \Omega^{(0)}_{\vec{q}'n'},\; V(0)A(0)\left(1+\left(i\frac{\partial}{\partial t}-H(t)\right)^{-1}V(0)\right)\Omega^{(0)}_{\vec{q},n}\right\rangle \tag{4.11}$$

where $\left(i\frac{\partial}{\partial t}-H(t)\right)^{-1}$ is a formal notation for the full Greens function. The other symbols have been defined above. The approximation of neglecting the coupling of the projectile to the laser allows us to replace $\chi_{\vec{q}}$ in (4.10) by an undistorted plane wave. The approximation of neglecting the laser-atom coupling in intermediate states allows us to replace H(t) in the Greens function by H, the Hamiltonian of the particles alone. Then there is no time dependence remaining in the operator of (4.11) except for the time derivative in the Greens function. But the initial and final states will now all have simple exponential time dependences which are eigenfunctions of the time derivative so the time derivative will be replaced by its eigenvalues and the S matrix is then simply expressible as an energy conserving delta function times a T matrix. For example if neither n or n' in (4.11) is 0 or 1 then the laser does not participate at all and so the T matrix is that in the absence of the laser. A more interesting case occurs when n=+ or - and n' > 1. This is

$$\begin{aligned} S(\vec{q}'n';\vec{q}\pm) = {} & -2\pi i\,\delta\!\left(\epsilon_{q'}+W_{n'}-\epsilon_q-W_0\mp\frac{|\Lambda|}{2}e^{\mp\mu}\right)\frac{e^{\pm\mu/2}}{(2\cosh\mu)^{1/2}}\times \\ & \langle\vec{q}'n'|\,T\!\left(\epsilon_q+W_0\pm\tfrac{1}{2}|\Lambda|e^{\mp\mu}\right)|\vec{q},0\rangle \\ & -2\pi i\,\delta\!\left(\epsilon_{q'}+W_{n'}-\epsilon_q-W_1\mp\tfrac{1}{2}|\Lambda|e^{\pm\mu}\right)(\pm)\frac{e^{\mp\frac{\mu}{2}-i\theta}}{(2\cosh\mu)^{1/2}}\times \\ & \langle\vec{q}'n'|\,T\!\left(\epsilon_q+W_1\pm\tfrac{1}{2}|\Lambda|e^{\pm\mu}\right)|\vec{q},1\rangle . \end{aligned} \tag{4.12}$$

These are off-shell T matrices but the energy by which they are off-shell is of the order of $|\Lambda|$ which is usually small compared to the energy spread of the electron beam. If these are neglected, only on shell T matrices occur and the cross section for production of electrons whose final energy is $\epsilon_{q'} = \epsilon_q - W_{no}$ is, using (4.6)

$$\frac{d\sigma}{d\Omega}(n;\vec{q}'\vec{q}) = (1-P_1)\left(\frac{d\sigma}{d\Omega}(\vec{q}'n,\vec{q}0)\right)_{\alpha_0=0} \tag{4.13}$$

where $P_1=\frac{1}{2}(\cosh 2\mu)^{-1}$ can be shown to be the probability of finding u_1 in the pre-collision atomic state by Mollow's technique[14]. The interpretation of this is obvious: The observed cross section is the cross section for scattering from the ground state, in the absence of the laser, times the probability of finding the ground state $(1-P_1)$ in the pre-collision atom. We can also find the cross section for finding the electron with energy $\epsilon_{q'} = \epsilon_q - W_{n1}$. It turns out to be

$$\frac{d\sigma}{d\Omega}(n;\vec{q}'\vec{q}) = P_1\left(\frac{d\sigma}{d\Omega}(\vec{q}'n,\vec{q}1)\right)_{\alpha_0=0} \tag{4.14}$$

and a similar interpretation follows. Both of these results show that, in this approximation, the measured cross sections give only cross sections in the absence of the laser but (4.14) at least allows for the measurement of scattering from an excited state which is something that would be much more difficult to measure in the absence of the laser. Experiments of this kind have been performed[16].

We could also write the S matrix in the case where both n and n' are $(\pm)$. The four different S matrices are somewhat lengthly and will not be reproduced here. We shall merely give the result for elastic scattering, $\epsilon_{q'} = \epsilon_q$

$$\frac{d\sigma_0}{d\Omega}(\vec{q}',\vec{q}) = (1-P_1)\left(\frac{d\sigma}{d\Omega}(\vec{q}'0,\vec{q}0)\right)_{\alpha_0=0} + P_1\left(\frac{d\sigma}{d\Omega}(\vec{q}'1,\vec{q}1)\right)_{\alpha_0=0} \tag{4.15}$$

whose interpretation is obvious. The super elastic scattering, $\epsilon_{q'} = \epsilon_q + \omega$, cross section is

$$\frac{d\sigma_{-1}}{d\Omega}(\vec{q}',\vec{q}) = P_1\left(\frac{d\sigma}{d\Omega}(\vec{q}'0,\vec{q}1)\right)_{\alpha_0=0} \tag{4.16}$$

whose interpretation is also obvious.

There is a maximum of one photon transferred during the

collision. This is a direct result of the approximations which neglect (1) the direct coupling of the laser to the projectile and (2) the coupling of the laser to the atom during the collision. The first is an approximation that (4.9) is small while the second is that $|\Lambda| T_c$ is small where T_c is the collision time. The ratio of these two is of the order of

$$\frac{T}{R_y}\left(\frac{R_y}{\hbar\omega}\right)^2\left|\frac{\Delta q}{q}\right|$$

where T is the kinetic energy of the projectile. This will be somewhat larger than unity, particularly for higher projectile energies and large angle scattering. This says that the first improvement of the approximations should be directed at including the projectile-laser coupling which means including the state χ_q in (4.10) instead of replacing it by a plane wave. This can be done with very little difficulty and results in cross sections which contain the interference of different T matrices which is an effect not observable in the absence of the laser. The effect is however small with current experiments and so will not be discussed here.

One can extend the ideas above to situations in which the laser is resonant with a transition between a pair of states neither of which is the ground state. These states must be intermediate states that occur during the collision. They then must be states of the compound system since the bare atomic states will be significantly shifted by the interaction with the projectile so they are not relevant to resonances during the scattering process.

An experiment of this type has been performed by Langendam et al.[17]. Electrons were scattered off Ne at an energy of 16.20 ev which is a known resonance energy of Ne^-. The experiment takes place in a laser beam whose photon energy is tuned near 2.75 ev so that the 16.20 ev Ne^- state can absorb one photon to yield a Ne^- state of about 18.95 ev which is the energy of another known resonance state of Ne^-. The absorbtion was detected by the autoionization of the upper Ne^- state to an excited state of Ne which is energetically forbidden without the addition of the energy of at least one laser photon. This state decays radiatively and that radiation was detected as a function of the laser frequency. The experiment therefore detects an absorbtion of at least one laser photon between two Ne^- resonance states. The widths of these states are of the order of a few millivolts and they live only 10^{-12} to 10^{-13} sec. so that the absorbtion of more than one photon is extremely unlikely. The effect may therefore be treated by first order perturbation theory, but because of our previous discussion, we can also allow for the distortion of the projectile states with very little difficulty. This allows for the possibility of another resonant mechanism for the process: Instead of the absorbtion taking place between states of the compound system the incoming electron can absorb a photon so that it directly forms the upper Ne^- resonance at the collision. The two different processes interfere but the experiments[17] have been carried out at too low an intensity for the effect to have been observed.

The experiments do however reveal as unexpected richness of structure which the theory[18] (which is straightforward but too lengthy to be included here) interprets as pairs of resonant substates. That is, the formation of the upper Ne^- state is detected as a function of the laser frequency. Each peak in this curve is interpreted as a dipole transition between a pair of Ne^- states. Naive theory[18] associates the width of each peak with the width of the upper state but the inclusion[19] of the effect of the finite width in energy of the electron distribution brings in the width of the lower state. The result is that the observed width of each peak is the sum of the widths of the two states of Ne^-.

Finally, we also should mention a suggestion by R.N. Compton[20] to the effect that these ideas could be extended so that instead of an absorbtion from of a photon in a resonance state we could have an induced emission with a final state which is a stable negative ion. This would allow for the measurement of electron affinities with much higher precision than has been possible before. The theory[19], which is very similar to the theories described above has been given but we shall not present it here. The experiment is in progress[20].

Acknowlegement - This research was support by U.S. O.N.R. Contract No. N00014-76C-0014 and CUNY Faculty Research Award No. 13098.

References

1. See for example I. Bialynicki-Birula and Z. Bialynicka-Birula, Phys. Rev. A14 1101 (1976).
2. See for example J.I. Gersten and M.H. Mittleman in Electron and Photon Interactions with Atoms, Ed. H. Kleinpoppen and M.R.C. McDowell, Plenum Pub. Co., N.Y. (1976).
3. The treatment of N.M. Kroll and K.M. Watson, Phys. Rev. A8 804 (1973) is followed here. See also F.V. Bunkin and M.V. Fedorov, J.E.T.P. 22 844 (1966).

4. See for example F. Bloch and A. Nordsieck, Phys. Rev. 52, 54 (1937).
5. See for example W.C. Henneberger, Phys. Rev. Lett. 21 838 (1968).
6. M.H. Mittleman, Phys. Rev. A19 134 (1979).
7. H. Kruger and C. Jung, Phys. Rev. A17 1706 (1978).
8. A. Weingartshofer, J.K. Holmes, G. Caudle, E.M. Clarke and H. Kruger, Phys. Rev. lett. 39 269 (1977).
9. C. Jung, Phys. Rev. A20 1585 (1979).
10. A. Weingartshofer, E.M. Clarke, J.K. Holmes and C. Jung, Phys. Rev. A19 2371 (1979).
11. See for example L. Rosenberg, Phys. Rev. A20 1352 (1979).
12. See for example Optical Resonances and Two Level Atoms by L. Allen and J. Eberly, Wiley Interscience.
13. J.I. Gersten and M.H. Mittleman, Phys. Rev. A13 123 (1976).
14. B.R. Mollow, Phys. Rev. A12 1919 (1975).
15. M.H. Mittleman, Phys. Rev. A16 1549 (1977).
16. H.W., Hermann, I.V. Hertel, W. Reiland, A. Stamatovic and W. Stoll, J. Phys. B (London) 10 251 (1971) and N.D. Bhaskar, B. Jaduszliwer and B. Bederson, Phys. Rev. Lett 38 14 (1977).
17. P. Langendam, M. Gavrila, H. Kaandorp and M. Van der Wiel, J. Phys. B (London) 9 L453 (1976).
18. M.H. Mittleman, Phys. Rev. A18 685 (1978).
19. M.H. Mittleman, J. Phys. B (London) 12 1781 (1979).
20. R.N. Compton, private communication.

LECTURERS

Prof. J.N. Bardsley
Faculty of Arts and Sciences,
Dept. of Physics and Astronomy,
University of Pittsburgh,
Pittsburgh, PA 15260, U.S.A.

Prof. P.G. Burke
Queens University,
Belfast, N. Ireland and
Daresbury Laboratory, SRC,
Daresbury, Warrington WA4 4AD,
United Kingdom

Dr. M.S. Child
Theoretical Chemistry Dept.
1, South Parks Rd.,
Oxford OX1 3TG
United Kingdom

Prof. F.A. Gianturco
Istituto di Chimica Fisica,
Cittá Universitaria,
00189 Roma, Italy

Prof. C.J. Joachain
Physique Thèorique,
Faculté des Sciences,
Université libre de Bruxelles,
Brussels, Belgium

Prof. R.D. Levine
Dept. of Physical Chemistry
The Hebrew University
Jerusalem, Israel

Prof. R. McCarroll
Laboratoire d'Astrophysique,
Université de Bordeaux 1,
40, rue Lamartine,
33400 Talence, France

Prof. M.H. Mittleman
Department of Physics
The City College of the
City University of New York,
New York, N.Y. 10031, U.S.A.

Prof. I.C. Percival
Dept. of Applied Mathematics,
Queen Mary College,
University of London,
Mile End Rd, London E1 4NS,
United Kingdom

Dr. H. Van Regemorter
Observatorie de Paris, Section
d'Astrophysique Fondamentale,
92190 Meudon, France

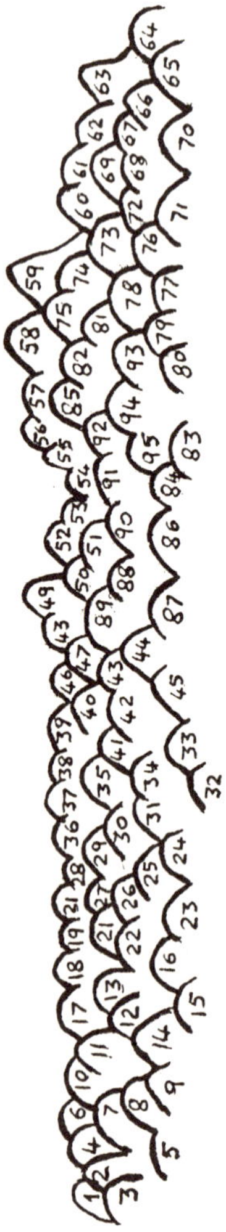

1. P.A. Amundsen
2. P. Palmieri
3. G. Bendazzoli
4. R. McCarroll
5. M.S. Child
6. D.C. Clary
7. M. Van Regemorter
8. H. Klar
9. L.A. Morgan
10. B.H. Bransden
11. J.M. Launay
12. J.P. Visticot
13. S. Rosendorff
14. F. Vecchiocattivi
15. A. Laganà
16. F. Pirani
17. C. Clark
18. L. Morgan
19. V. Aquilanti
20. K. Taulbjerg
21. F. Battaglia
22. P. Mandal
23. G. Grossi
24. M. Zarcone
25. M.R.C. Mc Dowell
26. J. Lin
27. K. Taylor
28. L. Ford
29. D. Malik
30. B. Andresen
31. N.C. Sil
32. S. Nuzzo
33. A.E. Orel
34. E. Vilallonga
35. D. Taylor
36. J. Hutson
37. J.E. Hansen
38. J. van den Biesen
39. S. Salvini
40. G. Drolshagen
41. H. Haugen
42. P. Burke
43. W. Bußert
44. F.A. Gianturco
45. C. Gianturco
46. T. Uzer
47. J. Jellinek
48. F. Borondo
49. E. Gerck
50. C. Mc Farlane
51. L. Avaldi
52. H.U. Kiefl
53. M. Mittleman
54. G. Akovali
55. G. Sinapius
56. C. Vasi
57. L. Errea
58. A. Jaquet
59. J. Vaaben
60. W. Burda
61. A. Defrance
62. M. Milazzo
63. A. Aquilar-Navarro
64. A. Maquet
65. L. Malegat
66. G.D. Fletcher
67. B. Piraux
68. S. Bliman
69. M. Mann
70. A.L. Roche
71. P. Villarreal
72. M.J. Jamieson
73. A.M.C. Moutinho
74. C.W. Newby
75. E. de Prunelé
76. S. Bivona
77. J.M. Alvariño
78. B. Spagnolo
79. J. Bastian
80. A.M. Cornille
81. W.D. Brewer
82. N.J. Allan
83. R. Daniele
84. M.C. Bacchus
85. D. Konowalow
86. I. Percival
87. C.J. Joachain
88. G. Delgado-Barrio
89. J.P. Ziesel
90. E. Fiordilino
91. A.D. Smith
92. N. Bardsley
93. E. van Dishoeck
94. C.J. Bocchetta
95. M.F. Laranjeira

PARTICIPANTS

BELGIUM

Mr. B. Piraux — Institut de Physique Corpusculaire
Bat. Sciences 1
B-1348 Louvain la Neuve (Belgium)

CANADA

Mr. R.D. Taylor — Dept. of Chemistry
Lash Miller Chemical Lab.
Toronto, Ontario M5S 1A1 (Canada)

DENMARK

Dr. B. Andresen — Physics Laboratory II
H.C. Ørsted Institute
Universitetsparken 5
DK-2100 København Ø (Denmark)

Mr. H.K. Haugen — Insitute of Physics
University of Aarhus
DK-8000 Aarhus C. (Denmark)

Dr. K. Taulbjerg — Institute of Physics
University of Aarhus
DK-8000 Aarhus C. (Denmark)

Dr. J. Vaaben A. — Physics Laboratory II
H.C. Ørsted Institute
Universitetsparken 5
DK-2100 København Ø (Denmark)

FRANCE

Ms. M.C. Bacchus	Laboratoire de Chimie Ecole Normale Superieure de Jeunes Filles 1, rue M. Arnoux 92120 Montrouge (France)
Ms. J. Bastian	Laboratoire de Ch. Physique Université P. et M. Curie 11, rue P. et M. Curie F-75231 Paris Cedex (France)
Prof. S. Bliman	Centre d'Etudes Nucléaires BP 95 Centre de TRI 38041 Grenoble Cedex (France)
Dr. M. Cornille	Dept. d'Astrophysique Fond. Observatoire de Paris 92190 Meudon (France)
Dr. A. Defrance	Lab. de Physique Moleculaire Université de Rennes Ave General Leclerc 35031 Rennes Cedex (France)
Dr. E. De Prunelé	Service de Physique Atomique CEN, Saclay B.P. 2 - 91190 Gif-sur-Yvette (France)
Dr. J.M. Launay	Dept. d'Astrophysique Fond. Observatoire de Paris 92190 Meudon (France)
Ms. L. Malegat	Laboratoire de Chimie Physique Université P. et M. Curie 11, rue P. et M. Curie F-75231 Paris Cedex (France)
Dr. J.P. Visticot	Service de Physique Atomique C.E.N., Saclay B.P. 2 - 91190 Gif-sur-Yvette (France)
Dr. J.P. Ziesel	Lab. des Collisions Atomiques et Moléculaires Bat. 351, Université Paris-Sud 91405 Orsay Cedex (France)

HOLLAND

Mr. J.E. Hansen	Zeeman Laboratory Plantage Muidergracht 4 1018 TV Amsterdam (The Netherlands)
Mr. J.J.H. van den Biesen	Huygens Laboratorium Wassenaarseweg 78 2300 RA Leiden (The Netherlands)
Ms. E.F. Van Dishoeck	Sterrewacht/Huygens Laboratorium Wassenaarseweg 78 2300 RA Leiden (The Netherlands)

ISRAEL

Mr. J. Jellinek	Dept. Chemical Physics Weizmann Institute of Science Rehovot (Israel)
Prof. S. Rosendorff	Dept. of Physics Technion IIT Haifa (Israel)

INDIA

Dr. P. Mandal	FOM Institute P.O.Box 41883 1009 DB Amsterdam (The Netherlands)
Prof. N.C. Sil	Indian Association for the Cultivation of Science Jadavpur, Calcutta-32 (India)

ITALY

Dr. L. Avaldi	Laboratorio raggi X Via Celoria, 16 Milano (Italy)
Dr. F. Battaglia	Max-Planck-Institut für Strömungsforschung Böttingerstraße 6-8 D-3400 Göttingen (West Germany)
Prof. G.B. Bendazzoli	Istituto di Chimica Fisica e Spettroscopia V. le Risorgimento, 4 Bologna (Italy)

Dr. S. Bivona	Istituto di Fisica Facoltà di Ingegneria Parco d'Orleans 90128 Palermo (Italy)
Dr. R. Daniele	Istituto di Fisica Via Archirafi 36 90123 Palermo (Italy)
Dr. E. Fiordilino	Istituto di Fisica Via Archirafi 36 90123 Palermo (Italy)
Dr. A. Laganà	Dipartimento di Chimica Università di Perugia 06100 Perugia (Italy)
Dr. M. Milazzo	Istituto di Fisica Via Celoria 16 Milano (Italy)
Prof. P. Palmieri	Istituto di Chimica Fisica V. le Risorgimento 4 Bologna (Italy)
Dr. S. Salvini	Dept. of Theoretical Physics and Applied Mathematics The Queen's University Belfast BT9 1NN Northern Ireland (U.K.)
Dr. B. Spagnolo	Istituto di Fisica Facoltà di Ingegneria Parco d'Orleans 90128 Palermo (Italy)
Dr. C. Vasi	Istituto di Fisica Università di Messina Messina (Italy)
Dr. B. Zarcone	Istituto di Fisica Via Archirafi, 36 90123 Palermo (Italy)

NORWAY

Dr. P.A. Amundsen	Institute of Physics University of Oslo P.O.Box 1048, Blindern Oslo 3 (Norway)

PORTUGAL

Prof. M.F. Laranjeira	Fac. de Ciencias e Tecnologia Universidade Nova de Lisboa Seminario dos Olivais Quinta do Cabeco Olivais, Lisboa (Portugal)
Dr. A.M.C. Moutinho	Centro de Fisica Molecular Complexo Interdisciplinar I.S.T. 1000 Lisboa (Portugal)

SPAIN

Dr. A. Aguilar-Navarro	Dpto Quimica Fisica Universidad de Barcelona Av. Diagonal 645 Barcelona 28 (Spain)
Dr. J.M. Alvariño	Dpto Quimica Fisica Universidad del Pais Vasco Apdo 644, Bilbao (Spain)
Dr. F. Borondo	Facultad de Ciencias C-XIV Universidad Autonoma Cantoblanco, Madrid 34 (Spain)
Dr. G. Delgado-Barrio	Istituto de Estructura de la materia Consejo Superior de Investigaciones Cientificas Serrano 119 Madrid-6 (Spain)
Dr. P. Villareal	Istituto de Estructura de la materia Consejo Superior de Investigaciones Cientificas Serrano 119, Madrid-6 (Spain)

TURKEY

Prof. G. Akovali	Dept. of Chemistry Middle East Tech. University Ankara (Turkey)
Dr. A.T. Uzer	Dept. of Theoretical Chemistry 1, South Parks Rd. Oxford OX1 3TG (U.K.)

UNITED KINGDOM

Mr. R.J. Allan	School of Physics The University Newcastle upon Tyne, NE1 TRU (U.K.)
Mr. C.J. Bocchetta	School of Chemistry University of Bristol Cantocks Close Bristol BS8 1TS (U.K.)
Prof. B. Bransden	Dept. of Physics University of Durham Science Laboratories South Rd, Durham, DH1 3LE (U.K.)
Dr. D.C. Clary	Dept. of Chemistry University of Manchester Institute of Science and Technology Manchester M13 9PL (U.K.)
Mr. J.M. Hutson	Physical Chemistry Lab. Oxford University South Parks Rd. Oxford OX1 3QZ (U.K.)
Dr. M.J. Jamieson	Dept. of Computing Science The University Glasgow, Scotland (U.K.)
Dr. C. Mc Farlane	Faculty of Mathematics The Open University Milton Keynes NK7 6AA (U.K.)
Dr. L.A. Morgan	Dept. of Statistics and Computer Science Royal Holloway College Egham Hill, Egham, Surrey (U.K.)
Mr. C.W. Newby	Dept. of Physics Science Labs, South Rd. Durham DH1 3LE (U.K.)
Mr. A.D. Smith	Dept. of Chemistry University of Manchester Manchester M13 9PL (U.K.)

Dr. K.T.A. Taylor — Daresbury Laboratory SRC
Daresbury, Warrington WA4 4AD
(U.K.)

USA

Dr. W.D. Brewer — Institut für Atom- und Festkörperphysik, FB Physik
Boltzmannstraße 20
1000 Berlin 33 (West Germany)

Dr. C.W. Clark — Daresbury Laboratory
Daresbury, Warrington WA4 4AD
(U.K.)

Dr. G.D. Fletcher — Physics Department
Yale University
P.O.Box 6666
New Haven CT 06511 (USA)

Prof. A.L. Ford — Physics Department
Texas A.-M. University
College Station, Texas 77843
(USA)

Prof. D.D. Konowalow — Dept. of Chemistry
State University of New York
Binghamton, NY 13901 (USA)

Dr. J. Lin — Dept. of Physics
University of Pittsburgh
Pittsburgh, PA 15260 (USA)

Dr. D.J. Malik — Dept. of Chemistry
Indiana U.-Purdue University
Indianapolis, IN 46205 (USA)

Dr. W.L. Morgan — Laser Theory Group L-472
Lawrence Livermore Lab.
P.O.Box 808
Livermore, Calif. 94550 (USA)

Ms. A.E. Orel — Dept. of Chemistry
University of California
Berkeley, Calif. 94720 (USA)

Dr. E.F. Vilallonga — Quantum Theory Project
University of Florida
Gainesville, Flo. 32611 (USA)

WEST GERMANY

Mr. W. Burda	Gesamthochschule Kassel Fachbereich 18 Physik D-3500 Kassel-Oberzwehren (BRD)
Mr. W. Bussert	Fachbereich Physik Universität Kaiserslautern D-6750 Kaiserslautern (BRD)
Mr. G. Drolshagen	M.P.I. für Strömungsforschung Böttingerstraße 6-8 D-3400 Göttingen (BRD)
Mr. E. Gerck	M.P.G. Projektgruppe für Laserforschung D-8046 Garching b. München (BRD)
Mr. H.U. Kiefl	Physikalisches Inst. d. Univ. Würzburg Am Hubland D-8700 Würzburg (BRD)
Dr. H. Klar	Fakultät für Physik A.Ludwigs-Universität D-7800 Freiburg i.Br. (BRD)
Dr. P. Hering	M.P.G. Projektgruppe für Laserforschung D-8046 Garching b. München (BRD)
Mr. U. Hefter	Fachbereich Physik Universität Kaiserlautern D-6750 Kaiserslautern (BRD)
Dr. R. Jaquet	Lehrstuhl für Theoretische Chemie Ruhr-Universität Bochum D-4630 Bochum (BRD)
Ms. M. Mann	Gesamthochschule Kassel Fachbereich 18 Physik D-3500 Kassel-Oberzwehren (BRD)
Dr. G. Sinapius	Fakultät für Physik Universität Bielefeld D-4800 Bielefeld 1 (BRD)

INDEX

Adiabatic
representation, 279, 294, 304
in ion atom collisions, 189-191
trial functions, 322, 323
Adiabatic Theory of e^--molecule collisions, 82
Abelian transformation, 292
Airy functions, 284, 291, 299
Alchemy code, 112
Angular distribution
e^--molecule collisions, 77, 78
molecular photoionization, 86, 87, 89
Yang's theorem, 89
Angular coupling, 294
Angular momentum transfer, 88
Atom
highly excited, 432
Rydberg, 432
atom collisions, 299, 301
Atom-molecule inelastic collisions
coupled representation(LAB), 337
rotational excitation, 335
coupled representation(BF), 348
vibrational excitation, 351
Atomic state expansion
in ion-atom collisions, 177-183
in charge exchange, 177-183

Bethe-Born approximation, 443
Body-fixed frame, 71
Bloch operator, 110
Born approximation in a laser field, 464
Born series, 13, 40, 178, 452
Born-Oppenheimer approximation, 318

Capture, electron (see charge exchange)
Charge exchange
atomic state expansion, 177-183
Born expansion, 174-176
capture in excited states, 218-221
classical model, 175, 188, 189
crossing of PES, 196, 198, 205, 206, 217-221
impact parameter (eikonal), 171-174, 176-184, 189-196, 224, 232, 233
molecular model (see perturbed stationary states)
momentum transfer (see translation factors)
multicharged ions, 166, 181, 215, 216, 226-231
nuclear-nuclear potential, 174, 179, 180
transfer to continuum (see ionization)
Chemical photophysics, 393
Classical
H atom, 437
path, 449
Coincidence e^--photon, 23
Combined theory, 460
Complex trajectory, 284, 297
Configuration interaction, 365, 294, 252, 258
Constraints, 422

Continuum multiple scattering, 99
Constrained phase space, 393
Correspondence principle, 432, 436, 441
Coordinate systems in rearrangement collisions, 168
Cosmic abundance, 243
Coulomb scattering, 10
Coupled equations, 178-181, 190, 192-193, 195, 199, 201-203
Coupled-state approximation, 358

Density of final states, 419, 443
Detailed balance, 128, 150, 396, 397, 237
Diabatic
  representation, 202-210, 279, 295, 296, 304
  trial functions, 325
Differential cross sections, 170, 172, 211-216, 283, 287, 290, 298
Dipole interaction, 451, 453
Dipole-length formula, 84
Dipole-velocity formula, 84
Direct potential in e^--molecule scattering, 75, 90
Direct scattering, 442
Discrete spectrum, 399
Dissociative attachment, 114
Dissociative continuum, 412
Distorted wave approximation, 176, 293, 297
Doorway states, 404
Diffraction oscillations, 279, 285, 286, 291, 302
Deflection function, 279, 281, 286, 293, 300
Demkov model, 279, 295, 296, 298, 302

Eigenstate expansion, 72
Eikonal approximation, 15, 46, 176, 177
Effective potential method, 378
Electron-molecule collisions
  angular distribution formulae,
Electron (continued)
  angular (continued)
    77, 78
  e^--CO, 91, 107
  e^--F_2, 107, 113
  e^--H_2, 96, 108, 113
  e^--HCl, 96
  e^--LiH, 91, 94
  e^--N_2, 91, 92, 101, 105, 107, 113, 116
  general theory, 71
  momentum transfer cross section, 88
  rotational excitation, 79
  vibrational excitation, 82
Electron capture (<u>see</u> charge exchange)
Electronic excitation, 411
Energy
  defect, 397
  disposal, 402
  randomization, 405
  requirements, 402
  transfer, 399
Entropy, 418, 424
Excitation rates, 237, 238
  of forbidden lines, 254-256
  in ion-atom collisions
    capture in excited states, 218-221
    inner-shell excitation, 231-234
    outer-shell excitation, 181-186, 224-226
Exchange potential, 75, 90
  approximate representation, 93
Exponential gap law, 197, 420

Feshbach operators, 38
First order perturbations, 451
Fixed-nuclei approximation
  in atom-molecule collisions, 373
  in electron-molecule collisions, 71
Fragmentation pattern, 414, 422
Frame-transformation theory, 82
  relation with R-matrix theory, 115
Free-electron gas exchange

Free-electron (continued)
potential, 93

Hara local exchange potential, 93
Heisenberg form of correspondence, 446
Hybrid theory in e⁻-molecule collisions, 82
HF-SCF approximation, 327
Hole burning, 407

Identical particles, 288
Impact parameter approximation
validity, 171
differential cross section, 173, 179
eikonal equations, 176-177
relationship with quantal formulation, 167, 171-174
Inner-shell processes
in ion-atom collisions
K-shell excitation, 181, 221-224
vacancy transfer, 224-226
Integral cross sections, 279, 287, 302, 303
Inversion
by unitarity, 291
Buck-Pauly, 290
distorted wave peeling, 293
Firsov, 293
IOS approximation in molecular collisions, 366
Invariant imbedding technique, 385
Integro-differential equations
numerical methods, 91
Ions
atmospheric, 125, 132
clusters, 142
in flames, 147
in fusion plasmas, 126
in lasers, 157
multiply charged, 129, 131, 133, 152
collisions with e⁻, 255, 258, 272
Ionization, collisional, 262

Ionization (continued)
in ion-atom collisions
direct ionization, 186, 187
charge transfer, 186-188
inner-shell, 221-226
Ionized regions, 433
Isotopic
enrichment, 399
shift, 400

K-matrix in e⁻-molecule collisions, 75

Lagrange multipliers, 422
Landau-Zener model, 144, 279, 297, 302
Laser
cavity, 399
low-frequency, 469
general use, 394, 395, 399
-induced-fluorescence, 399
Laboratory frame, 73, 81, 84
l-dominant decoupling, 376
L^2methods, 102
R-matrix method, 109
Stieltjes-Tchebycheff methods, 103
T-matrix method, 106
Local mode, 405

Maximum entropy, 422
Molecular frame, 71
Molecular state expansion (see Perturbed stationary state)
Momentum transfer cross sections, 78 (see also translation factors)
Monte-Carlo methods, 156, 439, 460
Multichannel scattering, 20
Multiphoton
absorption, 406
dissociation, 410, 422
ionization, 416, 422
Multicharged ions
charge exchange, 166, 215, 216, 226-231
capture in excited states, 218-221

Mutual neutralization, 143

Nikitin model, 279, 297, 302
Non-adiabatic transitions, 294-306
Nuclear-nuclear potential, 174, 179, 181

Optical potentials, 47, 137, 146
Orbiting limit, 293
Overtone spectra, 402

Partial wave expansion, 200-202
Perturbed stationary state (PSS) method, 189-196, 198-200
Phase shifts, 4, 281
Phase space, 418
Photoelimination process, 420, 421
Photodissociation, 413
Photoionization, 58
  of atoms, 256
  multiple, 61
  of molecules, 83
    angular distribution, 87, 88, 89
    $h\nu$ + CO, 97
    $h\nu$ + N_2, 97, 105
    Yang's theorem, 89
  resonances, 60
Photoisomerization, 405
Photoselective chemistry, 393
Planetary nebular
  atomic level population, 245, 248
  Balmer lines, 249
  charge transfer, 254
  collisions, 247, 248
  density diagnostics, 252, 268
  forbidden lines, 251
  important lines, 244, 250
  recombination theory, 244
  temperature diagnostics, 251, 252
Plasmas
  diagnostics, 239
  astrophysical, 235, 239
  optically thin, 246
  optically thick, 247
  transient, 271
Polarized electrons, 20
Polarization potential, 95, 99, 108
  pseudo states
    polarized, 72, 95
    in charge exchange, 183-185
Potential energy profile, 396
Potential energy surfaces, 412
  Li+ - H_2, 331
  H - H_2, 332
  H+ - H_2, 333
Potential scattering, 2, 281-294
  integral equation, 9
Power broadening, 407, 408
Prior distribution, 419

Quantum defect theory, 133
Quasicontinuum, 402

Rabi frequency, 408
Radial coupling, 196-198, 294
Radiationless transitions, 411
Radiorecombinations lines, 433
Radiative association, 148
Rainbow angle, 282, 287
  oscillations, 279, 285-287, 291, 302
Random phase approximation, 288, 302
Recombination
  atom-atom, 157
  collisional-dissociative, 153
  collisional-radiative, 151
  dielectronic, 129, 152, 263, 266
  dissociative, 134
  electron-ion, 166
  radiation emitted, 132
  radiative, 128
  rates, 126, 128, 130
Resonant charge transfer, 220-226
Resonant laser, 481
Resonance theory of vibrational excitation, 113, 129, 135, 149, 157
Rydberg states, 130, 132
RRKM theory, 405
R-matrix method
  applications, 133

R-matrix (continued)
applications (continued)
in at-m-molecule collisions, 36, 385
in e^--molecule collisions, 109
Bloch operator, 110
relation with FT theory, 115
vibrational excitation, 114
Rotational excitation, 79, 335
cross section, 81, 338
frame transformation theory, 82, 374

SA approximation in molecular collisions, 381
Satellite lines, 264
Scattering amplitude
in atom-atom collisions, 283
in atom-molecule collisions, 351
in e^--molecule collisions, 73, 77
in ion-atom collisions, 170
Selective excitation, 418
Selectivity, 418
Single-centre expansion, 89
convergence, 92
e^--molecule collisions, 89
molecular photoionization, 97
Single-m de laser, 462
Scaled hydrogenic basis set, 185-187
Symmetry oscillations, 210-216, 279, 287, 289, 290, 293
Solar corona
atomic level population, 266
dielectronic recombination, 263
density diagnostics, 268
ionization equilibrium, 261, 264-266
line spectrum, 259
satellite lines, 264, 270
solar flares, 271
S-matrix in e^--molecule collisions, 73, 77
Sum rules in a laser field, 473
Surface crossing, 413 (see also non-adiabatic crossing)
Surprisal, 414, 418, 421
Surprisal analysis, 419, 420
Spin-orbit coupling, 294
Sudden approximation, 452, 366
Strong coupling, 460
Sturmian state expansion, 182, 183
Stückelberg oscillations, 279, 299-302, 306
Static exchange approximation, 89
Static potential in e^--molecule collisions, 90
Stieltjes-Tchebysheff method, 103
Steric requirements, 399

Threshold laws, 126, 128, 146
Thermodynamical equilibrium, 235, 236
Time-dependent theories, 449
T-matrix in e^--molecule collisions, 77, 106
Translation factors, 191-196
Transition rates, 56
Total cross sections, 170, 174-180, 211-226

Uniform approximation, 283, 300

Validity, regions of, 459
Vacancy transfer, 181, 224-226
Vibrational energy disposal, 396
Vibrational excitation
in atom-molecule collisions, 351
in e^--molecule collisions, 79, 82
adiabatic theory, 83
hybrid theory, 83
R-matrix theory, 114
resonance theory, 114
shape resonances, 101